MW00844249

Optical Detection Theory
for Laser Applications

Optical Detection Theory
for Laser Applications

GREGORY R. OSCHE
Engineering Fellow
Raytheon Company
Tewksbury, Massachusetts

A JOHN WILEY & SONS, INC., PUBLICATION

Published by John Wiley & Sons, Inc., Hoboken, New Jersey.
Published simultaneously in Canada.

For general information on our other products and services please contact our Customer Care Department within the U.S. at 877-762-2974, outside the U.S. at 317-572-3993 or fax 317-572-4002.

Wiley also publishes its books in a variety of electronic formats. Some content that appears in print, however, may not be available in electronic format.

Library of Congress Cataloging-in-Publication Data Is Available

ISBN 0-471-22411-1

10 9 8 7 6 5 4 3 2

To Susan, Christopher, and Elizabeth
for their patience and support

To the memory of my parents
for their courage and perseverance

Contents

Preface xi

Chapter 1. Introduction and Background 1

1.1. Overview of Laser Systems 1
1.2. Review of Statistical Methods 4
1.2.1. Probability and Univariate Statistics 4
1.2.2. Statistical Moments 10
1.2.3. Bivariate Statistics 16
1.2.4. Transformation of Random Variables 21
1.2.5. Characteristic Function 26
1.2.6. Central Limit Theorem 30
1.2.7. Chi-Squared Distribution 34
1.2.8. Stationary and Ergodic Systems 36
1.2.9. Energy and Power Spectral Densities 38
1.2.10. Wiener–Khintchine Theorem 42
1.2.11. Fourier–Stieltjes Transform 44
1.2.12. Matched Filter Theory 46
1.3. Decision-Making Processes 47
1.3.1. Thresholding and Detection 48
1.3.2. Bayes' Decision Rule 54
1.3.3. Neyman–Pearson Decision Rule 58
1.3.4. Minimax Criterion 60
1.4. Optical Detection Techniques 61
References 66

Chapter 2. Signal and Noise Analysis 68

2.1. Introduction 68
2.2. Review of Diffraction Theory 68
2.3. Free-Space Propagation 74
2.3.1. Transformation of Gaussian Beams 78
2.3.2. Untruncated Focused Gaussian Beams 82
2.3.3. Multimode Beams 84
2.3.4. Beam Characterization 93
2.4. Truncated and Obscured Gaussian Beams 97
2.4.1. Optimum Beam Diameter 101
2.5. Fourier Optics and the Array Theorem 104
2.6. Antenna and Mixing Theorems 109

2.7. Analysis of Coherent Detection Systems 116
2.7.1. Untruncated Systems 117
2.7.2. Truncated Systems 125
2.8. Analysis of Direct-Detection Systems 129
2.8.1. Untruncated Systems 129
2.8.2. Truncated Systems 131
2.9. Receiver and Clutter Noise 136
2.9.1. Thermal Noise 136
2.9.2. Shot Noise 138
2.9.3. Dark Current and Excess Noise 140
2.9.4. Sources of Clutter 143
2.10. Power Signal-to-Noise-Ratio 145
References 148

Chapter 3. Random Processes in Beam Propagation **150**

3.1. Introduction 150
3.2. Review of Optical Coherence Theory 151
3.2.1. Coherence Properties of the Field 151
3.2.2. Van Cittert–Zernike Theorem 155
3.3. Surface Scattering 159
3.3.1. Scattering from a Rough Surface 160
3.3.2. Integrated Speckle Intensity 173
3.3.3. Speckle Correlation Diameter 176
3.4. Propagation through Turbulent Media 182
3.4.1. Atmospheric Model 183
3.4.2. Weak Turbulence Theory 187
3.4.2.1. Power Spectral Density 191
3.4.2.2. Correlation Function 194
3.4.2.3. Mutual Coherence Function 198
3.4.2.4. Statistics of the Turbulent Field 201
3.4.2.5. Aperture Averaging in Direct-Detection Systems 203
3.4.2.6. Turbulence-Limited Performance of Coherent Systems 208
3.4.2.7. Beam Wander 212
3.4.3. Strong Turbulence Theory 217
References 224

Chapter 4. Single-Pulse Direct-Detection Statistics **227**

4.1. Introduction 227
4.2. Single-Point Statistics of Fully Developed Speckle 228
4.3. Summed Statistics of Fully Developed Speckle 233
4.4. Poisson Signal in Poisson Noise 239

4.5. Negative Binomial Signal in Poisson Noise 245
4.6. Noncentral Negative Binomial Signal in Poisson Noise 255
4.6.1. Summed Statistics of Partially Developed Speckle 259
4.6.2. Single-Pulse Detection Statistics 263
4.7. Parabolic-Cylinder Signal in Gaussian Noise 267
4.8. Detection of Signals in APD Excess Noise 275
4.8.1. Poisson Signal 278
4.8.2. Negative Exponential Signal 288
4.8.3. Geiger-Mode APD Statistics 291
4.9. Detection in Atmospheric Turbulence 297
4.9.1. Poisson Signal in Turbulence 298
4.9.2. Bose–Einstein Signal in Turbulence 304
4.10. Detection in Atmospheric Clutter 306
4.11. Polarization Diversity 309
4.12. Multiple Uncorrelated Signals 313
References 315

Chapter 5. Single-Pulse Coherent Detection Statistics 318

5.1. Introduction 318
5.2. Constant-Amplitude Signal in Gaussian Noise 320
5.3. Rayleigh Fluctuating Signal in Gaussian Noise 328
5.4. One-Dominant-Plus-Rayleigh Signal in Gaussian Noise 331
5.5. Rician Signal in Gaussian Noise 334
5.6. Detection in Atmospheric Turbulence 338
5.6.1. Constant-Amplitude Signal in Weak Turbulence 339
5.6.2. Rayleigh Fluctuating Signal in Weak Turbulence 344
5.7. Coherent versus Noncoherent Performance 346
References 350

Chapter 6. Multiple-Pulse Detection 351

6.1. Introduction 351
6.2. Direct-Detection Systems 352
6.2.1. Poisson Signal in Poisson Noise 352
6.2.2. Negative Binomial Signal in Poisson Noise 356
6.2.3. Noncentral Negative Binomial Signal in Poisson Noise 358
6.2.4. Parabolic Cylinder Signal in Gaussian Noise 360
6.3. Coherent Detection Systems 363
6.3.1. Swerling Case 0 Model 366
6.3.2. Swerling Case I Model 371
6.3.3. Swerling Case II Model 375
6.3.4. Rician Signal Model 377

6.4. Binary Integration 381
6.4.1. Application to Geiger-Mode APD Detectors 384
References 389

Appendix A. Advanced Mathematical Functions **391**

A.1. Dirac Delta and Unit Step Functions 391
A.2. Gamma Function 393
A.3. Confluent Hypergeometric Function 395
A.4. Parabolic Cylinder Functions 397
A.5. Toronto Function 399
References 399

Appendix B. Additional Derivations **401**

B.1. Gamma Distribution 401
B.2. Burgess Variance Theorem 402
References 403

Index **405**

Preface

This book consists of notes and analyses that I have accumulated over the last 30 years while working in the field of laser radar and optical systems. The purpose of the book is to provide the electro-optics community with a comprehensive description of optical detection theory and associated phenomenologies. The current rapid development of active optical systems for both military and civilian applications has not been accompanied by a corresponding set of books that address the fundamental issues involved in laser system engineering. This is especially the case in the area of detection statistics, where the optical engineer must resort to journal articles or to books on microwave radar or communications theory in order to compile the necessary data for a detailed understanding of the field. A great deal of work has been accomplished by many authors in this field, but to my knowledge it has never been compiled and presented in a single coherent framework. With this motivation in mind, the book has been designed to be tutorial, deriving equations from first principles whenever possible so that the reader can establish a firm conceptual and mathematical understanding of the subject. A concerted effort has been made to present the theoretical discussions in a simple and straightforward manner for both clarity and comprehension by the interested reader. In this regard, I have tried to avoid highly generalized derivations and notation whenever possible.

The topics selected attempt to lay out the fundamental optical, statistical, and mathematical principles that are common to most, if not all, laser systems. Clearly, they are not all-inclusive but have been chosen based primarily on personal experience, the emphasis thus being on laser radar systems and to a lesser extent, communications systems. As will be seen, many of the topics have been derived from papers, textbooks, and monographs from other disciplines, such as radar, communications theory, statistical optics, diffraction theory, coherence theory, and atmospheric turbulence theory. Indeed, as with any technical book, there is very little new that is presented here except possibly the insights and viewpoints that are offered as an aid to an understanding of the material. To this end, I have included brief "Comments" or digressions from the main flow of the text to provide additional perspectives on the topics under discussion.

The book is basically a monograph or reference book for the professional scientist or engineer but may also be used as a textbook for a senior or graduate-level courses since it is relatively self-contained. For example, a review of some of the fundamental mathematical and physical concepts essential to a proper understanding of related optical phenomenologies comprises the first three chapters. Chapter 1 provides a brief review of the mathematical and statistical tools assumed in later chapters. These include, among others, mathematical statistics, matched filter theory, and statistical decision theory. Chapter 2 focuses on the deterministic processes that affect signal detection, beginning with the Huygens–Fresnel theory of diffraction, proceeding through both truncated and untruncated Gaussian beam theory, and ending with a formal development of the range equations for coherent and direct, truncated and untruncated laser systems. In Chapter 3 we consider the stochastic processes that affect signal detection, primarily those related to random fields generated by rough surfaces and turbulent atmospheres. Chapters 4, 5, and 6 are then devoted to the main subjects of the book: direct, coherent, and multiple-pulse detection statistics, respectively. The appendixes provide additional details on some of the more advanced mathematical functions used in the text. A comprehensive list of references is also provided at the end of each chapter for those who desire greater detail.

In the process of structuring the book, I have assumed that the reader has access to a modern high-speed personal computer with appropriate software and is therefore capable of performing calculations using the equations developed in the various chapters. For this reason, the need for extensive tabulation or graphical presentations of detection probabilities for example, as is usually found in classic radar books, has been avoided. A limited set of graphical examples are given in each case only as tools for conveying the underlying physical and mathematical trends that result from the theories derived.

I would like to thank Mike Fallica for his continued support during this project, as well as Frank Horrigan, Barbara Blyth, and Nancy DiMento Jerome for their efforts in proofreading the manuscript. A special thank you to Frank for his many helpful suggestions and thorough review of the mathematics. I would also like to thank Jeff Shapiro of the Massachusetts Institute of Technology for his helpful comments on scintillation statistics and Rick Marino of MIT Lincoln Laboratories for providing some important insights into APD Geiger-mode statistics. Finally, a word of acknowledgment to my colleagues at Raytheon Company, especially those from the old Sudbury EO group, for their pioneering work in the field of laser radar systems and for making my career an interesting and enjoyable one.

GREGORY R. OSCHE

Raytheon Company
Tewksbury, MA
January 2002

1

Introduction and Background

1.1. OVERVIEW OF LASER SYSTEMS

Everyone is familiar with the laser radar system developed for police organizations to reduce the ambiguities associated with conventional radar systems. These ambiguities have included multipath effects and poor target discrimination capabilities that frequently lead to unenforceable speed laws. This is a simple example of the advantages that can be gained by employing lasers to perform some of the well-known and well-established microwave radar functions. In this example the performance advantage arises as a result of the single most distinguishing feature of the laser, namely, a narrow beam, which in police use allows for the unambiguous selection of a single vehicle out of many in a heavy traffic environment. However, there are many more distinguishing features associated with laser systems that when considered alone or together lead to many more applications beyond that of simple range-Doppler measurements. Such features include operation in the infrared and visible portions of the electromagnetic spectrum, the capability to achieve high modulation bandwidths, and high Doppler sensitivities. These in turn lead to applications such as spectral probing of the atmosphere for known or unknown molecular species, high-resolution three-dimensional and Doppler imaging for use in robotic systems, and measurement of wind profiles using the backscatter from airborne aerosols and other naturally occurring particulates in the atmosphere. When combined with the usual range-Doppler measurement capabilities of a radar system, a multitude of system applications results, ranging from noninvasive medical diagnostic techniques to military remote sensing.

The laser radar system used by police is basically a *laser rangefinder* with some additional processing to extract the vehicle velocity. Laser rangefinders are commonly employed by the military, in geodetic surveying, and in space applications, as evidenced by the *Pathfinder* mission. Altimeters for low-flying aircraft and docking sensors for space vehicle rendezvous are obvious extensions of the rangefinder capabilities. Scanning laser rangefinders are generally classified as *laser radar systems*. These systems offer the unique capability of presenting high-resolution imagery in range, Doppler, or intensity formats, which, in turn, offers improved databases for processing of automatic target classification and identification algorithms. Commercial applications range from automatic part sorting based on shape to detailed mapping of man-made structures such as buildings and bridges for diagnostic and modeling purposes.

As mentioned earlier, operation in the infrared and visible portions of the spectrum implies a short wavelength that is on the order of the size of airborne particles in the atmosphere, such as dust and aerosols. Rayleigh and Mie scattering theories predict that the backscattered radiation under these conditions will be strong enough to be detectable with an optical receiver. Since airborne particles generally have a mean drift velocity equal to the local wind speed, scanning laser Doppler systems can be designed to map out the local three-dimensional wind vector. Systems that have been developed with these capabilities have included pollution monitoring systems, aircraft wake-vortex systems for airports, clear-air turbulence systems for commercial aircraft, and satellite-based systems designed to profile Earth's wind flow on a global scale.

The mere fact that laser systems operate in the infrared and optical spectrums makes them excellent candidates for spectral probing of the atmosphere. *Differential absorption lidar* (DIAL) *systems* are designed to have wavelength diversity over a given spectral band in order to map out the absorption profile of the atmosphere and therefore identify a particular airborne molecular species. Such systems have found application in monitoring carbon dioxide, water vapor, and ozone concentrations from airborne and space-based platforms. DIAL systems can be designed to operate against a hard target background to enhance the signal level while measuring the two-way integrated absorption at the probing wavelengths, or they can be designed to use the scattering properties of the atmosphere to generate the signal.

The optical receivers in all of these applications employ either *coherent* or *direct detection*, depending on the particular requirements of the system. Coherent detection allows for the extraction of frequency and phase information from the received signal in addition to amplitude. On the other hand, the *direct-detection receiver* is simply an energy-collection device that outputs the received pulse envelope only. Since phase is not measured, direct-detection is incapable of providing true Doppler information, although range rate can be estimated using two or more received pulses. However, the simplicity of the direct-detection receiver over that of the coherent detection receiver is significant, and usually leads to an interesting system trade-off between capability and cost.

Systems that operate against *soft* or *volume targets*, such as the atmosphere, may also employ coherent or direct-detection receivers. The former are generally used when Doppler information is required and are generally limited to the infrared because of difficulties in maintaining optical alignment and heterodyne losses due to atmospheric turbulence. They can be designed to operate in either a focused or a collimated beam mode, depending on range and aperture requirements, and are generally limited to either pulsed or continuous-wave (CW) waveforms. On the other hand, direct-detection receivers are generally used against soft targets when spectral measurements are required in the near-infrared or visible regions of the spectrum and Doppler shifts are not an issue. These systems offer lighter weight and lower cost for airborne or space-based applications and are generally limited to pulsed waveforms and collimated beams. (Some of these system limitations arise from fundamental optical processes discussed in later chapters.)

Systems that operate against *hard targets* may also use coherent or direct-detection receivers. However, in this case the coherent detection system has available to it a variety of range-Doppler waveforms that are either well known from the microwave radar community, such as FM-CW and pulse compression, or are unique to the laser transmitter and modulator technologies, such as the ultrashort mode-locked pulse waveform.

Optical communications has become a significant application for laser systems. The enormous growth potential for this market has spurred development of the photonics and fiber optics industries. On the other hand, long-range terrestrial-based free-space optical communications systems have not realized much success, due to obvious weather limitations. However, such limitations are less of an issue for short-range applications and certainly a nonissue for space-based inter-satellite and deep-space missions. Indeed, the high-bandwidth and narrow-beam characteristics of laser systems virtually assures their eventual use as high-data rate free-space links for global satellite communications while relying on lower frequencies for downloading data through the atmosphere to ground-based receivers. Short-range interbuilding free-space links may also prove to be an important application for relieving the "last mile" problem at city boundaries, where incoming long-range high-capacity fiber data transmissions require major improvements in the hardwired communication infrastructures for optimum distribution.

Finally, we should mention the promising new field of medical imaging using lasers and fiber optics for noninvasive or minimally invasive surgical and diagnostic procedures. This field, which may be considered a type of remote sensing, is already reporting on nonsurgical probing of living tissue using optical techniques, the theory of which is very similar to that of ground-penetrating radar. The simplest and best-known success in this area is the pulse oximeter, a small device placed on a patient's finger to measure heart rate and oxygen content of the blood. This is accomplished by measuring the difference in absorption at two wavelengths along a one-way path using light emitting diodes (LED) operating in the red and infrared regions of the spectrum. It is not inconceivable that in the future more sophisticated measurements of, say, phase delays through the tissue will reveal additional information on the health of the tissue or some other critical parameter.

In this book we focus on the fundamental detection processes that are inherent in most, if not all, laser applications that involve photodetection processes. The implications of using direct or coherent detection are addressed separately for each topic. In the subsections that follow in this chapter we review briefly some of the fundamental concepts of mathematical statistics, statistical decision theory, and optical detection techniques, topics that are basic to a proper understanding of optical detection theory. Those who are already versed in these topics may proceed directly to later chapters as desired.

1.2. REVIEW OF STATISTICAL METHODS

The need for an understanding of statistical analysis follows from the fact that many of the processes involved in optical detection are random. For example, receiver noise, target fluctuations, atmospheric perturbations, and clutter noise are all stochastic processes that can be described only from a statistical point of view. In this section we provide a brief overview of those aspects of mathematical statistics that are of particular importance in the description of phenomenologies associated with optical detection. Both continuous and discrete statistical processes are described, the latter being of particular importance in later chapters that address photoelectron counting statistics in direct-detection receivers. To those who wish a more detailed discussion of probability and statistics, the works of Kempthorne and Folks[1] and Papoulis[2] are highly recommended.

1.2.1. Probability and Univariate Statistics

The concept of probability follows naturally from considerations of the *frequencies of occurrence* of various *statistical events* compared to the total number of events that have occurred or are possible. The events themselves are described by a *random variable* that varies unpredictably from event to event but in accordance with some mathematical probability function. Determination of the form of this function is the ultimate goal of statistical theories since it is then possible to obtain as complete a description of the random process as possible, at least in a statistical sense.

A random variable can be associated with a *sample space* that contains all possible *outcomes* of a *random event or sample*. The events or samples themselves, which we denote with capital letters $A, B, C, \ldots$, may be continuous or discrete, with *sample spaces* that are also continuous or discrete. When the outcomes of two or more events are considered together, they are referred to as *joint events*, the probability of which is written $P(AB\ldots)$.

If the trial outcomes are *mutually exclusive*, only one outcome can occur for any given *observation* or *trial* of the random process, so that the probability of both A and B and ... occurring is $P(AB\ldots) = 0$. An example of a mutually exclusive process is the outcome of heads or tails in the flipping of a coin. The key word here is *or*, so that the probability of any one of the events occurring is $P(A \text{ or } B \text{ or } \ldots) = P(A) + P(B) + \cdots$. In the case of coin tossing, there are only two events, heads or tails, so that the sum equates to unity since there are no other possibilities.

If the events are *statistically independent*, the outcome of one event is independent of the outcomes of any other event. An example of statistically independent events that can be taken jointly is the flipping of two or more coins, a process in which coin 1 could yield heads, coin 2 could yield tails, and so on. The key word in this case is *and*, so that the probability of occurrence of any particular combination of heads and tails is $P(A, B, \ldots) = P(A \text{ and } B \text{ and } \ldots) = P(A)P(B)\ldots$.

If the events are *not mutually exclusive* [i.e., $P(AB\ldots) \neq 0$], the probability that at least one of the events will occur is given by the sum of their individual probabilities minus their joint probabilities. An example here is once again the flipping of two coins, but in this case we want the probability of obtaining heads and/or tails. The key phrase in this case is *and/or*, so that the probability of obtaining such an outcome becomes $P(A \text{ and/or } B) = P(A + B) = P(A) + P(B) - P(AB)$.

If two or more events are *statistically dependent*, the outcome of event A is dependent on the outcome of event B, and vice versa. In this case, their joint probability is not equal to the product of the individual probabilities [i.e., $P(A, B) \neq P(A)P(B)$]. Rather, it is dependent on the conditional probability that B will occur given that A has occurred times the probability that A will occur, and vice versa. This is written $P(A, B) = P(A \mid B)P(B) = P(B \mid A)P(A)$. Note that when the events are independent, $P(A \mid B) \equiv P(A)$ and $P(B \mid A) \equiv P(B)$. We have more to say about dependent random variables in the next section.

Mutually exclusive and joint events are easily visualized using Venn diagrams of set theory, as shown in Figure 1-1. Venn diagrams are graphical renditions of the various sample spaces involved. For example, the set of all possible outcomes of n events constitutes the entire sample space of the experiment denoted as Σ. Let the space A correspond to the probabilities associated with all possible outcomes of event A and similarly for space B and event B. Thus, if A and B are mutually exclusive, their regions do not overlap, as shown in Figure 1-1. On the other hand, if A and B are jointly related, the space of all possible outcomes of both A and B occurring jointly is represented by the shaded region labeled (A, B) in Figure 1-1. Finally, the probability of A and/or B occurring is represented by the shaded area in Figure 1-1.

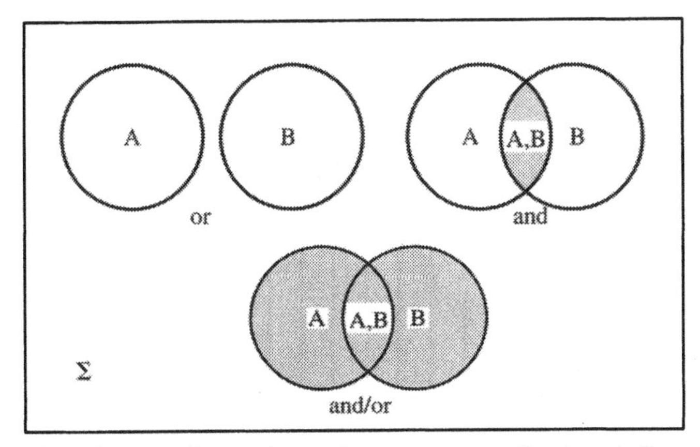

Figure 1-1. Venn diagrams of events in sample space corresponding to mutually exclusive (*or*) processes, statistically independent (*and*) processes, and combined mutual exclusive and statistically independent (*and/or*) processes.

Let us define the *frequency of occurrence* of a random variable x as $n(x)$ and the total number of occurrences as n. It is intuitively obvious that the *probability* that x will occur should then be defined as $P(x) = n(x)/n$. This is usually put in graphical form by plotting $P(x)$ as a function of x, where x takes the form of *frequency bins* of finite width Δx. As a specific example, consider the case of a golfer attempting a hole-in-one. Restricting the discussion to a single dimension for the moment, we could inquire as to the frequency with which the shots are placed in certain spatial intervals or bins of size Δx at various distances x from the hole. The relative probability is then $P(x) = n(x)/n$ of balls falling in spatial bins of width Δx located a distance x from the hole. However, notice that if we decrease the width Δx of the bins, the number of balls per bin also decreases such that as $\Delta x \to 0$, $n(x)/n \to 0$. To remove this dependency on the resolution and to establish the conceptual framework for some limiting cases, the magnitude of each bin can be redefined as the quantity $(1/\Delta x)[n(x)/n]$, which is a spatial density. The plot is therefore broken up into a series of rectangles of height $(1/\Delta x)[n(x)/n]$, width Δx, and area $n(x)/n$. The area $n(x)/n$ is then interpreted as the relative probability of event $n(x)$ occurring in the interval Δx located at position x. The resulting plot is called a *histogram* and is shown in Figure 1-2. Thus, in our example of the golfer, the probability of hitting a shot that lands within an interval $\pm\Delta x/2$ of some distance x from the hole is the area under the histogram plot between $x - \Delta x/2$ and $x + \Delta x/2$.

Consider the limit of letting $\Delta x \to 0$ and $n \to \infty$ in the histogram plot of Figure 1-2. The result is a curve, also shown in the figure, known as a *probability density function* or simply *density function*. If x is continuous over the interval $(-\infty, \infty)$, the function is referred to as a *continuous probability density function*. On the other hand, if x is discrete, with a finite or infinitely countable set of samples given by x_j, where j is an integer, the histogram becomes a discontinuous function known as a *discrete probability density function*. Hereafter denoting continuous density functions as $p(\cdot)$ and discrete density functions as

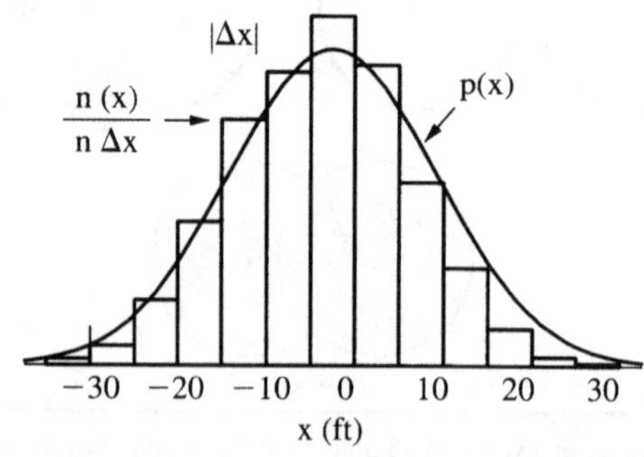

Figure 1-2. One-dimensional histogram of miss distances by a hypothetical golfer attempting a hole-in-one.

$q(\cdot)$, we have

$$
\begin{aligned}
p(x) &= \lim_{\substack{\Delta x \to 0 \\ n \to \infty}} \frac{n(x)/\Delta x}{n} && \text{continuous} \\
q(x) &= \sum_{j=1}^{\infty} P(x_j)\delta(x - x_j) && \text{discrete}
\end{aligned}
\tag{1-1}
$$

where $P(x_j) = n(x_j)/n$ is the probability of occurrence of x_j and $\delta(x - x_j)$ is the *Dirac delta* or *impulse function* (cf. Appendix A). For simplicity of notation, the discrete density function will be assumed to represent an infinitely countable sample space, that is, $j = 1, 2, 3, \ldots, \infty$. However, it should be recognized that this is not a necessary condition since the summation index j can be truncated at any value, depending on the form of the density function. Examples of continuous and discrete probability density functions are shown in Figure 1-3 and correspond to the *Gaussian or normal distribution* and the *binomial distribution*, respectively. Mathematically, they are written as

$$
p(x) = \frac{1}{\sqrt{2\pi\sigma^2}} e^{-(x-a)^2/2\sigma^2} \qquad -\infty \le x \le \infty \tag{1-2}
$$

where $a = 4$ and $\sigma^2 = 1$, and

$$
q(x) = \frac{n!}{j!\,(n-j)!} P^j (1-P)^{n-j} \delta(x - j) \qquad -\infty \le x \le \infty \tag{1-3}
$$

where $n = 7$ and $P = 0.5$. Equation (1-3) represents the probability of exactly j successes in n *Bernoulli trials*, where the term involving factorials represents the *binomial coefficients* and P the probability of success in a single trial. A Bernoulli trial is one that has only two mutually exclusive outcomes per trial, such as heads or tails in the flipping of a coin. These density functions are used frequently in later chapters.

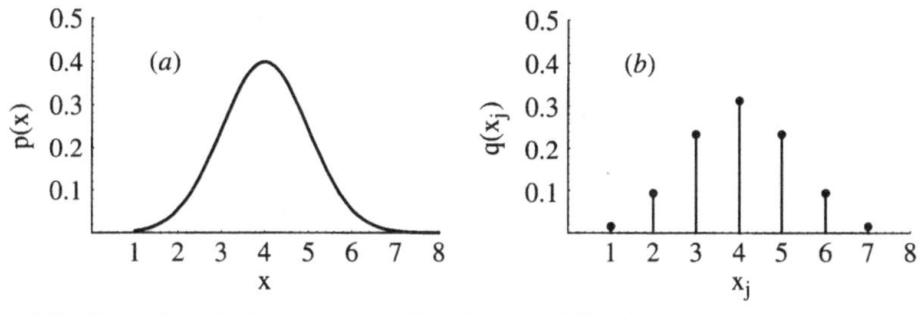

Figure 1-3. Examples of (*a*) a continuous Gaussian probability density function and (*b*) a discrete binomial probability density function for the case $p = \frac{1}{2}$ and $n = 7$.

With the definitions above, the probability P that the random number x will lie in the range $a \to b$ is found by integrating $p(x)$ or $q(x)$ over some specified limits, that is,

$$P(a \leq x \leq b) = \begin{cases} \displaystyle\int_a^b p(x)\,dx & \text{continuous} \\ \displaystyle\int_a^b q(x)\,dx = \sum_{j=1}^{\infty} P(x_j) \int_a^b \delta(x - x_j)\,dx & \\ \displaystyle\qquad = \sum_{j=m}^{n} P(x_j) & \text{discrete} \end{cases} \tag{1-4}$$

Here it should be noted that in the discrete case the order of summation and integration has been reversed and the integration region $a \to b$ is assumed to include the discrete points $m \leq j \leq n$. Thus we see from the definitions above that when $a \to b$, $p(x) \to 0$ for all x in the continuous case, and $q(x) \to q(x_j)$ in the discrete case. For P to be considered a true probability with a range $0 \leq P \leq 1$, the integrals above must equate to unity probability when extended over all sample space; that is,

$$\begin{aligned} P(-\infty \leq x \leq \infty) &= \int_{-\infty}^{+\infty} p(x)\,dx = 1, \qquad p(x) \geq 0 \qquad \text{continuous} \\ P(x_1 \leq x_j \leq x_\infty) &= \sum_{j=1}^{\infty} P(x_j) = 1, \qquad P(x_j) \geq 0 \qquad \text{discrete} \end{aligned} \tag{1-5}$$

Comment 1-1. Care must be exercised in the interpretation of the integrals associated with the discrete density functions when using the Dirac delta function formalism. Specifically, the integrals must totally include the discrete point or points in question and, to be formally correct, a notation should be used to indicate precisely the intended range of integration. For example, if $q(x)$ includes only two discrete points x_1 and x_2, and $P(x)$ in Eq. (1-4) is to include only the discrete point x_2, the integral in Eq. (1-4) might be written

$$P(x_1 < x \leq x_2) = \int_{x_1^+}^{x_2^+} q(x)\,dx = \sum_{j=1}^{\infty} P(x_j) \int_{x_1^+}^{x_2^+} \delta(x - x_j)\,dx = P(x_2) \tag{1-6}$$

where the $+$ superscript indicates values of x just beyond x_1 and x_2, as shown in Figure 1-4. In a similar fashion, if Eq. (1-4) is to include both discrete points x_1 and x_2, the integral would read

$$\begin{aligned} P(x_1 \leq x \leq x_2) &= \int_{x_1^-}^{x_2^+} q(x)\,dx = \sum_{j=1}^{\infty} P(x_j) \int_{x_1^-}^{x_2^+} \delta(x - x_j)\,dx \\ &= P(x_1) + P(x_2) \end{aligned} \tag{1-7}$$

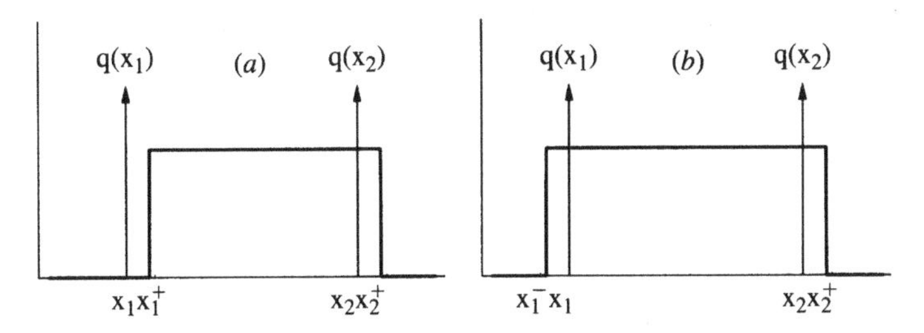

Figure 1-4. Range of integration for evaluating discrete probability density functions.

to indicate that the range of integration begins just before x_1 and ends just after x_2. However, such $+-$ notation is cumbersome and usually unnecessary if the range of integration is stated at the outset, and therefore is omitted in the remaining discussions. ■

The probability density function is a fundamental descriptor for statistical systems. From it other statistical measures can be derived, such as moments, in order to describe the behavior of a given random variable. Another such descriptor, the *cumulative distribution function*, or simply the *distribution function*, is defined as the probability that x will be less than or equal to some specified value. Letting $F(\cdot)$ and $G(\cdot)$ represent the continuous and discrete distribution functions, respectively, we have

$$F(x') = P(x \le x') = \int_{-\infty}^{x'} p(x)\,dx \qquad \text{continuous}$$
$$G(x_n) = P(x \le x_n) = \sum_{j=1}^{n} P(x_j)U(x - x_j) \qquad \text{discrete} \tag{1-8}$$

where x' and x_n are given samples of the random variable x and $U(x - x_j)$ is the unit step function (cf. Appendix A). Example distribution functions corresponding to the probability density functions of Figure 1-3 are shown in Figure 1-5. Alternatively, the density functions can be written in terms of the distribution functions,

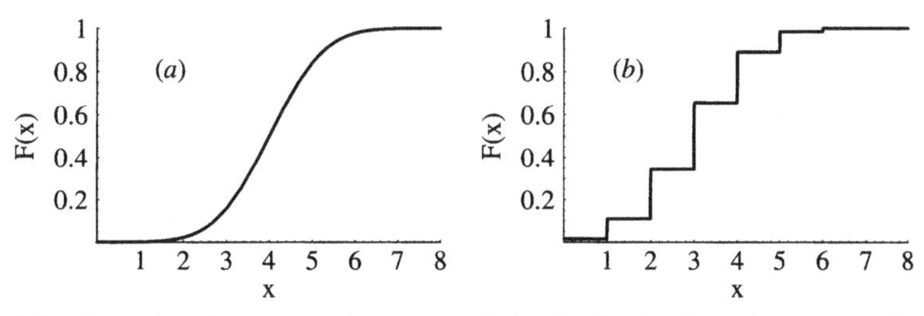

Figure 1-5. Examples of (*a*) the continuous cumulative distribution function corresponding to the density function (*a*) in Figure 1-3, and (*b*) the discrete distribution function corresponding to the density function (*b*) in Figure 1-3.

that is,

$$
\begin{aligned}
p(x) &= \frac{dF(x)}{dx} && \text{continuous} \\
q(x) &= \frac{dG(x)}{dx} = \sum_{j=1}^{\infty} P(x_j)\delta(x - x_j) && \text{discrete}
\end{aligned}
\tag{1-9}
$$

where the delta function and the unit step function are related by $\delta(x) = dU(x)/dx$.

Comment 1-2. The advantage of using the delta function in the discrete probability density functions is that it allows for a consistent formalism between the continuous and discrete cases. This is particularly useful in mixed probability density functions that contain both continuous and discrete components. In particular, if $p(x)$ and $q(x)$ represent the probability density functions of the continuous and discrete components of a single random variable x, then in order that they truly represent a probability, it is necessary that their *sum* satisfy the definition of probability, not their individual density functions, that is,

$$
\int_{-\infty}^{\infty} [p(x) + q(x)]\, dx = 1 \tag{1-10}
$$

■

Either the probability density function or the cumulative distribution function as defined above is sufficient to fully define the *first-order statistics* of a random variable; no other descriptors are required. Higher-order statistics are similarly defined by joint density functions or cumulative distributions, as will be shown later. It is for this reason that these functions play such a fundamental role in statistical theories.

1.2.2. Statistical Moments

The concept of statistical moments is analogous to the concept of moments in mechanics, where, for example, the first moment of a group of masses is just the *weighted* average location of these masses, the weights being the masses themselves. This is better known as the *center of mass* for the system and is written $r_c = \sum m_j r_j / M$, where the m_j are discrete masses located a radial distance r_j from the origin and M is the total mass of the system. The concept of center of mass is applicable to both discrete and continuous mass distributions. In direct analogy to the center of mass, a statistical *average* or *mean value* of a random variable x is defined as a *weighted average* of x, with the weighting functions being the probabilities of occurrence of x. Thus the *first moment* of x, denoted as m_1, is equal to the *expected value* of x, which is commonly written as $\langle x \rangle$, and is given by

$$m_1 = \langle x \rangle = \begin{cases} \displaystyle\int_{-\infty}^{\infty} x p(x)\, dx & \text{continuous} \\ \displaystyle\int_{-\infty}^{\infty} x q(x)\, dx = \sum_{j=1}^{\infty} P(x_j) \int_{-\infty}^{\infty} x\delta(x - x_j)\, dx & \\ \displaystyle\qquad = \sum_{j=1}^{\infty} x_j P(x_j) & \text{discrete} \end{cases} \tag{1-11}$$

[Other frequently used representations of the expected or mean value are $E(x)$ and $\overline{x}$, the latter being used interchangeably with $\langle x \rangle$ in later chapters for convenience of notation.] In the example of the golfer, $\langle x \rangle$ defines the center of the pattern of shots (assuming a symmetric distribution).

The next-higher moment in mechanics is the *moment of inertia*, defined as the sum of the products of the masses times the squares of their respective distances from the axis of rotation, that is, $I = \sum m_j r_j^2$. Once again the analogy with mechanics leads to the concept of the *second moment*, m_2, defined as the average of the square of the random variable x. Thus

$$m_2 = \langle x^2 \rangle = \begin{cases} \displaystyle\int_{-\infty}^{\infty} x^2 p(x)\, dx & \text{continuous} \\ \displaystyle\int_{-\infty}^{\infty} x^2 q(x)\, dx = \sum_{j=1}^{\infty} P(x_j) \int_{-\infty}^{\infty} x^2\delta(x - x_j)\, dx & \\ \displaystyle\qquad = \sum_{j=1}^{\infty} x_j^2 P(x_j) & \text{discrete} \end{cases} \tag{1-12}$$

The second moment is usually associated with the spread or width of the probability density function or, in the case of the golfer, dispersion in range or angle of the shot.

The moment definitions above can be generalized ad infinitum for any arbitrary random function of x such that the nth *moment* m_n of the function $f(x)$ becomes

$$m_n = \langle f(x)^n \rangle = \begin{cases} \displaystyle\int_{-\infty}^{\infty} f(x)^n p(x)\, dx & \text{continuous} \\ \displaystyle\int_{-\infty}^{\infty} f(x)^n q(x)\, dx & \text{discrete} \end{cases} \tag{1-13}$$

It is frequently desirable to be able to estimate the spread or width of a distribution relative to some point in the probability density function as a measure of the randomness of the processes involved. There are many possible definitions for such a parameter, the most commonly accepted one being the *variance* or mean-squared variation of x about the mean. The unique property of this definition is that it results in the minimum possible spread parameter. To show this, we define the spread about the point a as σ_x^2, where the subscript indicates the random variable, and write,

$$\sigma_x^2 = \langle (x-a)^2 \rangle = \langle x^2 - 2xa + a^2 \rangle = \langle x^2 \rangle - 2\langle xa \rangle + \langle a^2 \rangle \tag{1-14}$$

Assuming a unimodal distribution (i.e., one that has only a single peak), the minimum value can be obtained from

$$\frac{d\sigma_x^2}{da} = \frac{d}{da}\int_{-\infty}^{+\infty} (x-a)^2 p(x)\, dx = 2a \int_{-\infty}^{+\infty} p(x)\, dx - 2\int_{-\infty}^{+\infty} xp(x)\, dx = 0 \tag{1-15}$$

Using Eqs. (1-5) and (1-11), Eq. (1-15) yields

$$a = \langle x \rangle \equiv m_1 \tag{1-16}$$

Substituting Eq. (1-16) into Eq. (1-14), we find that

$$\sigma_x^2 = \langle x^2 \rangle - \langle x \rangle^2 \equiv m_2 - m_1^2 \tag{1-17}$$

The variance, which will occasionally be denoted as var(x), is also called the *second central moment*, μ_2. If the probability density function is symmetric and centered on the origin (i.e., $m_1 = 0$), the variance and second moment are equal (i.e., $\sigma_x^2 = m_2$). In all other cases, Eq. (1-17) is the proper definition. Note that the derivation above also holds for discrete distributions, as can be seen by replacing $p(x)$ by $q(x)$ in Eq. (1-15).

A parameter closely related to the variance is the *standard deviation* about the mean, σ_x, which is simply the positive square root of the variance. It is particularly useful because its units are those of the abscissa in the probability density function and may be measured in both positive and negative directions about the mean.

Other parameters of use are the median and mode of the distribution. The *median* is defined as the value of x that results in the distribution function F (or G) to be equal to $\frac{1}{2}$. It is generally equal to the middle value of the distribution, as opposed to the mean or average value given above. The *mode* of the distribution is defined as the value of x at which the peak of the distribution occurs and therefore corresponds to the *most probable value* of x.

Higher-order moments yield information about the symmetry properties of the distribution. For example, *skewness* is defined as

$$S = \frac{\mu_3}{\sigma_x^3} \tag{1-18}$$

where $\mu_3 = \langle (x - m_1)^3 \rangle$ and the denominator is used for normalization. Since the third central moment of a symmetric probability density function is zero, the skewness is a direct indication of the departure of the probability density function from perfect symmetry. The *kurtosis* is given by the fourth central moment normalized to the square of the variance, that is,

$$K = \frac{\mu_4}{\sigma_x^4} - K_G \tag{1-19}$$

where $\mu_4 = \langle (x - m_1)^4 \rangle$ and $K_G = \mu_4/\sigma_x^4 = 3$ corresponds to a Gaussian distribution. With this definition, K describes the "flatness" or "peakedness" of the distribution relative to a Gaussian or normal distribution. Thus $K = 0$ corresponds to a normal distribution, $K > 0$ a distribution that has a sharper peak than the normal distribution, and $K < 0$ a distribution that has a broader peak than the normal distribution.

Comment 1-3. Consider the continuous *Rayleigh probability density function* given by

$$p(x) = \frac{x}{\sigma^2} e^{-x^2/2\sigma^2} \qquad 0 \le x \le \infty \tag{1-20}$$

where σ^2 is the variance of the Gaussian distribution of Eq. (1-2) (cf. Comment 1-9). The various descriptors for this density function can be described as follows:

(a) The mode of the distribution occurs when $dp(x)/dx = 0$ or at $x = \sigma$. The amplitude at this point is $p(\sigma) = \sigma^{-1} e^{-1/2}$.

(b) The median is obtained by evaluating the distribution function for the case $F(x) = \frac{1}{2}$. Integrating Eq. (1-20) yields $F(x) = 1 - e^{-x^2/2\sigma^2}$. Solving for x yields $x_{\text{med}} = \sqrt{2 \ln 2}\, \sigma$.

(c) The mean of x is given by

$$\langle x \rangle = \frac{1}{\sigma^2} \int_{-\infty}^{+\infty} x^2 e^{-x^2/2\sigma^2} dx = \sqrt{\frac{\pi}{2}} \sigma \tag{1-21}$$

(d) The second moment, m_2, is given by

$$\langle x^2 \rangle = \frac{1}{\sigma^2} \int_{-\infty}^{+\infty} x^3 e^{-x^2/2\sigma^2} dx = 2\sigma^2 \tag{1-22}$$

(e) The variance follows from Eqs. (1-14), (1-21), and (1-22), that is,

$$\sigma_x^2 = \langle (x - \langle x \rangle)^2 \rangle = \langle x^2 \rangle - \langle x \rangle^2 = \left(2 - \frac{\pi}{2}\right) \sigma^2 \tag{1-23}$$

(f) The standard deviation is simply the positive square root of the variance, that is,

$$\sigma_x = \sqrt{2 - \frac{\pi}{2}} \sigma \tag{1-24}$$

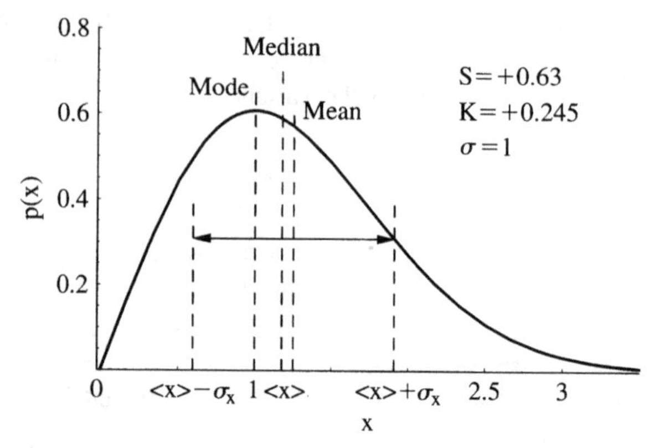

Figure 1-6. Parameters describing the statistics and shape of a Rayleigh probability density function.

(g) The skewness of the Rayleigh distribution is expected to be a positive number since it is asymmetric with a tail to the right of the mean. Thus

$$S = \frac{\mu_3}{\sigma_x^3} = \frac{\sigma^{-2}\displaystyle\int_{-\infty}^{+\infty}(x-m_1)^3 e^{-x^2/2\sigma^2}dx}{[(2-\pi/2)\sigma^2]^{3/2}} = \frac{(\pi-3)\sqrt{\pi/2}}{(2-\pi/2)^{3/2}} = 0.631 \tag{1-25}$$

(h) Finally, the kurtosis is given by

$$\begin{aligned} K &= \frac{\mu_4}{\sigma_x^4} - 3 = \frac{\sigma^{-2}\displaystyle\int_{-\infty}^{+\infty}(x-m_1)^4 e^{-x^2/2\sigma^2}dx}{[(2-\pi/2)\sigma^2]^2} - 3 \\ &= \frac{32-3\pi^2}{(4-\pi)^2} - 3 = 0.245 \end{aligned} \tag{1-26}$$

The fact that $K = 0.245$ is a positive number implies that the Rayleigh distribution is more peaked than the Gaussian or normal distribution. Notice also that S and K are both independent of σ. This is because the Rayleigh distribution maintains its shape as σ changes, thereby maintaining the skewness and kurtosis as constants. This is not the case for all density functions. Figure 1-6 shows the Rayleigh distribution plotted for the case of $\sigma = 1$ along with the various moments and shape parameters for the distribution. ■

Comment 1-4. For comparison, consider the case of the discrete binomial distribution introduced earlier. We have

$$q(x) = \frac{n!}{j!(n-j)!}P^j Q^{n-j}\delta(x-j) \tag{1-27}$$

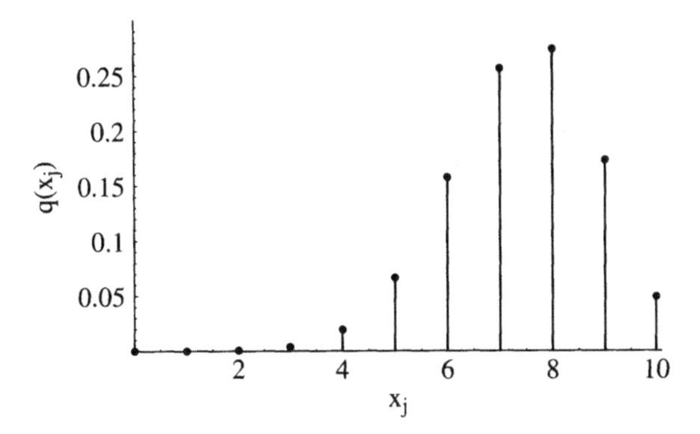

Figure 1-7. Discrete binomial distribution with parameters $n = 10$ and $p = 0.74$.

where P is the probability of success and $Q = 1 - P$ is the probability of failure. For example, letting $n = 10$ and $P = 0.74$, we obtain the asymmetric distribution shown in Figure 1-7. From Eq. (1-11) the mean of the distribution is

$$\langle x \rangle = \int_{-\infty}^{\infty} x q(x)\, dx = \sum_{j=0}^{n} x_j P(x_j) \tag{1-28}$$

Since x_j is an integer equal to j, we have

$$\begin{aligned}\langle x \rangle &= \sum_{j=0}^{n} j \frac{n!}{j!(n-j)!} P^j Q^{n-j} \\ &= \sum_{j=1}^{n} \frac{n!}{(j-1)!(n-j)!} P^j Q^{n-j} \\ &= n \sum_{j=1}^{n} \frac{(n-1)!}{(j-1)!(n-j)!} P^j Q^{n-j} \end{aligned} \tag{1-29}$$

where use was made of the fact that the $j = 0$ term is zero. Continuing, we have

$$\begin{aligned}\langle x \rangle &= nP \sum_{y=0}^{n-1} \frac{(n-1)!}{(y)!(n-1-y)!} P^y Q^{n-1-y} \\ &= nP(Q+P)^{n-1} = nP \end{aligned} \tag{1-30}$$

where $Q + P = 1$. The second moment may be calculated most easily by noting that $\langle x(x-1) \rangle + \langle x \rangle = \langle x^2 \rangle$. Therefore,

$$\langle x(x-1) \rangle = \int_{-\infty}^{\infty} x(x-1) q(x)\, dx = \sum_{j=0}^{n} x_j (x_j - 1) P(x_j)$$

$$= \sum_{j=2}^{n} \frac{n!}{(j-2)!(n-j)!} P^j Q^{n-j}$$

$$= n(n-1)P^2 \sum_{j=2}^{n} \frac{(n-2)!}{(j-2)!(n-j)!} P^{j-2} Q^{n-j}$$

$$= n(n-1)P^2 \sum_{y=0}^{n-2} \frac{(n-2)!}{y!(n-2-y)!} P^y Q^{n-2-y}$$

$$= n(n-1)P^2 (Q+P)^{n-2} = n(n-1)P^2 \tag{1-31}$$

Thus

$$\langle x^2 \rangle = n^2 P^2 - nP^2 + nP$$
$$= n^2 P^2 + nP(1-P) \tag{1-32}$$

The variance is therefore given by

$$\text{var}(x) = \langle x^2 \rangle - \langle x \rangle^2 = nPQ \tag{1-33}$$

Inserting the previous values for n, P, and Q in the moments above results in $\langle x \rangle = 7.4$, $\langle x^2 \rangle = 56.68$, and $\text{var}(x) = 1.924$. Thus we see that the moments themselves are, in general, not discrete. If it is necessary to have discrete quantities, the fractional values can be rounded off to the nearest integer. However, some descriptors, such as the mode and the cumulative distribution function, are necessarily discrete. For example, the mode is defined as that value of the random variable that has the greatest probability. For the discrete binomial distribution shown in Figure 1-7, it is equal to 8, which is necessarily an integer. General expressions for the S and K will not be derived here for the sake of conciseness, but are instead, calculated using the data shown in Figure 1-7. This yields $S = -0.34$ and $K - 3 = -0.0802$, thus indicating that the binomial distribution, with $n = 10$ and $P = 0.74$, has a tail to the left of the mean and is slightly less peaked than the Gaussian distribution. It is easy to show in this manner that $S < 0$ for $P < \frac{1}{2}$, $S = 0$ for $P = \frac{1}{2}$, and $S > 0$ for $P > \frac{1}{2}$.

It should be noted that in the remaining chapters discrete density functions are shown graphically as continuous functions. This allows several curves to be plotted within the same figure, as shown in Figure 1-8, for the purpose of showing trends due to parameter changes, a process that would produce unintelligible results if done with discrete lines and points. However, care must be exercised in extracting data from such curves since erroneous results can be obtained for certain descriptors, especially for low ranges of the count variable. As a simple example, note that the mode in Figure 1-8*c* appears to be located at $x \approx 7.62$, whereas the correct value, in strict conformance with the definition, is 8. ■

1.2.3. Bivariate Statistics

In Section 1.2.2 we described the statistics associated with a single random variable. This is usually referred to as *first-order* or *univariate statistics*. Most

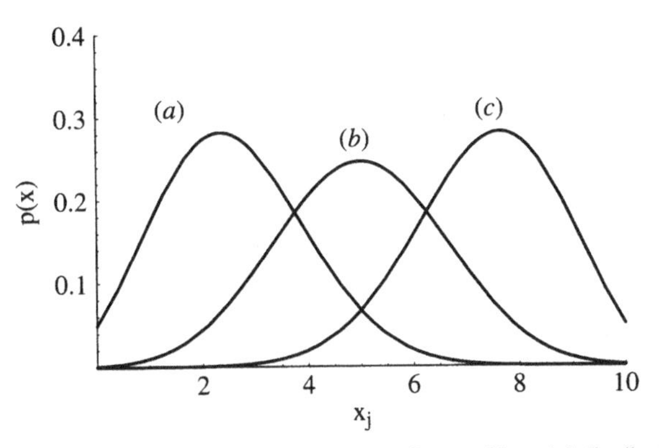

Figure 1-8. Continuous envelope representations of the discrete binomial distribution with parameters $n = 10$ and (*a*) $p = 0.74$, (*b*) $p = 0.5$, and (*c*) $p = 0.74$.

systems of interest involve many more variables, requiring at least *second-order* or *bivariate statistics* for an adequate description. Indeed, even our example case of the golfer attempting to hit a hole-in-one should be described by at least two independent variables, x and y (x being cross-range miss distance and y being down-range miss distance). The univariate probability density function is readily generalized to the *bivariate* case by requiring that the two-dimensional integral of the probability density function equates to the probability that the variables x and y will *jointly* lie between (x_a, x_b) and (y_a, y_b). That is,

$$P(x_a \le x \le x_b, y_a \le y \le y_b) = \begin{cases} \int_{x_a}^{x_b} \int_{y_a}^{y_b} p(x, y)\, dx\, dy & \text{continuous} \\ \int_{x_a}^{x_b} \int_{y_a}^{y_b} q(x, y)\, dx\, dy & \text{discrete} \end{cases} \tag{1-34}$$

where $p(x, y)$ and $q(x, y)$ represent continuous and discrete *joint probability density functions* of the random variables x and y. If x and y are dependent, their associated probabilities and frequencies of occurrence must be related. Thus, if $P(y) = n(y)/n$ represents the probability that y will occur, where n is the total number of events, the *probability* that x will occur under the *condition* that y occurs is given by $P(x \mid y) = n(x, y)/n(y)$. Here $n(x, y)$ is the number of times that x and y occur *jointly* out of the total number of trials n. Using these definitions, we can write for the *conditional probability* that x occurs given that y occurs as

$$P(x \mid y) = \frac{n(x, y)}{n(y)} = \frac{n(x, y)}{n} \frac{n}{n(y)} = \frac{P(x, y)}{P(y)} \tag{1-35}$$

or, in terms of the joint probability of x and y occurring,

$$P(x, y) = P(x \mid y) P(y) \tag{1-36}$$

Equation (1-36) constitutes the fundamental definition of *conditional* and *joint* probabilities. Since $P(x, y)$ is independent of the order of occurrence of x and y, we could also have written

$$P(y, x) = P(y \mid x)P(x) \tag{1-37}$$

Thus

$$P(x \mid y) = \frac{P(y \mid x)P(x)}{P(y)} \tag{1-38}$$

This is known as *Bayes' rule.*

The results above also hold for probability density functions. Thus

$$\begin{aligned} p(x, y) &= p(x \mid y)p(y) \qquad \text{continuous} \\ q(x, y) &= q(x \mid y)q(y) \qquad \text{discrete} \end{aligned} \tag{1-39}$$

where $p(x \mid y)$ and $q(x \mid y)$ are the *conditional probability density functions* for x given that y occurs. Note that if x and y are independent, $p(x \mid y) \rightarrow p(x)$ and $q(x \mid y) \rightarrow q(x)$, such that

$$\begin{aligned} p(x, y) &= p(x)p(y) \qquad \text{continuous} \\ q(x, y) &= q(x)q(y) \qquad \text{discrete} \end{aligned} \tag{1-40}$$

Equation (1-40) is a necessary condition for two variables to be independent since if it is violated, any probabilities calculated using this assumption will lead to unphysical results. Using the previous definitions of conditional probability, Eqs. (1-34) may now be written

$$P(x_a \leq x \leq x_b, y_a \leq y \leq y_b) = \begin{cases} \displaystyle\int_{x_a}^{x_b} \int_{y_a}^{y_b} p(x \mid y)p(y)\, dx\, dy & \text{continuous} \\ \displaystyle\int_{x_a}^{x_b} \int_{y_a}^{y_b} q(x \mid y)q(y)\, dx\, dy & \text{discrete} \end{cases} \tag{1-41}$$

Comment 1-5. Before proceeding, it is important to make clear how the Dirac delta function formalism, introduced earlier for discrete univariate statistics, generalizes to the case of bivariate statistics. Consider the case of two dependent random variables, x and y, each being associated with overlapping sample spaces of discrete points x_j and y_k. Then the joint probability density function of x and y is written

$$q(x, y) = \sum_{j=1}^{\infty} \sum_{k=1}^{\infty} P(x_j \mid y_k)P(y_k)\delta(x - x_j)\delta(y - y_k) \tag{1-42}$$

The *marginal probability density function* for x is then

$$q(x) = \int_{-\infty}^{\infty} q(x, y)\, dy = \sum_{j=1}^{\infty} \sum_{k=1}^{\infty} P(x_j \mid y_k) P(y_k) \delta(x - x_j) \tag{1-43}$$

and similarly for $q(y)$. The probability of x_j occurring is then

$$P(x_{j-1} < x \leq x_j) = \int_{x_{j-1}}^{x_j} q(x)\, dx = \sum_{k=1}^{\infty} P(x_j \mid y_k) P(y_k) \tag{1-44}$$

and of y_k occurring is

$$P(y_{k-1} < y \leq y_k) = \int_{y_{k-1}}^{y_k} q(y)\, dx = \sum_{j=1}^{\infty} P(y_k \mid x_j) P(x_j) \tag{1-45}$$

where the ranges of integration are understood to include only x_j and y_k, respectively, in accordance with Comment 1-1. ■

Using the results of Comments 1-1 and 1-5, the discrete density function in Eqs. (1-41) can be rewritten such that the pair now reads

$$P(x_a \leq x \leq x_b, y_a \leq y \leq y_b) = \begin{cases} \displaystyle\int_{x_a}^{x_b} \int_{y_a}^{y_b} p(x \mid y) p(y)\, dx\, dy & \text{continuous} \\ \displaystyle\sum_{j=m}^{n} \sum_{k=r}^{s} P(x_j \mid y_k) P(y_k) & \text{discrete} \end{cases} \tag{1-46}$$

Here the integration regions include the discrete points (m, n) and (r, s). If (m, n) and (r, s) extend over all sample space, then

$$\begin{aligned} &P(-\infty \leq x \leq \infty, -\infty \leq y \leq \infty) \\ &\quad = \int_{-\infty}^{\infty} \int_{-\infty}^{\infty} p(x \mid y) p(y)\, dx\, dy = 1 \qquad \text{continuous} \\ &P(x_1 \leq x_j \leq x_\infty, y_1 \leq y_k \leq y_\infty) \\ &\quad = \sum_{j=1}^{\infty} \sum_{k=1}^{\infty} P(x_j \mid y_k) P(y_k) = 1 \qquad \text{discrete} \end{aligned} \tag{1-47}$$

as required. For independent variables, Eq. (1-47) reduces to products of univariate probabilities, that is,

$$
\begin{aligned}
&P(-\infty \leq x \leq \infty, -\infty \leq y \leq \infty) \\
&\quad = \int_{-\infty}^{\infty} p(x)\, dx \int_{-\infty}^{\infty} p(y)\, dy = 1 \qquad \text{continuous} \\
&P(x_1 \leq x_j \leq x_\infty, y_1 \leq y_k \leq y_\infty) \\
&\quad = \sum_{j=1}^{\infty} P(x_j) \sum_{k=1}^{\infty} P(y_k) = 1 \qquad \text{discrete}
\end{aligned}
\tag{1-48}
$$

Similarly, the *joint cumulative distribution function* for the bivariate case is a straightforward generalization of the univariate case, that is,

$$
\begin{aligned}
F(x', y') &= \int_{-\infty}^{x'} \int_{-\infty}^{y'} p(x, y)\, dx\, dy \qquad \text{continuous} \\
G(x_n, y_s) &= \int_{-\infty}^{x_n} \int_{-\infty}^{y_s} q(x, y)\, dx\, dy \qquad \text{discrete}
\end{aligned}
\tag{1-49}
$$

Comment 1-6. Using Eq. (1-42) for the case of two random variables x and y, the probability of a golfer making exactly a birdie on a par-three hole can be calculated from the expression $P(x_1) = P(x_1 \mid y_1)P(y_1)$. Here the x_1 are the events related to putting the ball into the cup and the y_1 are the events related to driving the ball onto the green. Specifically, having (hypothetically) sampled the performance of our golfer over many trials, it is found that the probability of making a birdie on a particular par-three hole is dependent on whether or not the first shot lands on the green. It is found that the probability of putting the second shot in the cup, given that the first shot is on the green, is $P(x_1 \mid y_1) = 0.1$. Thus if the probability of a successful drive is, say, $P(y_1) = 0.75$, the probability of making a birdie is $P(x_1) = 0.75 \times 0.1 = 0.075$ or about 1 chance in 13 tries. ■

Comment 1-7. Consider now the joint probability of making a birdie given that the first shot does not land on the green. There are several possibilities for this case. For example, it might be found that the drive can land in the fairway with probability $P(y_2) = 0.1$, or in the rough with probability $P(y_3) = 0.1$, or in a bunker with probability $P(y_4) = 0.05$. The probability of the ball landing on the green is, from Comment 1-6, $P(y_1) = 0.75$, such that $\sum_{k=1}^{4} P(y_k) = 1$ as required, since there are no other (hypothesized) possibilities. Thus, if we assume for simplicity that the probability of making the second shot is about equally difficult in all the off-green cases, say $P(x_1 \mid y_k) = 0.01$, the probability of making a birdie, given that the drive is not on the green, is $P(x_1) = (0.1 +$

$0.1 + 0.05) \times (0.01) = 0.0025$, or about 1 chance in 400 tries! The total probability of all four events is $P(x_1) = (0.1 + 0.1 + 0.05 + 0.75) \times (0.01) = 0.01$, or 1 chance in 100 tries. Note that the total probability does not sum to 1. That is because other possibilities x_j have not been accounted for, such as making a par, a bogey, and so on. ■

Joint moments are straightforward generalizations of the univariate moments. Thus, for the moments

$$m_{jk} = \int_{-\infty}^{+\infty} \int_{-\infty}^{+\infty} x^j y^k p(x, y)\, dx\, dy \tag{1-50}$$

and the central moments

$$\mu_{jk} = \int_{-\infty}^{+\infty} \int_{-\infty}^{+\infty} (x - m_{10})^j (y - m_{01})^k p(x, y)\, dx\, dy \tag{1-51}$$

where m_{10} and m_{01} are defined by Eq. (1-11). By convention, μ_{20} and μ_{02} are usually referred to as the *variances* σ_1^2 and σ_2^2, respectively, of the random variables x and y. The discrete moments follow in accordance with the procedures outlined above.

μ_{11} is called the *covariance* of the two variables x and y and is zero when x and y are independent [i.e., when $p(x, y) = p(x)p(y)$ or $q(x, y) = q(x)q(y)$]. When x and y are dependent, μ_{11} becomes a measure of the degree of *correlation* of the variables with a magnitude determined by the units of measurement for x and y. If the correlation between x and y is such that a positive change in x (or y) corresponds to a positive change in y (or x), μ_{11} is positive. On the other hand, if a positive change in x (or y) corresponds to a negative change in y (or x), μ_{11} is negative. To remove the dependence on the measurement units and obtain a standardized measure of correlation between the variables, a *normalized covariance* function called the *correlation coefficient*, ρ, is defined as

$$\rho = \frac{\mu_{11}}{\sigma_1 \sigma_2} \tag{1-52}$$

where σ_1 and σ_2 are standard deviations of the random variables and $-1 \leq \rho \leq +1$. Thus if $\rho = \pm 1$, the variables are said to be *fully correlated*, either positively or negatively, and if $\rho = 0$, they are said to be *fully uncorrelated*.

1.2.4. Transformation of Random Variables

It is frequently the case that one must transform to a new set of variables, or some function of the original variables, in order to solve a problem of interest. For example, one may wish to convert the probability density function representing the amplitude, A, of the radiation field to the intensity, I, where $I = A^2$. This is a *one-to-one transformation*, which maps one set of coordinates onto another.

We consider the one- and two-dimensional cases, although the procedure can be generalized to any number of dimensions. For more detailed discussions, the reader is referred to the many texts on probability theory and integral calculus.

The simplest case is that of a one-dimensional transformation where the probability density function $p(x)$ of the random variable x is to be converted to a new probability density function $p(y)$ of the random variable y, where $y = f(x)$. This often occurs in detection theory as, for example, when the statistics of the detector output voltage is transformed to that of the output power. Consider first the case of $f(x)$ being a monotonically increasing differentiable function of x with a one-to-one or single-valued correspondence between x and y. Then, from Figure 1-9*a*, we find that $P(x \leq x') = P(y \leq y')$ or, in terms of distribution functions, $F(x') = F(y')$. Dropping the prime notation for the moment, the last expression can be differentiated with respect to y, yielding

$$\frac{dF(x)}{dy} = \frac{dF(x)}{dx}\frac{dx}{dy} = \frac{dF(y)}{dy} \tag{1-53}$$

Using the differential relationship between the probability density function and the cumulative distribution function given by Eq. (1-9), Eq. (1-53) becomes

$$p(x)\,dx = p(y)\,dy \tag{1-54}$$

This may be rewritten as

$$p(y) = p[f^{-1}(y)]\frac{dx}{dy} \tag{1-55}$$

where $f^{-1}(y)$ is the inverse of $y = f(x)$.

If we perform the same procedure for the case of a single-valued monotonically decreasing function of x, it follows from Figure 1-9*b* that $F(x') = 1 - F(y')$ and Eq. (1-55) becomes

$$p(y) = -p[f^{-1}(y)]\frac{dx}{dy} \tag{1-56}$$

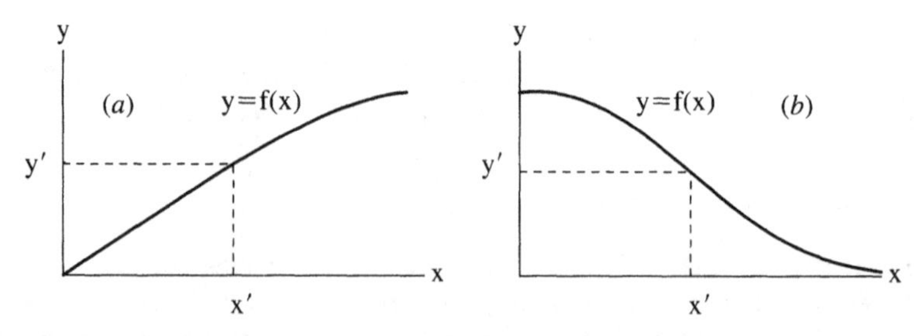

Figure 1-9. Relationship between the cumulative distribution functions $F(x')$ and $F(y')$ for single-valued functions $y = f(x)$ that are (*a*) monotonically increasing and (*b*) monotonically decreasing. In case (*a*), $F(x') = F(y')$; in case (*b*) $F(x') = 1 - F(y')$.

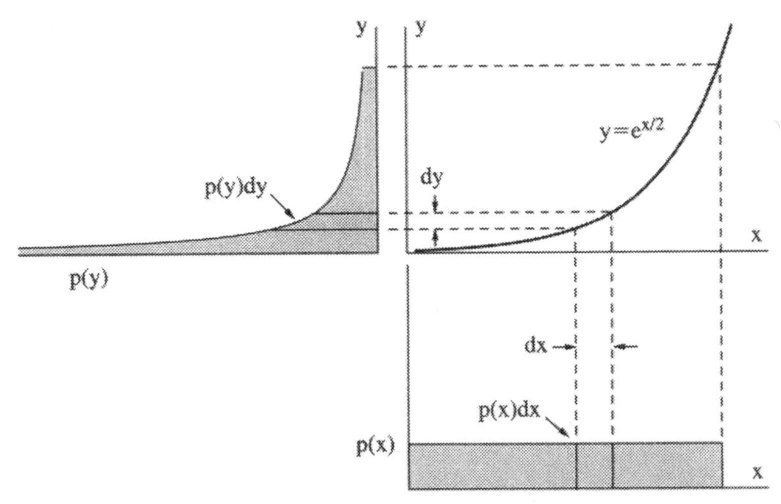

Figure 1-10. Graphical representation of the transformation of a uniform probability density function $p(x) = 1$ to $p(y) = 1/2y$, where the transformation is $y = e^{x/2}$.

But dx/dy is negative for a decreasing function of x, so that Eqs. (1-55) and (1-56) can be written as a single expression, namely,

$$p(y) = p[f^{-1}(y)]\left|\frac{dx}{dy}\right| \tag{1-57}$$

where the vertical bars indicate absolute value. When integrated over the appropriate limits, Eq. (1-54) ensures that the integrated area (i.e., probability) is the same in both x and y spaces.

Comment 1-8. Consider the case of a uniform distribution of the random variable x given by $p(x) = 1$ over the interval $0 \le x \le 1$. Assume a new random variable y related to x through the relationship $y = e^{x/2}$. Using Eq. (1-57), the transformed distribution becomes $p(y) = 1/2y$. This transformation is shown graphically in Figure 1-10. Note that the areas $p(x)\,dx$ and $p(y)\,dy$ are equal for any x in the interval (0, 1), which requires that $p(y) \to \infty$ as $x \to 0$. The variable y is referred to as a *log-normal random variable*. ■

In the case of multivalued functions, such as $y = x^2$, the transformation must take into account contributions from each monotonic or single-valued portion of the transformation function, since these represent mutually exclusive events. In such cases it is not too difficult to show that Eq. (1-57) generalizes to

$$p(y) = p(x_1)\left|\frac{dx_1}{dy}\right| + p(x_2)\left|\frac{dx_2}{dy}\right| + \cdots \tag{1-58}$$

where the x_j are the monotonically increasing or decreasing segments of the function $p(x)$.

Two-dimensional transformations take the form

$$x = f(u, v) \qquad y = g(u, v) \tag{1-59}$$

where the two-dimensional joint probability density function $p(x, y)$ is to be transformed to a new probability density function $p(u, v)$. Generalization of Eq. (1-54) to two dimensions yields

$$p(u, v)\, du\, dv = p(x, y)\, dx\, dy \tag{1-60}$$

Thus the transformation becomes

$$p(u, v) = p[f(u, v), g(u, v)]|J(x, y : u, v)| \tag{1-61}$$

where $|J(x, y : u, v)| = |\partial(x, y)/\partial(u, v)|$ is the absolute value of the Jacobian of the transformation. The Jacobian is defined by the determinant

$$J = \begin{vmatrix} \dfrac{\partial x}{\partial u} & \dfrac{\partial x}{\partial v} \\ \dfrac{\partial y}{\partial u} & \dfrac{\partial y}{\partial v} \end{vmatrix} \tag{1-62}$$

which is well known from the theory of integral calculus. This type of transformation may be generalized to n dimensions via an n-dimensional Jacobian, but this is not discussed here.

Comment 1-9. For the golf example mentioned earlier, it is instructive to calculate the most probable miss distance, the average miss distance, and the probability of making a hole-in-one if the standard deviation of the shots in the cross-range and down-range directions are ± 10 ft and the radius of the hole is 2 in.

Clearly, this problem must be described by a two-dimensional probability density function, x representing the cross-range error and y the down-range error. In order to simplify the problem, we assume that the distributions in x and y are both Gaussian with zero means and equal variances. The latter restrictions imply that this particular golfer has no biases in his shots in either the x or y directions ($\langle x \rangle = \langle y \rangle = 0$), that he has equal spreads in both cross-range and down-range directions ($\sigma_x = \sigma_y = \sigma$), and that there are no external influences, such as wind. The probability density function becomes

$$p(x, y) = p(x)p(y) = \frac{1}{2\pi\sigma^2} e^{-(x^2+y^2)/2\sigma^2} \tag{1-63}$$

where the individual density functions

$$p(x) = \frac{1}{\sqrt{2\pi\sigma^2}}e^{-x^2/2\sigma^2} \qquad p(y) = \frac{1}{\sqrt{2\pi\sigma^2}}e^{-y^2/2\sigma^2} \tag{1-64}$$

are considered independent. This is the same problem as throwing darts at a dartboard or firing bullets at a target. Equation (1-63) is shown in Figure 1-11. To calculate the various probabilities, it is more convenient to transform to the polar coordinates (r, θ), where $x = r\cos\theta$, $y = r\sin\theta$. Using these transformations in Eq. (1-61) (expressed in a slightly different but equivalent form) yields

$$p(r,\theta) = p(x,y)\begin{vmatrix} \dfrac{\partial x}{\partial r} & \dfrac{\partial x}{\partial \theta} \\ \dfrac{\partial y}{\partial r} & \dfrac{\partial y}{\partial \theta} \end{vmatrix}_{\substack{x\to r\cos\theta \\ y\to r\sin\theta}} \tag{1-65}$$

or

$$p(r,\theta) = \frac{re^{-r^2/2\sigma^2}}{2\pi\sigma^2}(\cos^2\theta + \sin^2\theta) = \frac{re^{-r^2/2\sigma^2}}{2\pi\sigma^2} \tag{1-66}$$

Assuming a uniform distribution for θ over the interval $0 \le \theta \le 2\pi$, we obtain the *marginal* probability density function for the random variable r, which is the Rayleigh distribution discussed in Comment 1-3, that is,

$$p(r) = \frac{re^{-r^2/2\sigma^2}}{\sigma^2} \tag{1-67}$$

Using the Rayleigh descriptors from Comment 1-3, the most probable value is clearly at the mode of the distribution, which is located at $r_p = \sigma = 10$ ft from the hole. Using Eq. (1-21), the average miss distance occurs at $\langle r\rangle = \sqrt{\pi/2}\,\sigma \approx$ 12.5 ft. Here one must be careful in interpreting r_p and $\langle r\rangle$. If enough golf balls

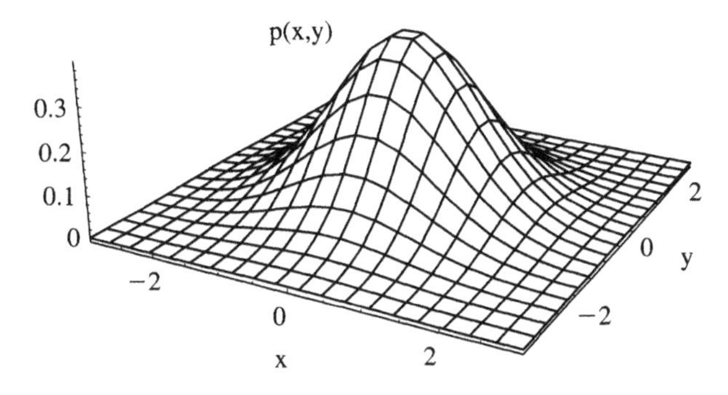

Figure 1-11. Two-dimensional joint Gaussian probability density function.

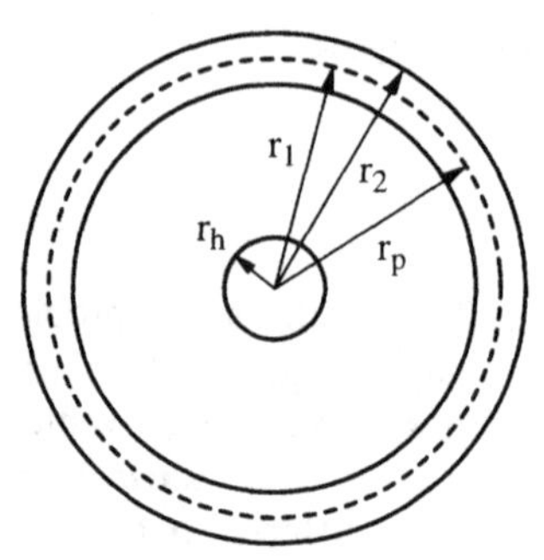

Figure 1-12. Regions of integration in calculation of the probability of a hole-in-one assuming Rayleigh statistics for the marginal probability density function for the radius r.

are hit, the resulting pattern becomes a two-dimensional Gaussian distribution with most of the balls near the hole as shown in Figure 1-11. However, r_p and $\langle r \rangle$ seem to indicate otherwise. This apparent contradiction can be understood when it is realized that the Rayleigh distribution, when integrated between r_1 and r_2, includes the area associated with an annular ring of width $r_1 - r_2$. Since the area of an annular ring of fixed width increases with radius while the Gaussian probability distribution decreases with radius, there will clearly be some radius $r > 0$ at which the probability of occurrence is a maximum, namely r_p. Similar arguments can be made about $\langle r \rangle$.

Assuming a uniform density function for θ, Eq. (1-67) can be integrated over the radius of the hole (i.e., $r_h = 2$ in.) to obtain the probability for a hole-in-one. From Figure 1-12 we have

$$P = \frac{1}{\sigma^2} \int_{r_1}^{r_2} r e^{-r^2/2\sigma^2} dr = e^{-r_1^2/2\sigma^2} - e^{-r_2^2/2\sigma^2} \tag{1-68}$$

With $r_1 = 0$ in. and $r_2 = 2$ in. and an assumed standard deviation of $\sigma = 10$ ft, $P = 2.77 \times 10^{-4}$ or 1 chance in 3600 tries! However, notice that for $r_1 = \sigma - 1$ in. and $r_2 = \sigma + 1$ in. which corresponds to a 2-in. ring around the hole at a mean radius of $(r_1 + r_2)/2 = \sigma$, the probability becomes $P = 0.0101$ or 1 chance in only 100 tries. ■

1.2.5. Characteristic Function

It is frequently desirable to obtain the density function of the sum of two or more *independent* random variables. For example, the statistics of a random variable $z = x + y$ may be desired, where x might represent the signal and y the noise in an optical receiver. The probability density functions for x and y are assumed known and given by $p(x)$ and $p(y)$, respectively. To find the statistics of z, one must find the probability density function of z [i.e., $p(z)$]. We can approach this problem by defining another set of variables:

$$z = x + y$$
$$\zeta = x \tag{1-69}$$

From Eq. (1-60) with $p(x, y) = p(x)p(y)$, we can write

$$\int_z \int_\zeta p(z, \zeta)\, dz\, d\zeta = \int_x \int_y p(x)p(y)\, dx\, dy \tag{1-70}$$

which can be rewritten as

$$\int_z \int_\zeta p(z, \zeta)\, dz\, d\zeta = \int_z \int_\zeta p(\zeta)p(z - \zeta)\, dz\, d\zeta \tag{1-71}$$

so that

$$p(z, \zeta) = p(\zeta)p(z - \zeta) \tag{1-72}$$

Finally, the desired probability density function $p(z)$ can be obtained by integrating over all values of ζ, that is,

$$p(z) = \int_{-\infty}^{\infty} p(\zeta)p(z - \zeta)\, d\zeta \tag{1-73}$$

This integral is referred to as a *convolution integral*. The arguments of the integral have the interesting and useful property that the Fourier transform of $p(z)$ may be obtained from the product of the Fourier transforms of $p(x)$ and $p(y)$. This is known as the *convolution theorem* and is written

$$G_z(iv) = G_x(iv)G_y(iv) \tag{1-74}$$

where

$$G_x(iv) = \int_{-\infty}^{+\infty} p(x)e^{ivx}dx$$
$$G_y(iv) = \int_{-\infty}^{+\infty} p(y)e^{ivy}dy \tag{1-75}$$

$G_x(iv)$ and $G_y(iv)$ are referred to as *characteristic functions* and correspond to the Fourier transforms of $p(x)$ and $p(y)$, respectively. $p(z)$ may then be obtained from the inverse transform of $G_z(iv)$. The complex number $i = \sqrt{-1}$ is included explicitly in the argument of G in order to make clear that the characteristic function is generally complex, whereas $p(x)$ is always real. Proof that Eq. (1-74) follows from Eq. (1-73) is easily shown by taking the Fourier transform of both sides of Eq. (1-73) followed by a change in the order of integration.

Comment 1-10. Depending on the sign conventions chosen, the characteristic function given by Eqs. (1-75) may correspond to either forward or reverse Fourier transforms. In particular, integral Fourier transforms may be defined in a variety of ways by appropriately defining the arbitrary parameters a and b in the generalized expressions

$$F(\omega) = \left[\frac{|b|}{(2\pi)^{1-a}}\right]^{1/2} \int_{-\infty}^{\infty} f(t)e^{-ib\omega t}\,dt \tag{1-76}$$

and

$$f(t) = \left[\frac{|b|}{(2\pi)^{1+a}}\right]^{1/2} \int_{-\infty}^{\infty} F(\omega)e^{ib\omega t}\,dt \tag{1-77}$$

In the chapters that follow we let $(a, b) = (1, 1)$ for the Fourier integrals and $(a, b) = (1, -1)$ for the characteristic functions, where $F(\omega) \to G(iv)$ and $f(t) \to p(x)$ in the latter case. Thus the characteristic function $G(iv)$ corresponds to a reverse or inverse Fourier transform [i.e., Eq. (1-77)] times 2π, a standard convention in probability theory. ■

From Eqs. (1-75) we see that the characteristic function can also be viewed as the average of e^{ivx} via the definition given by Eq. (1-11); that is, $G_x(iv) = \langle e^{ivx}\rangle$. Knowing this, the procedure can be generalized to the sum of an arbitrary number of *statistically independent variables*, n, (i.e., $z = x_1 + x_2 + x_3 + \cdots + x_n$). We have

$$\begin{aligned} G_z(iv) &= \langle e^{ivz}\rangle = \langle e^{iv(x_1+x_2+\cdots+x_n)}\rangle \\ &= \langle e^{ivx_1}\rangle\langle e^{ivx_2}\rangle \cdots \langle e^{ivx_n}\rangle \\ &= G_{x_1}(iv)G_{x_2}(iv)\cdots G_{x_n}(iv) \\ &= \prod_{i=1}^{n} G_{x_i}(jv) \end{aligned} \tag{1-78}$$

Equation (1-78) states that the characteristic function of the sum of n independent variables is equal to the product of the characteristic functions of the individual variables.

A power series expansion of $G_z(iv)$ shows that

$$G_z(iv) = \langle e^{ivz}\rangle = \left\langle \sum_{n=0}^{\infty} \frac{(ivz)^n}{n!} \right\rangle = \sum_{n=0}^{\infty} \frac{(iv)^n}{n!}\langle z^n\rangle = \sum_{n=0}^{\infty} \frac{(iv)^n}{n!} m_n \tag{1-79}$$

This implies that all the moments of a distribution may be obtained directly from an expansion of the characteristic function of that distribution. Indeed, a *moment generating function* can be defined, involving the derivative of the characteristic function evaluated at $v = 0$, which generates moments directly [i.e., without the associated coefficients apparent in Eq. (1-79)]. To show this, successive derivatives of $G_z(iv)$ are taken, yielding

$$\left.\frac{dG}{dv}\right|_{v=0} = i\int_{-\infty}^{\infty} xp(x)\,dx = im_1 \tag{1-80}$$

and

$$\left.\frac{d^2G}{dv^2}\right|_{v=0} = i^2\int_{-\infty}^{\infty} x^2p(x)\,dx = -m_2 \tag{1-81}$$

Generalizing these results, while taking into account the i^n factors, leads to a moment generating function given by

$$m_n = i^{-n}\frac{d^nG}{dv^n} \tag{1-82}$$

Other moment generating functions can be defined which do not include complex variables: for example,

$$G_x(t) = \int_{-\infty}^{\infty} p(x)e^{xt}dx \tag{1-83}$$

where

$$m_n = \frac{d^nG}{dt^n} \tag{1-84}$$

However, there is no guarantee that the integral in Eq. (1-83) exists, whereas the integrals in Eq. (1-75) always exist and will therefore be used throughout the book.

Comment 1-11. Equation (1-78) implies that the mean and variance of the sum of n independent variables is equal to the sum of the means and variances of the individual variables. To show this for the Gaussian probability density function, we let the random variable be given by $z = \sum_{j=1}^{n} x_j$ and the individual Gaussian probability density functions by

$$p_j(x_j) = \frac{1}{\sqrt{2\pi\sigma_j^2}}e^{-(x_j-\langle x_j\rangle)^2/2\sigma_j^2} \tag{1-85}$$

Then, by changing variables to $t^2 = [(x - \langle x\rangle)/\sigma - i\sigma v]$, we obtain for the Fourier transform of the probability density function of a single Gaussian variate,

$$\begin{aligned} G_{x_j}(iv) &= \int_{-\infty}^{+\infty} e^{ivx_j}p(x_j)\,dx_j \\ &= \frac{1}{\sqrt{2\pi}}e^{iv\langle x_j\rangle-\sigma_j^2v^2/2}\int_{-\infty}^{+\infty} e^{-t^2/2}dt \\ &= e^{iv\langle x_j\rangle-\sigma_j^2v^2/2} \end{aligned} \tag{1-86}$$

where $(2\pi)^{-1/2}\int_{-\infty}^{+\infty} e^{-t^2/2}dt = 1$. Hence

$$G_z(i\upsilon) = \prod_{j=1}^{n} G_{x_j}(i\upsilon)$$
$$= e^{i(\langle x_1\rangle+\langle x_2\rangle\cdots\langle x_n\rangle)\upsilon-(\sigma_1^2+\sigma_2^2\cdots\sigma_n^2)\upsilon^2/2}$$
$$= e^{i\langle z\rangle\upsilon-\sigma^2\upsilon^2/2} \tag{1-87}$$

Using Eq. (1-82) we find for the mean and variance of the summed random variable z:

$$\langle z\rangle = \sum_{j=1}^{n}\langle x_j\rangle \quad \text{and} \quad \sigma^2 = \sum_{j=1}^{n}\sigma_j \tag{1-88}$$

respectively. An inverse transformation on Eq. (1-87) will, by definition, produce a Gaussian distribution with mean $\overline{z}$ and variance σ^2, that is,

$$p(z) = \frac{1}{\sqrt{2\pi\sigma^2}}e^{-(z-\langle z\rangle)^2/2\sigma^2} \tag{1-89}$$

■

1.2.6. Central Limit Theorem

The *central limit theorem* is probably the most important theorem in statistics. The theorem states that the probability density function $p(z)$ representing the random variable $z = \sum_{j=1}^{n} x_j$ will converge to a Gaussian distribution function in the limit of large n, independent of the form of the probability density functions associated with the individual x_j. This assumes, of course, that all the random variables that are to be added have similar magnitudes such that no single density function will dominate the distribution. The convergence to a Gaussian distribution was clearly shown in Comment 1-11 for the case of a Gaussian distribution. However, the fact that the initial function was a Gaussian, whose Fourier transform is also a Gaussian, may not have convinced the reader of the generality of the theorem. Two additional examples will now be given that are less obvious in their conclusions and that will be useful in later chapters.

Consider the case of a set of random variables x_j each of which is described by a negative exponential probability density function given by

$$p_j(x_j) = \frac{1}{a_j}e^{-x_j/a_j} \tag{1-90}$$

The characteristic function may again be used to calculate the probability density function corresponding to n random variables that obey Eq. (1-90). Thus

$$G_j(i\upsilon) = \int_0^{\infty} a_j^{-1}e^{-x_j/a_j+i\upsilon x_j}dx_j$$
$$= \frac{1}{1-ia_j\upsilon} \tag{1-91}$$

The characteristic function for the sum of n of such random variables is

$$G^{(n)}(iv) = \prod_{j=1}^{n}(1 - ia_j v)^{-1} = (1 - iav)^{-n} \qquad \text{where all } a_j = a \tag{1-92}$$

The inverse transformation of Eq. (1-92) can be performed using a contour integration in the lower half of the complex plane. (cf. Appendix B). The result yields the *gamma distribution*, given by

$$p(z) = \begin{cases} \dfrac{a^n z^{n-1} e^{-az}}{\Gamma(n)} & z \geq 0 \\ 0 & z < 0 \end{cases} \tag{1-93}$$

where $\Gamma(n)$ is the *gamma function* defined by the integral $\Gamma(n) = \int_0^\infty t^{n-1} e^{-t} dt$ (cf. Appendix A) and a is a parameter. As will be seen in Chapter 4, the gamma distribution plays a fundamental role in the statistical description of radiation scattered from a rough surface. The question arises as to whether the gamma distribution can be approximated by a Gaussian distribution as n approaches infinity in accordance with the central limit theorem. To show that this is indeed the case, we first note that Eq. (1-92) equates to unity at $v = 0$ and decreases rapidly away from the origin as n is increased. Thus, for large n we can expand Eq. (1-92) about the origin, obtaining

$$G^{(n)}(iv) = 1 + ianv - \tfrac{1}{2}n(n-1)a^2v^2 + \cdots \tag{1-94}$$

Identifying the right side of Eq. (1-94) as the leading terms in the expansion of a Gaussian function yields

$$G^{(n)}(iv) \approx e^{inav - (1/2)n^2a^2v^2} \tag{1-95}$$

where the approximation $n(n-1) \approx n^2$ is used. But Eq. (1-95) is of the same form as Eq. (1-87) under the substitutions $\langle z \rangle \rightarrow na$ and $\sigma^2 \rightarrow n^2a^2$. Thus the inverse transformation must yield

$$p(z) \approx \frac{1}{\sqrt{2\pi n^2 a^2}} e^{-(z-na)^2/2n^2a^2} \tag{1-96}$$

and we see that the gamma distribution does approach a Gaussian distribution for large n as required by the central limit theorem. It must be kept in mind, however, that Eq. (1-96) is only an approximation that is valid for large n and small values of $z - na$. Equation (1-96) is plotted in Figure 1-13 for several

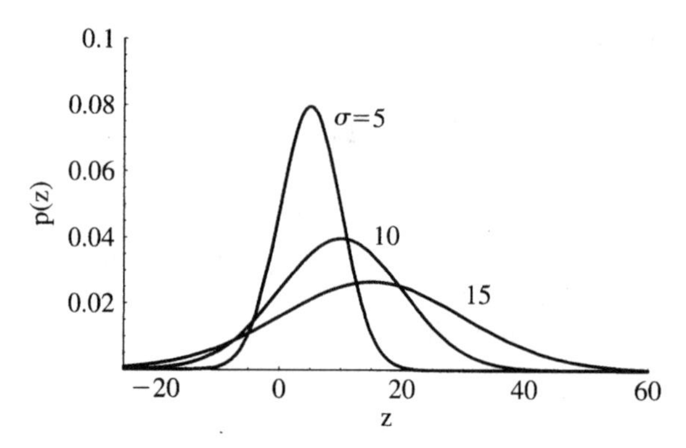

Figure 1-13. Gaussian approximations to the characteristic function for the gamma distribution.

values of $\sigma \equiv na$. Note that the mean is a function of σ as is the case for the gamma distribution.

The second example is that of the sum of n randomly phased sine waves. This has importance in modeling the noise properties of a narrowband filter, as discussed in Chapter 5. Consider the sine function

$$x = a \sin(\omega t + \phi) \tag{1-97}$$

where ϕ is a random variable. Letting $\theta = \omega t + \phi$ represent the randomly varying phase of the sine function, Eq. (1-97) can be written simply as

$$x = a \sin \theta \tag{1-98}$$

where θ is uniformly distributed over the interval $(0, 2\pi)$. The probability density function for θ is therefore given by $p(\theta) = 1/2\pi$. The density function for x is obtained by the procedures outlined in Section 1.2.4. Thus the probability density function for a randomly phased sine wave is

$$\begin{aligned} p(x) &= p(\theta) \frac{d\theta}{dx}\bigg|_{\theta=\sin^{-1}(x/a)} \\ &= \frac{1}{2\pi\sqrt{a^2 - x^2}} \end{aligned} \tag{1-99}$$

For the sum of n such sine waves, the random variable becomes

$$y = \sum_{j=1}^{n} a_j \sin \theta_j \tag{1-100}$$

where the θ_j are statistically independent and the a_j are distributed in a deterministic fashion (i.e., the a_j are not random variables). The characteristic function for the jth sine wave can be most easily obtained by averaging over θ_j, that is,

$$G_j(iv) = \int_0^{2\pi} p(\theta) e^{i a_j v \sin\theta_j} d\theta_j = \frac{1}{2\pi} \int_0^{2\pi} e^{i a_j v \sin\theta_j} d\theta_j \tag{1-101}$$

But this last form may be recognized as the definition of the *Bessel function of the first kind, zeroth order*. Thus

$$G_j(iv) = J_0(a_j v) \tag{1-102}$$

The series expansion for J_0 is given by

$$J_0(x) = \sum_{k=0}^{\infty} \frac{(-x^2/4)^k}{(k!)^2} = 1 - \frac{x^2}{4} + \cdots \tag{1-103}$$

which to order x^2 can be approximated by

$$J_0(x) \approx e^{-(1/4)x^2} \tag{1-104}$$

The characteristic function for the random variable y can therefore be written

$$G^{(n)}(iv) = \prod_{j=1}^{n} G_j(iv) \approx \prod_{j=1}^{n} e^{-(1/4)a_j^2 v^2} = e^{-(1/2)\sigma^2 v^2} \tag{1-105}$$

where $\sigma^2 = n a_j^2/2$. Thus the inverse transformation once again leads to a Gaussian probability density function given by

$$p(y) = \frac{1}{\sqrt{2\pi\sigma^2}} e^{-y^2/2\sigma^2} \tag{1-106}$$

It can be shown that the procedures above, which do not constitute a rigorous proof of the central limit theorem, may be generalized to the case of any probability density function[3] and is therefore a strong indication of the validity of the theorem. In general, it is found that the convergence to the Gaussian distribution is more rapid near the mean of the distribution than in the wings. Indeed, density functions with long tails, such as the log-normal distribution, tend to converge more slowly than more compact functions. In the case of the log-normal distribution, which is important for later chapters, it has been shown[4] that the convergence to the Gaussian distribution for large n is so slow that it is usually more correct to consider the limiting form as remaining log-normal. This behavior has been observed in numerous physical experiments, including optical propagation in a turbulent atmosphere and radio-wave propagation in the ionosphere. More will be said about the log-normal distribution in Chapters 3, 4, and 5.

1.2.7. Chi-Squared Distribution

The χ^2 *distribution* occurs frequently in detection theory and represents the distribution of the *sum of squares* of n normally distributed random variables. A random variable x is said to have a χ^2 distribution if its density function is of the form

$$p(x) = \begin{cases} \dfrac{x^{\alpha/2-1}e^{-x/2}}{2^{\alpha/2}\Gamma(\alpha/2)} & x \geq 0 \\ 0 & x < 0 \end{cases} \tag{1-107}$$

where the $\Gamma(\alpha/2)$ is the gamma function and α is referred to as the *number of degrees of freedom* of the density function $p(x)$. The characteristic function for $p(x)$ is given by

$$\begin{aligned} G(iv) &= \int_0^\infty e^{ivx} p(x)\, dx \\ &= \frac{2^{-\alpha/2}}{\Gamma(\alpha/2)} \int_0^\infty e^{-(x/2)(1-2iv)} x^{\alpha/2-1} dx \end{aligned} \tag{1-108}$$

Letting $z = (1 - 2iv)x$, Eq. (1-108) becomes

$$G(iv) = \frac{1}{(1-2iv)^{\alpha/2}} \int_0^\infty \frac{e^{-(z/2)} z^{\alpha/2-1}}{2^{\alpha/2}\Gamma(\alpha/2)} dz \tag{1-109}$$

But the expression under the integral sign is just the definition of the χ^2 distribution, which integrates to 1. Thus

$$G(iv) = \frac{1}{(1-2iv)^{\alpha/2}} \tag{1-110}$$

To show that the χ^2 distribution represents the sum of squares of n normally distributed random variables, consider the single random variable $y = x^2$, where x is normally distributed with zero mean and unity variance:

$$p(x) = \frac{e^{-x^2/2}}{\sqrt{2\pi}} \tag{1-111}$$

The characteristic function for $p(y)$ is

$$\begin{aligned} G_y(iv) &= \frac{1}{2\sqrt{2\pi}} \int_{-\infty}^\infty \frac{e^{ivy} e^{-y/2}}{\sqrt{y}} dy \\ &= \frac{1}{\sqrt{2\pi}} \int_{-\infty}^\infty e^{ivx^2} e^{-x^2/2} dx \\ &= \frac{1}{(1-2iv)^{1/2}} \end{aligned} \tag{1-112}$$

where a change of variables $y = x^2$ was used. Now for α such random variables, where $y = \sum_{j=1}^{\alpha} x^2$, we have

$$G_y^{(\alpha)}(i\upsilon) = \prod_{j=1}^{\alpha} G_{yj}(i\upsilon)$$

$$= \frac{1}{(1 - 2i\upsilon)^{\alpha/2}} \tag{1-113}$$

which agrees with Eq. (1-110).

Thus far we have shown that the sum of squares of α independent Gaussian random variables with *zero mean* and unity variance yields the χ^2 distribution. We will now show that the sum of squares of α independent Gaussian random variables with *nonzero* mean and unity variance yields the *noncentral* χ^2 *distribution*.

Consider the characteristic function of a single normally distributed random variable with mean μ and unity variance:

$$G_y(i\upsilon) = \frac{1}{\sqrt{2\pi}} \int_{-\infty}^{\infty} e^{i\upsilon x^2} e^{-(x-\mu)^2/2} dx \tag{1-114}$$

Completing the square in the exponent, we have

$$G_y(i\upsilon) = \frac{1}{\sqrt{2\pi}} \int_{-\infty}^{\infty} e^{-(x^2-2x\mu+\mu^2)^2/2+i\upsilon x^2} dx$$

$$= \frac{1}{\sqrt{2\pi}} e^{(\mu^2/2)(1/b^2-1)} \int_{-\infty}^{\infty} e^{-(bx-\mu/b)^2/2} dx \tag{1-115}$$

where $b^2 = 1 - 2i\upsilon$. Letting $z = bx$, the integral is readily solved to yield

$$G_y(i\upsilon) = \frac{e^{-(\mu^2/2)[1-1/(1-2i\upsilon)]}}{(1 - 2i\upsilon)^{1/2}} \tag{1-116}$$

Finally, the characteristic function for n variables, where $y = \sum_{j=1}^{n} x^2$, is simply

$$G_y^{(n)}(i\upsilon) = \prod_{j=1}^{n} G_{yj}(i\upsilon)$$

$$= \frac{e^{-n(\mu^2/2)[1-1/(1-2i\upsilon)]}}{(1 - 2i\upsilon)^{n/2}} \tag{1-117}$$

Letting $\lambda = n\mu^2/2$ in Eq. (1-117) yields

$$G_y^{(n)}(i\upsilon) = \frac{e^{-\lambda[1-1/(1-2i\upsilon)]}}{(1 - 2i\upsilon)^{n/2}} = \frac{e^{2i\upsilon\lambda/(1-2i\upsilon)}}{(1 - 2i\upsilon)^{n/2}} \tag{1-118}$$

The parameter λ is frequently referred to as the *noncentrality factor* for the distribution.

Some insight into Eq. (1-118) can be obtained using the series expansion $\exp(x) = \sum_{j=1}^{\infty} x^j/j!$. The first form of Eq. (1-118) can then be written as

$$
\begin{aligned}
G_y^{(n)}(i\upsilon) &= \frac{e^{-\lambda} e^{\lambda/(1-2i\upsilon)}}{(1-2i\upsilon)^{n/2}} \\
&= e^{-\lambda} \sum_{j=1}^{\infty} \frac{1}{j!} \left[\frac{\lambda}{(1-2i\upsilon)} \right]^j \frac{1}{(1-2i\upsilon)^{n/2}} \\
&= \sum_{j=1}^{\infty} \frac{\lambda^j e^{-\lambda}}{j!} \frac{1}{(1-2i\upsilon)^{(n+2j)/2}} \qquad (1\text{-}119)
\end{aligned}
$$

The coefficients of the terms involving υ are known as *Poisson probabilities*, which results in the series being a *convex sum* for $\lambda > 1$. A convex sum is defined as the sum of a series of terms whose largest values are not at the beginning of the series but at some value of j lying in the range $0 < j < \infty$. (Poisson probabilities and statistics are discussed in detail in Chapter 4.)

If we now inverse transform Eq. (1-119), we obtain

$$
\begin{aligned}
p(x) &= \frac{1}{2\pi} \int_{-\infty}^{\infty} e^{i\upsilon x} G_y^{(n)}(i\upsilon)\, d\upsilon \\
&= \sum_{j=1}^{\infty} \frac{\lambda^j e^{-\lambda}}{j!} \left[\frac{1}{2\pi} \int_{-\infty}^{\infty} \frac{e^{i\upsilon x}}{(1-2i\upsilon)^{(n+2j)/2}} d\upsilon \right] \qquad (1\text{-}120)
\end{aligned}
$$

But the term in brackets in Eq. (1-120) is, from Eq. (1-113), just the inverse transformation of the characteristic function for the χ^2 distribution with $n + 2j$ degrees of freedom. Thus, using Eq. (1-107) with $\alpha = n + 2j$, we obtain the series form for the noncentral χ^2 distribution,

$$
p(x) = \sum_{j=1}^{\infty} \frac{\lambda^j e^{-\lambda}}{j!} \frac{x^{(n+2j)/2-1} e^{-x/2}}{2^{(n+2j)/2} \Gamma[(n+2j)/2]} \qquad (1\text{-}121)
$$

Notice that for $\lambda = 0$, the distribution reduces to the χ^2 distribution.

1.2.8. Stationary and Ergodic Systems

It is frequently the case that random processes need to be considered over extended periods of time. Physical processes whose statistics do not vary over time are called *stationary random processes*. The assumption of stationarity is important in the formulation of useful statistical theories since the results assume a more general character than for nonstationary systems. However, nonstationary systems can and do frequently arise in the real world. One such example that is discussed in Chapter 3 is atmospheric turbulence, which is dependent on ambient conditions such as temperature, humidity, wind, and so on, all of which can vary significantly with time.

Stationary processes may be defined in the spatial domain as well, although they are usually not referred to as such. Random processes whose statistics are invariant over spatial distances can be treated using the same mathematical formalism as the time domain under the replacement $t \leftrightarrow x$. Indeed, the case of atmospheric turbulence is an example that may (or may not) exhibit both temporal and spatial stationarity, depending on ambient conditions.

There are several types of stationary processes with varying types of restrictions. A *strictly stationary process* is one in which all the joint probability density functions are independent of the space and time origins. Thus, with $u(x, t)$ representing the random variable,

$$\begin{aligned} &p(u_1, u_2, \ldots, u_n; t_1, t_2, \ldots, t_n; x_1, x_2, \ldots, x_n) \\ &\quad = p(u_1, u_2, \ldots, u_n; t_1 + \tau, t_2 + \tau, \ldots, t_n + \tau; \\ &\quad x_1 + \xi, x_2 + \xi, \ldots, x_n + \xi) \end{aligned} \tag{1-122}$$

which, of course, implies that the case of a single random variable obeys the relation $p(u, t, x) = p(u, t + \tau, x + \xi)$. It also implies that all the moments associated with the joint density functions are constant over space and time, which constitutes a *complete* statistical description of the random process.

Strictly stationary processes are usually unnecessarily restrictive for the description of most physical systems. A less restrictive definition is that of the *wide-sense stationary process*. In this case, a complete statistical description is usually not possible, so that the description is limited to first- and second-order statistics that satisfy the conditions

$$\langle u(x, t)\rangle = \text{constant} \tag{1-123}$$

and

$$\langle u(x_1, t_1), u(x_2, t_2)\rangle = R(x_1 - x_2; t_1 - t_2) \tag{1-124}$$

where $R(x_1 - x_2; t_1 - t_2)$ is the autocorrelation function of the random variable u (cf. Section 1.2.9).

It can also be the case that wide-sense stationary processes may be too restrictive to allow for an analytical description of some phenomena. In such a case, some of the restrictions implicit in the wide-sense processes may be removed. One way to do this is to limit the time over which the process is observed. Thus a function $g(x_1, x_2, t_2, t_1) = f(x_2, t_2) - f(x_1, t_1)$ is considered *stationary in increments* if g is strictly stationary for all x_1, x_2, t_1, and t_2. Such processes find application in stochastic processes that are slowly varying over space or time such that the definition can be applied locally. It will be shown in Chapter 3 that the concept of stationarity in increments allows for development of the atmospheric structure function, $D_\alpha(\rho) = \langle[\alpha(x_1, t_1) - \alpha(x_2, t_2)]^2\rangle$, which can be used to describe both the temporal and spatial statistics of a random variable α of the turbulent field.

The remaining and most restrictive class of random process is the *ergodic hypothesis*. The ergodic hypothesis states that for a strictly stationary temporal process, the joint statistics of a single *statistical realization* of a random process are equal to the joint statistics of an ensemble of realizations of the process. By *realization* is meant a "single throwing of the dice" for the statistical process at hand. In principle, there are an infinite number of potential realizations of the statistical process, all of which constitute the statistical ensemble. The function that describes a single realization of a statistical process is referred to as a *sample function* or, in the case of time functions, a *sample waveform*. A trivial example of an ergodic hypothesis is the flipping of a coin to obtain heads or tails. The same statistics will be obtained whether a single coin is flipped many times or an ensemble of coins is flipped once. Once again, the definition may be applied to the spatial domain as well.

However, not all processes that are strictly stationary are ergodic. An example is the noise voltage of a single optical receiver. If the noise voltage is truly ergodic, the joint statistics will be the same, whether they are calculated from a single sample waveform or from an ensemble of sample waveforms. An example of an ensemble of voltage waveforms is shown in Figure 1-14. On the other hand, the process is likely to become nonergodic if the ensemble of sample waveforms is taken from an ensemble of different receivers, say on a production line. In this case, slight differences in the characteristics of each receiver will probably result in statistics that are different than those measured in any single receiver. An example of such an ensemble is shown in Figure 1-15, where the variance of the noise itself is a random variable when taken over the ensemble. In this case, each independent sample waveform is strictly stationary, but the joint statistics do not agree with those of the ensemble, and therefore the process is not ergodic.

1.2.9. Energy and Power Spectral Densities

Determination of the signal-to-noise ratio of an optical receiver usually requires specification of its bandwidth. This follows from the fact that the wider the bandwidth, the greater the noise level that is ultimately added to the signal. Therefore, to characterize the various noise sources that may exist, knowledge of the frequency characteristics of the noise waveforms would be useful. Clearly, one can associate an energy and an average power with the sample waveforms shown in Figures 1-14 and 1-15. In the event that the waveforms are ergodic, the average power and energy can be calculated as space or time integrals over a particular sample waveform or as ensemble averages over an ensemble of waveforms. To characterize their frequency dependence, we must resort to some type of Fourier analysis. Consider a sample waveform $u(t)$ that represents an arbitrary real time-limited function. The energy associated with this waveform is

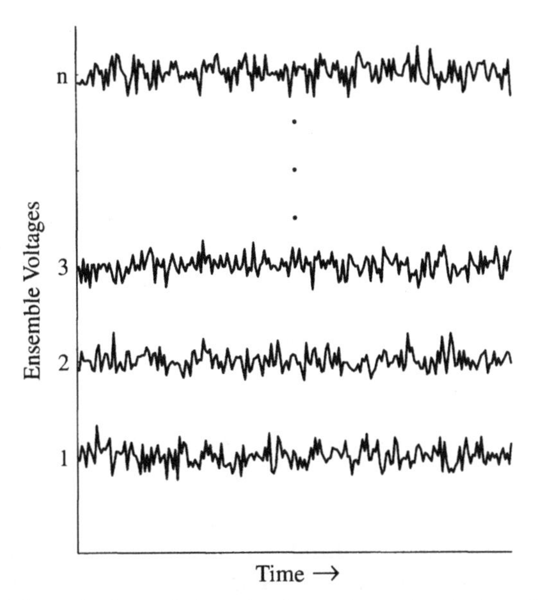

Figure 1-14. Multiple realizations of the noise fluctuations of a single optical receiver as an example of an ergodic process.

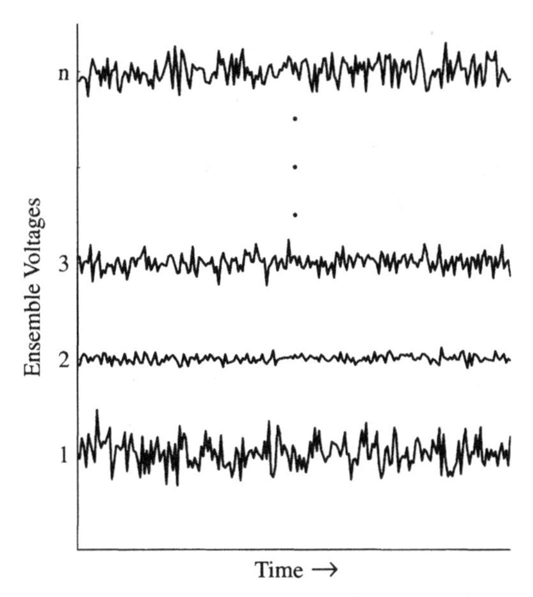

Figure 1-15. Single realizations of the noise fluctuations of multiple optical receivers as an example of a nonergodic process.

given by

$$E = \int_{-\infty}^{\infty} u^2(t)\, dt \tag{1-125}$$

Using Parseval's theorem from Fourier transform theory leads to

$$\int_{-\infty}^{\infty} u^2(t)\, dt = \int_{-\infty}^{\infty} |F(\omega)|^2\, df = 2\int_{0}^{\infty} |F(\omega)|^2\, df \tag{1-126}$$

where $\omega = 2\pi f$ and $u(t)$ and $F(\omega)$ are Fourier transform pairs given by

$$\begin{aligned} F(\omega) &= \int_{-\infty}^{\infty} u(t)e^{-i\omega t}\, dt \\ u(t) &= \frac{1}{2\pi}\int_{-\infty}^{\infty} F(\omega)e^{i\omega t}\, d\omega \end{aligned} \tag{1-127}$$

The function $|F(\omega)|^2 = F(\omega)F^*(\omega)$ has the dimensions of *energy spectral density*, which will be denoted as $\mathcal{E}(\omega) \equiv |F(\omega)|^2$. [Parseval's theorem is easily demonstrated by substituting the first of Eqs. (1-127) for $F(\omega)$ in the second term of Eq. (1-126), interchanging the order of integration, and noting that $|F(\omega)|^2$ is an even function of ω for real $u(t)$.]

If $u(t)$ is a periodic function of time, implying that it is defined for all time T, the limits of the integral in Eq. (1-125) must be finite (otherwise, the energy would be infinite). Thus

$$E = \int_{-T/2}^{T/2} u^2(t)\, dt \tag{1-128}$$

which represents the total energy in the interval T.

The average power in the time T can be obtained from Eq. (1-128) as

$$\langle P \rangle = \frac{1}{T}\int_{-T/2}^{T/2} u^2(t)\, dt \tag{1-129}$$

In the case of a spectrum defined for both positive and negative frequencies, a simple Fourier series analysis can be used to show that

$$\langle P \rangle = \sum_{n=-\infty}^{\infty} S(\omega_n)\Delta f \tag{1-130}$$

where $\Delta f = 1/T$ and $S(\omega_n) = |c_n|^2/T$ is defined as the *power spectral density* of the waveform $u(t)$. Here the c_n are the discrete Fourier coefficients of the waveform where $|c_n|^2 = c_n c_n^* = c_n c_{-n}$. If the spectral density is defined for only positive frequencies, then

$$\langle P \rangle = \sum_{n=1}^{\infty} S^+(\omega_n)\Delta f \tag{1-131}$$

where the relationship $S^+(\omega_n) = 2S(\omega_n)$ ensures conservation of energy between the two representations.

To find the power spectral density of random waveforms with continuous spectra, the Fourier integral representation of the analysis above is required. To do this, we must evaluate Eq. (1-130) in the limit of $T \to \infty$ and $\Delta f \to 0$, that is,

$$\begin{aligned} \langle P \rangle &= \lim_{\substack{T\to\infty \\ \Delta f \to 0}} \frac{1}{T} \sum_{n=-\infty}^{\infty} |c_n|^2 \, \Delta f \\ &= \int_{-\infty}^{\infty} S(\omega)\, df \end{aligned} \tag{1-132}$$

where

$$S(\omega) = \lim_{T\to\infty} |F(\omega)|^2 \tag{1-133}$$

is the power spectral density.

Care must be exercised when using Eq. (1-133), for if $u(t)$ is a time-limited waveform, Eq. (1-132) vanishes as $T \to \infty$. Thus the concept of power spectral density breaks down for time-limited waveforms. However, the true utility of the formalism is in the description of random processes that remain finite for $T \to \infty$. In particular, for a random process having a uniform power spectrum (i.e., white noise), the power spectral density is simply a constant N_0, that is,

$$S(\omega) = \frac{N_0}{2} \qquad \text{watts/Hz} \tag{1-134}$$

The factor of $\frac{1}{2}$ is again due to the fact that $S(\omega)$ is defined for positive and negative frequencies.

In Section 1.2.5 it was shown that the solution to a convolution integral could be obtained using Fourier transforms. In that case, the transforms were used in a special way to define characteristic functions. However, the relationship between convolution integrals and Fourier transforms is a far more general one, as evidenced by its widespread use in linear filter theory. In a manner similar to that shown in Section 1.2.5, it may be shown that the output of a linear filter may be obtained from the inverse transform of the product of the transforms of the filter function and the input waveform, that is,

$$F_o(\omega) = H(\omega) F_i(\omega) \tag{1-135}$$

Squaring both sides, dividing by T, and taking the limit as $T \to \infty$ yields

$$\lim_{T\to\infty} \frac{1}{T} |F_o(\omega)|^2 = \lim_{T\to\infty} \frac{1}{T} |H(\omega)|^2 \, |F_i(\omega)|^2 \tag{1-136}$$

or

$$S_o(\omega) = |H(\omega)|^2 \, S_i(\omega) \tag{1-137}$$

Thus the input power spectral density is related to the output power spectral density via the square of the filter transfer function. Thus, for white noise

$$S_o(\omega) = \frac{N_0}{2}|H(\omega)|^2 \tag{1-138}$$

1.2.10. Wiener–Khintchine Theorem

The sample waveforms shown in Figure 1-14 are ergodic and therefore strictly stationary. The definition of strictly stationary was given in terms of the joint statistics of the random variables, but very few specifics were given about their properties. Consider then the second-order statistics as represented by the average of the products of the sample waveforms evaluated at two instants of time t_1 and t_2 (or positions in space x_1 and x_2, as the case may be). If the process is ergodic, the average can be calculated as a temporal or spatial average using a single realization of the statistical process or by sampling many realizations of the process over the entire ensemble. Limiting our discussions to the time domain for the moment, the former averaging process becomes

$$R(\tau) = \lim_{T\to\infty} \frac{1}{T} \int_0^T u(t)u(t+\tau)\,dt \tag{1-139}$$

where $t_1 = t$ and $t_2 = t_1 + \tau = t + \tau$. $R(\tau)$ is known as the *time autocorrelation function* of the waveform over the interval $\tau = t_2 - t_1$. For the latter averaging process, we have

$$m_{11} = \int_{-\infty}^{\infty}\int_{-\infty}^{\infty} u_1 u_2 p(u_1, u_2)\,du_1\,du_2 \tag{1-140}$$

where u_1 and u_2 are the values of the waveform amplitude at the times t_1 and t_2 and $p(u_1, u_2)$ is the joint probability density function for the random variables u_1 and u_2. Since the process is ergodic,

$$R(\tau) = m_{11} \tag{1-141}$$

so that m_{11} may be identified as the *ensemble autocorrelation function*, frequently referred to simply as the *autocorrelation function*.

An important and very useful theorem in statistics is the *Wiener–Khintchine theorem*. This theorem states that the autocorrelation function and the power spectral density of a wide-sense stationary process form a Fourier transform pair. This can be written

$$S(\omega) = \int_{-\infty}^{\infty} R(\tau)e^{-i\omega\tau}\,d\tau \tag{1-142}$$

and

$$R(\tau) = \frac{1}{2\pi}\int_{-\infty}^{\infty} S(\omega)e^{i\omega\tau}\,d\omega \tag{1-143}$$

where S is the power spectral density of the waveform and ω is the angular frequency $2\pi f$. As will be shown in Chapter 2, Fourier transform theory may also be applied in the spatial domain. Thus an analogous Fourier transform pair may be defined for the *spatial autocorrelation function* $R(x_2 - x_1)$ and the *spatial power spectral density* $S(\omega_x)$, where $\omega_x = 2\pi f_x$ and f_x is a *spatial frequency* measured along the x-axis. More will be said on this topic in Section 2.5.

There are several important and useful properties associated with the autocorrelation functions. These are summarized below.

1. When $\tau \to 0$, the autocorrelation function reduces to the mean-squared value or average power of the waveform; that is,

$$R(0) = \lim_{T\to\infty} \frac{1}{T} \int_0^T |u(t)|^2 \, dt = \langle P(t) \rangle \tag{1-144}$$

2. $R(\tau)/R(0) \leq 1$. This follows from the fact that a phase factor does not affect the absolute value of a quantity. Thus, using Eq. 1-143, it can be seen that

$$\frac{|R(\tau)|}{R(0)} = \frac{\left| \int_{-\infty}^{\infty} S(\omega) e^{i\omega\tau} \, d\omega \right|}{\int_{-\infty}^{\infty} S(\omega) \, d\omega} \leq 1 \tag{1-145}$$

3. $R(\tau) = R(-\tau)$. This follows directly from the definition of the autocorrelation function. Thus, for stationary processes that are, by definition, independent of the time origin

$$\begin{aligned} R(\tau) &= \lim_{T\to\infty} \frac{1}{T} \int_0^T u(t) u(t+\tau) \, dt \\ &= \lim_{T\to\infty} \frac{1}{T} \int_{-\tau}^{T-\tau} u(t) u(t+\tau) \, dt \\ &= \lim_{T\to\infty} \frac{1}{T} \int_0^T u(t'-\tau) u(t') \, dt' \\ &= R(-\tau) \end{aligned} \tag{1-146}$$

where a change in variables $t' = t + \tau$ was used.

4. The autocorrelation function of a periodic function is also periodic. This can be shown as follows. Let a waveform be given by $u(t) = u_0 \sin(\omega t + \phi)$. Inserting this into Eq. (1-139) yields

$$R(\tau) = \lim_{T\to\infty} \frac{u_0^2}{T} \int_0^T \sin(\omega t + \phi) \sin(\omega t + \phi + \omega\tau) \, dt \tag{1-147}$$

Using the trigonometric identity $\sin A \sin B = [\cos(A - B) - \cos(A + B)]/2$ leads to

$$R(\tau) = \lim_{T\to\infty} \frac{u_0^2}{2T} \int_0^T [\cos \omega\tau - \cos(2\omega t + 2\phi + \omega\tau)]\, dt$$

$$= \frac{u_0^2}{2} \cos \omega\tau - \lim_{T\to\infty} \frac{u_0^2}{2T} \frac{\sin(2\omega T + 2\phi + \omega\tau) - \sin(2\phi + \omega\tau)}{2\omega}$$

$$= \frac{u_0^2}{2} \cos \omega\tau \qquad (1\text{-}148)$$

Notice that the properties of the autocorrelation function discussed previously are satisfied by this last result, namely, $R(0) \geq R(\tau)$, $R(\tau) = R(-\tau)$, and $R(0) = u_0^2/2$, the latter representing the average power of the waveform.

1.2.11. Fourier–Stieltjes Transform

Fourier transform theory is based on the idea that an arbitrary function $f(t)$ can be represented as a linear superposition of harmonic functions that, taken together, comprise a frequency spectrum given by $F(\omega)$, where $f(t)$ and $F(\omega)$ are related through integral transforms such as those of Eqs. (1-76) and (1-77). The frequency spectrum so defined may be considered fully determined by $f(t)$ such that the various spectral components, whether discrete or continuous, are fully correlated in amplitude and phase.

Consider now the case where $f(t)$ is a stationary random function. In this case the ordinary Fourier transform cannot be applied to the problem because a stationary random function does not satisfy the convergence criteria required for Fourier transforms[5], that is

$$\int_{-\infty}^{\infty} |f(t)|\, dt < \infty \qquad (1\text{-}149)$$

However, we may represent a single realization of such a function as the sum of harmonic functions, but in this case, the spectral components assume random amplitudes and phases. The integral transform that relates $f(t)$ with its spectrum is known as the *Fourier–Stieltjes transform*,[6] given by

$$f(t) = \frac{1}{\sqrt{2\pi}} \int_{-\infty}^{\infty} e^{i\omega t} dN(\omega) \qquad (1\text{-}150)$$

where $dN(\omega)$ represents the random amplitudes of the spectral components of the sample function. Note that in the case of $f(t)$ being real [i.e., $f(t) = f^*(t)$], letting $\omega \to -\omega$ in $f^*(t)$ leads to the important relation $dN(\omega) = dN^*(-\omega)$.

Consider the one-dimensional case of a stationary random function $f(t)$ with the properties

$$\langle f(t) \rangle = 0 \tag{1-151}$$

and

$$\langle f(t_1) f^*(t_2) \rangle = R(t_1 - t_2) \tag{1-152}$$

where $R(t_1 - t_2)$ is the autocorrelation function for the process. Note that Eqs. (1-151) and (1-152) satisfy the criteria for a wide-sense stationary process given by Eqs. (1-123) and (1-124).

Substituting Eq. (1-150) into Eq. (1-151) yields

$$\langle f(t) \rangle = \frac{1}{\sqrt{2\pi}} \int_{-\infty}^{\infty} e^{i\omega t} \langle dN(\omega) \rangle = 0 \tag{1-153}$$

where ω and t are nonrandom. From this we clearly have

$$\langle dN(\omega) \rangle = 0 \tag{1-154}$$

Substituting Eq. (1-150) into Eq. (1-152) yields

$$\langle f(t_1) f^*(t_2) \rangle = \frac{1}{2\pi} \int_{-\infty}^{\infty} e^{-i\omega_1 t_1 + i\omega_2 t_2} \langle dN(\omega_1)\, dN^*(\omega_2) \rangle \tag{1-155}$$

Now in order that $\langle f(t_1) f^*(t_2) \rangle$ may be considered an autocorrelation function, it must be a function only of the time difference $t_1 - t_2$. This requires that the averaging term in Eq. (1-155) take the form

$$\langle dN(\omega_1)\, dN^*(\omega_2) \rangle \equiv S(\omega_1) \delta(\omega_1 - \omega_2)\, d\omega_1\, d\omega_2 \tag{1-156}$$

where $S(\omega_1)$ is the power spectral density of the random function $f(t)$. Note that the definition given by Eq. (1-156) implies that

$$\langle dN(\omega_1)\, dN^*(\omega_2) \rangle = 0 \qquad \omega_1 \neq \omega_2 \tag{1-157}$$

which states that different spectral components in the spectrum of $f(t)$ are uncorrelated. Thus, using Eq. (1-156) in Eq. (1-155), we arrive at the familiar Fourier transform pair

$$\begin{aligned} R(\tau) &= \frac{1}{2\pi} \int_{-\infty}^{\infty} S(\omega) e^{i\omega\tau}\, d\omega \\ S(\omega) &= \int_{-\infty}^{\infty} R(\tau) e^{-i\omega\tau}\, d\tau \end{aligned} \tag{1-158}$$

which defines the Wiener–Khintchine theorem of Section 1.2.10.

The spectral representation of a random variable is particularly useful in describing turbulent fields and will therefore be seen again in Chapter 3, where it will be applied in the spatial domain for both two and three dimensions.

1.2.12. Matched Filter Theory

An optical receiver may be viewed as a filter function that performs in accordance with certain design criteria. The design criteria may be the preservation of the signal fidelity or in radar terms, the pulse shape, or it may be the maximization of the signal-to-noise ratio. In general, these two criteria are incompatible since preservation of pulse shape implies a wide bandwidth that necessarily degrades the signal-to-noise ratio. From a detection point of view, preservation of the pulse shape is unimportant but maximizing the signal-to-noise ratio is. Thus, our goal is to find a filter function for the receiver that maximizes the signal-to-noise ratio.

In Section 1.2.8 it was mentioned that the response of a linear filter with transfer function $H(\omega)$ to an input waveform with Fourier transform $F(\omega)$ is given by

$$f_o(t) = \frac{1}{2\pi} \int_{-\infty}^{\infty} F(\omega) H(\omega) e^{i\omega t} \, d\omega \tag{1-159}$$

where $\omega = 2\pi f$. For a given input pulse energy E_0, the signal-to-noise ratio at the peak of the waveform t_0 can be maximized by maximizing the ratio

$$\mathrm{SNR}_{\mathrm{peak}} = \frac{|f(t_0)|^2}{N_f} = \frac{\left| \int_{-\infty}^{\infty} F(\omega) H(\omega) e^{i\omega t_0} \, df \right|^2}{(N_0/2) \int_{-\infty}^{\infty} |H(\omega)|^2 \, df} \tag{1-160}$$

where N_f is the mean noise power at the output of the filter, given by

$$N_f = \frac{N_0}{2} \int_{-\infty}^{\infty} |H(\omega)|^2 \, df \tag{1-161}$$

Now Schwartz's inequality states that for any two complex functions $x(t)$ and $y(t)$, the following inequality holds:

$$\left| \int_{-\infty}^{\infty} x(t) y(t) \, dt \right|^2 \le \int_{-\infty}^{\infty} |x(t)|^2 dt \int_{-\infty}^{\infty} |y(t)|^2 dt \tag{1-162}$$

Thus with $x = F(\omega)e^{i\omega t_0}$ and $y = H(\omega)$ we can immediately write

$$\mathrm{SNR}_{\mathrm{peak}} = \frac{|f(t_0)|^2}{N_f} \le \frac{2E_0}{N_0} \tag{1-163}$$

where $E_0 = \int_{-\infty}^{\infty} |F(\omega)|^2 df$. The ratio is a maximum when $H(\omega) = KF^*(\omega) e^{-i\omega t_0}$, where K is a constant. Filters exhibiting the characteristic above are known as *matched filters* and have the following properties:

1. The matched filter optimizes the signal-to-noise ratio.
2. The matched filter does *not* preserve the input waveform shape.
3. If $f(t) = f(-t)$, then $h(t) = Kf(t - t_0)$.
4. If $f(t) = -f(-t)$, then $h(t) = Kf(-t - t_0)$.

The first two properties are self-explanatory. The third property is simply a statement of the fact that a filter matched to a symmetric input waveform has an impulse response that is a delayed replica of itself. The fourth property implies that for an asymmetric input waveform, the impulse response of the filter occurs in negative time. This last result is clearly unphysical, which suggests that matched filters are, in general, not realizable in practice. However, they do constitute an idealized filter function for use in comparison with other more practical filters, such as the low-pass filter or the Gaussian filter. Such comparisons are not carried out here but can be found in the works of Skolnik[7] and Schwartz.[8]

1.3. DECISION-MAKING PROCESSES

In optical detection systems, signals are received that must be discriminated against competing receiver noise and background clutter. Depending on the system, transmitted waveforms may provide many types of information. These might include range, Doppler, amplitude, frequency, or phase. In the case of a communication receiver, it might consist of a simple series of binary digits, possibly quantized in amplitude for additional information. All of these measurable quantities fall into the category of *information parameters* and the actual measured values, *parameter estimates*. There are two types of parameter estimation, one that determines whether a signal is present or not present and one that measures the parameters to some accuracy. For example, one may wish to determine solely that a signal was present or absent, regardless of the quality or form of the signal. This is a two-state process that is known as *binary detection* or simply, *signal detection*. On the other hand, one may wish to know some of the features of the signal (e.g., its phase, amplitude, Doppler, etc.) to an arbitrary degree of accuracy or precision. This process is known as *parameter estimation*.

The performance of an optical radar system is fundamentally the same as a conventional microwave radar system in that radiation is reflected off a target and must be detected in the presence of receiver noise and background clutter. In the chapters that follow, a problem will repeatedly arise as to the development of probability density functions that represent either noise alone or signal plus noise. These density functions may be discrete, as in the case of photon counting direct detection receivers, or they may be continuous, as in the case of coherent detection receivers, where photon numbers are large and therefore treated as continuous. The density function for noise alone is important in establishing the false alarm probability for the receiver, while the density function for signal plus noise is important in establishing the detection probability of the true signal in the presence of additive noise. As will be seen, a threshold level must be established that allows only a small fraction of the noise fluctuations to be detected (i.e., cross threshold) while satisfying requirements for detection probability.

Although the binary detection problem appears simple to define in a clutter-free environment, and the process of establishing a suitable threshold for detection generally easy to calculate, it is worthwhile to understand the criteria for detection from a more general decision-theoretic perspective. This is provided by *statistical decision theory*. We consider three of these: the Bayes criterion, the Neyman–Pearson criterion, and the minimax criterion. It will be seen that the latter two criteria are simply special cases of the Bayes criterion, the Neyman–Pearson criterion providing the detection criteria that are assumed in much of the remaining text. For more detailed treatments, the reader is referred to the textbooks written by DiFranco and Rubin[9] and McDonough and Whalen[10] or the original works on decision theory by Wald.[11]

1.3.1. Thresholding and Detection

Any decision-making process requires that a hypothesis be tested as to whether various alternatives are true or not true. These are then confirmed, or not confirmed, based on observations of real data. The overall process is referred to as *hypothesis testing*. In the case of signal detection the alternatives are usually classified as *simple hypotheses*; that is, either a signal is present or it is not, and therefore the situation reduces to a one-parameter binary problem. There are more complex hypotheses, however, in which each hypothesis has multiple parameters. These are called *composite hypotheses* and are particularly important in parameter estimation. However, for the binary detection problem, simple hypotheses are usually sufficient.

The decision as to whether a target is present or not is confronted with four possible outcomes:

1. A signal is not detected when a signal is absent.
2. A signal is detected when a signal is present.
3. A signal is detected when a signal is absent.
4. A signal is not detected when a signal is present.

Detection theory provides a formal construct from which the most likely outcome of the four possibilities above may be chosen.

The theory may be put on a more quantitative foundation with the following definitions. Let H_0 be the *null hypothesis* that a signal is absent and H_1 the *alternative hypothesis* that a signal is present. Then the probabilities $P(H_0)$ and

$P(H_1)$ represent the probabilities that H_0 and H_1 will occur, respectively, where $P(H_0) + P(H_1) = 1$. Let us also define the decision D_0 as the selection of H_0 as true and D_1 the selection of H_1 as true. Our four possible outcomes can then be written as the conditional probabilities

$$\begin{aligned} &(1)\ P(D_0 \mid H_0) \\ &(2)\ P(D_1 \mid H_1) \\ &(3)\ P(D_1 \mid H_0) \\ &(4)\ P(D_0 \mid H_1) \end{aligned} \tag{1-164}$$

where from the conservation of probability

$$\begin{aligned} P(D_0 \mid H_0) + P(D_1 \mid H_0) &= 1 \\ P(D_0 \mid H_1) + P(D_1 \mid H_1) &= 1 \end{aligned} \tag{1-165}$$

and

$$P(D_0, H_0) + P(D_1, H_1) + P(D_1, H_0) + P(D_0, H_1) = 1 \tag{1-166}$$

The first two decisions in Eq. (1-164) constitute correct decisions, and the last two constitute errors. Specifically, in the latter two cases, $P(D_1 \mid H_0)$ is referred to as an *error of the first kind* and corresponds in radar terminology to a *false alarm probability*. On the other hand, $P(D_0 \mid H_1)$ is referred to as an *error of the second kind* and corresponds to a *miss probability*. It, of course, follows that the *detection probability* is given by $P(D_1 \mid H_1) = 1 - P(D_0 \mid H_1)$.

It was mentioned earlier that the goal of decision theory was to develop a criterion by which the most likely outcome could be selected. In any statistical problem, everything that is known about a random variable is contained in the probability density function (or, equivalently, in the cumulative distribution function). Therefore, it is important that the theory be formulated to take advantage of these functions. Consider a sample point in observation space defined by the probability density function $p(x)$. Here x is a random variable that might represent, for example, a voltage waveform at the output of a receiver. The a priori density function for the null hypothesis is therefore written $p(x|H_0)$ and the simple alternative as $p(x \mid H_1)$. In the case of an optical receiver employing envelope detection, these might take the form of the Gaussian distributions shown in Figure 1-16. Although these distributions are continuous, the processes apply to discrete distributions as well. Note that the height and width of the distributions are in general different and since x represents the envelope of the waveform, are confined to the positive x-axis. Thus, if hypothesis H_0 is true, x is equal to the receiver noise n and is described by $p(x \mid H_0)$, while if H_1 is true, x is equal to the signal plus noise and is described by $p(x \mid H_1)$.

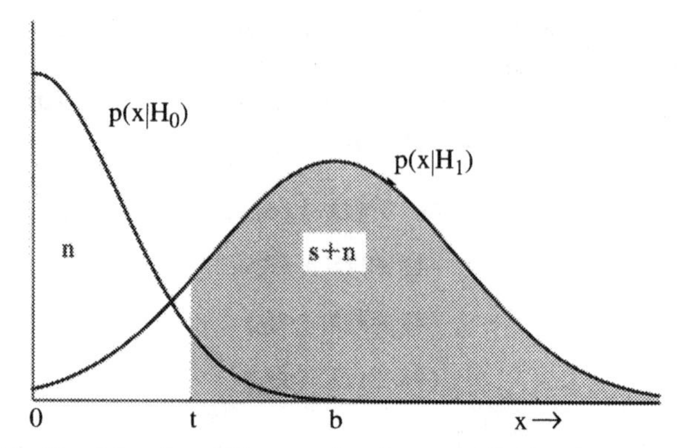

Figure 1-16. Likelihood functions with unequal variances confined to the positive real axis due to an energy detection process. The sample space is divided by the threshold t into regions where H_0 is most likely to occur (unshaded area) and H_1 is most likely to occur (shaded area).

Now the following question arises: If a single observation of x is made, which of the hypotheses can be assumed to be true? This can be written as a pair of conditional probabilities,

$$P(H_0 \mid x) = \text{conditional probability that } H_0 \text{ occurs given sample } x$$

$$P(H_1 \mid x) = \text{conditional probability that } H_1 \text{ occurs given sample } x$$

Here x is referred to as the *test statistic*. It should be noted that $P(H_0)$ and $P(H_1)$ are *a priori probabilities*, that is, probabilities that are assigned as a result of prior knowledge. (Webster's definition of *a priori* is "to arrive at through reasoning, not through experience or observation.") On the other hand, $P(H_0 \mid x)$ and $P(H_1 \mid x)$ are *a posteriori probabilities*, that is, probabilities that are assigned after observations are made. (In this case the definition is "to arrive at through observation or experience, not through pure reasoning.")

A decision rule can now be formulated as follows:

$$P(H_0 \mid x) > P(H_1 \mid x) \qquad H_0 \text{ is selected} \tag{1-167}$$

or

$$P(H_1 \mid x) \geq P(H_0 \mid x) \qquad H_1 \text{ is selected} \tag{1-168}$$

From the definition of conditional probability [cf. Eq. (1-36)] we can write

$$P(H_0 \mid x)P(x) = P(x \mid H_0)P(H_0) \tag{1-169}$$

and

$$P(H_1 \mid x)P(x) = P(x \mid H_1)P(H_1) \tag{1-170}$$

Since the sample x is represented by a probability density function $p(x)$, Eqs. (1-169) and (1-170) can be written as

$$P(H_0 \mid x)p(x) = p(x \mid H_0)P(H_0) \tag{1-171}$$

and

$$P(H_1 \mid x)p(x) = p(x \mid H_1)P(H_1) \tag{1-172}$$

respectively, where $p(x \mid H_j)$ is the probability density function for x given that H_j is true. Dividing Eq. (1-172) by Eq. (1-171) and using the decision rule given by Eq. (1-168) leads to

$$\begin{aligned} \ell(x) = \frac{p(x \mid H_1)}{p(x \mid H_0)} &\geq \frac{P(H_0)}{P(H_1)} \qquad H_1 \text{ selected} \\ &< \frac{P(H_0)}{P(H_1)} \qquad H_0 \text{ selected} \end{aligned} \tag{1-173}$$

Here $\ell(x)$ is called, by convention, the *likelihood ratio* for the estimate and $p(x \mid H_1)$ and $p(x \mid H_0)$ are the *likelihood functions*. If $p(x \mid H_1)$ and $p(x \mid H_0)$ are both monotonic functions of x, the inequality in Eq. (1-173) can always be rewritten in terms of x. That is,

$$x = \ell^{-1}\left[\frac{p(x \mid H_1)}{p(x \mid H_0)}\right] \geq \ell^{-1}\left[\frac{P(H_0)}{P(H_1)}\right] \tag{1-174}$$

where $\ell(x) = p(x \mid H_1)/p(x \mid H_0)$. This implies that there is a threshold value of $x \equiv t$ above which H_1 is chosen and below which H_0 is chosen. This division of the real axis into two distinct regions is shown graphically in Figure 1-16 for the assumed Gaussian probability density functions $p(x \mid H_0)$ and $p(x \mid H_1)$.

To be more specific, the Gaussian probability density functions in Figure 1-16 are described by

$$p(x \mid H_0) = \frac{1}{\sqrt{2\pi\sigma_n^2}} e^{-x^2/2\sigma_n^2} \tag{1-175}$$

and

$$p(x \mid H_1) = \frac{1}{\sqrt{2\pi\sigma_{s+n}^2}} e^{-(x-b)^2/2\sigma_{s+n}^2} \tag{1-176}$$

where σ_n^2 is the noise variance, σ_{s+n}^2 is the signal plus noise variance (here, for simplicity, we have assumed that the mean noise level is zero, which may not be true in the most general case), and b represents the mean signal level. The likelihood ratio becomes

$$\ell(x) = \frac{\sigma_n}{\sigma_{s+n}} e^{(x^2/2\sigma_n^2)-[(x-b)^2/2\sigma_{s+n}^2]} \geq \frac{P(H_0)}{P(H_1)} \tag{1-177}$$

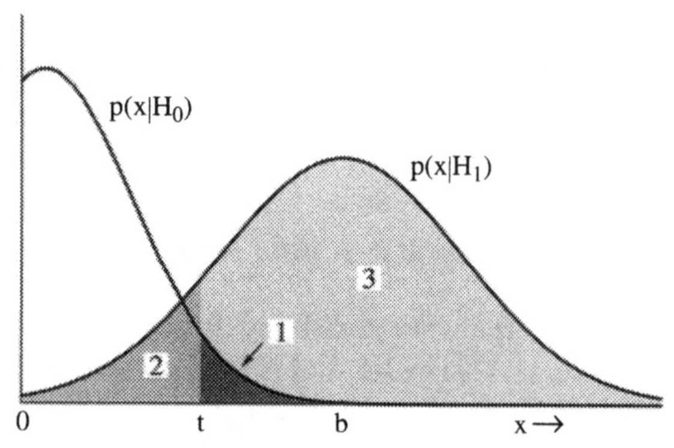

Figure 1-17. Likelihood functions with unequal variances confined to the positive real axis due to an energy detection process. Numbered regions correspond to (1) false alarm probability $P(D_0 \mid H_0)$, (2) miss probability $P(D_0 \mid H_1)$, and (3) detection probability $P(D_1 \mid H_1)$.

Consider the case of $\sigma_n = 1$ and $\sigma_{s+n} = 2$. Solving Eq. (1-177) for x results in a transcendental equation with two solutions. Thus the inequality becomes $x \geq t$, where

$$t = -\frac{b}{3} + \frac{2}{3}\sqrt{b^2 + 6 \ln \frac{2P(H_0)}{P(H_1)}} \tag{1-178}$$

and for positive values of b, only the positive sign in front of the radical is physically meaningful. Therefore, given the a priori probabilities $P(H_0)$ and $P(H_1)$, if $x \geq t$, the decision rule is to assume that H_1 is true (i.e., a signal is present) and if $x < t$, H_0 is true (i.e., a signal is not present). The parameter t therefore functions as a *threshold level* that resolves the sample space into three physically meaningful regions described by cases (2) through (4) of Eq. (1-164). These regions are shown in Figure 1-17 and can be seen to correspond to the false alarm probability $P(D_1 \mid H_0)$ in region 1, the miss probability $P(D_0 \mid H_1)$ in region 2, and the detection probability $P(D_1 \mid H_1)$ in region 3. If the a priori probabilities are known to be equal, as in a communication system where $P(H_0)/P(H_1) = 1$, the threshold t corresponds to the point where the likelihood functions are equal and the criterion is referred to as the *ideal observer test.* If $P(H_0)/P(H_1) > 1$ or < 1, the threshold moves to the right or left, respectively. It should be noted that the analysis above constitutes a one-sided test of the hypothesis under the assumption that x is always positive. If x can assume both positive and negative values, it may be necessary to perform a two-sided test with somewhat different results, but this is not discussed here.

Comment 1-12. Consider a simple binary optical communications system in which pulses are transmitted to a receiver at well-defined time intervals. When a pulse is received, it is decoded as a 1, and when no pulse is received, it is decoded as a 0. The detection problem in this case is one of quantitatively determining the reliability of such a system (i.e., its error rate when operated in the presence of additive noise). In a fiber link, the additive noise usually consists primarily of receiver noise, although other effects, such as transmitter leakage, must also be taken into account if present. However, in a free-space link, there are additional noise sources that can affect system performance besides receiver noise. These include, among others, atmospheric scintillation, beam wander, and solar background radiation.

The reliability of a communications link is characterized by its *error rate* or, for digital systems, its *bit error rate*. Abbreviated BER, the latter is defined as the number of error pulses N_e divided by the number of transmitted pulses, N_t:

$$\text{BER} = \frac{N_e}{N_t} \tag{1-179}$$

Typical bit error rates are on the order of 10^{-6} or less for a good system.

In a binary communication system, the bit error rate is determined by the probability density functions for the noise alone, $p(x \mid H_0)$, and the signal plus noise, $p(x \mid H_1)$. Thus the total error probability can be written as the sum of the individual errors weighted by their a priori probabilities:

$$\text{BER} = P(H_0) \int_t^{\infty} p(x \mid H_0)\, dx + P(H_1) \int_{-\infty}^{t} p(x \mid H_1)\, dx \tag{1-180}$$

where t is the threshold voltage. For the ideal observer case, $P(H_0) = P(H_1) \equiv \frac{1}{2}$, which means that as many 1's were transmitted as 0's. If we now assume the Gaussian distributions given by Eqs. (1-175) and (1-176) to be representative of the density functions in the off and on states, respectively, for $t = b/2$ and a likelihood ratio of $\ell = 1$ for optimum performance, Eq. (1-180) becomes

$$\text{BER} = \frac{1}{2}\left[1 - \text{erf}\left(\frac{b}{2\sqrt{2}\sigma}\right)\right] \tag{1-181}$$

Since b represents the mean signal amplitude and σ the root-mean-square (rms) noise voltage, we see that the bit error rate is very dependent on the mean signal-to-noise ratio as defined by b/σ.

The assumption of symmetric probability density functions for the on and off states above is somewhat simplistic. In practice, most noise sources have asymmetric density functions, as for example in the case of multiplication noise in avalanche photodiodes (cf. Section 4.8.1) or atmospheric scintillation (cf. Section 4.9). In such instances the optimum threshold level will necessarily be different than $t = b/2$ to equalize the error probabilities for the two states. An illustrative example of this is shown in Figure 1-18, where the threshold levels required by symmetric and asymmetric signal distributions are compared. For purposes of discussion, the symmetric signal is assumed to be Gaussian distributed with parameters $b = 3$ and $\sigma = 1$, while the asymmetric signal is assumed to be Rayleigh distributed with the same mean value. As shown earlier, the latter has a positive skewness, which implies a mean value that lies to the

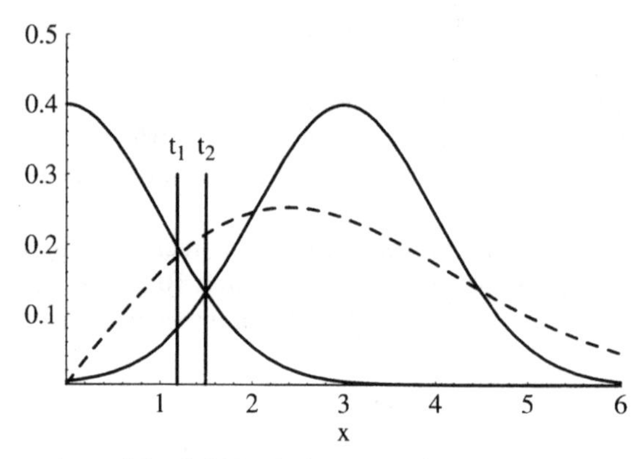

Figure 1-18. Comparison of threshold levels for symmetric (t_1) and asymmetric (t_2) signal distributions for the ideal observer case. The mean values of the symmetric (Gaussian) and asymmetric (Rayleigh) distributions are the same and equal to 3.

right of the mode. Assuming again that $P(H_0) = P(H_1)$ and distributions that are confined to the positive real axis, the threshold level is given by that value of t that satisfies the relationship

$$\int_t^{\infty} p(x \mid H_0)\, dx = \int_0^t p(x \mid H_1)\, dx \qquad (1\text{-}182)$$

Note from Figure 1-18 that the Rayleigh threshold level has shifted to the left of $t = b/2$ and does not coincide with the intersection of the two curves. ■

1.3.2. Bayes' Decision Rule

In the discussion above, the decision criterion was to maximize the a posteriori probabilities given the a priori probabilities. However, additional information concerning the relative importance of the two hypotheses H_0 and H_1 may have to be taken into account in forming the best decision strategy. For example, communication systems generally give equal weight to 1's and 0's since they have no significance except in the context of other 1's and 0's. However, radar systems might consider the consequences of a *type I error* (false alarm) as much more tolerable than a *type II error* (missed detection), although quantifying the *costs* of these errors may be very difficult. The introduction of cost or loss into the decision process allows us to take into account these "subjective" losses or gains in a quantitative way.

If we associate a cost C_{jk} with each of the joint probabilities in Eq. (1-166), an *average cost* can be defined as

$$\begin{aligned}\overline{C} &= C_{00}P(D_0, H_0) + C_{11}P(D_1, H_1) + C_{01}P(D_0, H_1) + C_{10}P(D_1, H_0) \\ &= C_{00}P(D_0 \mid H_0)P(H_0) + C_{11}P(D_1 \mid H_1)P(H_1) \\ &\quad + C_{01}P(D_0 \mid H_1)P(H_1) + C_{10}P(D_1 \mid H_0)P(H_0)\end{aligned} \qquad (1\text{-}183)$$

The problem is to minimize the average cost expressed by Eq. (1-183) and eventually obtain a likelihood ratio that includes cost. Thus, using the relationships given by Eqs. (1-165), Eq. (1-183) reduces to

$$\begin{aligned}\overline{C} &= C_{10}P(H_0) + C_{11}P(H_1) \\ &\quad + (C_{01} - C_{11})P(H_1)P(D_0 \mid H_1) \\ &\quad - (C_{10} - C_{00})P(H_0)P(D_0 \mid H_0)\end{aligned} \tag{1-184}$$

Note that the C_{jk} can assume any value depending on the particular problem. Since a "cost" is also a "loss," the convention is to assign a positive number to a loss and a negative number to a gain.

To obtain the desired likelihood ratio, Eq. (1-184) needs to be written in terms of the probability density functions $p(x \mid H_0)$ and $p(x \mid H_1)$. Now since $p(x \mid H_0)$ represents the sample distribution associated with the null hypothesis H_0 and $p(x \mid H_1)$ represents the simple alternative, the probability that H_0 is chosen correctly is $P(D_0 \mid H_0)$ and the probability that H_1 is not chosen is $P(D_0 \mid H_1)$. $P(D_0 \mid H_0)$ and $P(D_0 \mid H_1)$ correspond to the shaded areas under the two curves in Figure 1-16 and are given by the integrals over the corresponding probability density functions from minus infinity to the threshold value t:

$$P(D_0 \mid H_0) = \int_{-\infty}^{t} p(x \mid H_0)\,dx \quad \text{and} \quad P(D_0 \mid H_1) = \int_{-\infty}^{t} p(x \mid H_1)\,dx \tag{1-185}$$

Therefore,

$$\begin{aligned}\overline{C} &= C_{10}P(H_0) + C_{11}P(H_1) \\ &\quad + \int_{-\infty}^{t} [(C_{01} - C_{00})P(H_1)p(x \mid H_1) \\ &\quad - (C_{10} - C_{00})P(H_0)p(x \mid H_0)]\,dx\end{aligned} \tag{1-186}$$

The likelihood ratio at $x = t$ can be found by setting the $d\overline{C}/dt = 0$. This leads to

$$(C_{01} - C_{00})P(H_1)p(x \mid H_1) - (C_{10} - C_{00})P(H_0)p(x \mid H_0) = 0 \tag{1-187}$$

Therefore,

$$\ell(x = t) = \frac{p(x \mid H_1)}{p(x \mid H_0)} = \frac{C_{10} - C_{00}}{C_{01} - C_{11}} \frac{P(H_0)}{P(H_1)} \tag{1-188}$$

where $P(H_0) + P(H_1) = 1$. Thus if

$$\begin{aligned}\ell(x) &\geq \ell(t) \qquad H_1 \text{ is selected} \\ &< \ell(t) \qquad H_0 \text{ is selected}\end{aligned} \tag{1-189}$$

Note that the inequalities given by Eq. (1-189) are identical to those of Eq. (1-173) except for the addition of cost terms.

Consider once again the Gaussian density functions $p(x \mid H_0)$ and $p(x \mid H_1)$ defined by Eqs. (1-175) and (1-176). When inserted in Eq. (1-174) with $\sigma_n = \sigma_{s+n} = \sigma$, we obtain the threshold condition

$$t' = \frac{b}{2} + \frac{\sigma^2}{b} \ln\left[\frac{C_{10} - C_{00}}{C_{01} - C_{11}} \frac{P(H_0)}{P(H_1)}\right] \tag{1-190}$$

This can be rewritten as

$$\begin{aligned} t' &= \frac{b}{2} + \frac{\sigma^2}{b} \ln\left[\frac{P(H_0)}{P(H_1)}\right] + \frac{\sigma^2}{b} \ln\left[\frac{C_{10} - C_{00}}{C_{01} - C_{11}}\right] \\ &= t + \frac{\sigma^2}{b} \ln\left[\frac{C_{10} - C_{00}}{C_{01} - C_{11}}\right] \end{aligned} \tag{1-191}$$

The impact of cost on the threshold t can readily be obtained from Eq. (1-191). Since the cost of making an error is greater than the cost of a correct decision, it should always be possible to scale the costs such that $C_{01} - C_{00} > 0$ and $C_{10} - C_{11} > 0$. (Indeed, it is then possible to let $C_{00} = C_{11} = 0$ without loss of generality.) If $C_{10} - C_{11} = C_{01} - C_{00}$, $t' = t$, and there is no change in the threshold level compared to the maximum a posteriori test of Section 1.3.1. If, in addition, the a priori probabilities are known to be equal, the criterion corresponds to the *ideal observer test*.

Consider now the radar scenario mentioned earlier where C_{10} is the cost associated with false alarms and C_{01} is the cost associated with missed detections. If $C_{10} - C_{11} < C_{01} - C_{00}$, then $t' < t$ and the threshold moves to the left in Figure 1-17. This means that the costs specified have allowed the false alarms to be increased in order to achieve a higher detection probability. Similarly, if $C_{10} - C_{11} > C_{01} - C_{00}$, as, for example, in a scenario where false alarms are more costly than missed detections (possibly due to operator overload), then $t' > t$ and the threshold moves to the right or is increased relative to the ideal observer case. This reduces the false alarms but at the expense of detection probability. Once again, application of the Bayes criterion to radar scenarios usually reduces to the problem of assigning specific values to the losses and gains, although it should be remembered that these need not imply monetary values but can be other measurable quantities, such as time or energy.

Comment 1-13. A specific example is always helpful. Assume an automated assembly line that uses a laser system to perform a sorting process on randomly oriented parts that have to be integrated into a larger system. The parts include two types of products that are identical in shape but have different reflectivities. We can label these parts A and B, respectively. The laser sorting system is designed such that there is only one part in the laser beam at any given time. Measurements show that the reflectivity varies with orientation and that part A has a higher mean reflectivity than part B. Measurements also show that the probability density functions $p(x \mid H_1)$ and $p(x \mid H_0)$ representing reflectivity as

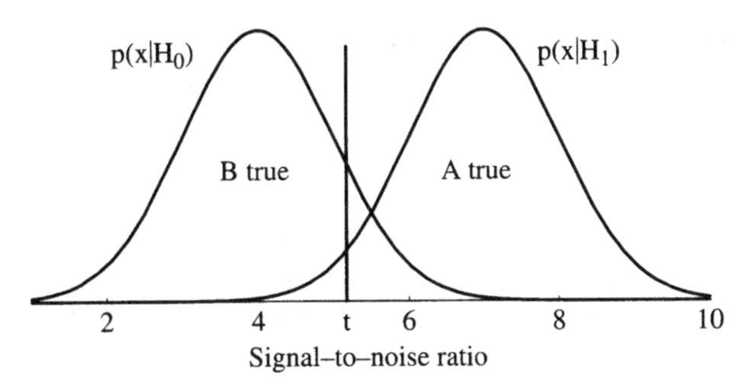

Figure 1-19. Threshold estimate t for an automated laser sorting process using the Bayes criterion.

a function of receiver signal-to-noise ratio are Gaussian, as shown in Figure 1-19, with means equal to 7 and 4, respectively, and variances equal to unity. Studies also show that the manufacturing process results in 1.5 times as many B parts as A parts. Thus the a priori probabilities are $P(H_1) = 0.6$ and $P(H_0) = 0.4$, respectively. Now the price for part A in the system is \$10 and that for part B is \$6. However, if the sorting process selects the wrong part for integration into the system, the part is destroyed during the integration process and has to be sold as scrap metal for \$2. Using the signal-to-noise ratio as the test statistic, the problem is to find the optimum threshold signal-to-noise ratio that will minimize the loss due to misclassification of the objects. The relevant costs are determined as follows:

(a) If the part is assumed to be B, and object B occurs, the manufacturer gains \$6. Thus $C_{00} = -6$.

(b) If the part is assumed to be A, and object B occurs, the manufacturer loses \$6 but gains \$2. Thus $C_{10} = 6 - 2 = 4$.

(c) If the part is assumed to be B, and object A occurs, the manufacturer loses \$10 but gains \$2. Thus $C_{01} = 10 - 2 = 8$.

(d) If the part is assumed to be object A, and object A occurs, the manufacturer gains \$10. Thus $C_{11} = -10$.

These results can be put in the form of a cost matrix,

$$C = \begin{pmatrix} C_{00} & C_{10} \\ C_{01} & C_{11} \end{pmatrix} = \begin{pmatrix} -6 & +4 \\ +8 & -10 \end{pmatrix} \tag{1-192}$$

Using these cost values, the a priori probabilities given above, and the parameters for the Gaussian density functions mentioned above, Eq. (1-188) predicts a threshold signal-to-noise ratio $t = 5.17$. Thus, if a signal-to-noise ratio measurement is greater than t, the laser system automatically labels the part as A, and if it is less than t, it labels it as B. Notice that the decision rule greatly favors the A decision. This is due to the high cost of a missed detection $P(D_0 \mid H_1)$, namely, $C_{01} = 8$. ■

1.3.3. Neyman–Pearson Decision Rule

In the case of a communication system, the a priori probabilities $P(H_0)$ and $P(H_1)$ at the receiver are usually assumed to be equal to $\frac{1}{2}$, that is, as many zeros are received on the average as are ones. This is based on a priori knowledge derived from information theory as applied to binary-coded messages. However, in the case of radar, a priori knowledge of such signal statistics are generally unknown. Also, costs are generally difficult to assign in the radar problem. The question therefore arises as to how to determine the optimum threshold setting for a receiver if only the probability density functions of the noise and signal plus noise are known. Neyman and Pearson[12] solved this problem in 1933 prior to the formal development of statistical decision theory. The Neyman–Pearson rule does not require a priori probabilities or cost functions and has the distinct advantage of maximizing the detection probability under the constraint of a prescribed false alarm probability. It is the decision process employed in most radar systems and is used as the baseline detection criterion in the remaining chapters.

The decision rule to be developed minimizes the total type II error probability subject to the constraint of a fixed total type I error probability. Since the type II error probability is given by $P(D_0 \mid H_1)$, the problem is, in reality, one of maximizing the detection probability $P(D_1 \mid H_1)$, where $P(D_1 \mid H_1) = 1 - P(D_0 \mid H_1)$. Now the type I error is the false alarm probability $P(D_1 \mid H_0)$, which is chosen to be a constant determined by the noise statistics in the absence of a signal and the desired threshold setting. The average error probability, P_e, to be minimized is then

$$P_e = P(D_0 \mid H_1) + \mu P(D_1 \mid H_0) \tag{1-193}$$

Here the second term is a constant and does not affect the minimization process. μ in Eq. (1-193) is a Lagrange multiplier[13] that has yet to be determined. However, the right side of Eq. (1-193) is also equal to the average cost given by Eq. (1-183) under the substitutions $C_{00} = C_{11} = 0$, $P(H_1)C_{01} = 1$, and $P(H_0)C_{10} = \mu$. Using the solution given by Eq. (1-188) with the values of C_{jk} above yields

$$\ell(t) = \frac{p(x \mid H_1)}{p(x \mid H_0)} = \mu \tag{1-194}$$

Once again,

$$\begin{aligned} \ell(x) &\geq \ell(t) \qquad H_1 \text{ is selected} \\ &< \ell(t) \qquad H_0 \text{ is selected} \end{aligned} \tag{1-195}$$

Here we see that the likelihood ratio at t is equal to the Lagrange multiplier μ, while the threshold setting t follows simply from the false alarm probability

$P_{\text{fa}} \equiv P(D_1 \mid H_0)$ given by

$$P_{\text{fa}} = \int_t^\infty p(x \mid H_0)\, dx \tag{1-196}$$

The corresponding detection probability $P_d \equiv 1 - P(D_0 \mid H_1) = P(D_1 \mid H_1)$ is given by

$$P_d = \int_t^\infty p(x \mid H_1)\, dx \tag{1-197}$$

Consider the Gaussian probability density functions shown in Figure 1-16 and described by Eqs. (1-175) and (1-176). The false alarm probability is, using Eq. (1-196),

$$\begin{aligned} P_{\text{fa}} &= \frac{1}{\sqrt{2\pi\sigma^2}} \int_t^\infty e^{-x^2/2\sigma^2} dx \\ &= \frac{\sigma}{2}\left[1 - \text{erf}\left(\frac{t}{\sqrt{2}\sigma}\right)\right] \end{aligned} \tag{1-198}$$

where $\text{erf}(z)$ is the error function defined by

$$\text{erf}(z) = \frac{2}{\sqrt{\pi}} \int_0^z e^{-y^2} dy \tag{1-199}$$

The corresponding detection probability is given by

$$\begin{aligned} P_d &= \frac{1}{\sqrt{2\pi\sigma^2}} \int_t^\infty e^{-(x-b)^2/2\sigma^2} dx \\ &= \frac{\sigma}{2}\left[1 - \text{erf}\left(\frac{t-b}{\sqrt{2}\sigma}\right)\right] \end{aligned} \tag{1-200}$$

Knowledge of the threshold level t then allows one to calculate the likelihood ratio from Eq. (1-177):

$$\ell(t) = \frac{p(t \mid H_1)}{p(t \mid H_0)} = e^{tb(1-b)/\sigma^2} \tag{1-201}$$

Comment 1-14. Assume a desired false alarm probability of $P_{\text{fa}} = 10^{-4}$, a mean signal level $b = 4$, and a noise variance of $\sigma = 1$. Then the threshold setting from Eq. (1-198) is $t \approx 3.72$. Hence if the observation is greater than or equal to 3.72, choose H_1, and if it is less than 3.72, choose otherwise. The probability of detection based on this criterion is, from Eq. (1-200), $P_d = 0.61$. The likelihood ratio evaluates to $\exp(6.88) = 972.6$. ■

1.3.4. Minimax Criterion

In Section 1.3.2 it was shown that the average cost of a Bayes decision was given by Eq. (1-183). Let us investigate the meaning of this equation in more detail. Letting $C_{00} = C_{11} = 0$, Eq. (1-183) can be written

$$\begin{aligned}\overline{C} &= C_{10}P(H_0)P(D_1 \mid H_0) + C_{01}P(H_1)P(D_0 \mid H_1) \\ &= C_{10}P(H_0)P_{\text{fa}} + C_{01}[1 - P(H_0)]P_m \end{aligned} \tag{1-202}$$

where from Eqs. (1-196) and (1-197), the false alarm probability is given by $P_{\text{fa}} = P(D_1 \mid H_0)$ and the miss probability by $P_m = 1 - P_d = P(D_0 \mid H_1)$. Now if we assume Gaussian distributions for the likelihood functions, the Bayes criterion was shown to yield a threshold level given by Eq. (1-190), which in this case becomes

$$t' = \frac{b}{2} + \frac{\sigma^2}{b} \ln\left[\frac{C_{10}}{C_{01}} \frac{P(H_0)}{1 - P(H_0)}\right] \tag{1-203}$$

Thus for given values of b, σ, C_{10}, and C_{01}, the average cost can be written solely in terms of the a priori probability $P(H_0)$. To be more specific, let $b = 1$ and $\sigma = 1$. Then

$$t' = \frac{1}{2} + \ln\left[\frac{C_{10}}{C_{01}} \frac{P(H_0)}{1 - P(H_0)}\right] \tag{1-204}$$

The average cost therefore becomes

$$\overline{C} = C_{10}P(H_0) \int_{t'}^{\infty} \frac{e^{-x^2/2}}{\sqrt{2\pi}} dx + C_{01}[1 - P(H_0)] \int_{t'}^{\infty} \frac{e^{-(x-1)^2/2}}{\sqrt{2\pi}} dx \tag{1-205}$$

Equation (1-205) is plotted in Figure 1-20 for $C_{10} = 1$ and $C_{01} = 0.5$, 1, and 2. (Letting $C_{10} = 1$ is equivalent to letting $\overline{C}$ be normalized to C_{10}.) It can be seen that a maximum occurs for any value of C_{01}. This is the maximum Bayes risk or cost. If, in the general case, the a priori probability $P(H_0)$ is known, the minimum Bayes risk can be found from data such as that shown in the Figure 1-19. However, if $P(H_0)$ is not known, an alternative method must be considered.

One such method is the *minimax criterion*. As shown by Wald[14], this criterion consists of assuming the *least favorable* statistics for the random variable in a Bayes decision rule. For systems in which the likelihood functions are not known, this amounts to choosing the least favorable probability density functions representing the signal. In cases where the density function is known but the a priori statistics $P(H_0)$ are not, it amounts to choosing the least favorable value for $P(H_0)$ in the average Bayes cost. In the latter case, the criterion is then to choose the value of $P(H_0)$ that corresponds to the peaks of the curves shown in Figure 1-20. The advantage of such a criterion is that it minimizes any possible excessive risk that may result from incorrect assumptions about the unknown a priori statistics.

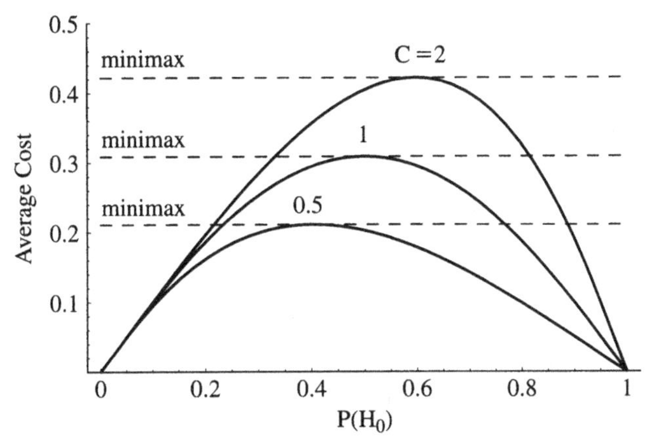

Figure 1-20. Minimax detection corresponds to the maximum Bayes criteria for any given cost. Examples are shown for $C = 0.5$, 1, and 2.

Comment 1-15. An observable in optical radar, as in microwave radar, that is best dealt with using the minimax criterion is the phase of the signal. The argument is usually made in the case of radar that the phase of the signal is uncertain to within the measurement accuracy of the round-trip time of flight to the target. It is therefore reasonable to assume that the phase measurement accuracy, when taken over an ensemble of such measurements, is uniformly distributed over the interval $(0, 2\pi)$. Clearly, the assumption is even more valid at optical frequencies where the wavelength is several orders of magnitude smaller. Since the uniform distribution corresponds to the least favorable distribution that one can assume for any random variable, the minimax criterion provides the *minimum* possible *maximum* cost, hence the name *minimax.* ■

1.4. OPTICAL DETECTION TECHNIQUES

Before concluding this chapter, we will review some basic concepts in optical detection for a better understanding of the succeeding chapters. The most fundamental consideration is that of the *photodetector*, the device that converts light into an electrical current. There are a variety of detector technologies that show high performance in the visible and infrared portions of the spectrum for use by laser radar and communications systems. For reasons of compactness and performance, most applications rely on modern semiconductor devices based on the *photoelectric effect* for the generation of detectable photocurrents. These photocurrents are comprised of photo-generated primary electrons and holes in the depletion region of the detector. Modest gains of 10 to 100 can be achieved in some detectors through *avalanche processes*, as in *photomultiplier tubes* (PMT) and *avalanche photodiodes* (APD). These devices typically exhibit an excess noise generated by the multiplication or avalanche process, which must be taken into account in the receiver design (cf. Section 2.9.3). Still higher gains can be achieved using APD devices operating in a *Geiger mode*, whereby the detector is biased well beyond avalanche breakdown, resulting in very high gains ($\sim 10^6$), ultra-fast rise-times (picoseconds), and sensitivity to single-photon events. It has been shown that in all cases, the primary photoelectron statistics are identical to the Poisson statistics of the impinging photon stream, while more complex statistical models are required to describe avalanche processes (cf. Chapter 4).

The use of photodetectors in optical receivers can be accomplished in two fundamental ways, either through *direct detection* or *coherent detection*. Direct detection may be considered a simple energy collection process that only requires a photodetector placed at the focal plane of a lens, followed by an electronic amplifier for signal enhancement. In comparison, coherent detection requires the presence of an *optical local oscillator* beam to be mixed with the signal beam on the photodetector surface. The coherent mixing process imposes stringent requirements on signal and local oscillator beam alignment in order to be efficient (cf. Section 2.6) and can be implemented in two fundamentally different ways. If the signal and local oscillator frequencies are different and uncorrelated, the process is referred to as *heterodyne detection*, and if they are the same and correlated, as *homodyne detection*. Figure 1-21 shows a generic optical heterodyne configuration in which the signal and local oscillator beams are generated by separate lasers of different, uncorrelated frequencies. They are combined at a beam splitter (BS) that is designed to have a reflectivity high enough to minimize signal loss but low enough to provide sufficient power for use as the local oscillator. Figure 1-22 shows a possible homodyne arrangement in which a small portion of the transmit beam is used for the local oscillator, thereby satisfying the requirement for correlated frequencies.

In homodyne laser radar applications, correlation between the signal and local oscillator frequencies over the round trip time-of-flight to the target τ_R is usually maintained by using a laser transmitter that has a coherence time that is longer than τ_R. In addition, the transmitter and local oscillator frequencies can be the

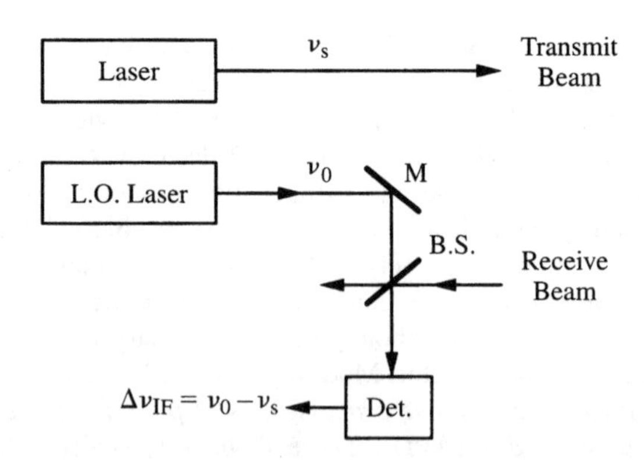

Figure 1-21. A generic heterodyne detection optical configuration.

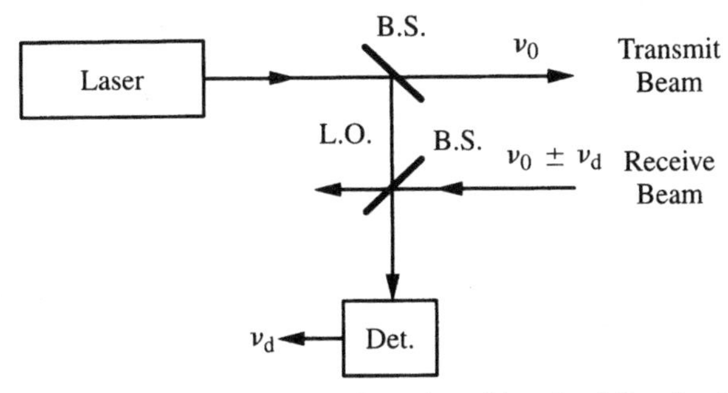

Figure 1-22. A generic homodyne optical configuration with a signal Doppler shifted away from baseband by an amount ν_d.

same or different, depending on whether frequency translators are employed in the optical system. Single-laser frequency offset systems are sometimes referred to as *offset-homodyne systems*. Frequency offsets can also occur for signals that have been Doppler-shifted away from baseband by an amount ν_d, where $\nu_d = \pm 2\,|v|\,/\lambda$ is the *Doppler frequency*. The negative sign corresponds to motion away from the source so that the Doppler frequency folds over at baseband to produce a positive frequency at ν_d. These processes may still be viewed as homodyne because of the inherent correlation between the transmitted and received frequencies.

From a system point of view there are several advantages to using homodyne detection over heterodyne detection. First, baseband homodyne detection is potentially a factor of 2 more sensitive than heterodyne detection because of the narrower (folded) bandwidth which contains the same amount of energy in half the bandwidth of an unfolded signal, as shown in Figure 1-23. Second, frequency-stabilized lasers are usually unnecessary to achieve a narrowband IF in (monostatic) homodyne systems due to the inherent correlation of the transmitted and received frequencies, whereas frequency-stabilized and frequency-locked lasers are usually necessary in heterodyne systems. Third, a single laser system usually offers more compactness and simplicity of design than a two-laser system.

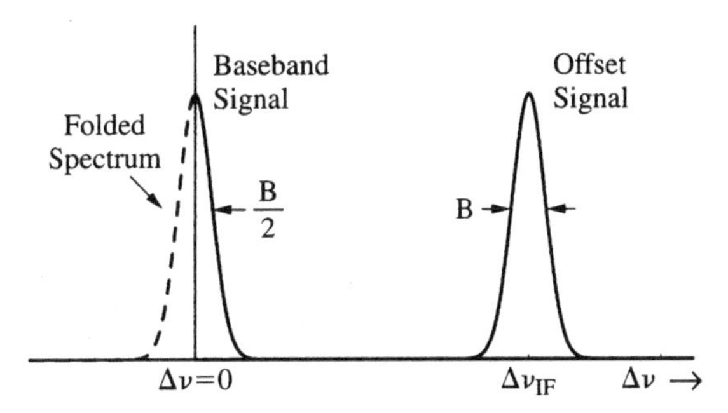

Figure 1-23. Typical IF spectra from a homodyne system operating at baseband and a heterodyne or offset-homodyne system operating at $\Delta\nu_{IF}$. Note the difference in bandwidth for the two cases.

Comment 1-16. Homodyne detection has some unique properties that have proven valuable in the field of quantum optics for demonstrating the possibility of achieving photon noise levels below the quantum limit[15]. Such noise levels are referred to as *squeezed states*, whereby the variance of amplitude or phase, but not both, is reduced below that given by *Heisenberg's uncertainty principle* for a *coherent state*. A coherent state is one in which the uncertainty principle is a minimum, otherwise known as laser light. These states are closely related to the *bunching and anti-bunching statistics* of the photon field, which will be referred to in Section 4.8.2 on APD detection statistics. ■

It becomes readily apparent that the versatility of coherent detection in terms of available modulation formats, and therefore the number of different parameters capable of being measured by the system (i.e., amplitude, frequency, or phase), is significantly greater than that of direct detection. Of course these advantages must be weighed against its higher degree of complexity and greater degree of susceptibility to phase perturbations due to atmospheric turbulence and surface scatter before a selection can be made (cf. Chapter 3). Indeed, one of the major advantages of direct-detection systems is the ability to smooth such beam perturbations via *aperture averaging* techniques (cf. Chapter 4).

Similar conclusions can be drawn for free-space optical communications systems, but with the additional problem of obtaining a coherent optical reference at the receiving end of the link to function as the local oscillator. For example, communication links that employ phase modulation must satisfy stringent requirements on the frequency and phase stability of the transmitter and local oscillator. Homodyne detection is particularly stressing in this regard since optimum performance is achieved when full phase correlation is achieved between the signal and local oscillator beams. (This is usually mitigated by using differential techniques, such as differential phase shift keying.)

Mathematically, we can describe direct-detection as a square-law process in which the photocurrent out of the detector is directly proportional to the square of the field on the detector surface, that is, the intensity. The field impinging on the detector surface may be written

$$E(t) = E_s(t)\cos[\omega_s t + \phi_s] \tag{1-206}$$

where both the signal frequency ω_S and the phase ϕ_S can, in general, be functions of time. The intensity is proportional to the square of the field, so that

$$\begin{aligned} E^2(t) &= E_s^2(t)\cos^2[\omega_s t + \phi_s] \\ &= \tfrac{1}{2}E_s^2(t)\{1 + \cos[\omega_s t + \phi_s]\} \end{aligned} \tag{1-207}$$

Since optical frequencies are not directly observable, a limitation imposed by quantum theory, and since there is no phase reference with which to measure phase, the cosine term can be dropped leaving

$$E^2(t) = \tfrac{1}{2}E_s^2(t) \equiv P(t) \tag{1-208}$$

where $P(t)$ is the incident power on the detector and the factor of 1/2 is identified as being the result of a time-average process.

The photocurrent out of the detector is proportional to the *detector responsivity* and is given by $\Re = e\eta_q/h\nu$ (amps/Watt), where e is the electronic charge, η_q is the detector quantum efficiency, h is Planck's constant, and $\nu = \omega/2\pi$ is the optical frequency. We therefore have

$$i(t) = i_s(t) = \frac{e\eta_q}{2h\nu} E_s^2(t) \tag{1-209}$$

where $i_s(t)$ is identified as the dc or slowly varying (compared to the optical frequency) component of the field.

In the case of coherent detection, a local oscillator beam is added to the signal on the detector surface so that the fields become

$$E(t) = E_0(t)\cos(\omega_0 t + \phi_0) + E_s(t)\cos(\omega_s t + \phi_s) \tag{1-210}$$

where the time dependencies of frequency and phase terms have been suppressed. The square-law device then yields

$$\begin{aligned} E^2(t) &= [E_0(t)\cos(\omega_0 t + \phi_0) + E_s(t)\cos(\omega_s t + \phi_s)]^2 \\ &= E_0^2(t)\cos^2(\omega_0 t + \phi_0) + E_s^2(t)\cos^2(\omega_s t + \phi_s) \\ &\quad + 2E_0(t)E_s(t)\cos(\omega_0 t + \phi_0)\cos(\omega_s t + \phi_s) \end{aligned} \tag{1-211}$$

Using the trigonometric identity $\cos x = (1 + \cos 2x)/2$, Eq. (1-211) becomes

$$\begin{aligned} E^2(t) &= \tfrac{1}{2}E_0^2(t) + \tfrac{1}{2}E_s^2(t) + E_0(t)E_s(t)\cos[(\omega_0 - \omega_s)t + \phi_0 - \phi_s] \\ &\quad \times \tfrac{1}{2}E_0^2(t)\cos(2\omega_0 t + 2\phi_0) + \tfrac{1}{2}E_s^2(t)\cos^2(2\omega_s t + 2\phi_s) \\ &\quad + E_0(t)E_s(t)\cos[(\omega_0 + \omega_s)t + \phi_0 + \phi_s] \end{aligned} \tag{1-212}$$

Here the last three terms on the right side of Eq. (1-212) correspond to unobservable processes and can be neglected. We are therefore left with

$$E^2(t) = \tfrac{1}{2}E_0^2(t) + \tfrac{1}{2}E_s^2(t) + E_0(t)E_s(t)\cos[(\omega_0 - \omega_s)t + \phi_0 - \phi_s] \tag{1-213}$$

or in terms of the detector current

$$i(t) = i_0(t) + i_s(t) + 2\sqrt{i_0(t)i_s(t)}\cos\{[\omega_0 - \omega_s]t + \phi_0 - \phi_s\} \tag{1-214}$$

where $i_0 = \Re E_0^2/2$ and $i_s = \Re E_s^2/2$ are the dc contributions from the local oscillator and signal beams. It is usually the case that $i_0 \gg i_s$, that is, the local oscillator power is much greater than the signal power. Since the third term in Eq. (1-214) contains i_0 as well as i_s, increasing the former has the effect of amplifying the latter above various circuit noises in the receiver. [Of course once the photon- or shot-noise limit has been reached, there is no further advantage in increasing i_0 (cf. Section 2.10)].

Equation (1-214) is the fundamental equation representing coherent detection. We see that the detected signal consists of a dc plus a harmonic component, the frequency and phase of the latter corresponding to the frequency and phase *differences* of the signal and local oscillator beams. The difference or *intermediate frequency* (IF) usually resides in the radio frequency (RF) or microwave regions of the spectrum.

In most applications, the dc terms are filtered out and the local oscillator power is constant, resulting in

$$i(t) = 2\sqrt{i_0 i_s(t)}\cos\{[\omega_0 - \omega_s]t + \phi_0 - \phi_s\} \quad (1\text{-}215)$$

Equation (1-215) represents the IF signal that is characteristic of heterodyne and offset-homodyne detection in which $\omega_0 \neq \omega_s$. On the other hand, there is no intermediate frequency for a CW homodyne signal when $\omega_0 = \omega_s$, so that Eq. (1-215) reduces to

$$i(t) = 2\sqrt{i_0 i_s(t)}\cos[\phi_0 - \phi_s] \quad (1\text{-}216)$$

Since the frequency and phase terms in Eqs. (1-215) and (1-216) may be made time-dependent through appropriate *optical modulation techniques*, we see that amplitude, frequency, or phase modulation may be used with coherent detection whereas, according to Eq. (1-209), only amplitude modulation can be supported with direct detection.

The relative sensitivities of coherent and direct detection are discussed in Section 2.10 and further clarified in the context of detection statistics in Section 5.7. In the latter discussion it is shown that environmental and system issues, such as atmospheric turbulence and surface scatter, can strongly impact the tradeoff between the two approaches.

REFERENCES

[1]O. Kempthorne and L. Folks, *Probability, Statistics, and Data Analysis*, Iowa State University Press, Ames, IA, 1971.

[2]A. Papoulis, *Probability, Random Variables, and Stochastic Processes*, McGraw-Hill Book Company, New York, 1965.

[3]A. M. Mood and F. A. Graybill, *Introduction to the Theory of Statistics*, 2nd ed., McGraw-Hill Book Company, New York, 1963, p. 149.

[4]R. L. Mitchell, Permanence of the log-normal distribution, *J. Opt. Soc. Am.*, Vol. 58, Sept. 1962, pp. 1267–1272.

[5]V. I. Tatarskii, *The Effects of the Turbulent Atmosphere on Wave Propagation* (translated by Israel Program for Scientific Translations; originally published in 1967), U.S. Department of Commerce, National Technical Information Service, Springfield, VA, 1971, pp. 8–9.

[6]A. M. Yaglom, *An Introduction to the Theory of Stationary Random Functions*, translated and edited by R. A. Silverman, Prentice Hall, Upper Saddle River, NJ, 1962.

[7]M. I. Skolnik, *Introduction to Radar Systems*, McGraw-Hill Book Company, New York, 1962, p. 414.

[8]M. Schwartz, *Information Transmission, Modulation, and Noise*, McGraw-Hill Book Company, New York, 1959, p. 289.

[9]J. V. DiFranco and W. L. Rubin, *Radar Detection*, Artech House, Norwood, MA, 1980.

[10]R. N. McDonough and A. D. Whalen, *Detection of Signals in Noise*, 2nd ed., Academic Press, San Diego, CA, 1995, p. 190.

[11]A. Wald, *Statistical Decision Functions*, John Wiley & Sons, New York, 1950.

[12]J. Neyman and E. S. Pearson, The testing of statistical hypothesis in relation to probability a priori, *Proc. Cambridge Philos. Soc.*, Vol. 29, 1933.

[13]H. Goldstein, *Classical Mechanics*, Addison-Wesley, Reading, MA, 1959.

[14]Ibid., ref. 11.

[15]H. P. Yuen and J. H. Shapiro, "Optical communication with two-photon coherent states — Part III: Quantum measurements realizable with photoemissive detectors," *IEEE Trans. Inf. Theory*, Vol. IT-26, No. 1, Jan., 1980, p. 78.

2

Signal and Noise Analysis

2.1. INTRODUCTION

Optical radar and communication systems are usually chosen over radio-frequency systems because of either bandwidth or beamwidth advantages. The small beamwidths achievable with laser systems, even with apertures on the order of a few centimeters, allow one to overcome the relatively high photon and thermal noise characteristics of optical receivers by placing a significant amount of energy on target. Beamwidths of less than 100 μrad are readily achieved with the aperture sizes above, so that a significant fraction of the energy transmitted may be captured by a communications receiver or intercepted by a small reflecting target in a laser radar application. However, any process that degrades the power or quality of the propagating beam from the initial design parameters detracts from the overall performance of the system.

There are many processes both internal and external to a laser system that can influence detection performance. In this chapter we focus primarily on those processes that (1) determine a system's ability to put energy on target and (2) are the sources of noise that may compete with the signal. The former processes include beam expansion due to diffraction, beam truncation and obscuration, and beam quality as determined by the transmitter and internal optics. The latter processes include internally generated noise sources such as detector and receiver noise as well as external noise sources such as solar radiation and atmospheric backscatter. Analysis of these effects allows for the development of signal-to-noise ratio expressions for coherent and direct detection laser systems that represent optimum performance for a given aperture size.

2.2. REVIEW OF DIFFRACTION THEORY

Diffraction is the fundamental effect that causes electromagnetic waves to spread as they propagate. It is the underlying phenomenon that drives aperture size in microwave or optical systems. In the case of resolved targets, a common mode of operation for laser radar systems, it is important for image reconstruction to maximize the number of spatial resolution cells on target. For unresolved targets, it is important for achieving high detection probabilities to maximize the energy on target. In this section, some elementary concepts in *diffraction theory* are reviewed that will lay the foundations for discussions of Gaussian and higher-order beam modes to be discussed in later sections.

Diffraction effects are described by solutions to the *Helmholtz scalar wave equation* for the electromagnetic field U given by

$$\nabla^2 U + k^2 U = 0 \tag{2-1}$$

where $k = 2\pi/\lambda$, λ is the wavelength of light, and the index of refraction is assumed to be unity. In the limit of plane wave illumination of the aperture and small observation angles relative to the normal of the emitting aperture, it can be shown[1] that the solution to Eq. (2-1) is the well-known *Fresnel–Kirchhoff diffraction formula*

$$U(x, y, R) = \iint_s u(\xi, \eta)\Lambda(\delta)\frac{e^{iks}}{s}\, d\xi\, d\eta \tag{2-2}$$

Referring to Figure 2-1, $u(\xi, \eta)$ is the complex wave amplitude at the aperture; s is the distance from the illuminating point ξ, η at the aperture to the observation point x, y; R is the distance from the center of the aperture to the observation point x, y; and R' is the distance from the source of illumination to the center of the aperture. $\Lambda(\delta) = -(i\kappa/2\lambda)(1 + \cos\delta)$ is an *inclination factor* describing the angular dependence of the amplitude of the *secondary Huygens wavelets* propagating toward the observation point. For the small angles to be considered here we can let $\cos\delta = 1$. κ is a factor related to source brightness such that as $s \to \infty$, U is nonzero. We also assume that $\kappa = 1$ in the remainder of the chapter.

Equation (2-2) represents the fact that each point on the aperture is a source of Huygens wavelets which when summed over the aperture with the illuminating

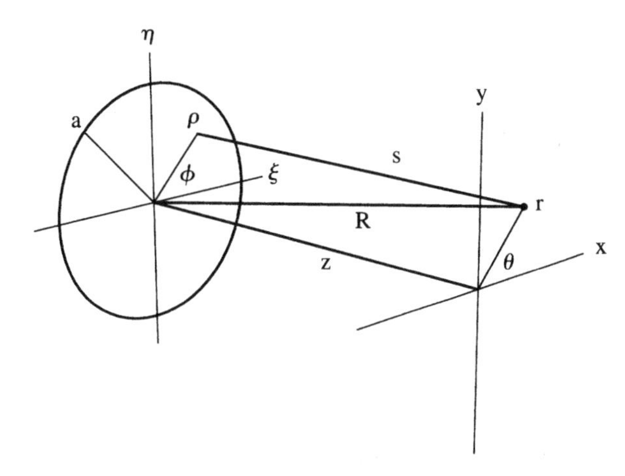

Figure 2-1. Geometry of Fraunhofer diffraction theory.

function yields the diffraction pattern. For the propagation conditions of interest here, the distance s in the denominator can be assumed to be slowly varying, compared to the exponential and can therefore be taken outside the integral. Thus Eq. (2-2) reduces to the *Huygens–Fresnel diffraction integral*, given by

$$U(x, y, z) = \frac{-i}{\lambda s} \iint_S u(\xi, \eta) e^{iks} d\xi d\eta \tag{2-3}$$

From Figure 2-1 we have

$$\begin{aligned} s &= [(x-\xi)^2 + (y-\eta)^2 + z^2]^{1/2} \\ &= z\left[1 + \frac{x^2+y^2}{z^2} + \frac{\xi^2+\eta^2}{z^2} - \frac{2(\xi x + \eta y)}{z^2}\right]^{1/2} \end{aligned} \tag{2-4}$$

Since most optical systems employ circular apertures with axially symmetric beams, we will assume azimuthal symmetry and expand s in polar coordinates. Thus, with $\xi = \rho\cos\phi$, $\eta = \rho\sin\phi$, $x = r\cos\theta$, and $y = r\sin\theta$,

$$s = z\left(1 + \frac{r^2+\rho^2}{z^2} - \frac{\rho\cos\phi r\cos\theta + \rho\sin\phi r\sin\theta}{z^2}\right)^{1/2} \tag{2-5}$$

Under the assumption that the distance z is large compared to either r or ρ, commonly referred to as the *paraxial approximation*, the square root can be expanded to yield

$$s \approx z + \frac{r^2+\rho^2}{2z} - \frac{r\rho\cos(\theta-\phi)}{z} + \cdots \tag{2-6}$$

where terms of order z^{-2} and higher are ignored. With the changes above, Eq. (2-3) becomes

$$U(r, \theta, z) = \frac{-ie^{ikz}}{\lambda z} \int_0^a u(\rho) e^{ik(r^2+\rho^2)/2z} \left[\int_0^{2\pi} e^{-ik\rho r\cos(\phi-\theta)/z} d\phi\right] \rho\, d\rho \tag{2-7}$$

where a is the radius of the aperture and $u(\rho)$ is the complex amplitude and phase of the source radiation at the aperture. But the integral in brackets can be identified with the integral representation of the Bessel functions:

$$J_n(x) = \frac{i^{-n}}{2\pi} \int_0^{2\pi} e^{ix\cos\delta} e^{in\delta} d\delta \tag{2-8}$$

which for $n = 0$ yields for Eq. (2-7),

$$U(r, z) = Ke^{ikr^2/2z} \int_0^a u(\rho) J_0 \frac{kr\rho}{z} e^{ik\rho^2/2z} \rho\, d\rho \tag{2-9}$$

where

$$K = \frac{-2\pi i e^{ikz}}{\lambda z} \tag{2-10}$$

Equation (2-9) is a frequently used form of the *Huygens–Fresnel formula* for the field distribution at an observation plane located at a distance z from the source aperture.

Equation (2-9) can be generalized to the case of spherical waves by assuming a convergent wavefront exiting the aperture with a radius of curvature equal to $-f$. (Here, radius of curvature is defined as positive for a surface that is concave when viewed from $z = -\infty$.) This condition is achieved by placing a positive lens of focal length f just after the aperture. It is accounted for in Eq. (2-9) by including a phase term in $u(\rho)$ of the form $u(\rho) = u_0(\rho)\exp(-ik\rho^2/2f)$, where $u_0(\rho)$ is the real part of the aperture distribution. This yields

$$U(r,z) = Ke^{ikr^2/2z}\int_0^a u_0(\rho)J_0\frac{kr\rho}{z}e^{(ik\rho^2/2)(1/z-1/f)}\rho\,d\rho \tag{2-11}$$

Here the form of the phase term may be understood from the geometry of Figure 2-2, where, for small displacements, the phase shift of the edge of the beam relative to the center is $\Delta z \approx \rho^2/2f$.

The well-known *Fraunhofer formula* for the far-field distribution of a circular aperture may now be obtained for either the case of a collimated beam, by letting $f \to \infty$ and $z \to \infty$, or a focused beam, by letting $z = f$. Thus, in either case, we obtain

$$U(\alpha) = K\int_0^a u(\rho)J_0(k\alpha\rho)\rho\,d\rho \tag{2-12}$$

where $\alpha = r/z$ and $K(z) \to K(f)$ in the focused beam case. The distribution obtained in the focused beam case is frequently referred to as the *simulated far-field pattern* since it does not occur in the far field of the aperture.

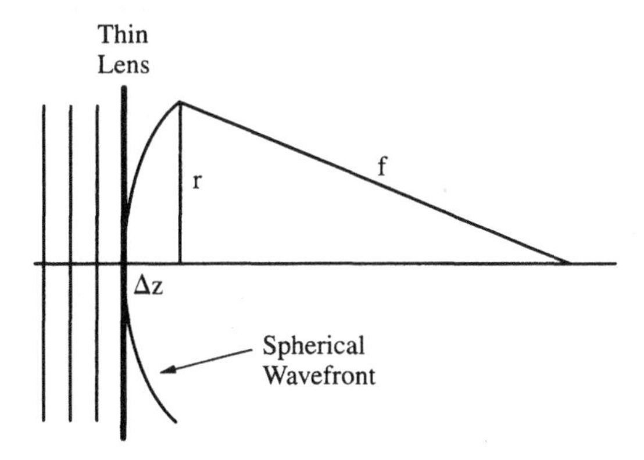

Figure 2-2. Phase shift as a function of radius for a focused spherical wavefront.

One can interpret Eq. (2-11) as stating that the Fraunhofer condition occurs at only one location, namely the geometric focal plane of the lens. (A collimated beam corresponding to $f = \infty$.) However, one must keep in mind that this condition occurs only for a uniform phase front at the input to the aperture. Any other phase front must be accounted for with an additional phase term in $u(\rho)$. Thus, for a spherical divergent wavefront input to the aperture-lens assembly from the left, $u(\rho) = u_0(\rho)\exp[ik\rho^2(1/2R_c - 1/2f)]$, where $u_0(\rho)$ is again the real part of $u(\rho)$, R_c is the input radius of curvature, and f is the focal length of the lens. Equation (2-11) then becomes

$$U(\alpha) = K\int_0^a u_0(\rho)J_0(k\alpha\rho)e^{(ik\rho^2/2)(1/z+1/R_c-1/f)}\rho\,d\rho \tag{2-13}$$

The Fraunhofer condition is therefore achieved for a nonplanar input wavefront when

$$\frac{1}{f} = \frac{1}{z} + \frac{1}{R_c} \tag{2-14}$$

But this is just the *lens formula*, which states that the Fraunhofer far-field pattern occurs, in the more general sense, at the *focal range* $z_f = f/(1 - f/R_c)$ of the focusing lens, the focal range being defined as the center of curvature of the exit rays from the aperture. It occurs at the focal plane of the lens $z = f$ only in the restricted case of a uniform phase front at the aperture (i.e., $R_c \to \infty$). It is also important to realize that all locations along the z-axis other than at the focal range produce near-field patterns, including the location $z = \infty$ in the focused beam case.

In addition, it is not too difficult to show that the Fraunhofer formula corresponds to a spatial Fourier transform of the aperture distribution and that a lens system performs nothing other than a Fourier transform of the input distribution when viewed at the focal plane of the lens. These facts are fundamental to the field of Fourier optics, which professes a completely analogous set of distance-spatial frequency transfer functions to the time-frequency transfer functions of temporal signals (cf. Section 2.5).

An estimate of the distance from the aperture at which the far-field approximation becomes valid for plane-wave illumination can be obtained from the quadratic phase term in Eq. (2-9) (i.e., when $k\rho^2/2z \ll 2\pi$ or $z \gg a^2/2\lambda$), where a is the radius of the aperture. This is the usual criterion for separating near- and far-field distances.

Comment 2-1. As an example, consider the case of a uniformly illuminated circular aperture with and without central obscuration. The Fraunhofer formula for a uniform plane-wave input becomes

$$U(\alpha) = K\int_0^a u(\rho)J_0(k\alpha\rho)\rho\,d\rho \tag{2-15}$$

A well-known recurrence relation

$$\int_0^x J_n(x')x'^{n+1}dx' = x^{n+1}J_{n+1}(x) \tag{2-16}$$

may be used such that with $x' = k\alpha\rho$, we obtain

$$\begin{aligned} U(\alpha) &= \frac{Ku_0}{k^2\alpha^2}\int_0^{k\alpha a} J_0(x')x'dx' \\ &= \frac{-ie^{ikz}}{\lambda z}Au_0\frac{2J_1(k\alpha a)}{k\alpha a} \end{aligned} \tag{2-17}$$

Here $A = \pi a^2$ and Eq. (2-10) was used for K. Recognizing that $u_0^* u_0$ must equate to an intensity so that $u_0 = (P/A)^{1/2}$, where P is the power out of the aperture, the intensity $I = U^* U$ becomes

$$I = I_0\left[\frac{2J_1(k\alpha a)}{k\alpha a}\right]^2 \tag{2-18}$$

where $I_0 = PA/\lambda^2 z^2$ is the intensity at the center of the diffraction pattern. Equation (2-18) represents the well-known *Airy pattern* shown in Figure 2-3. The first minimum occurs when $J_1(x) = 0$ at $x = 3.8328$. Hence $\alpha = 3.8328/\pi \times \lambda/D = 1.22\lambda/D$.

The centrally obscured aperture is solved in an identical manner except that the limits of integration are from b to a, where b is the radius of the obscuration. It is left to the reader to show that

$$I = I_0\left[\frac{2J_1(k\alpha a)}{k\alpha a} - \frac{2FJ_1(k\alpha b)}{k\alpha b}\right]^2 \tag{2-19}$$

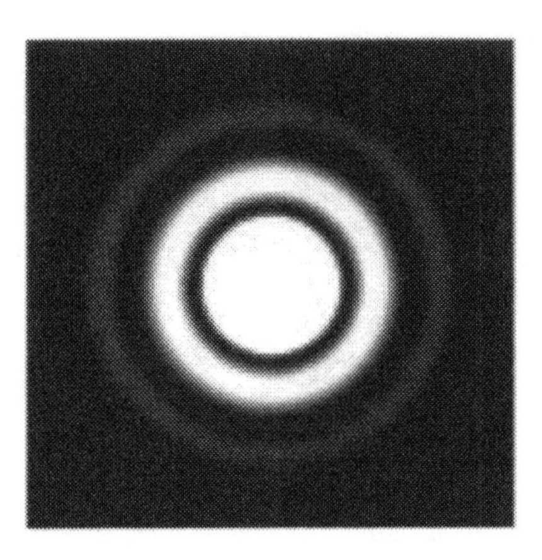

Figure 2-3. Far-field Airy diffraction pattern due to a circular aperture.

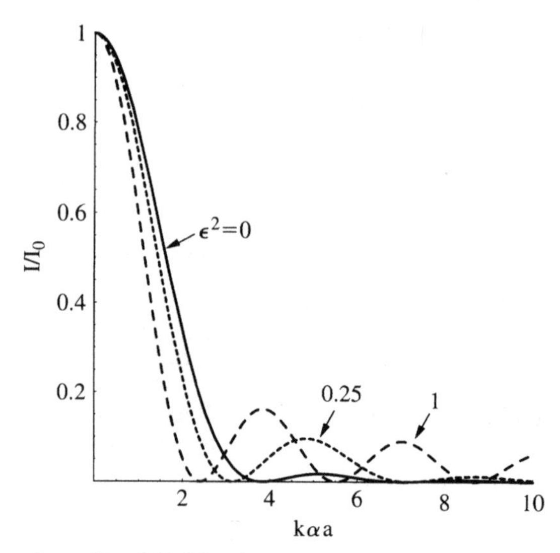

Figure 2-4. Comparison of far-field diffraction patterns of circular apertures with varying amounts of obscuration. $\varepsilon = b/a$, where b and a are the inner and outer radii of the aperture, respectively.

where $I_0 = PA_a/\lambda^2 z^2$, A_a is the total aperture area $= \pi a^2$, $F = A_b/A_a$, and A_b is the total obscuring area $= \pi b^2$. The centrally obscured aperture with a uniform plane-wave input is a particularly clear example of the Huygens–Fresnel principle for the superposition of diffracting waves. Indeed, Eq. (2-19) represents the superposition of two diffracting waves, one of which consists of a uniform "top-hat" distribution of outer area A_a and another which consists of an out-of-phase uniform top-hat distribution of inner area A_b. Since the forms of the two expressions in the brackets are identical, the patterns are scaled replicas of each other, the *obscuring* pattern being the larger pattern in the far field since its effective aperture diameter is smaller. Given a little thought, one realizes that the larger angular spread of the *effective aperture* A_b, when subtracted from the narrower pattern of A_a, results in a slight *narrowing* of the central lobe of the overall pattern compared to that produced by A_a alone. In this case, the minima correspond to the roots of $J_1(x) - \sqrt{F} J_1(x\sqrt{F}) = 0$, the first of which occurs at $x = 3.144$. Hence $\alpha = 3.144/\pi \times \lambda/D \approx 0.5\lambda/D$. This can be seen in Figure 2-4. ■

2.3. FREE-SPACE PROPAGATION

Modern optical systems frequently employ lasers for transmission of energy or information to the target. As will be shown in Section 2.3.3, an arbitrary field distribution can be described using an orthonormal set of *wavefunctions* derived from Helmholtz's wave equation under the constraint of boundary conditions corresponding to a laser cavity of infinite transverse extent. The resultant wavefunctions or *beam modes* have discrete resonant frequencies in directions both longitudinal and transverse to the laser cavity. It can be shown that these modes are also modes of free space, having the characteristic property of propagation without distortion, except for expansion due to diffraction, in an aberration-free environment.

A laser system that requires maximum energy density at the target should be designed with a beam profile that maximizes the on-axis intensity. A key design requirement to achieving this is to employ an optical system that operates in the *diffraction limit*. The diffraction-limited concept applies to both optical components and beams and is usually defined in terms of the minimum divergence (or focal spot diameter) that can be achieved for a given wavelength and aperture diameter. To achieve such a minimum, the beam must have a relatively smooth phase profile across the aperture. For example, an aperture that is uniformly illuminated with a plane or spherical phase front results in a diffraction-limited spot in the far field. On the other hand, an aperture illuminated by a multimode beam, or even a single high-order beam mode, that has several distinct phase discontinuities across the aperture (i.e., where the intensity goes to zero) results in *non-diffraction-limited performance*. Since the resultant beam divergence (or focal spot size) is some multiple n times the minimum possible divergence, a non-diffraction-limited beam is frequently characterized as being n times diffraction limited.

Some laser applications can tolerate high-order or even multimode beams, such as those employing direct detection receivers, while others, such as coherent or heterodyne systems, require diffraction-limited beams for optimal performance. Multimode beams can be generated in a variety of ways and are frequently the result of compromise trade-offs between transmitter efficiency and beam quality. Stable laser resonators will generally operate in multiple transverse modes unless apertured to restrict the lasing volume to a single mode. Unfortunately, this approach makes inefficient use of the lasing volume and results in reduced electrical efficiency. Other cavity configurations can yield high on-axis brightness in the far-field despite little or no aperturing, such as the unstable resonator, which is well suited to low-aspect-ratio cavities, where single-mode operation is inefficient and difficult to sustain. Such resonators can always be described in terms of orthonormal free-space modes.[2] We consider such a set, specifically the Hermite–Gaussian wavefunctions, in Section 2.3.3. We also explore a simplified method for characterizing multimode beams that takes advantage of the more tractable and intuitive parameters of the fundamental Gaussian mode. Presently, however, we focus on developing the key equations related to the fundamental Gaussian beam.

Gaussian beam theory follows from a straightforward solution of Helmholz's equation under the assumption of a slowly varying beam radius.[3,4] A general solution to Eq. (2-1) for a wave traveling to the right is

$$U = a_o \psi(x, y, z) e^{-ikz} \tag{2-20}$$

where a_o has the dimensions of square root of intensity. Substitution into Eq. (2-1) gives

$$\nabla^2\psi - 2ik\frac{\partial \psi}{\partial z} = 0 \tag{2-21}$$

A slowly varying beam profile along the propagation direction z can be assumed by neglecting $\partial^2\psi/\partial z^2$ in ∇^2 and postulating a solution of the form

$$\psi = e^{-i(p+kr^2/2q)} \tag{2-22}$$

where $r^2 = x^2 + y^2$ and $p \equiv p(z)$ and $q \equiv q(z)$. Substituting Eq. (2-22) into Eq. (2-21) yields

$$-2k\left(p' + \frac{i}{q}\right)\psi - \frac{k^2r^2}{q^2}(q' - 1)\psi = 0 \tag{2-23}$$

This yields the solutions

$$q' = 1 \quad \text{and} \quad p' = -\frac{i}{q} \tag{2-24}$$

where the primes indicate differentiation with respect to z. Integrating q' yields

$$q(z_2) = q(z_1) + z \tag{2-25}$$

Hence a *complex beam parameter* q can be defined which propagates linearly along the z-axis. As will be shown, it is directly analogous to the radius of curvature as defined in ray optics theory, with the additional feature of accounting for diffraction effects. With some foresight, two real beam parameters, R_g and ω, will be introduced, which are related to the complex beam parameter q by

$$\frac{1}{q} = \frac{1}{R_g} - i\frac{\lambda}{\pi\omega^2} \tag{2-26}$$

Letting $R_g \to \infty$, we find that $q = i\pi\omega_0^2/\lambda \equiv q_0$, where ω_0 is defined as the radius of the beam at the *beam waist* location, that position along z where the radius ω of the beam is a minimum and the phase front is flat. Hence $q(z) = z + q_0$, which when substituted into Eq. (2-26) yields

$$\omega^2(z) = \omega_0^2\left[1 + \left(\frac{\lambda z}{\pi\omega_0^2}\right)^2\right] \tag{2-27}$$

and

$$R_g(z) = z\left[1 + \left(\frac{\pi\omega_0^2}{\lambda z}\right)^2\right] \tag{2-28}$$

The significance of the parameters R_g and ω becomes clearer when q is substituted into Eq. (2-22):

$$\psi = e^{-ip-r^2/\omega^2(z)-ikr^2/2R_g(z)} \tag{2-29}$$

We see that the wavefunction ψ is described by a Gaussian amplitude distribution with beam radius $\omega(z)$ and a complex phase term having a spherical radius of curvature $R_g(z)$. Here R_g follows the usual sign convention of a positive number for a concave phase front when viewed from $z = -\infty$.

Notice from Eq. (2-28) that $R_g \to \infty$ at the beam waist location $z = 0$. At all other locations the radius of curvature $R_g \neq z$ (except for the limiting case of $z = \infty$), meaning that the beam waist location is, in general, not at the center of curvature of the phase front, as shown in Figure 2-5. It is easy to show that the difference between R_g and z is greatest when the radius of curvature is a minimum (i.e., where $dR_g/dz = 0$ or at $z_0 = \pi\omega_0^2/\lambda$). At this point $R_g = 2z_0$ and $\omega(z) = \sqrt{2}\,\omega_0$, where z_0 is frequently referred to as the *Rayleigh range* of an infinite Gaussian beam. It may also be identified as that range where the near field begins to evolve into the far field. (Since a pure mode does not change its intensity distribution as it propagates, the concept of far field has little meaning. However, the concept will gain meaning when multimode beams are considered in Section 2.3.3.) The form of the expression for $\omega(z)$ given by Eq. (2-27) is that of a hyperbola with asymptotes that define the far-field diffraction angle of the beam (i.e., $\theta_B = \lambda/\pi\omega_0$), where θ_B is the half-angle beam divergence. This is also shown in Figure 2-5.

The factor containing p is arrived at as follows. Substituting for q at the beam waist in the equation for p' [i.e., Eq. (2-24)], we obtain

$$p' = \frac{-i}{z + q_0} = \frac{-i}{z + i(\pi\omega_0^2/\lambda)} \tag{2-30}$$

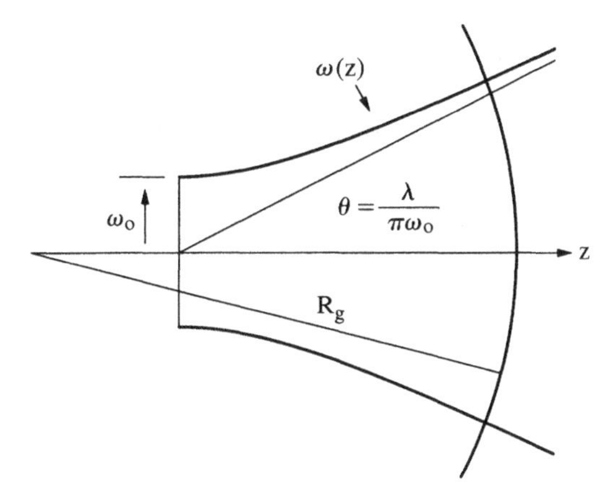

Figure 2-5. Evolution of Gaussian beam phase profile as a function of propagation distance z.

where the beam waist has been positioned at $z = 0$. Integrating Eq. (2-30) yields

$$p(z) = -\tan^{-1}\frac{\lambda z}{\pi\omega_0^2} - i\ln\sqrt{1 + \left(\frac{\lambda z}{\pi\omega_0^2}\right)^2} \tag{2-31}$$

Equation (2-20) now becomes

$$\begin{aligned} U(r,z) &= a_o e^{i\tan^{-1}(\lambda z/\pi\omega_0^2) - \ln\sqrt{1+(\lambda z/\pi\omega_0^2)^2} - r^2/\omega(z)^2 - ikr^2/2R_g} \\ &= \frac{a_o\omega_0}{\omega(z)} e^{-i(kz-\phi)-(r^2/\omega(z)^2 + ikr^2/2R_g)} \end{aligned} \tag{2-32}$$

where

$$\phi = \tan^{-1}\frac{\lambda z}{\pi\omega_0^2} \tag{2-33}$$

The intensity of a Gaussian beam is therefore given by

$$I = U^*(r,z)U(r,z) = \frac{a_o^2\omega_0^2}{\omega^2(z)} e^{-2r^2/\omega^2(z)} \tag{2-34}$$

The constant a_o is obtained by integrating the intensity over all space and setting it equal to the total optical power P. Hence

$$\frac{a_o^2\omega_0^2}{\omega^2}\int_0^{2\pi}\int_0^{\infty} e^{-2r^2/\omega^2} r\,dr\,d\theta = \frac{\pi a_o^2\omega_0^2}{2} = P \tag{2-35}$$

and Eq. (2-34) becomes

$$I = I_0 e^{-2r^2/\omega^2(z)} \tag{2-36}$$

where $I_0 = u^2 = 2P/\pi\omega^2(z)$ is the on-axis intensity of a Gaussian beam at position z and u is the amplitude of the beam. Using Eq. (2-35), the complex wave amplitude therefore becomes

$$U(r,z) = u(z)e^{-i(kz-\phi)-r^2/\omega^2(z)-ikr^2/2R_g(z)} \tag{2-37}$$

where $u = \sqrt{2P/\pi\omega^2(z)}$.

2.3.1. Transformation of Gaussian Beams

The q-parameters introduced above transform in a manner identical to spherical waves in geometric optics. For example, a spherical wavefront with radius of curvature R_1 at position 1 transforms into a spherical wavefront with radius of curvature R_2 at position 2 according to the relation

$$R_2 = R_1 + z \tag{2-38}$$

Similarly, an ideal thin positive lens transforms a spherical wavefront according to the lens formula:

$$\frac{1}{R_2} = \frac{1}{R_1} - \frac{1}{f} \tag{2-39}$$

where R_2 is the radius of curvature just after the lens and R_1 is the radius of curvature just before the lens. Finally, a general inhomogeneous media transforms an input ray at position x and slope x' within the paraxial ray approximation in accordance with the *ABCD ray transfer matrix* given by

$$\begin{pmatrix} x_2 \\ x_2' \end{pmatrix} = \begin{pmatrix} A & B \\ C & D \end{pmatrix} \begin{pmatrix} x_1 \\ x_1' \end{pmatrix} \tag{2-40}$$

The elements of the *ABCD* matrix can be found from the focal length f and principal planes h_1 and h_2 of the lenslike media through the relationships

$$h_1 = \frac{D-1}{C} \qquad h_1 = \frac{A-1}{C} \qquad f = -\frac{1}{C} \tag{2-41}$$

where $AD - BC = 1$. Here the principal planes[5] are defined by the intersection points of the collimated rays and focused rays of a lenslike medium, as shown in Figure 2-6. Equation (2-40) also holds for the simpler case of lenses and lens combinations. Indeed, since the rays are normal to the wavefront of a spherical wave, the ray parameters x and x' can always be related to the radius of curvature of a spherical wave by $R = x/x'$, so that Eq. (2-40) becomes

$$R_2 = \frac{AR_1 + B}{CR_1 + D} \tag{2-42}$$

where R_1 and R_2 are the input and output radii of curvatures, respectively.

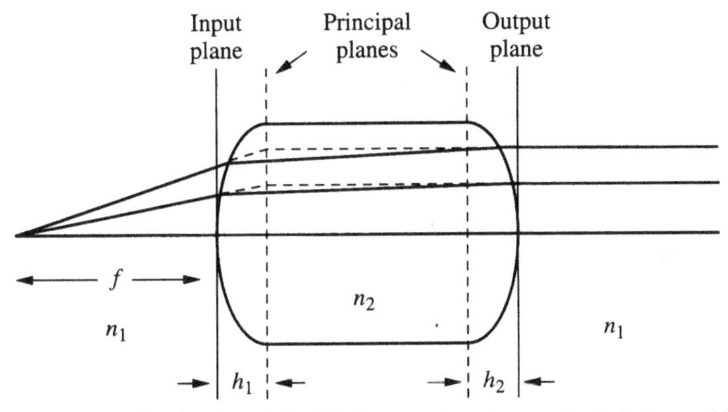

Figure 2-6. Ray trace showing the relationship between the primary principal plane and input plane of a lenslike medium with input focal length f and index of refraction n_2. A similar set of rays can be constructed for the output focal length using the secondary principal plane and output plane.

Since the radius of a Gaussian beam is unchanged in traversing a thin lens, but its radius of curvature R_g changes sign, the form of Eq. (2-39) must also hold for transformation of the Gaussian q-parameters. Similar arguments can be made for Eq. (2-42) such that with Eq. (2-25), a direct analogy between Eqs. (2-38), (2-39), and (2-40) and a corresponding set of q-transformations can be postulated. This correspondence is

$$\begin{aligned} R_2 = R_1 + z &\longrightarrow q_2 = q_1 + z \\ \frac{1}{R_2} = \frac{1}{R_1} - \frac{1}{f} &\longrightarrow \frac{1}{q_2} = \frac{1}{q_1} - \frac{1}{f} \\ R_2 = \frac{AR_1 + B}{CR_1 + D} &\longrightarrow q_2 = \frac{Aq_1 + B}{Cq_1 + D} \end{aligned} \tag{2-43}$$

Comment 2-2. There are only two matrices from which the transform properties of a sequence of thin lenses separated by arbitrary distances may be derived. These are the translation matrix over a distance d and the focusing matrix of a lens of focal length $\pm f$, where the plus and minus signs refer to positive and negative lenses, respectively. These matrices are given by

$$M_d = \begin{pmatrix} 1 & d \\ 0 & 1 \end{pmatrix} \quad \text{and} \quad M_f = \begin{pmatrix} 1 & 0 \\ \pm\dfrac{1}{f} & 1 \end{pmatrix} \tag{2-44}$$

It is easy to show that an expanding spherical wave of geometric optics having a center of curvature located at $2f$ is transformed by a positive lens of focal length f into a convergent wave with a center of curvature located at $-2f$. The q-parameters of a Gaussian beam transform similarly, except that the beam waist locations and center of curvatures of the wavefronts are not coincident. To show this, the translation formula and the lens formula in the Eqs. (2-43) will first be shown to follow from the matrices given by Eqs. (2-44). Thus

$$\begin{aligned} R_2 &= \frac{M_{d11}R_1 + M_{d12}}{M_{d21}R_1 + M_{d22}} = R_1 + d \\ \frac{1}{R_3} &= \left(\frac{M_{f11}R_2 + M_{f12}}{M_{f21}R_2 + M_{f22}}\right)^{-1} = \frac{1}{R_2} - \frac{1}{f} \end{aligned} \tag{2-45}$$

Hence for $R_1 = 0$ and $d = 2f$ we have $R_2 = 2f$ and $R_3 = -2f$. For the Gaussian beam case, we have

$$q_2 = \frac{M_{d11}q_1 + M_{d12}}{M_{d21}q_1 + M_{d22}} = q_1 + d \tag{2-46}$$

and

$$\frac{1}{q_3} = \left(\frac{M_{f11}q_2 + M_{f12}}{M_{f21}q_2 + M_{f22}}\right)^{-1} = \frac{1}{q_2} - \frac{1}{f} \tag{2-47}$$

From Eq. (2-25), $q_1 = q_0 = i\pi\omega_0^2/\lambda$ at $z = 0$ and $d = 2f$, so that $q_2 = 2f + i\pi\omega_0^2/\lambda$. Using q_2 in Eq. (2-26), while equating real and imaginary parts, the beam radius of curvature just before the lens is given by

$$R_2 = 2f\left[1 + \left(\frac{\pi\omega_0^2}{2\lambda f}\right)^2\right] \tag{2-48}$$

Similarly, using q_2 in Eq. (2-47) and solving for R_3 yields

$$R_3 = -2f\frac{[1 + (\pi\omega_0^2/2\,\lambda f)^2]}{[1 + (\pi\omega_0^2/\sqrt{2}\,\lambda f)^2]} \tag{2-49}$$

Equations (2-48) and (2-49) show that $|R_2|$ and $|R_3|$ can equal $2f$ only if $f = \infty$. This implies that the beam waist location at $z = 0$ and the center of curvature of the phase front cannot be coincident. Indeed, if we let $d = f$ rather than $2f$, we obtain

$$R_2 = f\left[1 + \left(\frac{\pi\omega_0^2}{\lambda f}\right)^2\right] \tag{2-50}$$

and

$$R_3 = -f\frac{\lambda f}{\pi\omega_0^2}\left[1 + \left(\frac{\pi\omega_0^2}{\lambda f}\right)^2\right] \tag{2-51}$$

In this case, R_2 and R_3 can be made symmetric about the lens plane if the input beam radius is chosen to be $\omega_0 = \sqrt{\lambda f/\pi}$.

Hence the diffractive effects in the Gaussian beam formalism result in some marked differences from the spherical waves of geometric optics. In particular, we have found that the center of curvature of a Gaussian beam phase front does

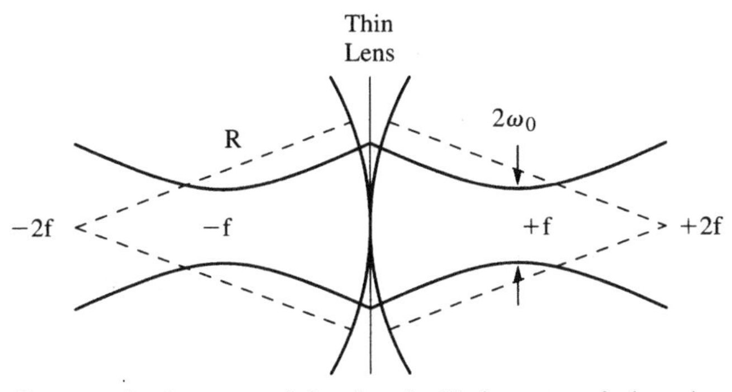

Figure 2-7. Geometry showing symmetric location of radii of curvatures for incoming and outgoing Gaussian beams propagating through a thin lens.

not coincide with the beam waist location where the intensity is a maximum. In addition, to achieve symmetric input and output beam parameters with a positive lens, the input beam waist must be located at the input focal plane $z = f$ with a radius equal to $\omega_0 = \sqrt{\lambda f/\pi}$. Figure 2-7 shows the geometry of the latter transformation. ■

2.3.2. Untruncated Focused Gaussian Beams

It will be shown in a later section that coherent detection systems have enhanced sensitivity when focused. This fact is exploited in atmospheric wind measurement systems that detect the backscattered radiation from weakly scattering particulates in the atmosphere. We first explore the beam-focusing capabilities of untruncated, unobscured Gaussian beams in order to gain insight into the optical limitations of such systems. Later the effects of truncation and obscuration are included.

A focused optical system usually employs a final telescope that defines the clear aperture for the system. This can be described using the beam translation and focusing equations of Section 2.3.1. For a beam waist located at the input to the focusing lens, the output q-parameters follow from

$$q = \frac{f}{(f/q_0) - 1} + z \tag{2-52}$$

where $q_0 = i\pi\omega_0^2/\lambda$ and f is the focal length of the lens. Letting $z_0 = \pi\omega_0^2/\lambda$, we obtain from Eqs. (2-26) and (2-52),

$$\frac{1}{R_g(z)} - \frac{i\lambda}{\pi\omega^2(z)} = \frac{z - (z_0^2/f)(1 - z/f) - iz_0}{z^2 + z_0^2(1 - z/f)^2} \tag{2-53}$$

Equating real and imaginary parts, the radius of curvature and beam radius at z become

$$R_g(z) = f\frac{z^2 + z_0^2(1 - z/f)^2}{zf - z_0^2(1 - z/f)} \tag{2-54}$$

and

$$\omega^2(z) = \omega_0^2\left[\left(\frac{1 - z}{f}\right)^2 + \left(\frac{z}{z_0}\right)^2\right] \tag{2-55}$$

At the focal plane $z = f$ we find that $R_g(f) = f$ and $\omega(f) = \lambda f/\pi\omega_0$. However, these are not the beam parameters at the beam waist, the latter being of more interest for optical system design since the intensity is highest at this point. Thus the *focal range* is defined as the distance from the lens aperture to the beam waist position, which, by definition, corresponds to a minimum in the axial beam radius. This minimum can be found from

$$\frac{d\omega}{dz} = \frac{1}{2}\frac{\omega_0}{\omega(z)}\left[\frac{2}{f}\left(\frac{z}{f} - 1\right) + \frac{2z}{z_0^2}\right] = 0 \tag{2-56}$$

which leads to

$$z_\omega = z_0 \left(\frac{z_0}{f} + \frac{f}{z_0} \right)^{-1} \tag{2-57}$$

Equation (2-57) states that for a given Rayleigh range z_0 (i.e., for a given input beam waist ω_0 and wavelength λ) there is a maximum output beam waist location at $z_\omega(\max) = f/2 = z_0/2$ for which the focal length $f = z_0$. Simply stated, an ideal Gaussian beam cannot be focused beyond half the focal distance f of a lens. This is shown in Figure 2-8, where z_ω is normalized to the input beam Rayleigh range z_0. Note that for $f/z_0 \ll 1$, the focal range is approximately equal to the geometric focal length f (i.e., $z_\omega \approx f$), while for $f/z_0 = 1$, $z_\omega = z_\omega(\max)$, and as $f/z_0 \to \infty$, $z_\omega \to 0$.

Now the corresponding size of the beam waist ω_0' at z_ω is found by inserting Eq. (2-57) into Eq. (2-55), which leads to

$$\omega_0'^2 = \omega_0^2 \left[1 + \left(\frac{z_0}{f} \right)^2 \right]^{-1} \tag{2-58}$$

Notice that as $f \to \infty$, $\omega_0' \to \omega_0$ and when $f = z_0$, ω_0' is located at $z_\omega = z_\omega$ (max) and is given by $\omega_0'(\max) = \omega_0/\sqrt{2}$. Similarly, it is easy to show by letting $z = z_\omega$ in Eq. (2-54) that $R_g(z_\omega) = \infty$ for all beam waist locations.

It will be shown in a later section that focused coherent systems operating against *soft* targets such as the atmosphere generate half of the signal within $\pm z_0'$ of the focal range z_ω. This region of a focused beam is, by convention, called the *range resolution* of the beam, which together with the mean beam diameter, constitutes the *focal volume* of the beam. Using Eq. (2-58), z_0' becomes

$$z_0' = \frac{\pi \omega_0'^2}{\lambda} = f \left(\frac{z_0}{f} + \frac{f}{z_0} \right)^{-1} \tag{2-59}$$

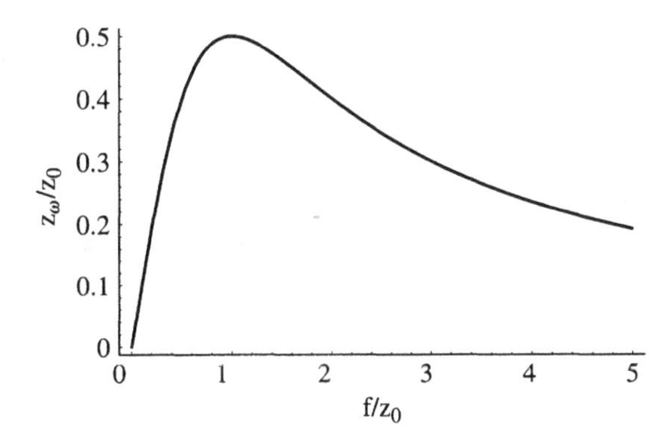

Figure 2-8. Normalized beam waist location plotted as a function of normalized lens focal length.

Equations (2-57), (2-58), and (2-59) are fundamental in the design of focused optical systems and are used later in the development of signal-to-noise ratio expressions for a heterodyne detection systems operating against soft targets.

It is frequently useful to express Eq. (2-55) in terms of the focused beam parameters z_ω and z_0'. Expanding the bracketed term in Eq. (2-55) results in

$$\omega^2(z) = a\omega_0^2\left(z^2 - \frac{b}{a}z + \frac{1}{a}\right) \tag{2-60}$$

where $a = f^{-2} + z_0^{-2}$ and $b = 2/f$. Completing the square and rearranging leads to

$$\omega^2(z) = \frac{\lambda^2}{\pi S}\left[1 + \left(\frac{z - z_\omega}{z_0'}\right)^2\right] \tag{2-61}$$

where

$$S = \frac{\lambda}{f}\left(\frac{z_0}{f} + \frac{f}{z_0}\right) \tag{2-62}$$

Note that S is a dimensionless quantity that increases as f decreases. Thus, at the beam waist $\omega(z_\omega) = \lambda/\sqrt{\pi S}$ and the focal volume can be approximated by $V_f = 2z_0' \times \omega(z_\omega) = 2\sqrt{\lambda/\pi}\, z_0'^{3/2}$. For $f \ll z_0$, $V_f \approx 2\sqrt{\lambda/\pi}\, f^3$, so that the focal volume decreases with the cube of the focal length f. We will see later that this f^3 dependence allows for significant signal enhancement in coherent detection systems employing tightly focused beams.

2.3.3. Multimode Beams

There are many laser transmitter configurations that can and will, despite the best efforts to the contrary, produce poor-quality beams, depending on how the laser cavity is constructed. For example, a stable laser resonator may preferentially operate in the lowest-order Gaussian mode due to either intentional or unintentional truncation effects occurring within the cavity. Less constrained cavities may oscillate in high-order transverse modes, possibly with random mode orientation and polarization relative to the cavity transverse axes. In addition to the transverse modes, the cavity may also support multiple longitudinal modes each having a different frequency separated by $\Delta\nu = c/2L$ and each associated with a complete set of transverse modes of yet different frequencies. However, in general, longitudinal modes do not affect beam propagation and can usually be constrained to a single wavelength with suitable wavelength-selective devices such as intracavity Fabry–Perot etalons. As will be seen later, longitudinal modes are of more concern in systems employing coherent detection where spectral purity is essential to good performance.

It has been shown by Boyd and Gordon[6] that laser cavities exhibiting rectangular and cylindrical geometries can be well approximated by transverse eigensolutions to the paraxial wave equation in the form of Hermite–Gaussian and

Laguerre–Gaussian functions, respectively. For simplicity we restrict our discussions to the Hermite–Gaussian functions, in particular their propagation characteristics and their impact on system design. The restriction to rectangular coordinates does not affect the generality of the results and allows for simple one-dimensional illustrations of basic concepts. The interested reader is referred to Casperson and Shekhani[7] for details on the Laguerre–Gaussian modes associated with cylindrical coordinates.

Solutions to the two-dimensional paraxial scalar wave equation, Eq. (2-21), are generalizations of Eq. (2-22) to include explicit functions of x and y as multiplicative factors, that is,

$$\psi = g(x)h(y)e^{-i[p(z)+kr^2/2q(z)]} \tag{2-63}$$

When inserted into Eq. (2-21), one obtains second-order differential equations that can be identified with the Hermite differential equation[8]

$$H_n''(\xi) - 2xH_n'(\xi) + 2nH_n(\xi) = 0 \tag{2-64}$$

where $H_n(\xi)$ are the well-known Hermite polynomials of order n. One of the many generating functions for these polynomials is

$$H_n(\xi) = (-1)^n e^{x^2} \frac{d^n}{dx^n} e^{-x^2} \tag{2-65}$$

which yields for the first few polynomials

$$\begin{aligned} H_0(\xi) &= 1 \\ H_1(\xi) &= 2\xi \\ H_2(\xi) &= 4\xi^2 - 2 \\ H_3(\xi) &= 8\xi^3 - 12\xi \end{aligned} \tag{2-66}$$

As in the Gaussian beam case, the higher-order solutions yield $q' = 1$ or $q_2 = q_1 + z$, while $q, \omega(z)$, and $R_g(z)$ are related through Eqs. (2-26), (2-27), and (2-28). However, the solutions for the phase propagation $p(z)$ become mode dependent:

$$\phi_{mn}(z) = (m + n + 1)\tan^{-1}\frac{z}{z_0} \tag{2-67}$$

where m and n are the x and y transverse mode numbers of the propagating beam. The normalized Hermite–Gaussian functions that satisfy the scalar wave equation then become

$$\begin{aligned} \psi_{mn}(x, y, z) = {} & N_m N_n H_m\left(\frac{\sqrt{2}x}{\omega(z)}\right) \\ & \times H_n\left(\frac{\sqrt{2}y}{\omega(z)}\right) e^{-(x^2+y^2)/\omega^2(z)} e^{-i[k(x^2+y^2)/2R_g]-i[kz-\phi_{mn}(z)]} \end{aligned} \tag{2-68}$$

where

$$N_n = \left(\frac{2}{\pi\omega^2(z)}\right)^{1/4}\left(\frac{1}{n!2^n}\right)^{1/2} \quad (2\text{-}69)$$

is the normalization factor for x with $n \to m$ for y. Examples of the first three wavefunctions are shown in Figure 2-9.

Notice that $R_g(z), \omega(z), \omega_0$, and z_0 are all identical to the Gaussian beam parameters defined earlier. This implies that the higher-order modes $\psi_{mn}(x, y, z)$, commonly referred to as *transverse electromagnetic* (TEM $_{mn}$) *modes*, propagate with the same radius of curvature and scale with distance z via the same quadratic equation as the TEM_{00} Gaussian mode. The *ABCD* beam transformation rules discussed earlier for the Gaussian beam may be applied to the higher-order modes as well, the only difference being the increased spot sizes at any location z along the propagation axis. These modes not only constitute eigensolutions to stable laser resonators but also constitute modes of free space. The latter implies that they do not change shape as they propagate but, instead, simply scale with distance in accordance with Eq. (2-68).

However, differences in mode propagation do arise due to the phase term $\phi_{mn}(z)$ in Eq. (2-68), which is mode dependent. This variation in phase velocity with mode number causes different modes of a stable laser cavity to oscillate at different frequencies in order to satisfy the boundary conditions at the cavity mirrors. The specific frequencies associated with the various TEM_{mn} transverse modes is strongly dependent on cavity design (i.e., mirror radius of curvature and separation), but these cavity issues will not be investigated here. It might be mentioned, however, that they are more closely spaced in frequency than the longitudinal modes (MHz versus 100 MHz for typical cavity parameters) and can appear on either side of the longitudinal mode frequencies, once again depending on cavity design. The reader is referred to the many excellent textbooks on laser theory for more detail.[9,10]

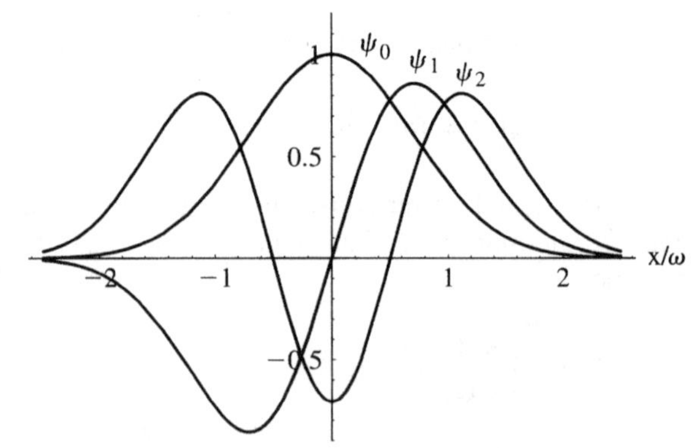

Figure 2-9. First three one-dimensional wavefunctions corresponding to the Hermite–Gaussian modes.

The intensity $I_{mn}(x, y)$ of the higher-order modes can be found, as in the Gaussian beam case, by equating the integral of $\psi_{mn}^2(x, y)$ over all x, y to the total power P. We have

$$P = b_o^2 \int_{-\infty}^{\infty}\int_{-\infty}^{\infty} \psi_{mn}^2(x, y)\, dx\, dy$$

$$= b_o^2 \int_{-\infty}^{\infty}\int_{-\infty}^{\infty} \psi_m^2(x)\psi_n^2(y)\, dx\, dy \tag{2-70}$$

where b_o is a constant to be determined. From the general orthogonality relation, we find that $b_o = P^{1/2}$. Hence

$$I_{mn}(x, y) = \frac{2P}{\pi\omega^2(z)} \frac{1}{m!n!2^{m+n}} H_m^2\left(\frac{\sqrt{2}\,x}{\omega(z)}\right) H_n^2\left(\frac{\sqrt{2}\,y}{\omega(z)}\right) e^{-2(x^2+y^2)/\omega^2(z)} \tag{2-71}$$

Relative intensity distributions along the $y = 0$ axis are shown in Figure 2-10 of the first three TEM_{m0} modes. From $dI_{10}/dx = 0$, we find that the peak intensity of the TEM_{10} mode is located at $x/\omega = \pm 1/\sqrt{2}$, so that $I_{01}/I_{00} = 2e^{-1} = 0.735 = -1.33$ dB. Similarly the peak intensities of the TEM_{20} mode occur at $x/\omega = 0$ and $\pm\sqrt{5}/2$, so that $I_{20}/I_{00} = 8e^{-5/2} = 0.656 = -1.83$ dB.

The $\psi_{mn}(x, y, z)$ constitute an orthonormal basis set of wavefunctions that satisfy the orthonormality condition

$$\iint_{x,y} \psi_{mn}(x, y)\psi_{pq}^*(x, y)\, dx\, dy = \delta_{mp}\delta_{nq} \tag{2-72}$$

where the δ are Kroneker delta functions defined by $\delta_{jk} = 1$ for $j = k$, 0 for $j \neq k$. Any arbitrary optical field $f(x, y)$ may be described in terms of this basis set in the usual manner, that is,

$$f(x, y) = \sum_m \sum_n c_{mn}\psi_{mn}(x, y) \tag{2-73}$$

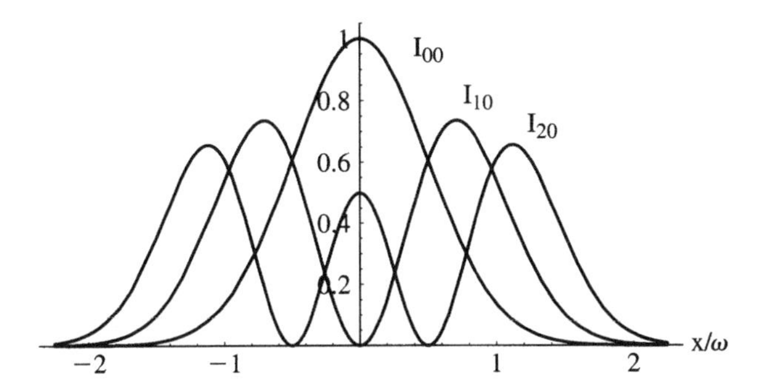

Figure 2-10. Intensity profiles of the first three Hermite–Gaussian modes.

where the expansion coefficients are obtained from the overlap integral

$$c_{mn} = \iint_{x,y} f(x, y)\psi_{mn}^*(x, y)\,dx\,dy \qquad (2\text{-}74)$$

and the integration is performed over the extent of $f(x, y)$.

Comment 2-3. Consider the one-dimensional case of a top-hat aperture function as shown in Figure 2-11 and located at $z = 0$:

$$f(x) = \begin{cases} a^{-1/2} & \text{for } -a \le x \le a \\ 0 & \text{for } a > |x| \end{cases} \qquad (2\text{-}75)$$

This is similar to the field distribution that we have seen before in two dimensions in solving the uniformly illuminated circular aperture. In this case we wish to expand $f(x)$ in the one-dimensional orthonormal basis set $\psi_n(x)$, find the expansion coefficients c_n, and plot the function $f(x, z)$ for several values of z. Therefore, we write

$$f(x, z) = \sum_{n=0}^{N} c_n \psi_n(x, z) \qquad (2\text{-}76)$$

where

$$\psi_n(x, z) = N_n H_n\left(\frac{\sqrt{2}x}{\omega(z)}\right) e^{-x^2/\omega^2 - ikx^2/2R_g - i[kz - \phi_n(z)]} \qquad (2\text{-}77)$$

$\phi_n(z) = (n + \frac{1}{2})\tan(z/z_0)$, and N_n is given by Eq. (2-69).

We begin by expanding $f(x)$ at the beam waist $z = 0$ of the basis set where $R_g = \infty$. The basis set $\psi_n(x, 0)$ then reduces to a set of purely real functions such that the c_n are all real. Also, it may have been noticed that we have set $f(x) = a^{-1/2}$ in the region between $-a$ and a rather than a dimensionless number

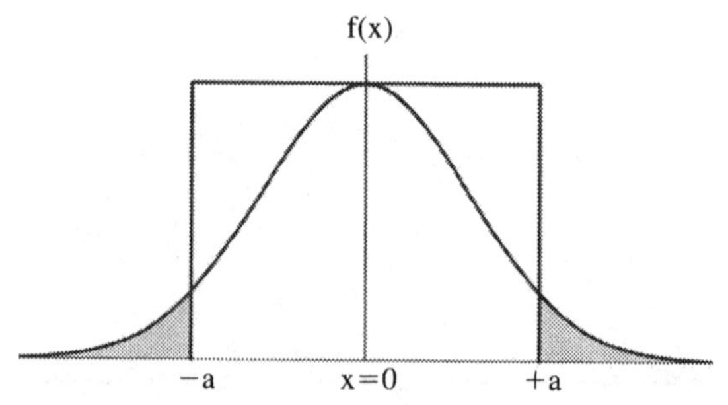

Figure 2-11. One-dimensional top-hat aperture function showing the truncated regions of the lowest-order Gaussian mode (shaded). (After A. E. Siegman, *Appl. Opt.*, Vol. 13, No. 12, Dec. 1974.)

such as 1. The reason for this is to obtain dimensionless c_n [i.e., $c_n \equiv c_n(a/\omega)$]. One could just as well define $f(x) = 1$, in which case $c_n \equiv c_n(a/\omega, \omega)$, requiring specification of ω and a individually rather than the ratio a/ω. We find for c_0,

$$c_0 = \frac{N_0}{\sqrt{a}} \int_{-a}^{a} e^{-x^2/\omega^2}\, dx$$
$$= (2\pi)^{1/4} \sqrt{\frac{\omega(z)}{a}}\, \operatorname{erf}\left(\frac{a}{\omega(z)}\right) \tag{2-78}$$

and for c_1,

$$c_1 = \frac{N_1}{\sqrt{a}} \int_{-a}^{a} H_1\left(\frac{\sqrt{2}x}{\omega(z)}\right) e^{-x^2/\omega^2}\, dx \tag{2-79}$$

But $H_1(x)$ is an odd function and therefore integrates to zero over even limits. Therefore, since all $H_n(x)e^{-x^2}$ are odd functions for n odd, all $c_n = 0$ for n odd.

For $n \geq 2$, the following recursive relations for the Hermite polynomials may be used to derive a recursive relation[11] for the c_n,

$$H_{n+1}(x) = 2x H_n(x) - 2n H_{n-1}(x) \tag{2-80}$$

and

$$\frac{d}{dx} H_n(x) = 2n H_{n-1}(x) \tag{2-81}$$

Some modest algebra results in

$$c_n = \left(\frac{n-1}{n}\right)^{1/2} c_{n-2} - \frac{2\omega(z)}{\sqrt{na}} \psi_{n-1}(a) \tag{2-82}$$

[Similar expressions can be developed for the asymmetric case of $f(x) = a^{-1/2}$ for $0 \leq x \leq a$ whereby $c_n \neq 0$ for n odd or even.]

We can now construct $f(x, z)$ using various c_n and $\psi_n(x, z)$ and investigate the propagation of the field $f(x, z)$ or the intensity $I(x, z)$. The first few terms of $f(x, z)$ are

$$\begin{aligned} f(x, z) &= c_0 \psi_0(x, z) + c_2 \psi_2(x, z) + \cdots \\ &= \frac{1}{\sqrt{a}} \left\{ \sqrt{2}\, \operatorname{erf}\left(\frac{a}{\omega}\right) e^{i\phi_0(z)} + \frac{1}{2\sqrt{2}} \left[\operatorname{erf}\left(\frac{a}{\omega}\right) - \frac{4a}{\sqrt{\pi}\,\omega} e^{-a^2/\omega^2} \right] \right. \\ &\quad \left. \times \left(\frac{8x^2}{\omega^2} - 2\right) e^{i\phi_2(z)} + \cdots \right\} e^{-x^2/\omega^2 - ikx^2/2R_g - ikz} \end{aligned} \tag{2-83}$$

Ultimately, the c_n are limited by the ability of the Hermite–Gaussian functions to fit into $f(x)$ without significant truncation. This leads to an approximate upper

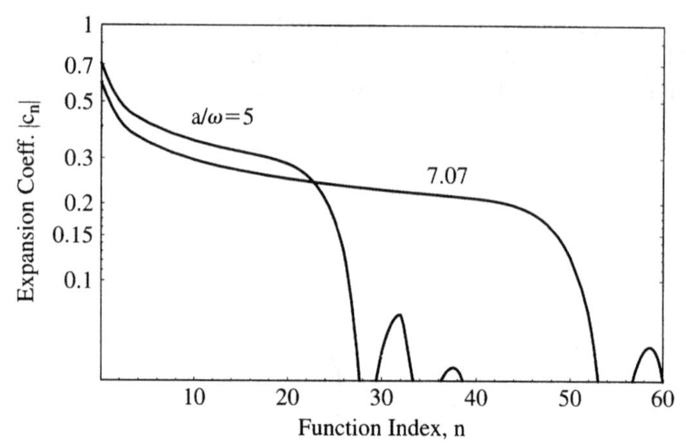

Figure 2-12. Expansion coefficients $|c_n|$ versus function mode number n for the case of an aperture extending from $-a$ to a about the origin. Aperture-to-beam radius ratios of $a/\omega = 5$ and 7.07 are shown. (After A. E. Siegman, *Appl. Opt.*, Vol. 13, No. 12, Dec. 1974.)

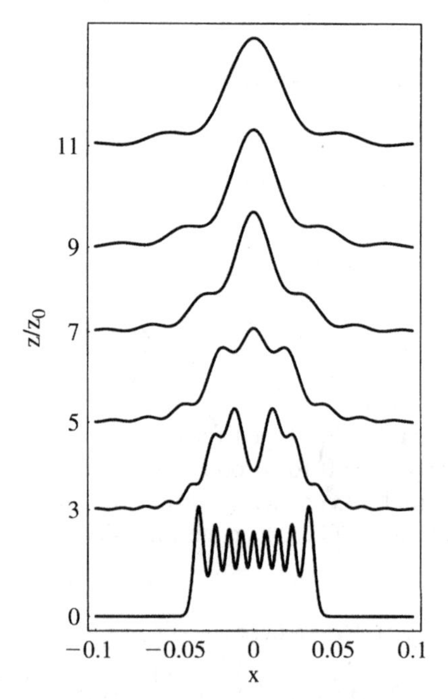

Figure 2-13. Propagation characteristics of a top-hat function generated by the coherent superposition of even-order Hermite–Gaussian modes to $n = 16$.

limit for n given by a^2/ω^2. Clearly, the smaller the initial choice of beam radius ω for a given aperture radius a, the greater the number of modes that will contribute to the expansion. Figure 2-12 supports these conclusions by showing a fast drop in $|c_n|$ beyond $n \approx a^2/\omega^2$ for two different values of a/ω.

To investigate the propagation effects on the intensity $I(x, z)$, we construct $I(x, z)$ using the first $n = 16$ terms in the expansion of $f(x)$ and plot $I(x, z)$ for several values of z/z_0 and $a/\omega_0 = 5$. It can be seen in Figure 2-13 that $f(x, z)$ starts out at $z = 0$ as constructed [i.e., a 16-term expansion (actually, 8 since all odd-n are zero)] in the Hermite–Gaussian functions. The intensity distribution then evolves through a near-field region at intermediate values of z/z_0, eventually stabilizing to a far-field pattern that approximates the distribution of a one-dimensional uniformly illuminated rectangular aperture. The difference between this case and the uniformly illuminated aperture is, of course, the fact that in the present problem the field is not uniformly distributed, due to the finite number of terms used in the expansion of $f(x)$. As indicated earlier, increasing n must be associated with an increase in the ratio a/ω_0, in order that the higher-order modes fit into the aperture. ■

The previous Fresnel-like propagation characteristics for the Hermite–Gaussian wavefunctions were a consequence of two effects, diffraction due to the real part of the wavefunctions and interference due to the mode-dependent phase term $\phi_n(z)$ in the imaginary part. In the case of a laser oscillating simultaneously in a set of TEM_{mn} modes, the cavity boundary conditions together with $\phi_n(z)$ usually force the modes to have well-defined but different frequencies (unless the cavity happens to be degenerate, in which case all modes have a common frequency). In this case, at any given plane z along the propagation path, the mode superposition can be treated as if incoherent [i.e., the intensities $I_{mn}(x, y)$ can simply be added to obtain the time-averaged resultant intensity]. As a consequence, the modes are assumed to propagate without coherent interference, and therefore since the modes all scale with distance according to the same hyperbolic equation, the resultant time-averaged pattern at the transmitting aperture does not change shape as it propagates.

To show this, we form the sum of intensities as follows:

$$\begin{aligned} I_{\text{inc}}(x, y, z) &= \sum_{mn} c_{mn} I_{mn}(x, y, z) \\ &= \sum_{mn} c_{mn} [\psi'_m(x, z)\psi'^*_m(x, z)] \times [\psi'_n(y, z)\psi'^*_n(y, z)] \\ &= \sum_m \sum_n c_{mn} N'^2_m N'^2_n \; H^2_m\left(\frac{\sqrt{2}x}{\omega(z)}\right) \\ &\quad \times H^2_n\left(\frac{\sqrt{2}y}{\omega(z)}\right) e^{-(x^2+y^2)/\omega^2(z)} \end{aligned} \tag{2-84}$$

where $\psi'_m = N'_m \psi_m$, $N'_m = P^{1/4} N_m$ and the c_{mn} are real coefficients that determine the power in each mode. [It should be noted that the set of functions $\psi_{mn}(x, y)$ do not constitute an orthogonal set, although they are normalized. Hence one cannot use them as a basis set for the expansion of an arbitrary function as in Eq. (2-73).]

Since the c_{mn} are independent of z, there is no loss in generality in letting all $c_{mn} = 1$. Figure 2-14 shows an incoherent summation of $n = 6$ modes in one dimension with all odd-n modes equal to 0. The significant difference between Figures 2-13 and 2-14 is the very large beam expansion associated with the incoherent summation compared to the coherent expansion. Indeed, the beam expansion is so large in the latter case, equal to that of the highest-order mode in the summation, that the distance z in the plot had to be limited to $z/z_0 = 3$ in order to resolve the details of the $z = 0$ distribution. The astute reader may have anticipated this result based on the fact that no interference effects can limit the growth of the high-order beams in an incoherent summation, hence the beam spread should approximate the growth of the highest-order beam in the summation. Also, the coherent expansion approximates a uniformly illuminated aperture in the far field which is known to produce the narrowest possible beam in the far field. This fact is demonstrated in Section 2.4. The actual expansion rate and characterization of multimode beams is the subject of Section 2.3.4, which follows.

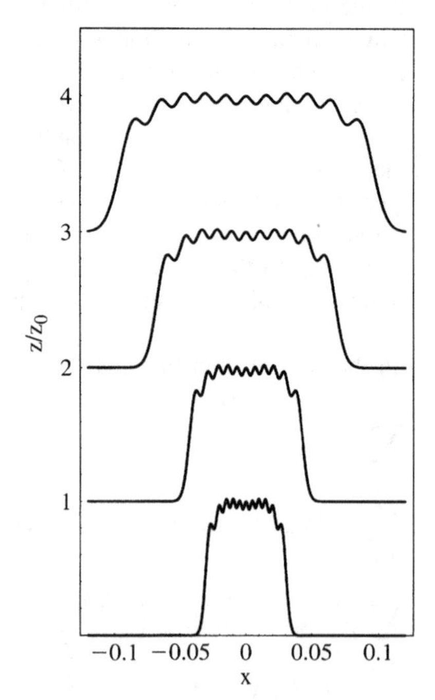

Figure 2-14. Propagation characteristics of incoherently summed even-order Hermit–Gaussian modes to $n = 10$.

2.3.4. Beam Characterization

Considerable time was spent developing the multimode concepts of Section 2.3.3 so that the reader could better appreciate the developments of the present section. As was stated earlier, the use of multimode beams in optical systems is not uncommon, so that it is important to be able to characterize the divergence properties of these types of beams as well. Since an orthonormal set of functions is always a convenient tool for describing an arbitrary aperture distribution, an understanding of the divergence properties of individual high-order beam modes will help in the characterization of more arbitrary distributions. In this section we develop the concept of laser beam divergence in greater detail based on the analysis of previous sections.

In Section 2.3.2 it was shown that the beam divergence of the lowest-order Gaussian mode may be defined as the angle of the e^{-2} asymptote of the propagation hyperbola with the beam axis, namely

$$\theta_b \underset{z\to\infty}{=} \frac{\omega(z)}{z} = \frac{\lambda}{\pi\omega_0} \tag{2-85}$$

It will become important in later sections to be able to quantitatively characterize the departure of beam quality from diffraction-limited performance. Here the *diffraction limit* is identified with the lowest-order Gaussian mode propagating through an aberration-free environment. (It should be noted, however, that such a beam does not necessarily produce the narrowest far-field beam pattern for a given aperture. As will be shown in the next section, the uniformly illuminated aperture produces the maximum on-axis intensity of any beam profile, assuming equal energies through the aperture, including the untruncated Gaussian.)

The definition above for the beam spread of a Gaussian beam is based on a spot size being defined as the e^{-2} points of the intensity profile. This is a reasonable and somewhat natural selection, based on the mathematical form of the Gaussian function. However, the definition is not so obvious for higher-order modes. For example, one could select the outermost peaks or minima of the distribution, or even the outermost e^{-2} points in analogy with the Gaussian. However, the more accepted criterion is to choose twice the mean-squared displacement of the intensity as the definition of spot size (frequently referred to as the *variance of the intensity distribution*, in analogy with statistical moments). Considering the x-dimension for the moment, we have

$$(\Delta x_n)^2 = \int_{-\infty}^{\infty} \psi_n^*(x, z)(x - \overline{x})^2 \psi_n(x, z)\, dx \tag{2-86}$$

where $\overline{x}$ is the mean of the distribution. A similar expression can be written for the y-coordinate. For simplicity we assume beams centered at $x = 0$ such that $\overline{x} = 0$.

Now it is convenient to express the Hermite–Gaussian functions in terms of the parameter $\alpha = \sqrt{2}/\omega$:

$$\psi_n(x, z) = N_n(\alpha)H_n(\alpha x)e^{-\alpha^2 x^2/2} \tag{2-87}$$

where $N_n = (\alpha/\sqrt{\pi}n!2^n)^{1/2}$. Applying the recursion relation of Eq. (2-80) twice leads to the expression $x^2 H_n(x) = \frac{1}{4}H_{n+2}(x) + (n + \frac{1}{2})H_n(x) + n(n-1)H_{n-2}(x)$. Thus, inserting Eq. (2-87) into Eq. (2-86) while invoking the orthogonality relation for the Hermite–Gaussian functions results in

$$(\Delta x_n)^2 = \frac{2n+1}{2\alpha^2} \tag{2-88}$$

The standard deviation of the distribution is therefore

$$\Delta x_n = \frac{\omega}{2}(2n+1)^{1/2} \tag{2-89}$$

From the form of Eq. (2-89), we can define a generalized spot size of any mode of order n as $(2n+1)^{1/2}$ times the corresponding Gaussian beam spot radius $\omega(z)$ [i.e., $\omega_n(z) = 2\Delta x_n = (2n+1)^{1/2}\omega(z)$]. Substituting for $\omega(z)$ in Eq. (2-27), results in

$$\begin{aligned} \omega_n^2(z) &= \omega_0^2(2n+1)\left[1 + \left(\frac{\lambda z}{\pi\omega_0^2}\right)^2\right] \\ &= \omega_{0n}^2\left[1 + \left(\frac{\lambda z}{\pi\omega_0^2}\right)^2\right] \end{aligned} \tag{2-90}$$

where $\omega_{0n} = \omega_0(2n+1)^{1/2}$ is the beam waist radius of the nth-order mode. The definition of modal spot size above improves with increasing mode number, that is, a greater fraction of the energy is contained within the bounds $\pm\omega_n(z)$ for the higher-order modes. It is left to the reader to verify that the TEM_{00} mode contains 91% of the energy, the TEM_{01} mode 94.5%, and the TEM_{11} mode 98.5%, and so on, within the rectangular spots defined by Eq. (2-87).

Equation (2-89) implies that the higher-order modes expand by a factor of $(2n+1)^{1/2}$ more than the Gaussian beam mode in each direction. From the large-z limit of Eq. (2-90), we obtain for the half-angle beam spread of the nth-order mode

$$\theta_{bn} = (2n+1)^{1/2}\frac{\lambda}{\pi\omega_0} \tag{2-91}$$

where ω_0 is the radius of the Gaussian beam waist.

The implications of Eq. (2-89) can be explored further by taking the Fourier transform of the Hermite–Gaussian functions, thereby obtaining their distributions in angle space. It can be shown that the Fourier transform of a Hermite–Gaussian function is also a Hermite–Gaussian function[12] but with a different parameter α denoted as $\widehat{\alpha}$. Specifically,

$$\begin{aligned}\widehat{\psi}_n(k_x) &= \int_{-\infty}^{\infty} \psi_n(x) e^{-2\pi i k_x x}\, dx \\ &= (-i)^n N_n(\widehat{\alpha}) H_n(\widehat{\alpha}\, k_x) e^{-(1/2)\widehat{\alpha}^2 k_x^2} \\ &= (-i)^n \psi_n(\widehat{\alpha}\, k_x)\end{aligned} \tag{2-92}$$

where $\widehat{\alpha} = \sqrt{2}\pi\omega$ and $k_x = \sin\theta_x/\lambda \approx \theta_x/\lambda$.

Defining a corresponding variance in k as

$$(\Delta k_x)^2 = \int_{-\infty}^{\infty} \widehat{\psi}_n^*(k_x)(k_x - \overline{k}_x)^2\, \widehat{\psi}_n(k_x)\, dk_x \tag{2-93}$$

With $\overline{k}_x = 0$, procedures similar to those used in obtaining Eq. (2-88) can be applied again to obtain

$$(\Delta k_x)^2 = \frac{2n+1}{2\tilde{\alpha}^2} = \frac{2n+1}{4\pi^2\omega^2} \tag{2-94}$$

or

$$\Delta k_x = \frac{(2n+1)^{1/2}}{2\pi\omega} \tag{2-95}$$

Now in complete analogy with the Hermite–Gaussian wavefunctions of the quantum mechanical harmonic oscillator, the product of the spreads in x and k_x becomes

$$\Delta x\, \Delta k_x = \frac{2n+1}{4\pi} \tag{2-96}$$

Equation (2-96) is the analog of the *uncertainty principle* of quantum mechanics as applied to classical electromagnetic waves. It states that the more confined the beam is at the aperture, the greater the *angular uncertainty* or beam spread due to diffraction. The *minimum uncertainty,* or more appropriately, the *minimum beam spread* condition, is achieved only for the lowest-order Gaussian mode ($n = 0$) and is given by

$$\Delta x\, \Delta k_x = \frac{1}{4\pi} \tag{2-97}$$

This relationship will be shown in Section 2.6 to be closely related to the antenna theorem for coherent detection systems.

We mentioned briefly at the beginning of Section 2.3 that single-mode operation in laser transmitters is sometimes difficult to obtain as well as being costly as far as transmitter efficiency and fabrication complexity is concerned. It is therefore usually advantageous, at least from a systems point of view, to design an optical system with the least requirements on diffraction-limited performance. As discussed in later sections, such a design goal is more easily accommodated using direct detection receivers than heterodyne receivers.

Arbitrary beam profiles can always be analyzed through an expansion in free-space modes, but it would be convenient for purposes of system analysis if a formalism existed that could describe such "real beams" in a fashion similar to that for the fundamental Gaussian beam mode. Siegman[13] has proposed just such a formulation by identifying the variance of the intensity distribution of all optical beams as obeying a quadratic propagation rule similar to Eq. (2-27) given by

$$\Delta x^2(z) = \Delta x^2(0) + \lambda^2 \Delta k_x^2(0) z^2 \tag{2-98}$$

where the beam waist location is at $z = 0$. We have already seen that orthonormal sets, such as the Hermite–Gaussian modes, obey such a rule and it is reasonable to expect that any optical beam follows this rule given the shape of the envelope of rays into and out of the focal plane of a lens. A *real-beam* quality factor M_x^2 may then be defined as

$$\begin{aligned} M_x^2 &= \frac{\Delta x \, \Delta k_x (\text{real beam})}{\Delta x \, \Delta k_x (\text{Gaussian beam})} \\ &= 4\pi \, \Delta x \, \Delta k_x (\text{real beam}) \end{aligned} \tag{2-99}$$

where Eq. (2-97) was used. Thus $M_x^2 = 1$ for a Gaussian beam and $M_x^2 > 1$ for all other beams. By *real beam* is meant any optical beam, not necessarily pure modes or modes possessing a high degree of transverse coherence. The advantage here is that one can describe complex and potentially ill-formed beams, such as those generated by unstable resonators, using descriptors similar to those used for Gaussian beams, namely, ω_0 and $\omega(z)$, plus a parameter M^2 which provides a measure of the departure from diffraction-limited performance.

Letting $W_x(z) = 2\Delta x(z)$ and $W_{0x} = 2\Delta x(0)$ represent the real-beam parameters, Eq. (2-90) can be generalized to arbitrary beam profiles by substituting Eq. (2-99) into Eq. (2-98), yielding

$$W_x^2(z) = W_{0x}^2 + \frac{\lambda^2 M_x^4}{\pi^2 W_{0x}^2} z^2 \tag{2-100}$$

Hence one can characterize a low-quality beam by specifying M_x^2, W_{0x}, and z_0, or, equivalently, by measuring $W_x(z)$ at any three points[14] along z and fitting to Eq. (2-100).

If we rewrite Eq. (2-90) in terms of $\omega_{0n} = \omega_0(2n+1)^{1/2}$, we obtain

$$\omega_n^2(z) = \omega_{0n}^2 + \frac{\lambda^2(2n+1)^2}{\pi^2\omega_{0n}^2}z^2 \tag{2-101}$$

and compare with Eq. (2-100), we see that for a pure Hermite–Gaussian mode, $M_x^2 = 2n+1$.

Equation (2-100) suggests that the propagation rule for a low-quality beam could be modeled as an effective or "embedded" Gaussian beam of wavelength $\lambda' = M^2\lambda$. For example, an $M^2 = 10$, $\lambda = 1$-μm wavelength beam will propagate through an optical system in the same manner as a pure Gaussian beam at a wavelength of 10 μm. This scaling relationship allows for simplified calculations of beam radii in any optical system given knowledge of M^2, $W_x(z)$, and λ. Indeed, from Eq. (2-101) we can write for the half-angle far-field beam divergence of an nth-order mode,

$$\theta_{bn} = M^2\frac{\lambda}{\pi\omega_{0n}} = \frac{\lambda'}{\pi\omega_{0n}} \tag{2-102}$$

which states that the nth-order mode diverges as would a beam of wavelength $\lambda' = M^2\lambda$ and beam waist diameter ω_{0n}.

The M^2 formalism has recently become very popular in characterizing less than diffraction-limited-quality laser beams, despite the fact that the theory has not yet been fully developed. It should be understood that the formalism says nothing about the internal structure of the beam, that is, whether there is a peak or a hole at the center or how many modes contribute to the value of M_x^2. Indeed, M_x^2 is ambiguous with regard to the latter since, for example, an $M_x^2 = \sum(2n+1) = 9$ beam could consist of a single TEM_{04} mode or a combination of TEM_{00}, TEM_{01}, and TEM_{02} modes, or even a more complex unstable resonator pattern.

2.4. TRUNCATED AND OBSCURED GAUSSIAN BEAMS

The propagation of an infinite Gaussian laser beam was discussed in Section 2.3.1. However, real systems do not have infinite apertures, so that the question arises as to the near- and far-field characteristics of truncated Gaussian beams. In addition to truncation, an optical train may be used which has beam obscuring components, such as the secondary mirror in a Cassegrain or Matsutov telescope.[15] In such cases it is necessary to calculate the diffraction and power losses for the transmitted beam. As will be seen later, calculation of the intensity distributions resulting from truncated and obscured Gaussian beams is of particular importance in the performance characterization of coherent and direct detection systems (cf. Sections 2.7.2 and 2.8.2).

For generality, we begin with the Huygens–Fresnel formula derived earlier, namely Eq. (2-9), which may be written

$$U(\alpha) = Ke^{ikr^2/2z} \int_b^a u(\rho) J_0(k\alpha\rho) e^{ik\rho^2/2z} \rho \, d\rho \tag{2-103}$$

where $\alpha = r/z$ and $K = -2\pi i e^{ikz}/\lambda z$. Here the integration is performed over the radial coordinate ρ in the plane of the aperture between the limits b and a, where a is the aperture outer radius and b is the radius of the central obscuration. The geometry is shown in Figure 2-15. $u(\rho)$ represents the amplitude and phase of the beam pattern at the aperture. If $u(\rho)$ is constant (i.e., a uniform plane-wave with no obscuration), it may be taken outside the integral and the solution becomes the Airy pattern discussed earlier. If the distribution is nonuniform, $u(\rho)$ must remain under the integral. In most cases the integral must be solved numerically (or with approximate analytical forms[16]) except for some special limiting cases, as will be shown.

Assume an infinite Gaussian beam with a beam waist ω_0 impinging on an aperture that is followed by a thin lens of focal length f. This is a typical beam arrangement employed in most optical systems. The Gaussian field exiting the lens then follows from Eq. (2-37) and is given by

$$u(\rho) = u_0 e^{-\rho^2/\omega_0^2 - ik\rho^2/2f} \tag{2-104}$$

Inserting Eq. (2-104) into Eq. (2-103) results in

$$U(\alpha) = Ku_0 e^{ikr^2/2z} \int_b^a J_0(k\alpha\rho) e^{-\rho^2/\omega_0^2 + ik\rho^2/2z(1-z/f)} \rho \, d\rho \tag{2-105}$$

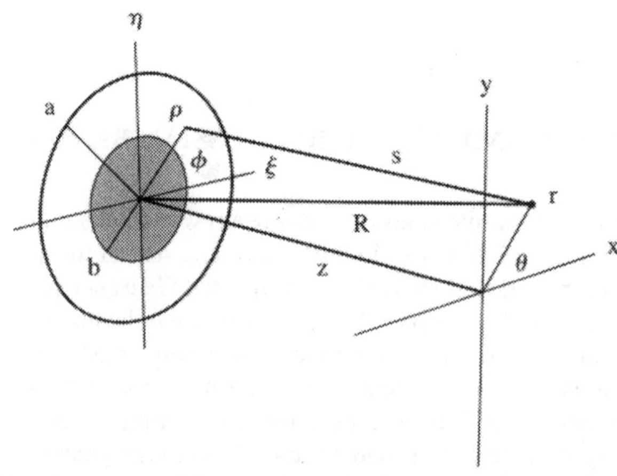

Figure 2-15. Coordinate system for calculation of far-field on-axis flux from a centrally obscured aperture.

At this point it is helpful to change variables. Defining $\xi = a/\omega_0$ as an aperture-to-beam radius ratio or truncation parameter, $\rho' = \rho/a$ as a normalized radial coordinate in the aperture plane, $\varepsilon = b/a$ as the ratio of inner to outer radii of the aperture, $\alpha' = 2a\alpha/\lambda$ as a normalized angle, and $u_0 = (2P/\pi\omega_o^2)^{1/2}$, Eq. (2-105) becomes

$$U(\alpha') = 2\left(\frac{2PA}{\lambda^2 z^2}\right)^{1/2} K'\xi \int_{\varepsilon}^{1} J_0(\pi\alpha'\rho')e^{-\xi^2\rho'^2+(i\rho'^2/2)(\beta_z-\beta_f)}\rho' d\rho' \qquad \text{(2-106)}$$

where $A = \pi a^2$ is the aperture area, $\beta_z = ka^2/z$, $\beta_f = ka^2/f$, $K' = -i \exp[ik(z+r^2/2z)]$ is a phase term that will be ignored in calculations of intensity, and P is the power of the input Gaussian beam. This is a convenient representation of the diffraction integral since all terms having physical dimensions appear in the multiplicative factor $PA/\lambda^2 z^2$, which is recognizable as the maximum on-axis intensity of a uniformly illuminated circular aperture in the far field [cf. Eq. (2-18)] and is therefore a meaningful normalization factor.

The intensity at any point in the longitudinal and transverse planes may now be obtained from

$$I = U^*(\zeta)U(\zeta) = \frac{PA}{\lambda^2 z^2}\Phi(\alpha') \qquad \text{(2-107)}$$

where Φ is the square of the dimensionless integral in Eq. (2-105):

$$\Phi(\alpha') = 8\xi^2 \left| \int_{\varepsilon}^{1} J_0(\pi\alpha'\rho')e^{-\xi^2\rho'^2+(i\rho'^2/2)(\beta_z-\beta_f)}\rho' d\rho' \right|^2 \qquad \text{(2-108)}$$

Figure 2-16 shows several radial diffraction patterns obtained from Eq. (2-108) in the Fraunhofer limit of $\beta_z = \beta_f$. The patterns include the following cases: (a) $\xi = 1$ (i.e., e^{-2} truncation), (b) $\xi = 1$ with obscuration $\varepsilon^2 = 0.1$ (i.e., $b/a = 0.316$), (c) plane-wave illumination with no obscuration ($\xi = 0$, $\varepsilon^2 = 0$), and (d) an untruncated, unobscured Gaussian beam having an e^{-2} beam waist radius ω_0 equal to the aperture radius a. Several interesting trends can be noticed. First, the truncated beam increases the central lobe width while the obscured beam decreases the central lobe width compared to a uniformly illuminated unobscured aperture. Second, the highest on-axis intensity is achieved with uniform illumination[17] followed closely by that of an untruncated Gaussian, although the untruncated Gaussian has the largest beam spread of those compared.

Equation (2-108) can be solved exactly for the on-axis case of $\alpha' = 0$. After some modest algebra, we obtain for the normalized on-axis intensity,[18] Φ_0:

$$\Phi_0(z) = \frac{2}{\xi^2}\frac{(e^{-\varepsilon^2\xi^2} - e^{-\xi^2})^2 + 4e^{-(1+\varepsilon^2)\xi^2}\sin^2[A(1-\varepsilon^2)(1-z/f)/2\lambda z]}{1 + [A(1-z/f)/\lambda z\xi^2]^2} \qquad \text{(2-109)}$$

Notice that for the far-field case of $z = f$, Eq. (2-19) simplifies to

$$\Phi_0(z = f) = \frac{2}{\xi^2}(e^{-\varepsilon^2\xi^2} - e^{-\xi^2})^2 \qquad \text{(2-110)}$$

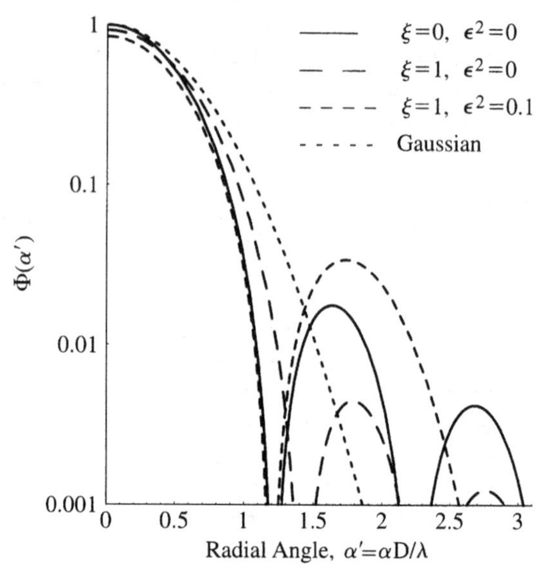

Figure 2-16. Comparison of far-field diffraction patterns corresponding to truncated obscured Gaussian beams and plane-wave illuminated circular apertures.

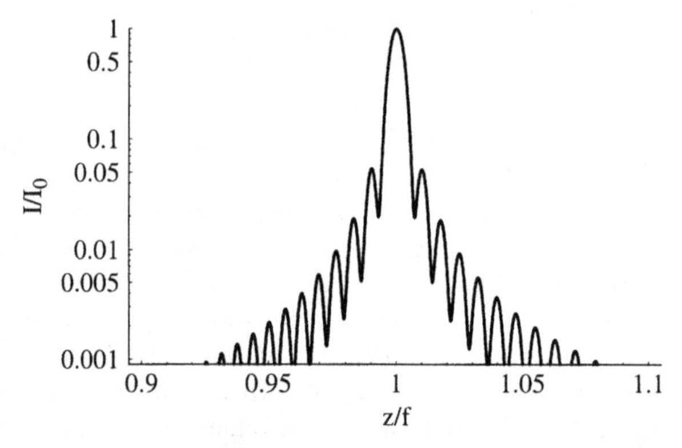

Figure 2-17. Normalized on-axis intensity of a circular aperture in the vicinity of the focal plane of a focused, truncated, obscured Gaussian beam shown as a function of the normalized variable z/f. Here $\varepsilon^2 = 0.1$, $\xi = 1$, and $A/\lambda f = 1000$.

Figure 2-17 shows the on-axis diffraction of a circular aperture near the focal plane of a focused, truncated ($\xi = 1$), obscured ($\varepsilon^2 = 0.1$) Gaussian beam as given by Eq. (2-109). A value for the ratio $A/\lambda f = 1000$ was used in the normalized expression $I/I_0 = f^2\Phi_0(z)/z^2\Phi_0(z = f)$. Note the strong oscillations along the z-direction. The amplitude of these oscillations, including the central peak, is a function of the amount of truncation and obscuration, whereas the frequency is dependent on the ratio $A/\lambda f$. The effect of these oscillations on the performance of a focused laser system is to cause a spatially dependent signal response that must be taken into account when interpreting the data.

Comment 2.4. We know intuitively that the received power P_R in a one-way data link should be proportional to the ratio of the receiver aperture area to the transmitted beam area. Indeed, we can write

$$P_R = I_0 A_R = \frac{2PA_R}{\pi\omega^2(z)} = 2P\frac{A_R}{A_B} \tag{2-111}$$

where I_0 is the intensity at the center of the beam, $\omega(z)$ the Gaussian beam radius at the receiver, A_R the receiving aperture area, and A_B the area of the transmitted beam at the receiver. Equations (2-109) and (2-110) are fully consistent with this formulation, as can be seen by letting $\xi \to \infty$ and $\varepsilon = 0$ in Eq. (2-110), which leads to

$$P_R = \frac{PA_T}{\lambda^2 z^2}\Phi_0 A_R = \frac{PA_T}{\lambda^2 z^2}\frac{2}{\xi^2}A_R = \frac{2PA_R}{\pi(\lambda z/\pi\omega_0)^2} = \frac{2PA_R}{\pi\omega^2(z)} = 2P\frac{A_R}{A_B} \tag{2-112}$$

where A_T is the untruncated transmit aperture area. ■

2.4.1. Optimum Beam Diameter

In real optical systems, the final clear aperture is usually constrained by volume or cost considerations. In such cases the question arises as to the size of the input Gaussian beam relative to the final aperture if maximum on-axis intensity or power is desired at the target. Clearly, if the beam is too large, significant energy is lost due to truncation by the limiting aperture, assumed here to be the final output aperture. On the other hand, if it is too small, diffraction spreading will reduce the energy density at the target. If the target is small compared to the beam, the amount of energy on target is reduced in both limits compared to the optimum beam diameter, which, by definition, maximizes the on-axis far-field intensity. In a later section we will see that similar optimizations can be made on the system level in order to maximize the system signal-to-noise ratio.

The optimum input beam diameter for a given aperture diameter can be found by maximizing Eq. (2-110) with respect to ξ. Since an analytical solution does not exist, the solution must be found numerically. Figure 2-18 shows a plot of Φ for several values of ε^2. It can be seen that for no obscuration (i.e., $\varepsilon^2 = 0$), $\xi_{\text{opt}} = 1.12$, whereas letting $\varepsilon^2 = 0.1$ results in $\xi_{\text{opt}} = 1.02$. These numbers are very close to $\xi = 1$, which corresponds to truncation at the e^{-2} point of the beam.

Alternatively, one may wish to optimize the output power rather than the intensity. Integrating the Gaussian intensity over the clear aperture yields

$$\begin{aligned} P_{\text{out}} &= 2\pi \int_{\varepsilon a}^{a} u_0^2 e^{-2\rho^2/\omega_0^2}\rho\, d\rho \\ &= 2A_T u_0^2 \int_{\varepsilon}^{1} e^{-2\xi^2 x^2} x\, dx \end{aligned} \tag{2-113}$$

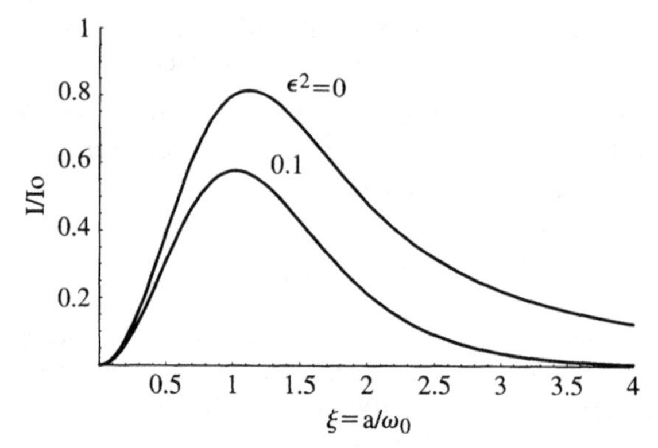

Figure 2-18. Optimized on-axis intensity of a truncated Gaussian beam for several values of obscuration ratio ε^2.

With $u_0^2 = I_0 = 2P_T/\pi\omega_0^2$, this becomes

$$P_{\text{out}} = P_T(e^{-2\varepsilon^2\xi^2} - e^{-2\xi^2}) \tag{2-114}$$

Taking the derivative of Eq. (2-114) and setting it equal to zero shows that a maximum occurs at ξ_{opt}, given by

$$\xi_{\text{opt}} = \left(\frac{\ln \varepsilon}{\varepsilon^2 - 1}\right)^{1/2} \tag{2-115}$$

The fact that an optimum value exists follows from the fact that losses occur at both the obscuring and truncating apertures. According to Eq. (2-115), when $\varepsilon^2 = 0$, then $\xi_{\text{opt}} = \infty$, which implies that maximum power transmission will occur only for an infinite aperture-to-beam diameter ratio, as would be expected.

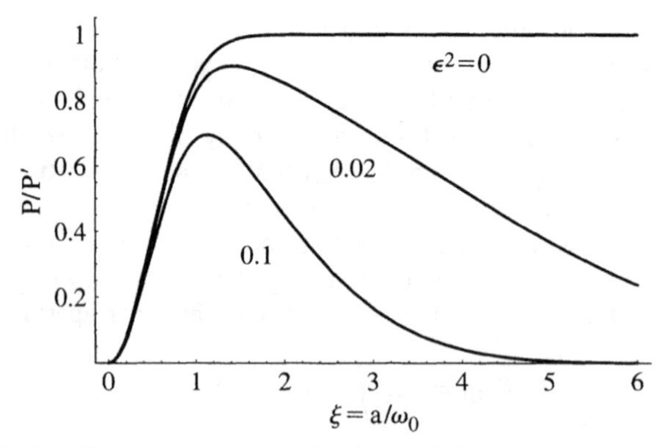

Figure 2-19. Optimized transmitted power of a truncated Gaussian beam for several values of obscuration ratio ε^2.

On the other hand, when ε^2 is nonzero, finite values of ξ_{opt} are obtained. For example, $\varepsilon^2 = 0.1$ yields $\xi_{\text{opt}} = 1.31$, which is somewhat larger than that found above for the optimized intensity (i.e., $\xi_{\text{opt}} = 1.02$). A plot of Eq. (2-115) is shown in Figure 2-19 for several values of obscuration ratio ε^2.

Comment 2.5. Based on the analysis above, we can calculate the required transmitter power of a laser communication link between earth-orbit satellites. Assume a collimated beam that is well centered on a receiver aperture that is located in the far field of the transmitted beam. Since for large propagation distances the received beam diameter will be much larger than the receiver aperture diameter, it is more important to maximize the on-axis intensity in the far field rather than the total transmitted power. Hence the optical part of the link budget can be written $P_R = \eta_o I_0 A_R \Phi_0(z = f)$, where η_o is the optical efficiency of the receiver and Φ_0 is given by Eq. (2-110). If we also assume unity optical efficiency for the receiver, we obtain for the received power

$$P_R = I_0 A_R \Phi_0(z = f) = \frac{P_T A_T A_R}{\lambda^2 z^2} \frac{2}{\xi^2} (e^{-\varepsilon^2 \xi^2} - e^{-\xi^2})^2 \tag{2-116}$$

where $A_R = \pi D_R^2/4$ is the area of the receive aperture and $A_T = \pi D_T^2/4$ is the area of the transmit aperture. Here P_T is the (peak) power input to the transmit aperture. Equation (2-116) can be rewritten as

$$P_T A_T = \frac{\lambda^2 z^2 P_R}{2 A_R} \xi^2 (e^{-\varepsilon^2 \xi^2} - e^{-\xi^2})^{-2} \tag{2-117}$$

Note that Eq. (2-117) can also be written in terms of the output power–aperture product by using Eq. (2-114), yielding

$$P_{\text{out}} A_T = \frac{\lambda^2 z^2 P_R}{2 A_R} \xi^2 \frac{e^{-\varepsilon^2 \xi^2} + e^{-\xi^2}}{e^{-\varepsilon^2 \xi^2} - e^{-\xi^2}} \tag{2-118}$$

Consider an operating wavelength of $\lambda = 1.5$ μm, a required mean signal power of $P_R = 10^{-9}$ W for a unity signal-to-noise ratio, an obscuration ratio of $\varepsilon^2 = 0.1$, and an optimum truncation ratio of $\xi_{\text{opt}} = 1.02$ in Eq. (2-117). We have

$$P_T \approx 3.9 \times 10^{-21} \frac{z^2}{A_R A_T} \qquad \text{watts} \tag{2-119}$$

Thus for an intersatellite distance of $z = 10^6$ m and equal transmit and receive aperture diameters of $D_R = D_T = 0.01$ m, a transmitter peak power of $P_T = 0.632$ W is required to achieve a minimum detectable signal of 1 nW. ■

2.5. FOURIER OPTICS AND THE ARRAY THEOREM

It was mentioned in Section 2.3.1 that Fourier optics is a powerful tool for describing far field patterns of arbitrary phase and amplitude distributions at an aperture. In this section we expand on this topic by describing briefly Fourier transform theory as applied to spatial distributions[19,20] and apply the theory to two-dimensional Gaussian beam arrays. Beam arrays are frequently employed in laser radar systems as a means for obtaining high power from inherently low-power sources, such as laser diodes, or reducing scan rates in systems limited by scanner capabilities or lag-angle effects. (*Lag angle* is the angular misalignment of the signal beam and receiver IFOV due to scan motion during the round-trip time of flight.)

Earlier sections showed that propagation of coherent, coaxial multimode beams results in a unique beam profile in the far field. From the form of the Fraunhofer integral it was shown that such free space propagation corresponds to a Fourier transform of the aperture distribution. The same argument can be made for two-dimensional beam arrays, arrays of similar but spatially separated aperture distributions. Intuitively we expect that n-beam arrays will produce n distinct spots in the far field if their angular separations are greater than their individual beam spreads, a result that is easily verified within the theory. However, for beam spreads that overlap in the far field, Fourier transform theory predicts unique far field patterns that depend specifically on the details of the array.

Fourier transform theory applied to spatial distributions is completely analogous to the corresponding theory for temporal distributions under the substitution of distance x for time t, and spatial frequency ω_x for temporal frequency ω. The one-dimensional reciprocal transforms become

$$F(\omega_x) = \int_{-\infty}^{\infty} f(x)e^{-i\omega_x x}\,dx$$

$$f(x) = \frac{1}{2\pi}\int_{-\infty}^{\infty} F(\omega_x)e^{i\omega_x x}\,d\omega_x \tag{2-120}$$

where $F(\omega_x)$ is the angular spectrum of radiation in the far field and $f(x)$ is the complex phase and amplitude distribution in the near field. The corresponding two-dimensional transforms are

$$F(\omega_x, \omega_y) = \int_{-\infty}^{\infty}\int_{-\infty}^{\infty} f(x, y)e^{-i(\omega_x x+\omega_y y)}\,dx\,dy$$

$$f(x, y) = \frac{1}{(2\pi)^2}\int_{-\infty}^{\infty}\int_{-\infty}^{\infty} F(i\omega_x, i\omega_y)e^{i(\omega_x x+\omega_y y)}\,d\omega_x\,d\omega_y \tag{2-121}$$

Noting that a unit amplitude plane-wave propagating with direction cosines $(\theta_x, \theta_y, \theta_z)$ is written

$$e^{-i(2\pi/\lambda)(\theta_x x+\theta_y y+\theta_z z)} \tag{2-122}$$

where $\theta_x^2 + \theta_y^2 + \theta_z^2 = 1$, the spatial angular frequencies ω_x and ω_y may be identified with the angular displacements θ_x and θ_y through the relationships

$$\omega_x = 2\pi f_x = 2\pi \frac{\theta_x}{\lambda}, \qquad \omega_y = 2\pi f_y = 2\pi \frac{\theta_y}{\lambda} \tag{2-123}$$

such that $\theta_z = \sqrt{1 - (\lambda f_x)^2 - (\lambda f_y)^2}$. [Note that as long as circularly symmetric beams are considered, the use of ω_x and ω_y for the spatial angular frequencies should not lead to confusion with the beam radius ω. In the more general case of asymmetric beams, problems of identifying variables can be avoided by writing Eq. (2-121) in terms of the frequencies $f_x = \omega_x/2\pi$ and $f_y = \omega_y/2\pi$.]

Consider first a beam array consisting of two equal-amplitude circularly symmetric Gaussian beams arranged symmetrically on the x-axis at $x = \pm a/2$, as shown in Figure 2-20. Using the Gaussian beam equations of Section 2.3.2, the aperture distribution is described by

$$\begin{aligned} u(x, y) = {} & exp\left\{-\frac{(x - a/2)^2 + y^2}{\omega^2} - ik\left[\frac{(x - a/2)^2 + y^2}{2R} + \alpha_1 x + \beta_1 y\right]\right\} \\ & + e^{i\phi} exp\left\{-\frac{(x + a/2)^2 + y^2}{\omega^2} - ik\left[\frac{(x + a/2)^2 + y^2}{2R} + \alpha_2 x + \beta_2 y\right]\right\} \end{aligned} \tag{2-124}$$

where $\alpha_{1,2}$ and $\beta_{1,2}$ are the angles that the individual beams make with the z-axis (assumed small), ω is the radii of the two beams at the aperture (assumed equal), and ϕ is an arbitrary phase factor between the two beams. Letting $\phi = 0$ and applying the first of Eqs. (2-121) leads to an angular spectrum given by

$$F(\omega_x, \omega_y) = \frac{2\pi exp\left[-\dfrac{\omega_x^2}{4(ik/2R + 1/\omega^2)} - \dfrac{\omega_y^2}{4(ik/2R + 1/\omega^2)}\right] \cos\left(\dfrac{a\omega_x}{2}\right)}{ik/2R + 1/\omega^2} \tag{2-125}$$

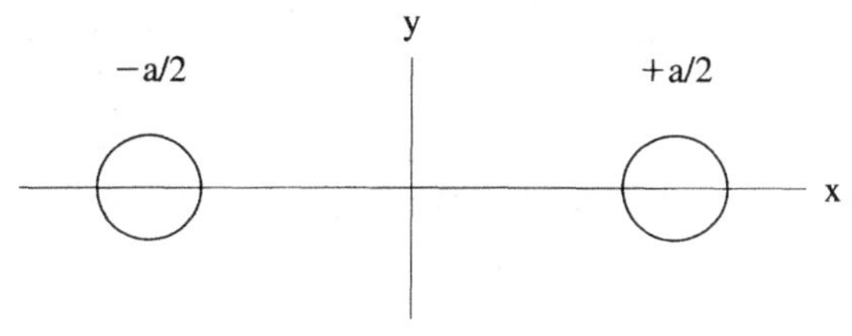

Figure 2-20. Aperture distribution consisting of two Gaussian beams located at $\pm a/2$ on the x-axis. Beam radii are $\omega = a/10$, where $a = 2$ cm and $\lambda = 1$ μm.

where $\alpha_{1,2}$, $\beta_{1,2}$ have been set equal to zero. The intensity distribution is obtained as the square of the absolute magnitude of Eq. (2-125). For simplicity, a radius of curvature $R = \infty$ will be chosen corresponding to a beam waist located at $z = 0$, while the angle variables will be normalized to the angle λ/a, resulting in $\omega_x = 2\pi\Theta_x/a$ and $\omega_y = 2\pi\Theta_y/a$, where $\Theta_x = a\theta_x/\lambda$ and $\Theta_y = a\theta_y/\lambda$. The intensity distribution in angle space therefore becomes

$$I(\Theta_x, \Theta_y) = 4\pi^2\omega^4 e^{-2\pi^2\omega^2/a^2(\Theta_x^2+\Theta_y^2)} \cos^2(\pi\Theta_x) \tag{2-126}$$

Equation (2-126) is plotted in Figure 2-21 for a nominal wavelength of $\lambda = 1\ \mu\text{m}$, a spacing of $a = 2$ cm, and a beam radius ω at the aperture of 2 mm. The cosinusoidal fringing contained in Eq. (2-126) is clearly evident. Note that the fringes are spaced by $\Theta_x = 1$ or $\theta_x = \lambda/a$. If additional beams were included in the array with regular spacing along the x-axis, the resultant pattern would exhibit intensity modulation in accordance with $I \propto [\cos(a\omega_x/2) + \cos(a\omega_x) + \cdots + \cos(na\omega_x/2)]^2$. Furthermore, if the outgoing beams were to be given small angular separations (i.e., $\alpha, \beta \neq 0$), little change would be found in the far field pattern, due to the fact that the beams are, to first order, being rotated about their centers of curvature so that the overlapping phase fronts experience little relative change. If α and β are made large, the beams separate into distinct Gaussian beams, as mentioned earlier.

The procedures above are a simple example of the *array theorem*, which states that the far-field distribution of n apertures having similar distributions is given by the product of the far-field distributions of one of those apertures and an array of identically distributed point sources. Since the far field distribution is given by the Fourier transform of the aperture distribution, the statement above

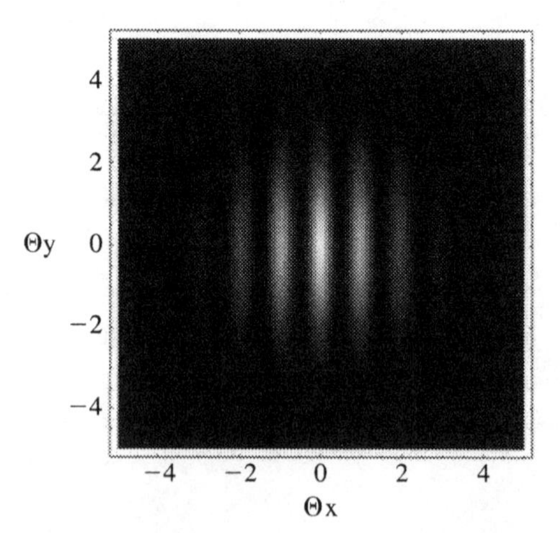

Figure 2-21. Far-field diffraction pattern of two coherently propagating Gaussian beams with a 2-cm separation, beam radii of 2 mm, and a wavelength of 1 μm.

corresponds to a convolution of a single aperture distribution with an array of point sources at the given locations. To show this mathematically, consider the one-dimensional case of an array of identical apertures $\psi(x)$ given by

$$\psi(x) = \sum_{n=1}^{N} \psi(x - x_n) \tag{2-127}$$

where n stands for the nth position in the array. From Eq. (2-120) the far field pattern for this array is given by

$$F(\omega_x) = \int_{-\infty}^{\infty} \psi(x)e^{-i\omega_x x}\,dx \tag{2-128}$$

Substituting Eq. (2-127) into Eq. (2-128) and interchanging the order of summation and integration results in

$$F(\omega_x) = \sum_{n=1}^{N} \int_{-\infty}^{\infty} \psi(x - x_n)e^{-i\omega_x x}\,dx \tag{2-129}$$

Now we can use the properties of the Dirac delta function to write $\psi(x - x_n)$ in terms of a convolution integral. Thus

$$\psi(x - x_n) = \int_{-\infty}^{\infty} u(x - x')\delta(x' - x_n)\,dx' \tag{2-130}$$

where $u(x)$ is the aperture distribution. But we know from the convolution theorem that the Fourier transform of the convolution of two functions is equal to the product of the Fourier transforms of the individual functions. Thus Eq. (2-129) becomes

$$\begin{aligned} F(\omega_x) &= \sum_{n=1}^{N} \int_{-\infty}^{\infty} u(x - x')e^{-i\omega_{x'}x'}\,dx' \int_{-\infty}^{\infty} \delta(x'' - x_n)e^{-i\omega_{x'}x''}\,dx'' \\ &= \int_{-\infty}^{\infty} u(x - x')e^{-i\omega_{x'}x'}\,dx' \int_{-\infty}^{\infty} \sum_{n=1}^{N} \delta(x'' - x_n)e^{-i\omega_{x'}x''}\,dx'' \\ &= \int_{-\infty}^{\infty} u(x - x')e^{-i\omega_{x'}x'}\,dx' \sum_{n=1}^{N} e^{-i\omega_{x'}x_n} \end{aligned} \tag{2-131}$$

The first transform in the last line of Eq. (2-131) is the far field pattern of an individual aperture function, while the latter is that associated with an array of point sources located at x_n. This is readily generalized to two dimensions, yielding

$$
\begin{aligned}
F(\omega_x, \omega_y) &= \int_{-\infty}^{\infty}\int_{-\infty}^{\infty} u(x - x')u(y - y')e^{-i(\omega_{x'}x' + \omega_{y'}y')}dx'dy' \\
&\quad \times \int_{-\infty}^{\infty}\int_{-\infty}^{\infty} \sum_{n=1}^{N}\sum_{m=1}^{M} \delta(x' - x_n)\delta(y' - y_n)e^{-i(\omega_{x'}x' + \omega_{y'}y')}dx'dy' \\
&= \int_{-\infty}^{\infty}\int_{-\infty}^{\infty} u(x - x')u(y - y')e^{-i(\omega_{x'}x' + \omega_{y'}y')}dx'dy' \\
&\quad \times \sum_{n=1}^{N}\sum_{m=1}^{M} e^{-i(\omega_{x'}x_n + \omega_{y'}y_m)}
\end{aligned} \tag{2-132}
$$

Consider the case of numerous small, co-propagating, randomly phased, equal-amplitude Gaussian beams randomly located throughout a square aperture. As the beam radii $\omega \to 0$ and the number of beams n becomes large, the array theorem states that the far-field pattern should consist of a Gaussian beam profile with an internal pattern that approaches the well-known *speckle pattern* observed with coherent light. As shown in Chapter 4, a speckle pattern results from the interference of a large number of randomly phased and randomly positioned radiators such as those resulting from the coherent illumination of a surface that is rough compared to a wavelength. Figure 2-22 shows an array of 100 Gaussian beam positions distributed randomly throughout a square aperture with the phases distributed uniformly over the interval $(0, 2\pi)$. Figure 2-23 shows the resultant far-field pattern obtained with a 128×128 point fast Fourier transform (FFT). Here the ratio of the beam radii to aperture width is 0.02, which is shown to scale in Figure 2-22. Note that the overall beam profile maintains a

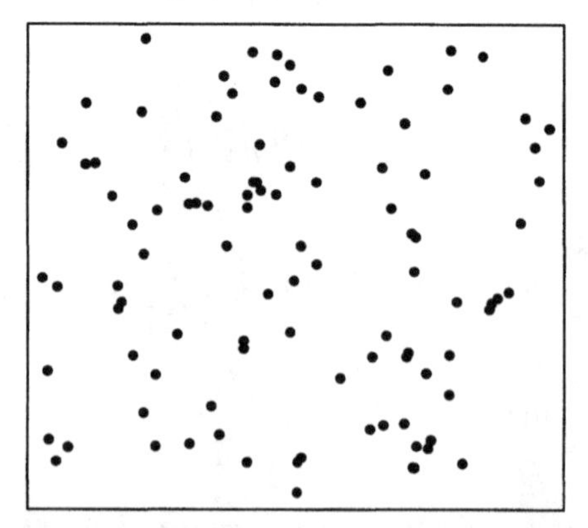

Figure 2-22. Aperture distribution of 100 randomly distributed equal-amplitude Gaussian beams used in Figure 2-23. The ratio of beam diameters to aperture width is as shown in the figure (~0.02).

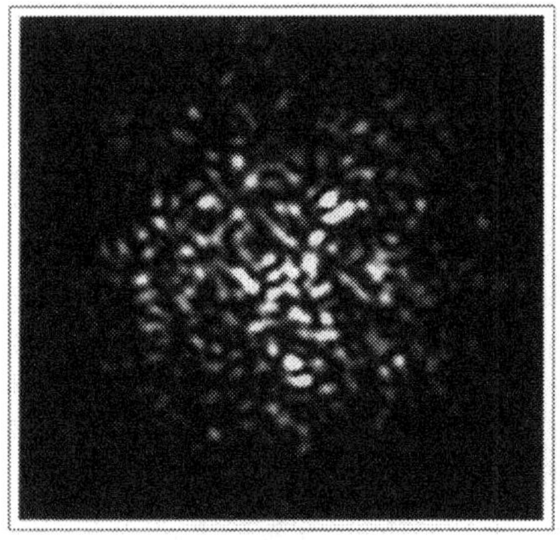

Figure 2-23. Simulated far-field speckle intensity pattern obtained with the aperture distribution shown in Figure 2-22. The beams are randomly phased and co-propagating with equal amplitudes at the aperture. A 128 × 128 FFT was used to generate the pattern.

Gaussian-like falloff with radius while the internal structure of the beam exhibits the random structure typical of real speckle. Indeed, in Chapter 4 it is shown that simulated speckle patterns can easily be constructed that exhibit all the statistical properties of real speckle. This is a consequence of the linear superposition of fields inherent in the scattering process, which is very amenable to computer simulation.

2.6. ANTENNA AND MIXING THEOREMS

In Section 2.3.4 we saw that a minimum spread condition could be developed for any pure mode based on the diffraction characteristics of the beam. For the fundamental Gaussian mode, this relationship took the form

$$\Delta x \Delta k_x = \frac{1}{4\pi} \tag{2-133}$$

where Δx is the beam diameter at the reference plane and Δk_x was the angular spread of the beam as expressed through the wave vector $k_x = 2\pi \sin\theta_x/\lambda$. In this section we show that a similar relationship holds for the field distribution and angular sensitivity of a coherent optical system. This is the well-known antenna theorem first put forth by Siegman[21] and is applicable to any coherent detection system independent of its wavelength or specific optical design. (It should be noted that the theorem does not apply to direct detection systems, where angular sensitivities are determined by aperture stops, detector sizes, and so on, not by amplitude and phase matching conditions on the detector surface.)

Consider a heterodyne detection system in which the signal and local oscillator beams are coaligned in what is usually referred to as the *maximum heterodyne efficiency configuration*. That is, the beams are aligned in polarization and phase such that little or no out-of-phase mixing occurs on the detector surface. Referring to the plane-wave example in Figure 2-24, we see that if the signal and local oscillator beams are misaligned by an amount $\theta = n\lambda/2d$, where n is an integer, complete cancellation of the signal by out-of-phase components on the detector surface can occur. Obviously, this condition is to be avoided in practice, but it does suggest an upper limit to the phase-matching condition of the signal and local oscillator beams. Thus, if we allow for misalignment in both directions from the normal, we obtain for the full-angle central lobe of the receiving antenna, $\theta \approx \lambda/d$. Assuming a square detector, this leads to

$$A_R \Omega \approx \lambda^2 \tag{2-134}$$

where $A_R = d^2$ is the effective area of the receiving aperture and $\Omega = \theta^2$ is the solid angle defined by the first zero of the phase-matching condition. As with Eq. (2-133), Eq. (2-134) expresses an inverse relationship between the effective aperture and the angular field of view of a heterodyne detection system and constitutes the fundamental statement of the antenna theorem. A little thought leads to the conclusion that Eq. (2-134) holds even if energy collection optical elements are placed in front of the detector. This follows from the fact that as A_R is increased by the insertion of magnifying optics, for example, the instantaneous field of view is reduced proportionately, thereby maintaining the product equal to λ^2. With this simplified picture in mind, we now consider a more general proof of the theorem.

Consider a heterodyne receiver that performs photomixing on a detector surface. Assuming identical polarization characteristics for the signal and local oscillator beams, we can write for the total scalar amplitude

$$u(x, y, t) = u_o(x, y)e^{i\omega_o t} + u_s(x, y)e^{i\omega_s t} \tag{2-135}$$

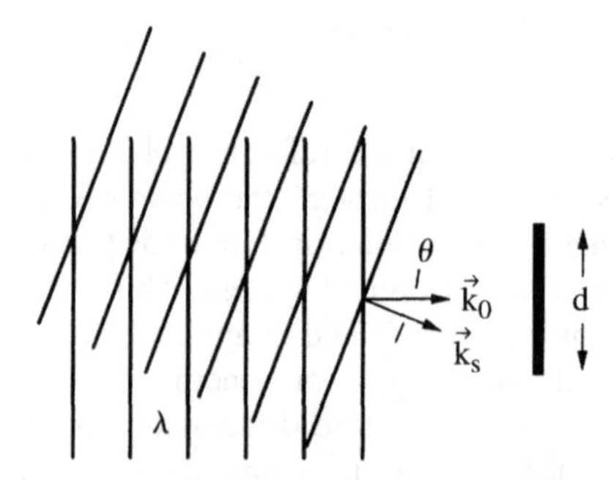

Figure 2-24. Mixing of local oscillator and signal waves on a photodetector surface of diameter d.

where u_o and u_s are the complex field amplitudes of the local oscillator and signal beams, respectively. The photomixing process then produces an incremental photocurrent given by

$$di(t) = \Re(x, y)|u(x, y, t)|^2\, dx\, dy \tag{2-136}$$

where $\Re(x, y) = q\eta_q(x, y)/h\nu$ is the spatially dependent *detector responsivity*, $\eta_q(x, y)$ is the quantum efficiency, and $q/h\nu$ is the electronic charge divided by the local oscillator photon energy. Substituting for u and integrating leads to

$$\begin{aligned} i &= \iint \Re(x, y)u(x, y, t)u^*(x, y, t)\, dx\, dy \\ &= \iint \Re(x, y)|u_o(x, y)|^2\, dx\, dy + \iint \Re(x, y)\, |u_s(x, y)|^2\, dx\, dy \\ &\quad + \iint \Re(x, y)\big[u_o(x, y)u_s^*(x, y)e^{i(\omega_o - \omega_s)t} \\ &\quad + u_o^*(x, y)u_s(x, y)e^{-i(\omega_o - \omega_s)t}\big]\, dx\, dy \end{aligned} \tag{2-137}$$

The time-independent terms in Eq. (2-137) can each be identified with a photocurrent as follows:

$$i_{\omega_o} = \iint \Re(x, y)\, |u_o(x, y)|^2\, dx\, dy \tag{2-138}$$

$$i_{\omega_s} = \iint \Re(x, y)\, |u_s(x, y)|^2\, dx\, dy \tag{2-139}$$

$$i_{\Delta\omega} = \iint \Re(x, y)u_o^*(x, y)u_s(x, y)\, dx\, dy \tag{2-140}$$

$$i_{\Delta\omega}^* = \iint \Re(x, y)u_o(x, y)u_s^*(x, y)\, dx\, dy \tag{2-141}$$

where $\Delta\omega = \omega_o - \omega_s$ and $\omega = 2\pi f$. Here Eqs. (2-138) and (2-139) correspond to the dc photomixing currents induced by the local oscillator and signal, respectively, and Eqs. (2-140) and (2-141) correspond to the complex phasor amplitudes of the intermediate frequency (IF). With these definitions we can write

$$i = i_{\omega_o} + i_{\omega_s} + \big[i_{\Delta\omega}e^{-i(\omega_o - \omega_s)t} + i_{\Delta\omega}^* e^{i(\omega_o - \omega_s)t}\big] \tag{2-142}$$

Now the IF signal power is given by

$$|i_{\Delta\omega}|^2 = \left| \iint \Re(x, y)u_o^*(x, y)u_s(x, y)\, dx\, dy \right|^2 \tag{2-143}$$

But Schwartz's inequality relating the integral of the products of complex functions states that

$$\left|\int f(x, y) g(x, y)\, dx\, dy\right|^2 \leq \int |f(x, y)|^2\, dx\, dy \int |g(x, y)|^2\, dx\, dy \tag{2-144}$$

Thus Eq. (2-143) can be written

$$|i_{\Delta\omega}|^2 \leq \iint \Re(x, y)\, |u_o(x, y)|^2\, dx\, dy \iint \Re(x, y)\, |u_s(x, y)|^2\, dx\, dy \tag{2-145}$$

Equations (2-145) and(2-143) can therefore be used to define a *mixing* or *heterodyne efficiency* given by

$$\eta_e = \frac{\left|\iint \Re(x, y) u_o^*(x, y) u_s(x, y)\, dx\, dy\right|^2}{\iint \Re(x, y)\, |u_o(x, y)|^2\, dx\, dy \iint \Re(x, y)\, |u_s(x, y)|^2\, dx\, dy} \tag{2-146}$$

where $\eta_e \leq 1$. The equality holds only if $u_o(x, y)$ and $u_s(x, y)$ have identical amplitude and phase variations over the detector surface. Equation (2-146) represents the *mixing theorem*[22], which states that for any mismatch in phase or amplitude between the local oscillator and signal beams, the IF signal current will be less than the optimum value, in accordance with the relationship above.

Comment 2-6. The functional dependence of heterodyne efficiency on angular misalignment of the signal and local oscillator beams can be calculated using Eq. (2-146). Other effects, such as polarization misalignment and beam quality, will be neglected. Assume that the signal beam is incident on the detector an angle θ in the x–z plane, as shown in Figure 2-24, and that the local oscillator is coaligned with the z-axis. Letting the responsivity be uniform across the detector, we have

$$\eta_e = \frac{\left|\iint u_o^*(x, y) u_s(x, y)\, dx\, dy\right|^2}{\iint |u_o(x, y)|^2\, dx\, dy \iint |u_s(x, y)|^2\, dx\, dy} \tag{2-147}$$

where the integration is taken over the detector surface. Since the y-dependence of the total field on the detector is constant, Eq. (2-147) reduces to

$$\eta_e = \frac{\left|\int_0^d u_o^*(x) u_s(x)\, dx\right|^2}{\int_0^d |u_o(x)|^2\, dx \int_0^d |u_s(x)|^2\, dx} \tag{2-148}$$

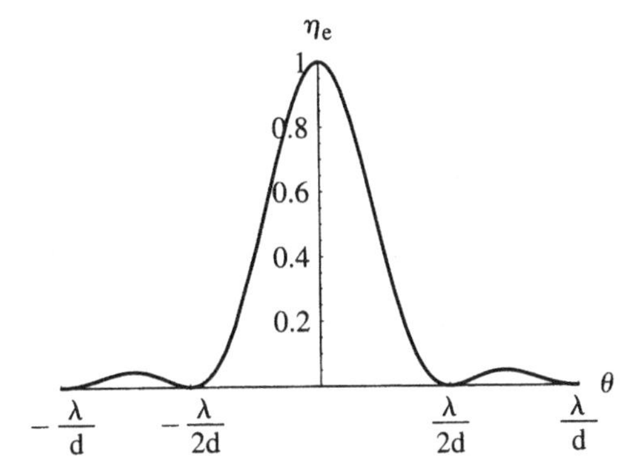

Figure 2-25. Heterodyne efficiency η_e as a function of misalignment angle θ.

Now the components of $u_o(x)$ and $u_s(x)$ along the x-direction are given by

$$u_o(x) = u_o e^{i\vec{k}_o \cdot \vec{x}} = u_o \tag{2-149}$$

and

$$u_s(x) = u_s e^{i\vec{k}_s \cdot \vec{x}} = u_s e^{ik_{sx} x \sin\theta} \tag{2-150}$$

where $\vec{k}_s = k_{sx}\,\widehat{i} + k_{sz}\,\widehat{k}$, $\vec{k}_o = k_{oz}\,\widehat{k}$, and $\vec{x} = x\,\widehat{i}$. Thus

$$\eta_e = \left| \frac{1}{d} \int_0^d e^{ik_{sx} x \sin\theta}\, dx \right|^2 = \left[\frac{\sin(k_{sx} d \sin\theta)}{k_{sx} d \sin\theta} \right]^2 \tag{2-151}$$

Note that for $\theta = 0$, $\eta_e = 1$, and for $k_{sx} d \sin\theta = n\pi$, $\eta_e = 0$. For small angles, this last condition can be written $\theta \approx n\lambda/2d$. Thus for a given detector size d, the signal out of a heterodyne detector will periodically go to zero as the misalignment angle θ is increased, the envelope of the function eventually approaching zero at large angles. This is shown in Figure 2-25. In practice, η_e is usually not so deterministic as indicated above, but is, rather, a consequence of a large number of unknown angle errors distributed throughout the system. ■

Returning to the antenna theorem and Eq. (2-140), we assume a signal wave given by

$$u_s(x, y, z) = a e^{-i\vec{k}\cdot\vec{r}} \tag{2-152}$$

where a is a complex phasor amplitude. For a detector located at $z = 0$, we have

$$u_s(x, y) = a e^{-i(k_x x + k_y y)} \tag{2-153}$$

and Eq. (2-140) becomes

$$i_{\Delta\omega} = a \iint \Re(x, y) u_o^*(x, y) e^{-i(k_x x + k_y y)} \, dx \, dy \tag{2-154}$$

To prove the antenna theorem, a relationship needs to be developed between the effective aperture A_R of the receiver and the photodetector current $i_{\Delta\omega}$. This is accomplished by noting that the IF signal current is given by $|i_{\Delta\omega}|^2 = i_{\omega_o} i_{\omega_s}$ and from a strictly dimensional argument we can write Eq. (2-139) as $i_{\omega_s} = \overline{\Re} \, |a|^2 A_R$, where $\overline{\Re} \, |a|^2$ is a current density and A_R has units of area. Thus

$$A_R = \frac{|i_{\Delta\omega}|^2}{i_{\omega_o} \overline{\Re} \, |a|^2} \tag{2-155}$$

Now $\overline{\Re}$ is the mean responsivity across the detector surface, defined by

$$\overline{\Re} \equiv \frac{\iint \Re(x, y) \, |u_o(x, y)|^2 \, dx \, dy}{\iint |u_o(x, y)|^2 \, dx \, dy} \tag{2-156}$$

Using Eqs. (2-138), (2-154), and (2-156) in Eq. (2-155), the effective receiving aperture takes the form

$$A_R(k_x, k_y) = \frac{\left| \iint \Re(x, y) u_o^*(x, y) e^{-i(k_x x + k_y y)} \, dx \, dy \right|^2}{\overline{\Re}^2 \iint |u_o(x, y)|^2 \, dx \, dy} \tag{2-157}$$

It can be seen that the numerator in Eq. (2-157) represents the power spectral density of the spatial distribution $\Re(x, y) u_o^*(x, y)$ in the space frequencies k_x and k_y. This fact suggests the use of the Wiener–Khintchine theorem to express A_R in terms of the average power of the surface distribution (cf. Section 1.2.10).

To accomplish this, we first assume a spherical coordinate system centered on the detector surface as shown in Figure 2-26, the incoming signal wave being assumed to be a plane wave propagating along the k_s-direction at angles θ and ϕ relative to the z and x axes, respectively. Then it is easy to show from Figure 2-26 that $k_x = k_s \sin\theta \cos\phi$ and $k_y = k_s \sin\theta \sin\phi$ such that

$$dk_x \, dk_y = J(k_x, k_y : \theta, \phi) = k_s^2 \cos\theta \sin\phi \, d\theta \, d\phi \tag{2-158}$$

where J is the Jacobian of the transformation and $k_s = 2\pi/\lambda$. Defining the solid angle about the z-direction as $d\Omega = \sin\theta \, d\theta \, d\phi$, we obtain for small angles with the z-axis,

$$dk_x dk_y \approx k_s^2 d\Omega \tag{2-159}$$

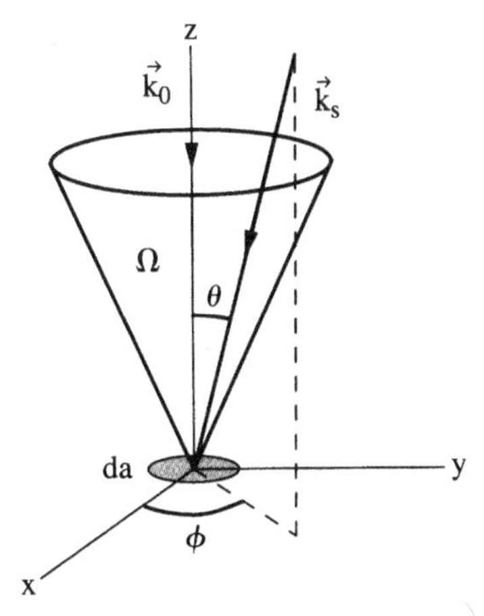

Figure 2-26. Spherical coordinate system for relating the solid angle Ω and the wave vectors $\vec{k}_o$ and $\vec{k}_s$.

We can therefore write

$$\iint A_R(\Omega)\, d\Omega \approx \frac{1}{k_s^2} \iint A_R(k_x, k_y)\, dk_x\, dk_y$$

$$= \left(\frac{2\pi}{k_s}\right)^2 \frac{\iint \Re^2(x, y)\, |u_o(x, y)|^2\, dx\, dy}{\overline{\Re}^2 \iint |u_o(x, y)|^2\, dx\, dy} \tag{2-160}$$

or

$$\iint A_R(\Omega)\, d\Omega \approx \frac{\overline{\Re^2}}{\overline{\Re}^2} \lambda^2 \tag{2-161}$$

For most detectors, $\overline{\Re^2}/\overline{\Re}^2 \approx 1$, so that Eq. (2-161) reduces to

$$\iint A_R(\Omega)\, d\Omega \approx \lambda^2 \tag{2-162}$$

In the simplified case of A_R = constant within the solid angle subtended by the receiver and zero outside, Eq. (2-157) further reduces to the angle–aperture product given by Eq. (2-134).

Thus far we have referred mainly to the receiver channel of the heterodyne system. However, the formulations above also say something about the transmit channel. In particular, the integral expression in the definition of $A_R(k_x, k_y)$ is identical in form to the Fraunhofer approximation to the Huygens–Fresnel diffraction integral developed in Section 2.2. This has several implications. First, if the local oscillator distribution were backpropagated out through the optical system to the far field, the resulting distribution would represent the far field antenna pattern for the system. Second, if this pattern were to be convolved with the far field distribution of the transmitted beam, a corresponding set of matching conditions for the amplitude and phase would be obtained that are identical to those obtained in the detector plane. Indeed, it can be shown that the mixing efficiency may be calculated at any point along the propagation path, as long as the local oscillator is backpropagated to the same plane as the transmitted beam.

These conclusions lead to severe constraints on the quality of the transmitted beam. For example, if a Gaussian local oscillator distribution is backpropagated through an infinite aperture diffraction-limited optical system, the resulting beam profile in the far field would still be Gaussian. Optimum heterodyne efficiency would then require that the transmitted far-field beam profile also be Gaussian, which, in turn, implies a near-field distribution that is Gaussian. Similar arguments can be made for uniform local oscillator beams where the far-field profile assumes the form of an Airy pattern. We conclude, therefore, that for most practical local oscillator distributions on the detector, diffraction-limited, or at least near diffraction-limited, transmit beams must be employed to achieve high heterodyne efficiencies.

2.7. ANALYSIS OF COHERENT DETECTION SYSTEMS

In this section we develop general expressions for the signal-to-noise ratio starting with the Huygens–Fresnel diffraction integral for two generic target types (i.e., the hard target and the soft or atmospheric target). The analyses will consider the more common case of Gaussian transmit and local oscillator beams in a monostatic application using either single- or dual-aperture configurations for the transmit and receive channels. A generic dual-aperture heterodyne configuration is shown in Figure 2-27. Exact and numerical solutions will be developed for

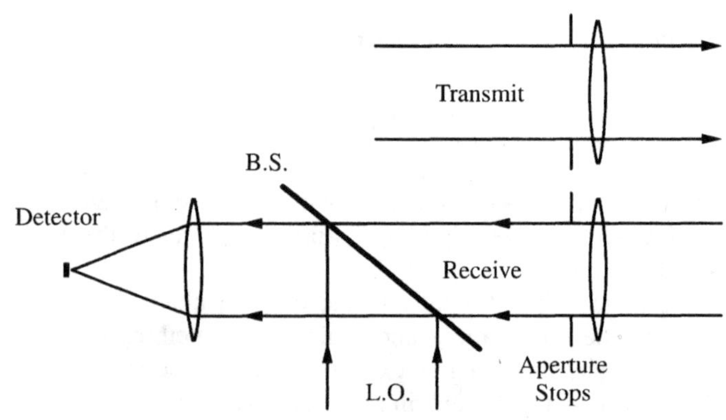

Figure 2-27. Generic heterodyne configuration showing beam truncation occurring at aperture stops in the beam-forming telescopes.

the untruncated and truncated beam configurations, respectively. Atmospheric losses due to scattering and absorption will be ignored for the moment, as will turbulence effects on the beam. Analysis of other beam profiles, such as the uniform local oscillator beam, may be obtained by straightforward generalization of the Gaussian beam results.

2.7.1. Untruncated Systems

The mixing efficiency of a coherent optical receiver was considered in Section 2.6. In this section, the effects of diffraction on the signal level of a coherent detection system are analyzed for the case of untruncated infinite Gaussian beams. Although this implies an unphysical situation involving infinite detector and aperture dimensions, the apertures need only be a few times larger than the Gaussian beam radius for the theory to be applicable. In addition, for Gaussian beams the mathematics can be done exactly,[23] thereby providing a convenient set of analytic equations for use in system analysis. Although large aperture-to-beam diameter ratios occur in many systems, especially during sensitivity measurements prior to insertion of the final aperture-limiting telescope, it will be shown in the next section that some truncation is necessary in order to achieve maximum sensitivity for a given aperture diameter.

The field distribution in the observation or target plane (r, θ) located at a distance z from an emitting aperture plane (ρ, ϕ) is given by Eq. (2-9), which is repeated here for reference:

$$U_T(r) = \frac{-2\pi i}{\lambda z} \int_0^a u(\rho) J_0\left(\frac{kr\rho}{z}\right) e^{ik(r^2+\rho^2)/2z} \rho\, d\rho \tag{2-163}$$

It was shown previously that the field of a Gaussian beam with its waist located at the emitting aperture followed by a lens of focal length f is described by

$$u(\rho) = u_T e^{-\rho^2/\omega_{0T}^2 - ik\rho^2/2f} \tag{2-164}$$

where $u_T = \sqrt{2P_T/\pi\omega_{0T}^2}$, ρ is a radial vector in the aperture plane, f is the focal length of the output optics, ω_{0T} is the beam waist radius located at the input to the focusing optics, and P_T is the power transmitted. Thus, as in the case of Eq. (2-105), the field at z becomes

$$U_T(r) = \frac{-2\pi i u_T}{\lambda z} e^{ikr^2/2z} \int_0^\infty J_0\left(\frac{kr\rho}{z}\right) e^{-\rho^2/\omega_{0T}^2 + ik\rho^2/2z(1-z/f)} \rho\, d\rho \tag{2-165}$$

Now the field at the receiver plane r' generated by a single scatterer located at r can be considered the source of a single Huygens' wavelet described by

$$U_R(r') = \frac{S_c U_T(r)}{z} e^{ik|\vec{r}-\vec{r}'|^2/2z} \tag{2-166}$$

where S_c is the scattering amplitude of the particle. This received field mixes with the local oscillator on the detector plane to produce an IF signal current. The local oscillator field may be written

$$U_{\mathrm{LO}}(r') = u_{\mathrm{LO}} e^{-r'^2/\omega_{0\mathrm{LO}}^2 - ikr'^2/2f} \tag{2-167}$$

where frequency terms have been suppressed and $u_{\mathrm{LO}} = \sqrt{2P_{\mathrm{LO}}/\pi\omega_{0\mathrm{LO}}^2}$. Here $\omega_{0\mathrm{LO}}$ is the local oscillator beam radius at the detector and P_{LO} is the local oscillator power. The IF signal current is therefore

$$i_{\Delta\omega} = 2\Re \iint U_R(\vec{r}')U_{\mathrm{LO}}(\vec{r}')d^2\vec{r}' \tag{2-168}$$

where $\Re = \eta_q q/h\nu$ is the detector responsivity assumed uniform across the detector surface. Inserting Eq. (2-167) into Eq. (2-168), performing the integration over θ, and rearranging yields

$$i_{\Delta\omega} = 2i\Re S_c \lambda U_T(r) \left[\frac{-2\pi i u_{\mathrm{LO}}}{\lambda z} e^{ikr^2/2z} \right.$$
$$\left. \times \int_0^\infty J_0\left(\frac{krr'}{z}\right) e^{-(r'^2/\omega_{0\mathrm{LO}}^2)+(ikr'^2/2z)(1-z/f)} r'\,dr' \right]. \tag{2-169}$$

But the term in brackets is identical in form to $U_T(r)$ under the replacement $r' \to \rho, \omega_{0\mathrm{LO}} \to \omega_{0T}$, and $u_{\mathrm{LO}} \to u_T$. Hence this expression can be considered the field at the target due to a *backpropagated local oscillator*. Such a concept is frequently referred to as a *virtual LO*. Thus, designating this field as $U_{\mathrm{VLO}}(r)$, the IF signal current can be written

$$i_{\Delta\omega} = -2i\Re S_c \lambda U_T(r) U_{\mathrm{VLO}}(r) \tag{2-170}$$

From the form of Eq. (2-170) we see that knowledge of the transmitted and backpropagated local oscillator distributions at the target is sufficient, at least for a single scatterer, to provide the solution. In the case of an untruncated system, the integral solutions for $U_T(r)$ and $U_{\mathrm{VLO}}(r)$ can be solved exactly, at least in the case of Gaussian beams. Indeed, as we know from Section 2.3, Eq. (2-37), which represents a focused Gaussian beam with beam parameters $R_g(z)$ and $\omega(z)$ given by Eqs. (2-54) and (2-55), is the solution to the Huygens–Fresnel integral given by Eq. (2-165).

The quantity of interest in calculating the power signal-to-noise ratio is the mean-squared current. Thus, for a single scatterer,

$$\overline{i_{\Delta\omega}^2} = 2\Re^2 S_c^2 \lambda^2 \left|U_T(r)\right|^2 \left|U_{\mathrm{VLO}}(r)\right|^2 \tag{2-171}$$

where division by a factor of 2 accounts for the time-averaging process.

Comment 2-7. The averaging process alluded to here may seem mysterious, but that is only because we have suppressed the time dependence of $U_T(r)$ and $U_{\mathrm{VLO}}(r)$. If these were made explicit, the averaging process in Eq. (2-171) would lead to terms such as $\overline{\cos^2 \Delta\omega t} \equiv \frac{1}{2}$. ■

The contributions from all scatterers is then the integral over the volume of the target times the density of scatterers ρ_a on the target surface S, which is assumed uniform:

$$\overline{i^2_{\Delta\omega}} = 2\Re^2 \rho(\pi)\lambda^2 \iint_S |U_T(r)|^2 \, |U_{\mathrm{VLO}}(r)|^2 \, r \, dr \, d\theta \tag{2-172}$$

where $\rho(\pi) = \rho_a S_c^2$ is the *inverse steradian power reflectivity* of the target, expressed in units of inverse steradians (sr^{-1}). (The definition of a diffuse Lambertian surface scatterer is one that scatters radiation in accordance with the relationship $\rho(\pi) = \rho \cos\phi/\pi$, where ρ is the reflectivity of the surface. For more detail, see Section 3.3.1). In the case of a volume target, such as the atmosphere, this relationship becomes $\beta(\pi) = \rho_v S_c^2$, which has units of inverse meters-steradians ($\mathrm{m}^{-1}\ \mathrm{sr}^{-1}$).

From Section 2.3, the intensity distributions at the target plane can therefore be written

$$|U_T(r)|^2 = \frac{2P_T}{\pi\omega_T^2(z)} e^{-2r^2/\omega_T^2(z)} \tag{2-173}$$

and

$$|U_{\mathrm{VLO}}(r)|^2 = \frac{2P_{\mathrm{LO}}}{\pi\omega_{\mathrm{VLO}}^2(z)} e^{-2r^2/\omega_{\mathrm{VLO}}^2(z)} \tag{2-174}$$

Substituting Eqs. (2-173) and (2-174) into Eq. (2-172) yields

$$\overline{i^2_{\Delta\omega}} = 2\Re^2 \lambda^2 \rho(\pi) \left(\frac{2}{\pi}\right)^2 \frac{P_T P_{\mathrm{LO}}}{\omega_T^2(z)\omega_{\mathrm{VLO}}^2(z)} \iint_S e^{-2r^2/\omega_T^2(z) - 2r^2/\omega_{\mathrm{VLO}}^2(z)} r \, dr \, d\theta \tag{2-175}$$

For the case of perfectly matched beam parameters, $\omega_T = \omega_{\mathrm{VLO}} = \omega$, Eq. (2-175) reduces to

$$\overline{i^2_{\Delta\omega}} = 2\Re^2 \lambda^2 \rho(\pi) P_T P_{\mathrm{LO}} \left(\frac{2}{\pi\omega^2(z)}\right)^2 \iint_S e^{-4r^2/\omega^2(z)} r \, dr \, d\theta \tag{2-176}$$

For a circular or "disk" target oriented perpendicular to the beam, Eq. (2-176) integrates to

$$\overline{i^2_{\Delta\omega}} = 2\Re^2 \rho(\pi) P_T P_{\mathrm{LO}} \frac{\lambda^2}{\pi\omega^2(z)} [1 - e^{-4r_t^2/\omega^2(z)}] \tag{2-177}$$

Equation (2-177) can be rewritten as

$$\overline{i_{\Delta\omega}^2} = 2\Re^2 P_{\mathrm{LO}} P_R = 4\Re^2 \rho(\pi) P_T P_{\mathrm{LO}} \frac{\lambda^2}{\pi \omega^2(z)} \eta_T \eta_D \tag{2-178}$$

where P_R is the received optical power, to be defined below, and

$$\eta_T = 1 - e^{-2r_t^2/\omega^2(z)} \tag{2-179}$$

and

$$\eta_D = \tfrac{1}{2}\left[1 + e^{-2r_t^2/\omega^2(z)}\right] \tag{2-180}$$

η_T is easily identified as the fractional power intercepted by a uniform circular target in a Gaussian beam of radius $\omega(z)$ and is generally referred to as the *capture efficiency* of the target–beam interaction. It is defined mathematically as the convolution of the target reflectivity and the beam intensity profile over the geometric extent of the target. Thus, for targets of a more general nature, Eq. (2-179) can be generalized to

$$\eta_T = \iint_S f_\rho(r, \theta) e^{-2r^2/\omega^2(z)} r\, dr\, d\theta \tag{2-181}$$

where $f_\rho(r, \theta)$ is the reflectivity function of the target surface.

η_D assumes values between $\frac{1}{2}$ and 1 as r goes from infinity to zero. The case of $r = \infty$ corresponds to an infinite diffuse target, while the limiting case of $r \to 0$ can be viewed as the fractional power intercepted by a target that is much smaller than the beam. These interpretations and results are consistent with results obtained from independent considerations of heterodyne efficiency associated with glint and diffuse targets.[24]

At this point, an expression for the power signal-to-noise ratio (SNR) of a coherent optical system can be developed. The mean-squared noise power out of a shot-noise-limited photodetector is given by $\overline{i_N^2} = 2q\Re P_{\mathrm{LO}} B$, where B is the IF bandwidth of the receiver. It will be shown in Section 2.10 that the signal-to-noise ratio can be written

$$\mathrm{SNR}(z) = \frac{\overline{i_{\Delta\omega}^2}}{\overline{i_N^2}} = \frac{P_R}{\mathrm{NEP}} \tag{2-182}$$

where $\mathrm{NEP} = qB/\Re$ is the *noise equivalent power* of the receiver, $\Re = \eta_q q/h\nu$ is the detector responsivity, η_q is the detector quantum efficiency, $h\nu$ is the photon energy, and

$$P_R = 2\eta_p \rho(\pi) P_T \frac{\lambda^2}{\pi \omega^2(z)} \eta_T \eta_D \tag{2-183}$$

is the received optical power. Here

$$\omega^2(z) = \omega_0^2\left[\left(\frac{1-z}{f}\right)^2 + \left(\frac{z}{z_0}\right)^2\right] \tag{2-184}$$

is the beam spot radius at the target and η_p is a factor introduced into Eq. (2-183) to account for all power losses internal and external to the system. (η_p does not include the quantum efficiency since this is included in the expression for NEP.) For example, $\eta_p = \eta_o\eta_e\eta_a$, where η_o represents the internal optical losses of the system, including power loss due to truncation (to be discussed later), η_e represents the heterodyne or mixing efficiency discussed in Section 2.6, and $\eta_a = \exp(-2\alpha z)$ represents a two-way atmospheric loss due to absorption and scattering. Equation (2-182) is the fundamental power SNR expression for untruncated coherent optical systems and is valid for both near- and far field ranges. It is discussed in greater detail in Section 2.10, where it will be shown to correspond to the high signal limit of the direct detection signal-to-noise ratio.

Note that for an infinite target (i.e., $\eta_T \to 1$ and $\eta_D \to \frac{1}{2}$), the received power is inversely proportional to the area of the transmitted beam at the target [i.e., $\pi\omega^2(z)$]. For a small target of radius $r_t \ll \omega(z)$, Eqs. (2-179) and (2-180) can be expanded in Taylor series to give $\eta_T \approx 2r_t^2/\omega^2$ and $\eta_D \approx \frac{1}{2}$. This results in an inverse fourth-power dependence on beam radius for the received power, that is,

$$P_R \approx \eta_p \rho(\pi) P_T \frac{2r_t^2\lambda^2}{\pi\omega^4(z)} \tag{2-185}$$

Comment 2.8. For collimated beams, $f = \infty$ in Eq. (2-184). With $z_0 = \pi\omega_0^2/\lambda$, it is easy to show that the far field limits of Eqs. (2-183) and (2-185) can be written

$$P_R = \begin{cases} \eta_p \rho(\pi) P_T \dfrac{\pi D^2}{4z^2} & \text{(infinite target)} \quad (2\text{-}186) \\[2ex] 2\pi\eta_p \rho(\pi) P_T \dfrac{\pi D^2}{4z^2}\dfrac{A_t}{A_b} & \text{(small target)} \quad (2\text{-}187) \end{cases}$$

where $D = 2\omega_0$ is identified with the diameter of the receive aperture, $A_t = \pi r_t^2$ is the area of a circular target on the axis of the beam, and $A_b = \pi(\lambda z/\pi\omega_0)^2$ is the area of the beam at the target. Note that Eq. (2-187) has a z^{-4} range dependence when A_b is expanded out and that the ratio $2A_t/A_b$ is the first-order approximation to the capture efficiency η_T defined earlier. In radar terminology, Eqs. (2-186) and (2-187) are usually referred to as the R^{-2} and R^{-4} range equations for targets that are larger than or smaller than the beam, respectively. ■

The SNR expression given by Eq. (2-182) can be put in a particularly intuitive form as follows. Letting $\eta_T = 1, \eta_D = \frac{1}{2}$, and inserting Eqs. (2-183) and (2-184) into Eq. (2-182) yields

$$\begin{aligned} \mathrm{SNR}(z) &= \frac{\eta}{h\nu B}\rho(\pi)P_T\frac{\lambda^2}{\pi\omega^2(z)} \\ &= \frac{\eta}{h\nu B}\rho(\pi)P_T\frac{\lambda^2}{\pi\omega_0^2}\left[\left(\frac{1-z}{f}\right)^2+\left(\frac{z}{z_0}\right)^2\right]^{-1} \end{aligned} \tag{2-188}$$

where $\eta = \eta_q\eta_p$. Using the results of Section 2.3.2, the last two terms can be rewritten in terms of focused beam variables:

$$\mathrm{SNR}(z) = \frac{\eta}{h\nu B}\rho(\pi)P_T\frac{S}{1+[(z-z_\omega)/z_0']^2} \tag{2-189}$$

where

$$S = \frac{\lambda}{f}\left(\frac{z_0}{f}+\frac{f}{z_0}\right) \tag{2-190}$$

$$z_\omega = z_0\left(\frac{z_0}{f}+\frac{f}{z_0}\right)^{-1} \tag{2-191}$$

and

$$z_0' = f\left(\frac{z_0}{f}+\frac{f}{z_0}\right)^{-1} \tag{2-192}$$

Equations. (2-191) and (2-192) can be recognized from Section 2.3.2 as the focal range and Rayleigh range of the focused beam, respectively, and are shown in Figure 2-28. The quantity S corresponds to the maximum possible signal, which from the form of Eq. (2-189), must occur at $z = z_\omega$, where the beam spot size is a minimum. That is, using Eq. (2-190) in Eq. (2-189) yields

$$\mathrm{SNR}_{\max} = \frac{\eta}{h\nu B}\rho(\pi)P_T\frac{\lambda^2}{\pi\omega_0'^2} \tag{2-193}$$

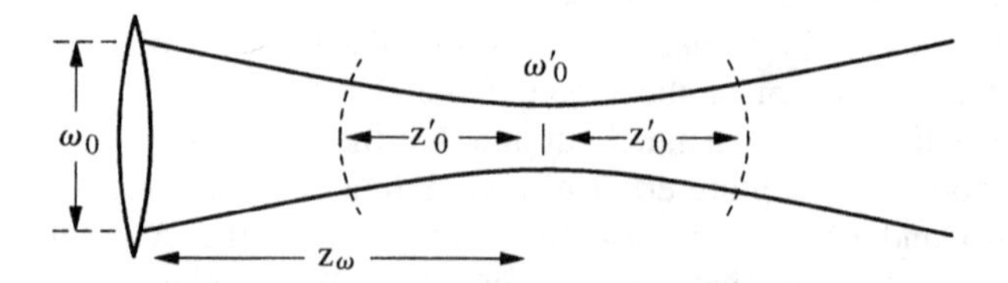

Figure 2-28. Gaussian beam parameters for a focused coherent optical system.

where $\omega_0'^2 = \omega_0^2/[1 + (z_0/f)^2]$ is the beam waist radius at the focal range z_ω. Note also that the maximum value of z_ω is $2\,z_0$, twice the Rayleigh range of the transmitted beam, and that for a collimated system (i.e., $f \to \infty$ and $z_\omega \to 0$), the last term in Eq. (2-193) becomes simply $\lambda^2/\pi\omega_0^2$.

Comment 2-9. The last term in Eq. (2-189) may be considered a range sensitivity function for the system that peaks at $z = z_\omega$. Some insight into its behavior may be obtained by rewriting it in terms of the ratio z_0/f:

$$\Phi(z_0, f) = \frac{S}{1 + [(z - z_\omega)/z_0']^2} = \frac{(\lambda/z_0)(1 + z_0^2/f^2)}{1 + [(z/z_0)(1 + z_0^2/f^2) - z_0/f]^2} \tag{2-194}$$

The quantity $\Phi z_0/\lambda$ is plotted in Figure 2-29 as a function of the normalized range z/z_0 with z_0/f as a parameter. It can be seen that as the system becomes more focused, the width of the high-sensitivity region, defined by $2z_0'$, narrows, while the peak response, located at z_ω, increases in magnitude. Also note that the maximum focal range for the system is one-half the Rayleigh range z_0, as can easily be seen from the $z_0/f = 1$ curve. For $f \to \infty$, maximum sensitivity occurs at $z = 0$ and is finite (i.e., it does not go to infinity as in z^{-2} range equations that do not take into account near-field effects). ■

Calculation of the actual SNR requires definition of the target characteristics. Consider the interesting case of a volume target such as the atmosphere. In this case the surface reflectivity $\rho(\pi)$ of a hard target must be replaced by the volume reflectivity $\beta(\pi)$ of a soft target. With this change, we can integrate Eq. (2-189) over z, obtaining

$$\mathrm{SNR} = \frac{\eta}{h\nu B}\beta(\pi)P_T \int_{z_1}^{z_2} \frac{S}{1 + [(z - z_\omega)/z_0']^2}\,dz \tag{2-195}$$

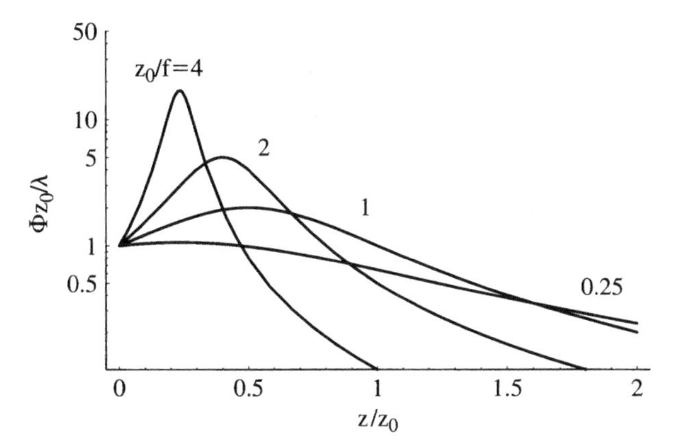

Figure 2-29. Normalized SNR versus normalized range for different values of $g = z_0/f$. Notice the enhanced but narrow peak response for large g, corresponding to a tightly focused system.

Since S is independent of z, this yields the exact solution

$$\mathrm{SNR} = \frac{\eta P_T}{h\nu B}\beta(\pi)\lambda\left(\tan^{-1}\frac{z_2 - z_\omega}{z_0'} - \tan^{-1}\frac{z_1 - z_\omega}{z_0'}\right) \tag{2-196}$$

where $Sz_0' = \lambda$.

Several interesting cases might be considered. The first is the SNR obtained from a *focused CW system* in the range interval $0 \le z \le \infty$. Assuming uniformly distributed scatterers as shown in Figure 2-30, we let $z_1 = 0$ and $z_2 = \infty$ in Eq. (2-196), obtaining

$$\mathrm{SNR} = \frac{\eta P_T}{h\nu B}\beta(\pi)\lambda\left(\frac{\pi}{2} + \tan^{-1}\frac{z_\omega}{z_0'}\right) \tag{2-197}$$

But for most tightly focused systems, the focal range is much larger than the Rayleigh range of the focused beam. Thus, with $z_\omega/z_0' \gg 1$, the second term in the parentheses on the right-hand side approximates to $\pi/2$, resulting in

$$\mathrm{SNR} \approx \frac{\eta P_T}{h\nu B}\lambda\pi\beta(\pi) \tag{2-198}$$

(This result would be exact if contributions to the signal extended to $z = -\infty$, which is clearly not possible for a finite aperture system.)

Consider now the SNR obtained from the focal volume only. In this case, $z_2 = z_\omega + z_0'$ and $z_1 = z_\omega - z_0'$, which results in

$$\mathrm{SNR} = \frac{\eta P_T}{h\nu B}\frac{\lambda\pi\beta(\pi)}{2} \tag{2-199}$$

Thus, half the signal energy arises from the focal volume. This feature allows for a certain degree of range resolution achievable with a system employing a CW waveform.

Finally, consider a collimated system as shown in Figure 2-31, where the beam waist $\omega_0' = \omega_0$ and $z_\omega = 0$. With $z_1 = 0$ and $z_2 = \infty$ in Eq. (2-196), we obtain

$$\mathrm{SNR} = \frac{\eta P_T}{h\nu B}\frac{\lambda\pi\beta(\pi)}{2} \tag{2-200}$$

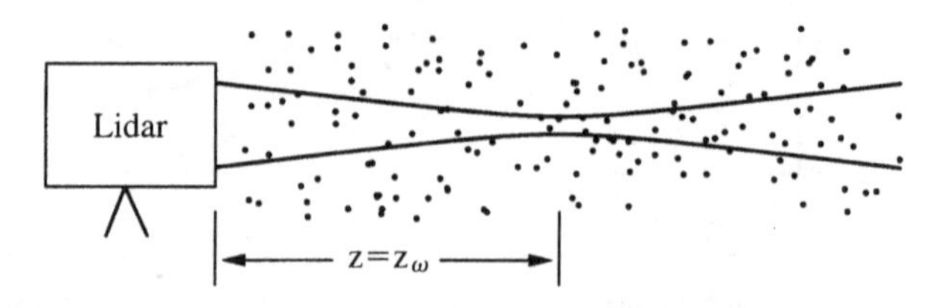

Figure 2-30. Focused beam lidar in a volumetric scattering medium.

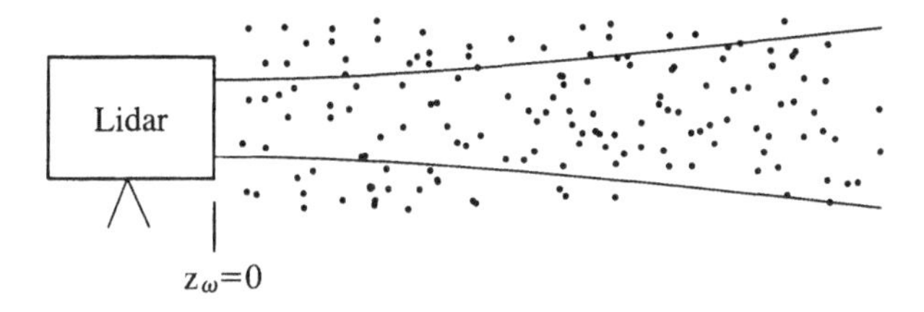

Figure 2-31. Collimated beam lidar in a volumetric scattering medium.

In this case, the factor of $\frac{1}{2}$ results from the fact that contributions to the signal arise from only one side of the beam waist, which is located at the aperture. A variety of other configurations might be considered, including the effects of pulsed waveforms on the system performance.

2.7.2. Truncated Systems

In this section the effects of beam truncation by the limiting aperture in a coherent optical system, assumed to be the final output aperture, are considered. It will be shown that an optimum aperture-to-beam diameter ratio exists similar to that which was found in Section 2.4.1 for Gaussian beams. In both cases it is assumed that the aperture is fixed by cost or volume constraints and the beam is adjusted for maximum signal. The fact that an optimum ratio exists for a coherent detection system is easily understood. Consider a common aperture monostatic system where the transmitter and local oscillator beams are optimally matched and totally intercepted by the detector. If they are also small compared to the clear aperture, the effective receive aperture, which is determined by the backpropagated local oscillator distribution, is smaller than that allowed by the output aperture and the signal is therefore less than optimum. If, on the other hand, the beam is oversized, truncation occurs and transmitted power is lost. In addition, there is increased diffraction spreading associated with the truncation process, thereby reducing the signal even further.

Returning to the case of hard targets, the optimum aperture-to-beam diameter ratio[25] for the case of a large diffuse target, we divide the mean-squared current given by Eq. (2-172) by the mean-squared noise current to obtain

$$\mathrm{SNR}_{\mathrm{TL}} = \frac{\eta\rho(\pi)\lambda^2}{h\nu B P_{\mathrm{LO}}} \int_0^{\infty} \int_0^{2\pi} |U_T(r)|^2 \, |U_{\mathrm{VLO}}(r)|^2 \, r \, dr \, d\theta \qquad \text{(2-201)}$$

where the subscripts in $\mathrm{SNR}_{\mathrm{TL}}$ refer to "truncated" and "large," respectively, and $\eta = \eta_q \eta_p$. $U_T(r)$ and $U_{\mathrm{VLO}}(r)$ are given by Eqs. (2-165) and the bracketed term in Eq. (2-169), respectively, but with the integration limits extending from zero to the aperture radius a rather than from zero to infinity. If the analysis is now restricted to a matched monostatic common aperture system, we can let the integral expressions in $U_T(r)$ and $U_{\text{VLO}}(r)$ be equal, so that Eq. (2-201) becomes, after integrating over θ,

$$\text{SNR}_{\text{TL}} = \frac{\eta P_T \rho(\pi)\lambda^2}{h\nu B} \frac{8}{\pi \omega_0^4} \frac{k^4}{z^4} \int_0^\infty |f(r)|^4 r\, dr \tag{2-202}$$

where

$$f(r) = \int_0^a J_0\left(\frac{kr\rho}{z}\right) e^{-(\rho^2/\omega_{0T}^2)+(ik\rho^2/2z)(1-z/f)} \rho\, d\rho \tag{2-203}$$

[Note that the complex term $-i\exp(ikr^2/2z)$ multiplying the integral in $f(r)$ can be ignored since its absolute value will be calculated.]

Normalizing variables to the aperture radius a by letting $\rho' = \rho/a$ and $\zeta = r/a$, Eq. (2-203) becomes

$$f(\zeta) = a^2 \int_0^1 J_0(\beta_z \zeta \rho') e^{-\xi^2 \rho'^2 + (i\rho'^2/2)(\beta_z - \beta_f)} \rho' d\rho' \tag{2-204}$$

where $\xi = a/\omega_0$ is the truncation parameter, $\beta_z = ka^2/z$, and $\beta_f = ka^2/f$. (Note that ka^2 equals $2\times$ the far field distance of the aperture.) Equation (2-202) can therefore be written as

$$\text{SNR}_{\text{TL}} = \frac{\eta P_T \rho(\pi)}{h\nu B} \left(\frac{\lambda}{a}\right)^2 \frac{8}{\pi} \beta_z^4 \xi^4 \int_0^\infty |f(\zeta)|^4 \zeta\, d\zeta \tag{2-205}$$

Some insight into the performance impact of beam truncation may be obtained by comparing the performance of a truncated system to that of an untruncated system. To do this, we normalize Eq. (2-205) to the signal-to-noise ratio of an untruncated system with an exit beam diameter equal to (1) the optimally truncated *beam* diameter, and (2) the system *aperture* diameter. As will be seen, the former occurs at $\xi_{\text{opt}} \approx 1.25$ while the latter corresponds to $\xi = 1$. We therefore obtain

$$\text{SNR}_{\text{CL}} = \frac{\text{SNR}_{\text{TL}}(\xi)}{\text{SNR}_{\text{UTL}}(\xi = \xi_0)} \tag{2-206}$$

where $\text{SNR}_{\text{UTL}}(\xi = \xi_0)$ is the untruncated signal-to-noise ratio given by Eq. (2-205) but with the integration limit in Eq. (2-204) extended to infinity and evaluated at the beam radius $\xi = \xi_0$. Here $\xi_0 = 1.25$ for case 1 and $\xi_0 = 1$ for case 2. Equation (2-206) is numerically integrated and shown in Figure 2-32 for the two cases above as a function of the parameters $\xi = a/\omega_0$, with $\beta_z = 0.1$ and $\beta_f = 0$. Note that since ka^2 is twice the far-field distance of an aperture of

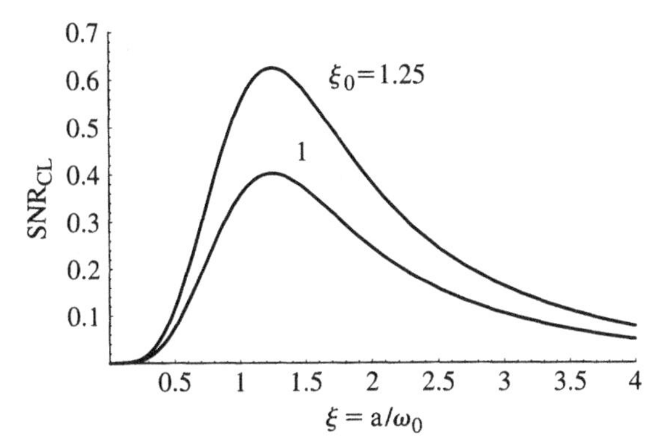

Figure 2-32. Normalized SNR versus truncation parameter $\xi = a/\omega_0$ for a coherent detection system operating against an infinite diffuse target. The optimum truncation ratio occurs at $\xi_{\rm opt} \approx 1.25$ for both collimated and focused beam systems. Normalization is relative to the SNR of an untruncated system with a beam radius equal to the optimally truncated beam radius ($\xi_0 = 1.25$) and the aperture radius ($\xi_0 = 1$), respectively.

radius a, a value of $\beta_z = 0.1$ implies a range that is well into the far field, while a value of $\beta_f = 0$ implies a collimated beam.

The value of $\xi_{\rm opt} \approx 1.25$ is clearly evident[26] in Figure 2-31. It can be seen that a truncated system is down by a factor of 0.624, or approximately -2 dB, relative to an untruncated system having a beam diameter $\omega_0 = 0.8a$. The result for case 2, where $\xi_0 = 1$, shows that the performance ratio is down by a factor of 0.4, or -3.95 dB, relative to an untruncated system having a beam radius $\omega_0 = a$. It should be noted that Figure 2-31 holds for focused beam systems as well. For example, an identical set of curves are obtained by letting $\beta_z = 4$ and $\beta_f = 4$, in which case the beam waist location $z_\omega \approx f$ is approximately one-half the far-field distance $ka^2/2$.

Now that we know the optimum truncation ratio, the signal loss due solely to truncation at the aperture can be calculated. Using Eq. (2-114) with $\varepsilon = 0$ while normalizing to the total power in the transmitted beam, P_T, leads to a one-way power loss given by

$$\eta_{1\rm w} = 1 - e^{-2\xi^2} \tag{2-207}$$

However, since the effective receive aperture is determined by the Gaussian local oscillator distribution on a matched detector, a two-way power loss occurs given by

$$\eta_{2\rm w} = (1 - e^{-2\xi^2})^2 \tag{2-208}$$

Thus, with $\xi = \xi_{\rm opt}$, $\eta_{2\rm w} = 0.906 \approx -0.42$ dB, and we find that very little actual power is lost at the optimum design point, so that the -2 dB or -4 dB loss found above may be attributed primarily to diffraction.

The solution for small targets can be found by evaluating the integral in Eq. (2-202) in the vacinity of the beam axis, where the intensity may be considered constant. Using the subscript "S" to indicate the small-target case, we have

$$\mathrm{SNR_{TS}} = \frac{\eta P_T \rho(\pi)\lambda^2}{h\nu B} \frac{8}{\pi\omega_0^4} \frac{k^4}{z^4} |f(r=0)|^4 \int_0^{r_t} r\,dr \tag{2-209}$$

Normalizing variables as in Eq. (2-205) yields

$$\mathrm{SNR_{TS}} = \frac{\eta P_T \rho(\pi)}{h\nu B} \left(\frac{\lambda}{a}\right)^2 \frac{4A_t}{\pi^2} \beta_z^4 \xi^4 |f(\zeta=0)|^4 \tag{2-210}$$

where $A_t = \pi r_t^2$ is the area of the target.

Once again we find that an optimum truncation ratio exists but this time obtaining $\xi_{\mathrm{opt}} \approx 1.12$. We therefore normalize to cases 1 and 2, as before, which leads to

$$\mathrm{SNR_{CS}} = \frac{\mathrm{SNR_{TS}}(\xi)}{\mathrm{SNR_{UTS}}(\xi=\xi_0)} \tag{2-211}$$

$\mathrm{SNR_{CS}}$ is plotted in Figure 2-33, where $\xi_0 = 1.12$ was used for case 1 and $\xi_0 = 1$ for case 2. Once again it is pointed out that the curves are valid for both collimated and focused beams. Note that the optimum truncation ratio of $\xi_{\mathrm{opt}} = 1.12$ is identical to the value obtained in Section 2.4.1 for the optimized on-axis intensity of a truncated (unobscured) Gaussian beam. It can also be seen that the loss has increased by about 4 dB (−2 dB and −4 dB to −6 dB and −8 dB, respectively, for $\xi_0 = 1.12$ and $\xi_0 = 1$). The two-way loss given by Eq. (2-208) is $\eta_{2w} = -0.737$ dB, which is almost twice that for an infinite diffuse target.

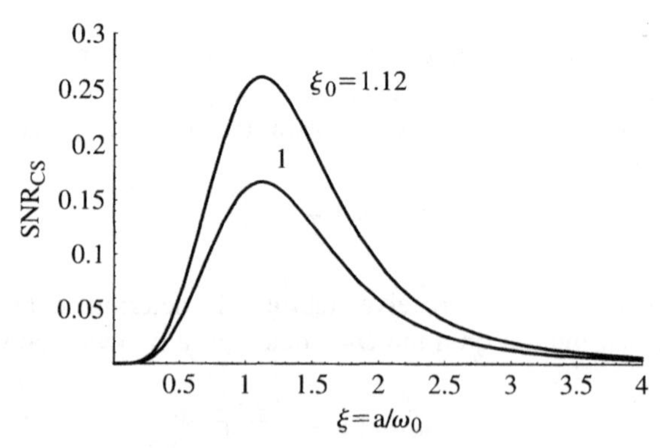

Figure 2-33. Normalized SNR versus truncation parameter $\xi = a/\omega_0$ for a coherent detection system operating against a diffuse target that is small compared to the beam. The optimum truncation ratio occurs at $\xi_{\mathrm{opt}} \approx 1.12$ for both collimated and focused beam systems. Normalization is relative to the SNR of an untruncated system with a beam radius equal to the optimally truncated beam radius ($\xi_0 = 1.12$) and the aperture radius ($\xi_0 = 1$), respectively.

2.8. ANALYSIS OF DIRECT-DETECTION SYSTEMS

The range performance of direct-detection systems can also be developed from the Huygens–Fresnel integral. In this case, the transmit and receive channels are not constrained by the antenna theorem, so that they can be designed and optimized somewhat independently. The word *somewhat* is used here since an optimal receiver design must take into account the transmit channel design, despite their being independent. For example, systems employing aperture averaging of the speckle fluctuations typically have transmit beam sizes that are small compared to the receive aperture dimensions. The angular fields of view of the two channels are then matched with a detector size of the appropriate dimensions. In other words, the receive channel need not be diffraction limited as in the case of heterodyne receivers, since there are no stringent mixing requirements. For simplicity, the following two sections consider only monostatic common aperture configurations, although the analysis can easily be extended to accommodate more general designs.

2.8.1. Untruncated Systems

Analysis of the direct-detection system begins with the equivalent of Eq. (2-139) for the signal current in the absence of a local oscillator, that is,

$$i_{\omega_s} = \iint \Re(x, y)\left|U_R(r')\right|^2 d^2\vec{r}' \tag{2-212}$$

where $U_R(r')$ is the field at the receiver plane r' due to a single scatterer at r as given by Eq. (2-166), which will be repeated here for convenience:

$$U_R(r') = \frac{S_c U_T(r)}{z} e^{ik|\vec{r}-\vec{r}'|^2/2z} \tag{2-213}$$

From Eq. (2-165),

$$U_T(r) = \frac{-2\pi i u_T}{\lambda z} e^{ikr^2/2z} \int_0^\infty J_0\left(\frac{kr\rho}{z}\right) e^{-(\rho^2/\omega_{0T}^2)+(ik\rho^2/2z)(1-z/f)} \rho\, d\rho \tag{2-214}$$

As was pointed out earlier, an exact solution can be found for Eq. (2-214) in the form of the propagation equation for Gaussian beams given by Eq. (2-37), such that the intensity is given by Eq. (2-173). We have

$$|U_T(r)|^2 = \frac{2P_T}{\pi\omega_T^2(z)} e^{-2r^2/\omega_T^2(z)} \tag{2-215}$$

Thus, inserting Eqs. (2-213) and (2-215) into Eq. (2-212) while assuming a uniform detector responsivity yields

$$
\begin{aligned}
i_{\omega_s} &= \frac{\Re S_c^2}{z^2} |U_T(r)|^2 \int_0^a \int_0^{2\pi} r' dr' d\theta' \\
&= \Re S_c^2 \frac{2P_T}{\pi \omega^2(z)} \frac{\pi D^2}{4z^2} e^{-2r^2/\omega^2(z)}
\end{aligned}
\tag{2-216}
$$

where $D = 2a$ is the diameter of the receive aperture. The current produced by a diffuse target is then obtained by integrating Eq. (2-216) over the target extent:

$$
i_{\omega_s} = \Re S_c^2 \frac{2P_T}{\pi \omega^2(z)} \frac{\pi D^2}{4z^2} \int_0^{r_t} \int_0^{2\pi} \rho_a e^{-2r^2/\omega^2(z)} r \, dr \, d\theta \tag{2-217}
$$

where ρ_a is the density of scatterers at the target plane and r_t is the target radius. Thus, with $\rho(\pi) = \rho_a S_c^2$, we have

$$
i_{\omega_s} = \Re \rho(\pi) P_T \frac{\pi D^2}{4z^2} [1 - e^{-2r_t^2/\omega^2(z)}] \tag{2-218}
$$

The mean signal power is proportional to the mean-square current, thus

$$
\overline{i_{\omega_s}^2} = \Re^2 P_R^2 = \Re^2 \rho^2(\pi) P_T^2 \left(\frac{\pi D^2}{4z^2} \right)^2 [1 - e^{-2r_t^2/\omega^2(z)}]^2 \tag{2-219}
$$

where P_R is the received optical power. Note that the averaging process does not result in a factor of $\frac{1}{2}$ as in coherent detection systems since the energy detection process does not involve an intermediate frequency.

Now, unlike coherent detection receivers in which local oscillator shot noise dominates all other noise contributions, the noise in a direct-detection receiver can arise from a variety of sources. These might include, among others, background radiation, signal shot noise, dark current noise, backscattered radiation, amplifier noise, excess noise, and thermal noise. Any one of these can dominate, depending on the receiver design and operating environment. They are discussed in detail in Section 2.10, but for the moment we consider, without loss of generality, the two representative cases of signal and background noise only. The mean-squared noise current is then given by

$$
\overline{i_N^2} = 2q\Re(P_R + P_b)B \tag{2-220}
$$

where P_R and P_b are the signal and background powers, respectively. The power signal-to-noise ratio may then be written

$$
\text{SNR}_p = \frac{\overline{i_{\omega_s}^2}}{\overline{i_N^2}} = \left(\frac{P_R}{\text{NEP}} \right)^2 \tag{2-221}
$$

where, in agreement with Eq. (2-219),

$$P_R = \eta_p \rho(\pi) P_T \frac{\pi D^2}{4z^2}[1 - e^{-2r_t^2/\omega^2(z)}] \tag{2-222}$$

and η_p accounts for all internal and external losses associated with a direct-detection system [cf. Eq. (2-183)]. The noise equivalent power (NEP) in Eq. (2-221) may be written

$$\text{NEP} = \frac{2q(P_R + P_b)B}{\Re} \tag{2-223}$$

Note that unlike the coherent detection case, the received power expression of a direct detection system is not a function of $\omega(z)$, except in the exponential term. This means that for large targets, where the exponential term vanishes, there is no signal dependence on the focal length of the transmitting optics. Hence there is no advantage in using a focused beam in a direct-detection system to enhance the sensitivity when operating against, say, atmospheric targets. However, when operating against small targets, the signal can clearly be enhanced by maintaining the ratio $r_t/\omega(z)$ large (i.e., by focusing the beam, assuming of course, that the target is within the focusing range of the optics).

The limiting case of an infinite diffuse target yields for the received power

$$P_R = \eta_p \rho(\pi) P_T \frac{\pi D^2}{4z^2} \tag{2-224}$$

while the limiting case of a small on-axis target in the far field can be put into the form

$$P_R = 2\pi \eta_p \rho(\pi) P_T \frac{\pi D^2}{4z^2} \frac{A_t}{A_b} \tag{2-225}$$

Here $A_t = \pi r_t^2$ is the area of the on-axis target and $A_b = \pi(\lambda z/\pi\omega_0)^2$ is the area of the beam at the target. It might be noted that Eq. (2-225) is identical in form to that found for the coherent detection case given by Eq. (2-187). This is to be expected since the received optical power is independent of the detection process at the detector.

2.8.2. Truncated Systems

We saw in Section 2.7.2 that for any finite aperture size, a maximum signal-to-noise ratio exists for a truncation ratio $\xi = a/\omega_0$ less than infinity. This was attributable to the fact that the effective receiver aperture is determined by the (backpropagated) local oscillator beam profile, which, for a matched common aperture system, is identical to the beam profile transmitted, via the mixing theorem. However, in the case of a direct-detection system, the transmit and receive apertures are not as coupled as in the coherent case and may be quite different, depending on system requirements such as aperture averaging or lag-angle compensation.

The problem of optimization of the on-axis intensity of a truncated Gaussian beam has already been solved in Section 2.4.1. There it was shown that a finite $\xi_{\text{opt}} = a/\omega_0$ exists for a nonzero value of obscuration parameter $\varepsilon = b/a$, where b and a are the inner and outer radii, respectively, and that for $\varepsilon = 0$, ξ_{opt} occurs at infinity. The corresponding received power and signal-to-noise ratio may be developed along similar lines. Consider the signal current produced by a monostatic direct-detection system employing a truncated and obscured Gaussian beam operating against a single scatterer. We have

$$i_{\omega_s} = \Re S_c^2 \frac{\pi D^2}{4z^2} |U_T(r)|^2 \tag{2-226}$$

where $U_T(r)$ is given by Eq. (2-165). The case of a diffuse target of radius r_t is given by

$$i_{\omega_s} = \Re \rho(\pi) \frac{\pi D^2}{4z^2} \int_0^{r_t} \int_0^{2\pi} |U_T(r)|^2 \, r \, dr \, d\theta \tag{2-227}$$

where the relationship $\rho(\pi) = \rho_a S_c^2$ was used. The mean-squared signal current is therefore

$$\overline{i_{\omega_s}^2} = \Re^2 P_R^2 = 4\pi^2 \Re^2 \rho^2(\pi) \left(\frac{\pi^2 D^2}{2z^2} \right)^2 \left[\int_0^{r_t} |U_T(r)|^2 \, r \, dr \right]^2 \tag{2-228}$$

where a circularly symmetric target is assumed. Assuming for the moment that all the energy collected at the aperture reaches the detector, the signal-to-noise ratio becomes

$$\text{SNR} = \left(\frac{P_R}{\text{NEP}} \right)^2 \tag{2-229}$$

where

$$P_R = 2\pi \rho(\pi) \frac{\pi D^2}{4z^2} \int_0^{r_t} |U_T(r)|^2 r \, dr \tag{2-230}$$

and the NEP is given by Eq. (2-223). Assuming a Gaussian input beam at the aperture given by Eq. (2-164), Eq. (2-229) can be recast in the form

$$P_R = 2\pi \rho(\pi) \frac{\pi D^2}{4z^2} u_T^2 \frac{k^2}{z^2} \int_0^{r_t} |f(r)|^2 r \, dr \tag{2-231}$$

where $f(r)$ is given by Eq. (2-203) and $u_T^2 = 2P_T/\pi\omega_0^2$. The signal-to-noise ratio for an infinite diffuse target then becomes

$$\text{SNR}_{\text{TL}} = \left[\frac{2\pi \rho(\pi)}{\text{NEP}} \frac{\pi D^2}{4z^2} u_T^2 \frac{k^2}{z^2} \int_0^{\infty} |f(r)|^2 r \, dr \right]^2 \tag{2-232}$$

where the integration has been extended over all space in the plane of the target. Letting $\rho' = \rho/a$ and $\zeta = r/a$, Eq. (2-232) becomes

$$\mathrm{SNR_{TL}} = \left[\frac{4\rho(\pi)P_T}{\mathrm{NEP}}\frac{\pi D^2}{4z^2}\beta^2\xi^2\int_0^\infty |f(\zeta)|^2\zeta\,d\zeta\right]^2 \tag{2-233}$$

Anticipating that $\xi_{\mathrm{opt}} = \infty$, we will normalize Eq. (2-233) to a single case, that of an untruncated system with $\omega_0 = a$. This yields

$$\mathrm{SNR_{DL}} = \frac{\mathrm{SNR_{TL}}(\xi)}{\mathrm{SNR_{UL}}(\xi = 1)} \tag{2-234}$$

where $\mathrm{SNR_{UL}}(\xi = 1)$ is the infinite aperture limit of Eq. (2-232) evaluated at $\xi = 1$.

Numerical integration of Eq. (2-233) is shown in Figure 2-34 as a function of $\xi = a/\omega_0$ for the case of a collimated transmit beam ($\alpha = 0$) and a far field range ($\beta_z = 0.1$). It can be seen that the optimum beam radius occurs at infinity; hence there is no penalty for a large aperture-to-beam ratio. This is consistent with the notion that the smaller the beam size, the less the truncation, such that all the energy transmitted reaches the target. However, introducing a central obscuration into the aperture will result in a finite value for ξ_{opt}. The reasons for this are the same as those in the related case of optimum power transmission of a truncated, obscured Gaussian beam discussed in Section 2.4.1 (cf. Figure 2-18).

The case of a small target located on the axis of the beam follows in the same manner as in the coherent detection case. Equation (2-234) becomes

$$\mathrm{SNR_{TS}} = \left[\frac{4\rho(\pi)P_T}{\mathrm{NEP}}\frac{\pi D^2}{4z^2}\beta^2\xi^2|f(0)|^2\int_0^{r_t/a}\zeta\,d\zeta\right]^2 \tag{2-235}$$

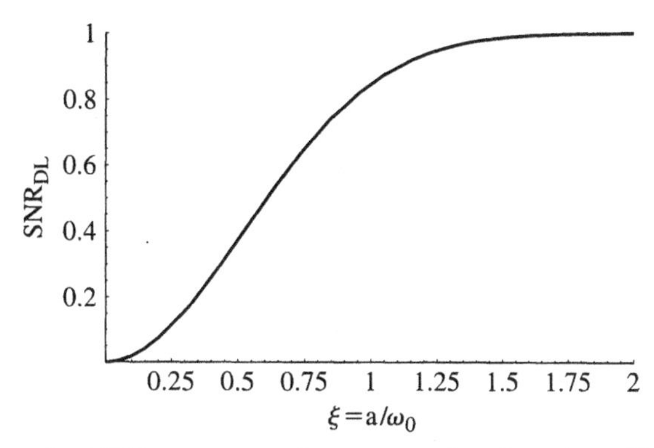

Figure 2-34. Normalized SNR versus truncation parameter for a $\xi = a/\omega_0$ collimated beam direct detection system operating against a diffuse target that is large compared to the beam. The optimum truncation ratio occurs at $\xi_{\mathrm{opt}} = \infty$. Normalization is relative to the SNR of an untruncated system with a beam radius equal to the aperture radius ($\xi_0 = 1$).

where $f(\zeta)$ is assumed constant in the vicinity of $\zeta = 0$. Thus

$$\text{SNR}_{\text{TS}} = \left[\frac{2}{\pi} \frac{\rho(\pi)}{\text{NEP}} \frac{\pi D^2}{4z^2} u_T^2 \frac{k^2}{z^2} |f(0)|^2 A_{tn}^2 \right]^2 \tag{2-236}$$

where $A_{tn} = r_t^2/a^2$ is the normalized area of the target. The normalized small-target direct-detection signal-to-noise ratio then becomes

$$\text{SNR}_{\text{DS}} = \frac{\text{SNR}_{\text{TS}}(\xi)}{\text{SNR}_{\text{US}}(\xi = \xi_0)} \tag{2-237}$$

Equation (2-237) is numerically integrated and shown in Figure 2-35. Once again an optimum value occurs at $\xi_{\text{opt}} = 1.12$, so that we let $\xi_0 = 1.12$, and $\xi_0 = 1$ in Eq. (2-237) as was done for the truncated coherent system. Note that $\xi_{\text{opt}} = 1.12$ is again in agreement with Figure 2-17 for the case of optimized on-axis intensity of a truncated Gaussian beam, which could have been anticipated. Note also that performance degradations of -2.9 dB and -3.8 dB occur for cases $\xi_0 = 1.12$ and $\xi_0 = 1$, respectively, due to the loss of power at the target.

The optimization procedure can be taken a step further by considering the detector size relative to the received spot size. Signal optimization in a monostatic system usually involves the selection of an aperture size that provides the desired detection capability or speckle averaging or both. The detector size is then chosen to optimally match the received beam spot size at the focal plane of the detector optics. The latter procedure limits the background radiation reaching the detector to the solid angle subtended by the transmitted beam and therefore maximizes the signal-to-background noise ratio. Clearly, a detector that is larger than the received spot size potentially introduces additional noise, while

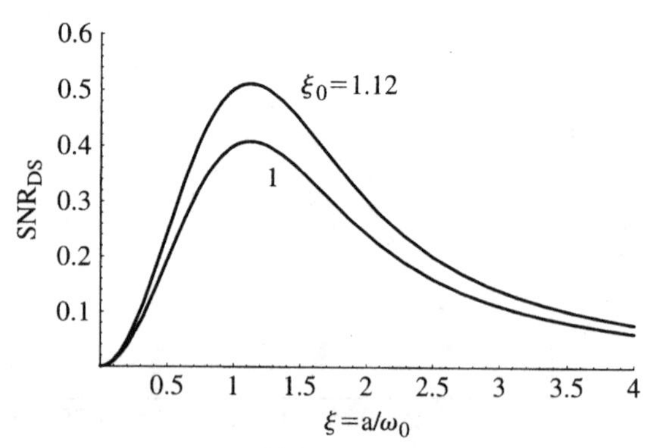

Figure 2-35. SNR versus truncation parameter $\xi = a/\omega_0$ for a collimated direct detection system operating against a diffuse target that is small compared to the beam. The optimum truncation ratio occurs at $\xi_{\text{opt}} = 1.12$. Normalization is relative to the SNR of an untruncated system with a beam radius equal to the optimally truncated beam radius ($\xi_o = 1.12$) and the aperture radius ($\xi_o = 1$), respectively.

a detector that is smaller truncates the received beam and sacrifices signal. The use of low-cost non-diffraction-limited optics can further aggravate the problem by increasing the received spot size and therefore the detector size.

A simple example highlights the detector optimization trade-offs noted above. Consider the case of background-limited noise given by Eq. (2-223) with $P_b \gg P_r$, where P_r and P_b are the signal and background powers that actually reach the detector. We have

$$\mathrm{SNR}_D = \frac{\eta_q P_r^2}{2h\nu B P_b} \tag{2-238}$$

For simplicity of argument, let us assume that the signal spot on the detector has a Gaussian-like beam profile. Then from Eq. (2-115) with $\varepsilon = 0$, P_r is related to the total received power P_R by

$$P_r = P_R(1 - e^{-2r_d^2/\omega_0^2}) \tag{2-239}$$

where ω_0 is the beam spot radius and r_d is the detector radius.

Consider first the case where the receiver field-of-view is much larger than the transmitted beam divergence such that the background radiation is essentially uniform over the detector surface. Equation (2-237) can then be written

$$\begin{aligned}\mathrm{SNR_D} &= \frac{\eta_q P_R^2}{2h\nu B I_b A_d}(1 - e^{-2r_d^2/\omega_0^2})^2 \\ &= \frac{\eta_q P_R^2}{2h\nu B I_b A_s}\frac{1}{\xi_d^2}(1 - e^{-2\xi_d^2})^2\end{aligned} \tag{2-240}$$

where I_b is the uniform background intensity at the detector, $A_d = \pi r_d^2$ the detector area, $A_s = \pi\omega_0^2$ the area of the spot at the detector plane, and $\xi_d = r_d/\omega_0$ the detector truncation parameter. The functional dependence of Eq. (2-240) on ξ_d can easily be investigated by normalizing to the multiplicative factor $\eta_q P_R^2/2h\nu B I_b A_s$ and denoting the resultant ratio as $\mathrm{SNR_{D}}_d$. $\mathrm{SNR_{D}}_d$ is plotted in Figure 2-36, where it can be seen that an optimum ratio of detector-to-beam radius occurs at $\xi_{d\mathrm{opt}} \approx 0.79$, at which point a loss of approximately 0.81 or -0.9 dB results relative to a detector of radius ω_0. Similar results would be obtained for other noise sources that are dependent on the detector area, such as dark current.

In the case of matched transmit and receive fields of view, the background radiation can be assumed to have the same spatial distribution as the signal, so that Eq. (2-239) becomes

$$\mathrm{SNR_D} = \frac{\eta_q P_R^2}{2h\nu B P_{0b}}(1 - e^{-2\xi_d^2}) \tag{2-241}$$

where $P_b = P_{0b}(1 - e^{-2\xi_d^2})$ and P_{0b} is the untruncated background power. Note that in this case the optimum ξ_d is at infinity. Hence in this idealized case the

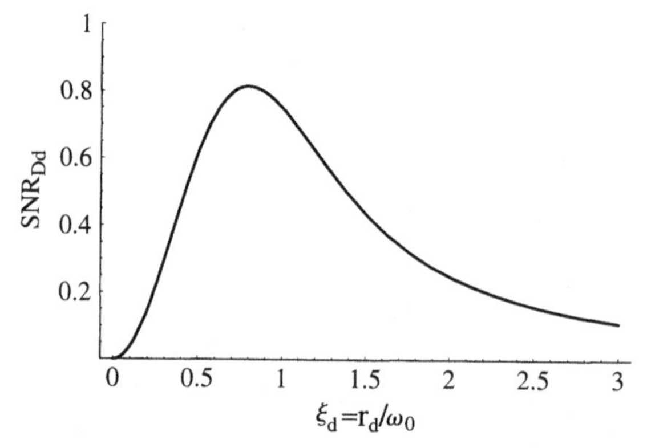

Figure 2-36. Normalized SNR versus detector truncation parameter $\xi_d = r_d/\omega_0$ for a direct detection receiver with a finite detector size. Here r_d is the detector radius and ω_0 is the assumed spot radius at the detector. The optimum detector size occurs for $\xi_{opt} = 0.79$ with an associated loss of 0.81 or -0.9 dB relative to a detector of radius ω_0.

detector can be made much larger than the spot size without any penalty in sensitivity. It should be remembered, however, that in real systems there are typically many more contributions to the receiver NEP than have been discussed here and that each spatial distribution must be included in the analysis to properly determine the optimum detector size.

2.9. RECEIVER AND CLUTTER NOISE

Earlier we discussed noise processes arising from sources external to the receiver, specifically those generated by randomizing effects in the received signal due to target roughness and atmospheric turbulence. There are, however, additional noise processes that must be taken into account if we are to fully characterize receiver performance. The primary noise contributors in an optical receiver can be categorized into three generic processes: thermal or Johnson noise associated with resistive circuit losses, shot noise associated with photon and photoelectron fluctuations, and excess noise associated with multiplication processes in avalanche photodetectors. In the first two sections we develop general expressions for the power spectral densities of thermal and shot noise, followed by a discussion of the concept of excess noise. These results are then used to define the *noise equivalent power* (NEP) of an optical receiver in terms of the power spectral densities of the specific noise sources common to such systems.

2.9.1. Thermal Noise

Johnson noise was first explained by Nyquist[27] using a variation on Planck's derivation of the blackbody radiation law applied to a one-dimensional ideal resistive circuit. By *ideal* we mean that the resistive elements are noninductive

Figure 2-37. Resistive transmission line used to describe thermal noise according to Nyquist's theory.

and noncapacitive, which is not a restriction to the generality of the problem, since power dissipation occurs only in resistive elements. Referring to Figure 2-37, the instantaneous voltage drops across resistors R_1 and R_2 due to noise voltages v_{n1} and v_{n2} generated in R_2 and R_1, respectively, are given by

$$\begin{aligned} V_1 &= v_{n2}\frac{R_1}{R_1 + R_2} \\ V_2 &= v_{n1}\frac{R_2}{R_1 + R_2} \end{aligned} \tag{2-242}$$

Thus the average power generated in R_1 and R_2 become

$$\begin{aligned} P_{n1} &= \frac{V_{n1}^2}{R_1} = \langle v_{n2}^2 \rangle \frac{R_1}{(R_1 + R_2)^2} \\ P_{n2} &= \frac{V_{n2}^2}{R_1} = \langle v_{n1}^2 \rangle \frac{R_2}{(R_1 + R_2)^2} \end{aligned} \tag{2-243}$$

where $\langle v_{n1}^2 \rangle$ and $\langle v_{n2}^2 \rangle$ are the mean-squared noise voltages generated by R_1 and R_2, respectively. In thermal equilibrium $P_{n1} = P_{n2}$, so that

$$\langle v_{n2}^2 \rangle R_1 = \langle v_{n1}^2 \rangle R_2 \tag{2-244}$$

As a further simplification, let $R_1 = R_2 = R$. The corresponding resistive impedances are then matched and power dissipation is equal in the two resistors.

At this point it is useful to view the resistive circuit as a transmission line of length L. The number of electromagnetic modes that can be supported by a transmission line of length L in a bandwidth Δf is

$$\Delta n = \frac{2L\Delta f}{v} \tag{2-245}$$

where v is the velocity of propagation. From Planck's radiation law, the number of photons per mode of vibration is given by

$$\frac{1}{e^{h\nu/k_B T} - 1} \approx \frac{k_B T}{h\nu} \tag{2-246}$$

where $k_B = 1.38 \times 10^{-23}$ J/K is Boltzmann's constant, T is the temperature in kelvin, $h = 6.62 \times 10^{-34}$ J-s is Planck's constant, and $h\nu$ is the energy of a single photon. It is easy to show that the approximation is quite valid for room temperatures and frequencies normally associated with receiver circuits. Thus the number of photons is given by

$$\Delta n \frac{kT}{h\nu} = \frac{2L}{v} \frac{kT}{h\nu} \tag{2-247}$$

and the total energy in the bandwidth Δf by

$$\Delta n \frac{k_B T}{h\nu} h\nu = \frac{2L}{v} k_B T \Delta f \tag{2-248}$$

Equation (2-248) represents the two-way flow of energy into the two resistors. Thus the average noise power flowing into one of the resistors is half of this, and the noise power per unit length per second is

$$P_n = \frac{1}{2}\left(\frac{v}{L}\right)\frac{2L}{v} k_B T \Delta f = k_B T \Delta f \tag{2-249}$$

It therefore follows that with $R_1 = R_2 = R$ in Eqs. (2-243),

$$P_n = \frac{v_n^2}{4R} = k_B T \Delta f \tag{2-250}$$

The mean-squared noise current in the bandwidth Δf is therefore

$$\langle i_n^2 \rangle = \frac{4k_B T \Delta f}{R} \tag{2-251}$$

and the thermal noise current

$$i_n = \left(\frac{4k_B T \Delta f}{R}\right)^{1/2} \tag{2-252}$$

From Eq. (2-250) and the definition of power spectral density given in Section 1.2.9, we can conclude that the one-sided power spectral density of thermal noise is simply by $S^+ = k_B T$, which is independent of frequency and therefore constitutes a *white noise process*.

2.9.2. Shot Noise

Shot noise is a consequence of the random occurrence of events that follow Poisson statistics. As will be seen, there are many noise sources that exhibit shot noise in optical receivers, including photon fluctuations, dark current, and photoemmissive processes. Rice[28] was first to develop a comprehensive treatment of the subject using Campbell's theorem, from which we will provide a simplified derivation for the power spectral density. Campbell's theorem was originally applied by Rice to the behavior of Poisson-distributed electrons in electron tubes but may be applied to any Poisson-distributed process.

Consider then a current that is measured at the output of a circuit that has at its input a stream of electrons obeying Poisson statistics. Assuming that the effects add linearly at the output, we can write for the instantaneous current at the output

$$I(t) = \sum_{k=-\infty}^{\infty} I(t - t_k) \tag{2-253}$$

where t_k corresponds to the arrival time of the kth electron. Campbell's theorem then states that the mean and mean-squared noise currents are given by

$$\langle i_n(t) \rangle = \lambda \int_{-\infty}^{\infty} I(t)\, dt \tag{2-254}$$

$$\langle i_n^2(t) \rangle = \lambda \int_{-\infty}^{\infty} I^2(t)\, dt \tag{2-255}$$

respectively, where $\lambda = \langle I \rangle / q$ is the mean rate of arrival of electrons in the measurement interval $\tau, q = 1.6 \times 10^{-19}$ C is the unit of electronic charge, and i_n is the noise current. It can be shown (cf. Chapter 4) that the variance equals the mean for Poisson statistics, which is implicit in the form of Eq. (2-255). In the case of a steady stream of Poisson-distributed electrons (or photons), the current at any instant is defined as the average current $\langle I \rangle$ during the measurement interval τ, as shown in Figure 2-38. Based on this model, it is reasonable to assume that $I(t) = \langle I \rangle = q/\tau$, which is a constant. Hence, limiting the integration to the measurement interval τ in Eqs. (2-254) and (2-255) results in $\langle i_n(t) \rangle = \lambda q = I$ and $\langle i_n^2(t) \rangle = \lambda q^2/\tau = qI/\tau$, respectively, where $I \equiv \langle I \rangle$. Applying Parseval's theorem to the integral in Eq. (2-255) results in

$$\langle i_n^2(t) \rangle = 2\lambda \int_0^{\infty} |F(\omega)|^2 df \tag{2-256}$$

where $F(\omega)$ is an even function. Evaluating $F(\omega)$ yields

$$\begin{aligned} F(\omega) &= \int_0^{\tau} I(t) e^{-i\omega t} dt = \frac{q}{\tau} \int_0^{\tau} e^{-i\omega t} dt \\ &= \frac{iq}{\omega \tau}(e^{-i\omega\tau} - 1) \end{aligned} \tag{2-257}$$

Thus

$$|F(\omega)|^2 = \frac{2q^2(1 - \cos \omega\tau)}{\omega^2 \tau^2} \tag{2-258}$$

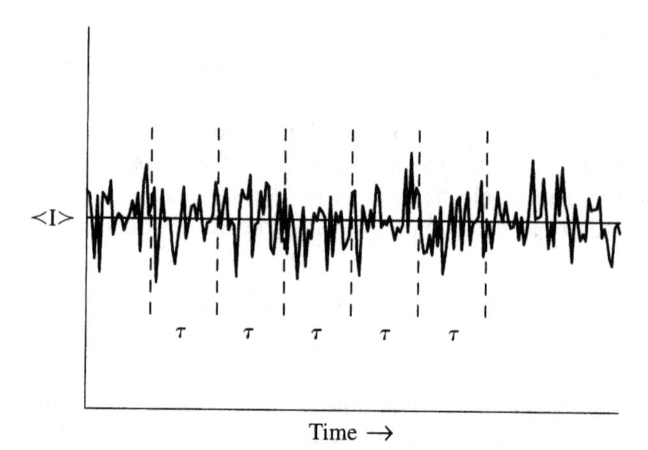

Figure 2-38. Measurement intervals τ in a stream of Poisson-distributed electrons of mean current $\langle I \rangle$.

For small measurement intervals, $\omega\tau \ll 1$, so that Eq. (2-258) reduces to $|F(\omega)|^2 \approx q^2$. Inserting this into Eq. (2-256) and integrating over the bandwidth Δf yields for the mean-squared current due to shot noise,

$$\langle i_n^2 \rangle = 2qI\,\Delta f \tag{2-259}$$

The average power in the bandwidth Δf is therefore

$$P_n = 2qIR\,\Delta f \tag{2-260}$$

Once again the one-sided power spectral density can be identified as a white noise process, in this case

$$S^+ = 2qIR \tag{2-261}$$

The corresponding two-sided spectrum is given by $S = qIR$ such that the average power remains $P_n = qIR \times 2\Delta f = 2qIR\,\Delta f$.

2.9.3. Dark Current and Excess Noise

Detectors that employ internal avalanche gain mechanisms to boost the signal above the thermal noise of amplifier stages of the receiver exhibit what is commonly referred to as *excess noise*. This class of detectors includes, among others, the *photomultiplier tube*, the conventional *avalanche photodiode* (APD), and the *quantum well* APD. We focus on the conventional APD devices, which includes semiconductor materials such as silicon and germanium, primarily because of their small size, low cost, and high performance, which makes them attractive for many laser applications. Excess noise in avalanche photodiodes is due to the multiplication process in the high-field region of the device, where each primary electron and hole can generate an additional electron and hole through impact ionization of bound electrons. These additional carriers can then create still additional carriers in a cascading or avalanche process. Here gain G is defined as the ratio of the mean multiplied current to the mean unmultiplied or primary photocurrent and can be as high as 200 in some semiconductor devices.

Dark current can also contribute to excess noise. There are two types of dark current that can generate excess noise, *surface dark current* and *bulk dark current.* Both types generate primary carriers that are Poisson distributed and therefore add yet another noise process that competes with the signal even without multiplication. Surface dark current, designated as I_{ds}, is usually negligible in well-designed devices that employ guard rings to bleed off the surface currents. On the other hand, bulk dark current, designated as I_{db}, may be generated throughout the depletion region and therefore can experience a wide spectrum of gains, thereby adding to the excess noise. As discussed by Hakim et al.,[29] under certain conditions dark excess noise can be larger than the photoinjected excess noise.

If all primary carriers were to be multiplied equally in an avalanche device, the mean-square current gain $\langle G^2 \rangle$ would equal the mean gain $G \equiv \langle G \rangle$ and the excess noise, defined as $F = \langle G^2 \rangle / G$, would be equal to 1. This is the case for unity-gain devices such as the p-i-n photodiode. However, due to the statistical nature of the multiplication process, F is always greater than 1 in avalanche devices. To be able to calculate F, it would be desirable to be able to express F in terms of measurable parameters of the material. This is achieved by relating the excess noise to the hole and electron *ionization rates* of the semiconductor material. Since this derivation is quite complex,[30] only the result is given here. Thus

$$F_e = \frac{k_2 - k_1^2}{1 - k_2} G_e + 2\left[1 - \frac{k_1(1 - k_1)}{1 - k_2}\right] - \frac{(1 - k_1)^2}{G_e(1 - k_2)} \tag{2-262}$$

and

$$F_h = \frac{k_2 - k_1^2}{k_1^2(1 - k_2)} G_h + 2\left[1 - \frac{k_2(1 - k_1)}{k_1^2(1 - k_2)}\right] + \frac{(1 - k_1)^2 k_2}{G_h(1 - k_2)k_1^2} \tag{2-263}$$

where k_1 and k_2 are weighted hole-to-electron ionization ratios given by

$$k_1 = \frac{\displaystyle\int_{\Delta x} \beta(x) g(x)\, dx}{\displaystyle\int_{\Delta x} \alpha(x) g(x)\, dx} \tag{2-264}$$

and

$$k_2 = \frac{\displaystyle\int_{\Delta x} \beta(x) g^2(x)\, dx}{\displaystyle\int_{\Delta x} \alpha(x) g^2(x)\, dx} \tag{2-265}$$

Here α and β are spatially dependent *ionization coefficients* for the medium defined in terms of the probabilities αdx and βdx that the electron or hole will experience an ionizing collision between x and $x + dx$. The integration is performed over the high-field region Δx, where $g(x)$ is defined[31] as

$$g(x) = e^{-\int_{\Delta x} [\alpha(x)-\beta(x)]\,dx} \qquad (2\text{-}266)$$

If one makes the simplifying assumption that k_1 and k_2 are constant and equal, which is generally the case for most practical devices, then Eqs. (2-262) and (2-263) can be reduced to

$$F_e = k_{\text{eff}} G_e + \left(\frac{2-1}{G_e}\right)(1 - k_{\text{eff}}) \qquad (2\text{-}267)$$

and

$$F_h = k'_{\text{eff}} G_h + \left(\frac{2-1}{G_h}\right)(1 - k'_{\text{eff}}) \qquad (2\text{-}268)$$

where k_{eff} and k'_{eff} are the electron and hole *effective ionization rate ratios*, respectively, given by

$$\begin{aligned} k_{\text{eff}} &= \frac{k_2 - k_1^2}{1 - k_2} \approx k_2 \\ k'_{\text{eff}} &= \frac{k_{\text{eff}}}{k_1^2} \approx \frac{k_2}{k_1^2} \end{aligned} \qquad (2\text{-}269)$$

Equation (2-267) is plotted in Figure 2-39 as a function of electron gain G_e and effective ionization ratio k_{eff}. It can be seen that low values of k_{eff} are desirable if excess noise is to be minimized. Photosensitive materials such as silicon can be fabricated with effective ionization ratios as low as $k_{\text{eff}} \approx 0.01$, which, as can be seen from the curves, results in an excess noise factor of $F_e \approx 3$ for a gain of 100. From Eqs. (2-264) and (2-265) it can be seen that if the ionization coefficients α and β are equal, the excess noise is at its maximum, whereas if β decreases, the excess noise also decreases such that in the limit of $\beta = 0$, F_e is at a minimum of $F_e \approx 2$.

The primary photogenerated carriers in an APD device obey Poisson statistics as in any photodetector. However, when multiplied in the avalanche process, the probability density function is modified significantly. The resulting distribution is discussed in Chapter 4. On the other hand, the generalization of the power spectral density derived in the preceding section for primary photoemmissive processes assumes a particularly simple form in the case of multiplied electrons. Specifically, Eq. (2-261) becomes

$$S^+ = 2qIR\langle G^2\rangle = 2qIRG^2F \qquad (2\text{-}270)$$

where $I \equiv \langle I \rangle$ and the last form follows from the definition of F given above. A formal derivation of Eq. (2-270) may be found in the works of Personick.[32]

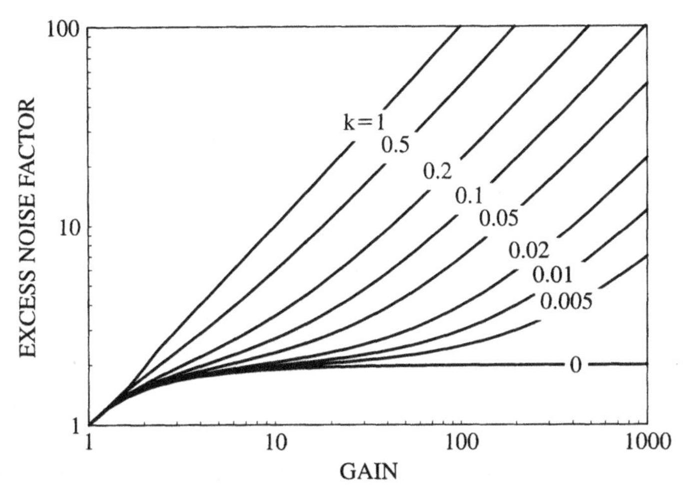

Figure 2-39. Excess noise factor versus avalanche gain for avalanche photodiode detectors. (From P. Webb, R. J. McIntyre, and J. Conradi, *RCA Rev.*, Vol. 35, June 1974.)

2.9.4. Sources of Clutter

As in radar systems, there are a variety of *clutter sources* that an optical system must contend with. The two most common are background radiation and backscattered radiation. Background radiation may consist of solar radiation scattered from either background objects or the target, or it can originate as thermal emission from the objects themselves. These effects are usually expressed in terms of radiometric quantities depending on whether the viewed source is treated as a reflector or as an emitter and are listed in various handbooks on optical and infrared phenomena.[33]

To minimize background radiation in direct detection systems, it is usually preferable to insert a narrowband spectral filter of width $\Delta\lambda$ in front of the detector, as shown in Figure 2-40. The background power reaching the detector may then be determined from

$$P_b = E_\lambda \rho(\pi) A\Omega\, \Delta\lambda \eta_f \eta_o \eta_a \tag{2-271}$$

or

$$P_b = N_\lambda A\Omega\, \Delta\lambda \eta_f \eta_o \eta_a \tag{2-272}$$

where E_λ is the *spectral irradiance* at the target (W m^{-2} μm^{-1}), defined as the radiant power per unit area per unit wavelength incident upon a surface, N_λ is the *spectral radiance* at the receiving aperture (W m^{-2} μm^{-1} sr^{-1}), defined as the radiant power per unit area per unit solid angle per unit wavelength incident upon a surface, $\rho(\pi)$ is the diffuse target reflectivity (sr^{-1}), A is the receiver aperture area (m^2), Ω is the receiver acceptance solid angle (rad^2), $\Delta\lambda$ is the *spectral filter bandpass* (μm), η_f is the target area fill factor in the receiver

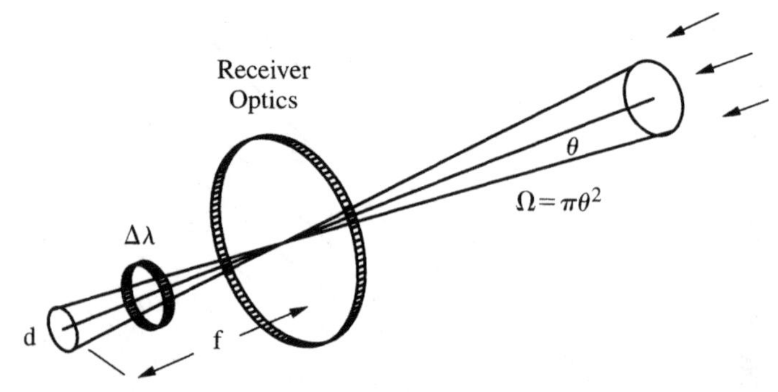

Figure 2-40. Geometry showing background irradiance accepted within the solid angle of a receiver.

field of view, η_o is the receiver optical efficiency, and η_a is the atmospheric transmission. The acceptance angle of an optical receiver is given by $\Omega = \pi\theta^2$, where θ is determined by the detector radius r_d and the focal length f of the optics through the relationship $\theta = r_d/f$. The reflectance of a diffuse Lambertian surface is given by $\rho(\pi) = \rho\cos\phi_s/\pi$, where ρ is the power reflectivity and ϕ_s is the scattering angle relative to the surface normal (cf. Chapter 3).

In the case of coherent detection receivers the incoherent nature of background radiation, together with the narrow IF bandwidth typical of such systems, usually prevents any background radiation from competing with the signal. The exception is, of course, the sun itself, which can easily saturate the detector if viewed directly.

Laser radiation backscattered from the intervening atmosphere may also be considered a source of clutter for some systems, while in others it may be considered the signal itself (e.g., in wind sensing and DIAL systems) (cf. Section 1.1). While background clutter can be reduced with narrowband spectral filters, the spectral width of backscattered radiation is too narrow to be optically filtered. Since most direct detection systems employ pulsed waveforms, the clutter power due to backscatter from suspended particles along the propagation path is given by

$$P_c = P_T A\beta(\pi)\frac{ct_p}{2z^2}g(\Omega)\eta_o\eta_a \qquad g(\Omega) = \begin{cases} 1 & \text{for } \Omega \geq \Omega_0 \\ \dfrac{\pi\theta^2}{\Omega_0} & \text{for } \Omega < \Omega_0 \end{cases} \tag{2-273}$$

where t_p is the pulse width (s), $\beta(\pi)$ is the atmospheric backscatter coefficient ($\text{m}^{-1}\ \text{sr}^{-1}$), z is the range (m), Ω is the solid-angle field of view of the receiver, Ω_0 is the solid angle subtended by the transmitted beam, and $c = 3 \times 10^8$ m/s. Note that $ct_p/2$ is the range resolution ΔR of the waveform such that an effective reflectivity $\rho_e(\pi)$ for the medium can be defined as $\rho_e(\pi) = \beta(\pi)ct_p/2 = \beta(\pi)\Delta R$.

The dependence of Eq. (2-274) on the pulse length implies that considerable clutter rejection can be achieved by using short pulses. (On the other hand, wind sensing or DIAL systems may want a long pulse.) Note also that in the case of common aperture systems, a receiver field of view that is larger than the transmitted beam divergence results in no further increase in backscattered power beyond that contained within Ω_0. This is not necessarily the case for bistatic or dual-aperture systems.

At this point it is worthwhile to mention briefly the fluctuation statistics associated with these clutter sources. As discussed in detail in Chapter 4, solar background radiation is incoherent, with the intensity obeying the same statistics as that of specularly reflected laser radiation. On the other hand, backscattered radiation is generated from coherent laser radiation and therefore obeys the same statistics as that of diffusely reflected laser radiation from a rough surface. Since Eq. (2-273) has a z^{-2} dependence and, as we have seen earlier in this chapter, speckle correlation size depends on range in accordance with $\rho_c \approx \lambda z/d_s$, improved clutter or speckle averaging might be expected at the near-in ranges. However, since the illuminating beam diameter also decreases with range, the speckle count at the receiver due to scatter in the far field of the transmitter remains relatively constant. On the other hand, the beam diameter stays relatively constant with range in the near field so that improved clutter averaging can occur in this region.

2.10. POWER SIGNAL-TO-NOISE-RATIO

In Section 2.8.1 it was shown that the electrical *power signal-to-noise ratio* of a laser receiver is the ratio of the mean-squared signal current to the mean-squared noise current. For direct detection receivers it was shown that the mean-squared signal current was proportional to the square of the product of the received power and the responsivity of the detector. Since the noise sources are uncorrelated, the total mean-squared noise current is given by the sum of the mean-squared noise currents, so that the power signal-to-noise ratio may be written

$$\mathrm{SNR} = \frac{\langle i_s^2 \rangle}{\langle i_n^2 \rangle} = \frac{\langle i_s^2 \rangle}{\langle i_p^2 \rangle + \langle i_b^2 \rangle + \langle i_c^2 \rangle + \langle i_{db}^2 \rangle + \langle i_{\mathrm{ds}}^2 \rangle + \langle i_t^2 \rangle} \tag{2-274}$$

where from left to right the various noise currents represent photon shot noise, background noise, clutter or backscatter noise, bulk dark current noise, surface dark current noise, and thermal or Johnson noise. Using the results of the preceding three sections in Eq. (2-274), the power signal-to-noise ratio of a direct detection receiver may be written

$$\mathrm{SNR} = \frac{G^2 \Re^2 P_R^2}{2q\Re(P_R + P_b + P_c)G^2 F B + 2q(I_{db}G^2 F + I_{\mathrm{ds}})B + 4kT_n B/R_L} \tag{2-275}$$

where $\langle i_s^2 \rangle = G^2\Re^2 P_R^2$ is the mean-squared signal current due to received power P_R, $B \equiv \Delta f$ is the matched bandwidth of the receiver, $\Re = \eta_q q/h\nu$ = detector responsivity (A/W), and R_L is the load resistor at the input to the amplifier. $T_n = T_d + (F_a - 1)T_a$ is the effective noise temperature of the detector/amplifier combination, where T_d is the temperature associated with the detector resistance, F_a the noise figure of the amplifier, and T_a the temperature of the amplifier load resistance. (It should be noted that the expression for amplifier noise is very simplistic and usually requires modification when transimpedance or other specialized amplifiers are employed.[34])

Dividing Eq. (2-275) by $G^2\Re^2$ results in

$$\mathrm{SNR} = \frac{P_R^2}{\left(\begin{array}{c}(2q/\Re)(P_R + P_b + P_c)F(G)B + (2q/\Re^2)(I_{db}F(G) \\ +I_{\mathrm{ds}}/G^2)B + 4kT_n B/\Re^2 G^2 R_L\end{array}\right)} \tag{2-276}$$

where the functional dependence of F on G is given by Eq. (2-267). In this form, the overall functional dependence of the signal-to-noise ratio on G can be seen to result in reduced values of SNR at low and high values of G, suggesting that an optimum value of G exists where the signal-to-noise ratio is a maximum. Depending on specific values of the noise contributions, the optimum value is usually found to occur in the range $10 < G < 100$.

Comparing Eq. (2-276) with the form of Eq. (2-221), the SNR can be expressed in terms of the received power and a noise equivalent (optical) power as

$$\mathrm{SNR} = \left(\frac{P_R}{\mathrm{NEP}}\right)^2 \tag{2-277}$$

where P_R is as described in Chapter 2. The NEP can be expressed in terms of *noise spectral densities* having units of watts per root hertz bandwidth:

$$\mathrm{NEP} = \mathrm{NEP}^*\sqrt{B} = \left(\sum_i \mathrm{nep}_i^{*2}\right)^{1/2} \sqrt{B} \tag{2-278}$$

where $\mathrm{NEP}^*(\mathrm{W}/\sqrt{\mathrm{Hz}})$ is the overall spectral density of the receiver and the $\mathrm{nep}_i^*(\mathrm{W}/\sqrt{\mathrm{Hz}})$ are the individual spectral densities of the various noise contributions. From Eq. (2-276) the latter are

$$\mathrm{nep}_s^* = \left[\frac{2qFP_R}{\Re}\right]^{1/2} = \text{shot noise due to the signal} \tag{2-279}$$

$$\mathrm{nep}_b^* = \left[\frac{2qFP_b}{\Re}\right]^{1/2} = \text{shot noise due to the background} \tag{2-280}$$

$$\mathrm{nep}_c^* = \left[\frac{2qFP_c}{\Re}\right]^{1/2} = \text{shot noise due to backscatter} \tag{2-281}$$

$$\text{nep}_d^* = \left[\frac{2q(I_{ds} + I_{db}G^2F)}{G^2\Re^2}\right]^{1/2} = \text{shot noise due to dark current} \quad (2\text{-}282)$$

$$\text{nep}_t^* = \left[\frac{4k_BT_n}{G^2\Re^2R_L}\right]^{1/2} = \text{thermal noise} \quad (2\text{-}283)$$

Inspection of Eq. (2-276) shows that for low receiver and clutter noise or high signal levels, the receiver can approach the *shot noise limit*, $\langle i_n^2\rangle \to 2qP_RFB/\Re$. In that case, Eq. (2-276) approaches the *quantum limit* given by

$$\text{SNR} = \frac{P_R^2}{2qP_RFB/\Re} = \frac{\eta_q P_R}{2h\nu FB} \quad (2\text{-}284)$$

This is very similar to the signal-to-noise ratio of a coherent receiver, as shown below, a consequence of the fact that a strong signal acts as its own local oscillator. In comparison, the signal-to-noise ratio of a coherent receiver is obtained from Eq. (2-274) by letting $\langle i_s^2\rangle = 2\Re^2G^2P_{\text{LO}}P_R$ and $\langle i_n^2\rangle = 2q\Re G^2P_{\text{LO}}FB$, which yields

$$\text{SNR} = \frac{\Re P_R}{qFB} = \frac{\eta_q P_R}{h\nu FB} \quad (2\text{-}285)$$

Note that Eq. (2-285) is twice that of Eq. (2-284). This is, of course, due to the factor of 2 that occurs in $\langle i_s^2\rangle = 2P_{\text{LO}}P_R$ because of the mixing process. In addition, for small signals or strong noise sources, direct detection receivers may not reach the shot noise limit and, consequently, may suffer additional performance penalties relative to the coherent receiver.

Equations (2-284) and (2-285) indicate that the presence of avalanche gain reduces the signal-to-noise ratio of both receiver types in proportion to the amount of excess noise that is generated. This is one of the reasons that avalanche detectors are considered nonoptimum for coherent detection. Another reason is that shot-noise-limited performance can be achieved via a strong local oscillator, thereby eliminating the need for a high-gain detector to overcome the thermal noise of the amplifiers.

Note that for $F = 1$, the last form of Eq. (2-285) can be written

$$\text{SNR} = \frac{P_R}{\text{NEP}} = \frac{\eta_q P_R}{h\nu B} \quad (2\text{-}286)$$

where $\text{NEP} = h\nu B/\eta_q$ is the noise equivalent power of a shot-noise-limited coherent detection receiver. Comparing Eq. (2-286) with Eq. (2-277) for the non-shot-noise-limited direct detection receiver, we see that for receivers employing matched filters, the signal-to-noise ratio of a coherent receiver is proportional to the received energy (i.e., $\text{SNR} \propto P_R/B = E_R$), whereas in a direct detection receiver it is proportional to an energy-peak power product (i.e., $\text{SNR} \propto P_R^2/B = E_RP_R$).

Finally, it should be remembered that the efficiency factor $\eta_p = \eta_o \eta_e \eta_a$ introduced in Section 2.7.1 to account for power losses internal and external to the system needs to be included in Eqs. (2-277) and (2-286). Recall that $\eta_o = \eta_t \eta_r$ represents optical losses internal to the system for both transmit and receive channels, and η_e accounts for any mismatch between the signal and local oscillator in the mixing process of a coherent receiver. In a real system, η_o and η_e must be obtained through precise measurements using calibrated targets and instruments. $\eta_a = \exp(-2\alpha z)$ accounts for atmospheric absorption and scattering, where $\alpha(m^{-1})$ is an *extinction coefficient* and z a one-way propagation path.

REFERENCES

[1]M. Born and E. Wolf, *Principles of Optics*, 3rd ed., Pergamon Press, Elmsford, NY, 1964, p. 375.

[2]A. E. Siegman and E. A. Sziklas, "Mode calculations in unstable resonators with flowing saturable gain, 1: Hermite–Gaussian expansion," *Appl. Opt.*, Vol. 13, No. 12, Dec. 1974, pp. 2775–2791.

[3]H. Kogelnik, "On the propagation of Gaussian beams of light through lenslike media including those with a loss or gain variation," *Appl. Opt.*, Vol. 4, No. 12, Oct. 1965, p. 1562.

[4]H. Kogelnik and T. Li, "Laser beams and resonators," *Appl. Opt.*, Vol. 5, No. 10, Oct. 1966, p. 1550.

[5]F. A. Jenkins and H. E. White, *Fundamentals of Optics*, 3rd ed., McGraw-Hill Book Company, New York, 1957, p. 64.

[6]G. D. Boyd and J. P. Gordon, "Confocal multimode resonator theory," *Bell Syst. Tech. J.*, Vol. 40, Mar. 1961, pp. 489–508.

[7]L. W. Casperson and M. S. Shekhani, "Mode properties of annular gain lasers," *Appl. Opt.*, Vol. 14, No. 11, Nov. 1975, pp. 2653–2661.

[8]G. Arfken, *Mathematical Methods for Physicists*, 2nd ed., Academic Press, San Diego, CA, 1970, p. 611.

[9]A. Yariv, *Quantum Electronics*, 3rd ed., John Wiley & Sons, New York, 1989.

[10]A. E. Siegman, *Lasers*, University Science Books, Mill Valley, CA, 1986.

[11]Ibid., ref. 8, p. 609.

[12]E. Titchmarsh, *Introduction to the Theory of Fourier Integrals*, 3rd ed., American Mathematical Society, Providence, RI, 1986, p. 81.

[13]A. E. Siegman, "New developments in laser resonators," SPIE/OE LSAE '90, Conference on Laser Resonators, Los Angeles, Jan. 1990.

[14]T. F. Johnston, "Beam propagation (M^2) measurement made easy as it gets: the four-cuts method," *Appl. Opt.*, Vol. 37, No. 21, July 20, 1998, p. 4840.

[15]Ibid., ref. 1, p. 246.

[16]G. Oluremi Olaofe, "Diffraction by Gaussian apertures," *J. Opt. Soc. Am.*, Vol. 60, No. 12, Dec. 1970, pp. 1654–1655.

[17]R. Barakat, "Solution of the Luneberg apodization problems," *J. Opt. Soc. Am.*, Vol. 52, No. 3, Mar. 1962, pp. 264–275.

[18]D. A. Holmes, P. V. Avizonis, and K. H. Wrolstad, "On-axis irradiance of a focused, apertured Gaussian beam," *Appl. Opt.*, Vol. 9, No. 9, Sept. 1970, pp. 2179–2180.

[19]J. W. Goodman, *Introduction to Fourier Optics*, McGraw Hill Book Company, New York, 1968.

[20]G. O. Reynolds, J. B. DeVelis, G. B. Parrent, Jr., and B. J. Thompson, *The New Physical Optics Notebook: Tutorials in Fourier Optics*, SPIE Optical Engineering Press, Bellingham, WA, 1989.

[21]A. E. Siegman, "The antenna properties of optical heterodyne detection receivers," *Proc. IEEE*, Vol. 54, 1966, p. 1350.

[22]A. H. M. Ross, "Optical heterodyne mixing efficiency invariance," *Proc. IEEE*, Oct. 1970, p. 1766.

[23]C. M. Sonnenschein and F. A. Horrigan, "Signal-to-noise relationships for coaxial systems that heterodyne backscatter from the atmosphere," *Appl. Opt.*, Vol. 10, No. 7, July 1971, p. 1600. This work was later refined by C. DiMarzio and K. Seeber in unpublished work.

[24]J. Y. Wang, "Detection efficiency of coherent optical radar," *Appl. Opt.*, Vol. 23, No. 19, Oct. 1, 1984, p. 3421.

[25]J. Y. Wang, "Optimum truncation of a lidar transmitted beam," *Appl. Opt.*, Vol. 27, No. 21, Nov. 1, 1988, p. 4470.

[26]This value differs from that given by Wang (ref. 25), due to a difference in the definitions of the e^{-1} beam radius. Wang's beam radius $\alpha = \omega_0/\sqrt{2}$ such that $r/\alpha = \sqrt{2}r/\omega_0 = \sqrt{2} \times 1.24 \approx 1.75$.

[27]H. Nyquist, "Thermal agitation of electric charge in conductors," *Phys. Rev.*, Vol. 32, July 1928, p. 110.

[28]S. O. Rice, "Mathematical analysis of random noise," *Bell Syst. Tech. J.*, Vol. 23, 1944, p. 282.

[29]N. Z. Hakim, B. E. A. Saleh, and M. C. Teich, "Generalized excess noise factor for avalanche photodiodes of arbitrary structure," *IEEE Trans. Electron Devices*, Vol. 17, No. 1, Mar. 1990, p. 599.

[30]P. P. Webb, R. J. McIntyre, and J. Conradi, "Properties of avalanche photodiodes," *RCA Rev.*, Vol. 35, June 1974, pp. 234–277.

[31]R. J. McIntyre, "Multiplication noise in uniform avalanche junctions," *IEEE Trans. Electron Devices*, Vol. ED-13, 1966, p. 164.

[32]S. D. Personick, "Receiver design for digital fiber optic communication systems, I," *Bell Syst. Tech. J.*, Vol. 52, 1973, p. 843.

[33]W. L. Wolfe and G. J. Zissis, Editors, *The Infrared Handbook*, 3rd ed., Environmental Research Institute of Michigan, Ann Arbor, MI, 1989.

[34]G. Keiser, *Optical Fiber Communications*, McGraw-Hill Book Company, New York, 1983, p. 302.

3

Random Processes in Beam Propagation

3.1. INTRODUCTION

The discussions in Chapter 2 considered the propagation of well-behaved deterministic beams. There are, however, other beam profiles that an optical system must deal with, especially in the receive channel. The most obvious case is that of scattered radiation from an object that is rough compared to a wavelength. This includes volumetric targets such as airborne aerosols, dust, or fog. As shown in this chapter, such objects produce radiation patterns that exhibit significant spatial granularity in the plane transverse to the direction of propagation. This granularity is commonly referred to as *target-induced speckle* or simply *speckle*. Another commonly encountered stochastic process that can randomize a transmitted beam profile is atmospheric turbulence. Once again the propagated beam exhibits a granularity that will be referred to as *turbulence-induced speckle*. In both cases, the laser receiver can experience deep fades in signal intensity that can seriously degrade the detection statistics. In addition, these phenomena are difficult to predict, since in the former case, they depend on the object's orientation and surface condition, and in the latter case, the weather. Thus a laser system must be designed for the worst-case conditions anticipated, while the phenomena themselves must be dealt with on a statistical basis.

An obvious mechanism for reducing both types of fading in direct-detection systems is to use a receiver aperture that is large compared to the correlation size of the pattern. This is a form of averaging commonly referred to as *speckle* or *aperture averaging*. Unfortunately in the case of coherent detection, aperture averaging of either type of pattern can seriously degrade the signal-to-noise ratio of the system. In addition, propagation of the correlation functions of the target- and turbulence-induced speckle are quite different, the former increasing in size with range and the latter decreasing in size with range. Thus, to properly quantify the effects of speckle averaging on system performance it will be useful to know the correlation properties of the field at the receiving aperture. As will be seen, optical coherence theory provides an ideal framework for accomplishing this, either in the spatial domain via ensemble averages or in the time domain via convolution integrals.

3.2. REVIEW OF OPTICAL COHERENCE THEORY

The interference of light beams requires for a full description a theoretical construct known as *optical coherence theory*.[1] The most general formulation of this theory must take into account the spectral width of the illuminating source since in the limit of *white light* (or *natural light* as it has historically been called), no interference is observed at all. In the following sections we dealing with quasimonochromatic light, light whose spectral width is small compared to its mean frequency. Laser light is well known to fall into this category and can even be considered fully monochromatic for many analytic computations. However, in the case of spatially randomized beams due to scattering or turbulent media, the spatial coherence properties can approach complete incoherence. This occurs despite the fact that the temporal coherence of the randomized field remains unchanged.

In the following section we briefly review the statistical descriptors used in coherence theory to describe quasimonochromatic light. It is not meant to be all-inclusive as far as topics are concerned but is only meant to review those concepts pertinent to the phenomenologies associated with optical detection theory. References are provided for the reader who wishes to explore the various topics in greater detail.

3.2.1. Coherence Properties of the Field

The instantaneous intensity of a radiation field has been defined in Chapter 2 as the quantity

$$I = U^*(\vec{z}, t)U(\vec{z}, t) = |U(\vec{z}, t)|^2 \tag{3-1}$$

where $U(\vec{z}, t)$ is the deterministic complex scalar field amplitude at the point defined by $\vec{z}$ and time t. If $U(\vec{z}, t)$ is a random function of time, one must resort to statistical moments to describe the behavior of the field. Thus, in the case of the intensity, we have for the first two moments

$$\begin{aligned} \langle I \rangle &= \langle |U(\vec{z}, t)|^2 \rangle \\ \sigma_I^2 &= \langle I(\vec{z}, t)^2 \rangle - \langle I(\vec{z}, t) \rangle^2 \end{aligned} \tag{3-2}$$

If U corresponds to a spatially and temporally stationary field, $\langle I \rangle$ and σ_I^2 are constants. These expressions constitute, in part, the *first-order statistics* of intensity. In this and later chapters we will frequently be concerned with the *second-order statistics* of the field, which are required to describe the coherence properties of scattered and scintillating fields. A fundamental descriptor in this process is the *cross-correlation function* (cf. Chapter 1), which when applied to the complex field amplitudes, is known as the *mutual coherence function*. Since we are dealing with second-order statistics, it is important that the fields be at least wide-sense stationary (i.e., they should possess the property that their mean values be constant and their autocorrelation functions be dependent only on space and time differences). With this restriction, the mutual coherence function Γ_{12} is then defined as the mean value of the product of the fields at a point Q in the observation plane located at a distance $\vec{z}_1$ and $\vec{z}_2$ from two points P_1 and P_2 in the source plane. The geometry is shown in Figure 3-1. Since the propagation times from P_1 and P_2 to Q are in general different and given by $t_1 = |\vec{z}_1|/c$ and $t_2 = |\vec{z}_2|/c$, we have

$$\Gamma_{12}(\tau) = \langle U^*(\vec{z}_1, t_1)U(\vec{z}_2, t_1 + \tau)\rangle \tag{3-3}$$

where $\tau = t_2 - t_1 = |\vec{z}_2 - \vec{z}_1|/c$. Here the angle brackets may represent ensemble or time averages, depending on the process being described, and if it is ergodic, they are equal and interchangeable.

Experimental verification of Eq. (3-3) can be accomplished in a variety of ways, the most common being that of Young's interference experiment. Young's experiment involves two overlapping beams in an observation plane located a distance $|\vec{z}|$ from an opaque screen containing two pinholes at P_1 and P_2 that sample the radiation from a source S. The intensity at a single point Q on the observation plane is then described by

$$\begin{aligned}
I(\tau) &= \langle [U^*(\vec{z}_1, t_1) + U^*(\vec{z}_2, t_1 + \tau)][U(\vec{z}_1, t_1) + U(\vec{z}_2, t_1 + \tau)]\rangle \\
&= \langle U^*(\vec{z}_1, t_1)U(\vec{z}_1, t_1)\rangle + \langle U^*(\vec{z}_2, t_1 + \tau)U(\vec{z}_2, t_1 + \tau)\rangle \\
&\quad + \langle U^*(\vec{z}_1, t_1)U(\vec{z}_2, t_1 + \tau)\rangle + \langle U^*(\vec{z}_2, t_1 + \tau)U(\vec{z}_1, t_1)\rangle \\
&= I_1 + I_2 + \Gamma_{12}(\tau) + \Gamma_{12}^*(\tau)
\end{aligned} \tag{3-4}$$

where $I_1 = \langle U^*(\vec{z}_1, t_1)U(\vec{z}_1, t_1)\rangle$ and $I_2 = \langle U^*(\vec{z}_2, t_2)U(\vec{z}_2, t_2)\rangle$.

If the measurements are made simultaneously (i.e., $\tau = 0$), the mutual coherence function reduces to the *mutual intensity*, commonly denoted as J_U. In this case, the time dependence is usually ignored, resulting in

$$J_U(\vec{z}_1, \vec{z}_2) \equiv \Gamma_{12}(0) = \langle U^*(\vec{z}_1)U(\vec{z}_2)\rangle \tag{3-5}$$

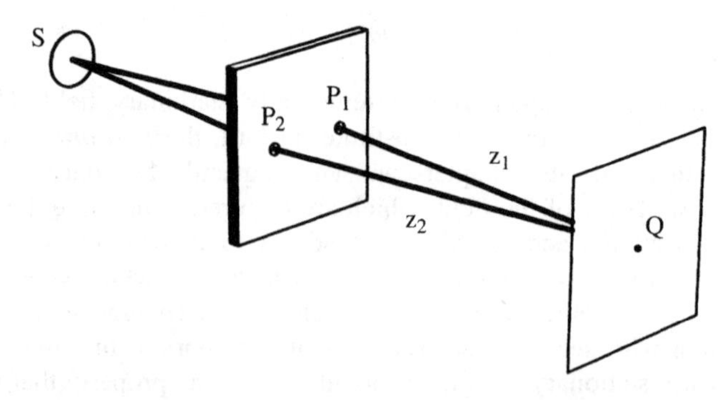

Figure 3-1. Young's two-beam experiment demonstrating the coherence properties of light.

$J_U(\vec{z}_1, \vec{z}_2)$ is seen to correspond to the autocorrelation function of the field at the two points $\vec{z}_1$ and $\vec{z}_2$. Similarly, if the measurements are made at a single point $\vec{z}_1 = \vec{z}_2 \equiv \vec{z}$, the z-dependence may be ignored and Eq. (3-3) written as

$$\Gamma_{11}(\tau) = \langle U^*(t_1)U(t_1 + \tau)\rangle \tag{3-6}$$

Here $\Gamma_{11}(\tau)$, sometimes written simply as $\Gamma(\tau)$, is referred to as the *temporal coherence function* of the field at a single point z. It is, once again, the autocorrelation function of the field evaluated at the times t_1 and t_2.

Normalization of Eqs. (3-3), (3-5), and (3-6) leads to several additional descriptors that are somewhat simpler to interpret. The normalization factor that will be chosen is the mutual coherence function when $\tau = 0$. Thus

$$\gamma_{12}(\tau) = \frac{\Gamma_{12}(\tau)}{\sqrt{\Gamma_{11}(0)\Gamma_{22}(0)}} = \frac{\langle U^*(\vec{z}_1, t_1)U(\vec{z}_2, t_1 + \tau)\rangle}{\sqrt{I_1 I_2}} \tag{3-7}$$

where $\gamma_{12}(\tau)$ is known as the *complex degree of coherence* or the *complex coherence factor* of the field evaluated at the space-time points $\vec{z}_1, t_1$ and $\vec{z}_2, t_2$. The absolute magnitude of $\gamma_{12}(\tau) \equiv |\gamma_{12}(\tau)|$ is denoted simply as the *degree of coherence*. In a similar fashion, the normalized mutual intensity becomes

$$\mu_{12} \equiv \gamma_{12}(0) = \frac{J_U(\vec{z}_1, \vec{z}_2)}{\sqrt{\Gamma_{11}(0)\Gamma_{22}(0)}} = \frac{\langle U^*(\vec{z}_1)U(\vec{z}_2)\rangle}{\sqrt{I_1 I_2}} \tag{3-8}$$

which is referred to as the *complex degree of spatial coherence* or the *spatial coherence factor*. Proceeding in this manner, the *complex degree of temporal coherence* or the *temporal coherence factor* is defined by letting $\vec{z}_1 = \vec{z}_2$ in Eq. (3-8), resulting in

$$\mu(\tau) = \frac{\Gamma_{11}(\tau)}{\Gamma_{11}(0)} = \frac{\langle U^*(t_1)U(t_1 + \tau)\rangle}{I_1} \tag{3-9}$$

If the field is statistically stationary in the spatial domain, the mean values at the positions $\vec{z}_1$ and $\vec{z}_2$ are equal, so that

$$\Gamma_{11}(0) = \Gamma_{22}(0) = \langle |U(\vec{z}_1, t_1)|^2\rangle = \langle |U(\vec{z}_2, t_2)|^2\rangle = \langle I\rangle \tag{3-10}$$

Equations (3-7), (3-8), and (3-9) represent normalized correlation functions and can therefore be used to describe the spatial and temporal correlation properties of the field. In particular, if the fields are well behaved (i.e., continuous over the domains of z and t), it is reasonable to expect that as $\vec{z}_2 - \vec{z}_1$ or $t_2 - t_1$ become large, $|\mu_{12}| \to 0$ and $|\mu(\tau)| \to 0$, respectively, in which case the fields are considered *spatially and temporally incoherent*. On the other hand, when $\vec{z}_2 = \vec{z}_1$ or $t_2 = t_1$, then $|\mu_{12}(0)| = 1$ and $|\mu(\tau)| = 1$ and the fields are considered to be *spatially and temporally coherent*. Thus the range of values for the degree of coherence is given by $0 \leq |\mu_{12}(\tau)| \leq 1$, with values in the range $0 < |\mu_{12}(\tau)| < 1$ representing *partially coherent light.*

As shown in Section 3.3.3, a *coherence area* and a *coherence time* may be defined for quasimonochromatic fields given by

$$S_c = \int_{-\infty}^{\infty} \int_{-\infty}^{\infty} |\mu_{12}(\Delta x, \Delta y)|^2 \, d\Delta x \, d\Delta y \tag{3-11}$$

and

$$\tau_c = \int_0^{\infty} |\mu(\tau)|^2 \, d\tau \tag{3-12}$$

The specific form of the coherence factors will depend on the physical processes involved but usually take the form of exponential functions such as $|\mu(\tau)|^2 = \exp(-\tau/\tau_c)$, such that S_c and τ_c are those points where the correlations reach their e^{-1} value.

If the radiation has a finite coherence time, it follows that the spectral width is nonzero. Thus the coherence properties of the field can be represented in the spectral domain as well. In this case a Fourier analysis of the random wavefunction $U(t)$ can be defined and by using the Wiener–Khintchine theorem, a power spectral density $S(\omega)$ thereby obtained. Proof of this follows procedures similar to those of Section 1.2.10 and are left to the reader. We have

$$\Gamma_{12}(\tau) = \int_{-\infty}^{\infty} S(\omega) e^{i\omega\tau} \, d\omega \tag{3-13}$$

and

$$S(\omega) = \frac{1}{2\pi} \int_{-\infty}^{\infty} \Gamma_{12}(\tau) e^{-i\omega\tau} \, d\tau \tag{3-14}$$

Comment 3-1. The physical significance of the mutual coherence function and its derivative functions can be made clearer by rearranging Eq. (3-4). We have

$$\begin{aligned} I(\tau) &= I_1 + I_2 + 2 \ \text{Re}[\Gamma_{12}(\tau)] = I_1 + I_2 + 2\sqrt{I_1 I_2} \, \text{Re}[\gamma_{12}(\tau)] \\ &= I_0 \left\{ 1 + \frac{2\sqrt{I_1 I_2}}{I_1 + I_2} \text{Re}[\gamma_{12}(\tau)] \right\} \end{aligned} \tag{3-15}$$

where $I_0 = I_1 + I_2$ and $\Gamma_{12}(\tau) + \Gamma_{12}^*(\tau) = 2 \ \text{Re}[\Gamma_{12}(\tau)]$. If we now let $\gamma_{12}(\tau)$ be a damped harmonic process, such as might occur in Young's experiment with a finite coherence time source, given by $\gamma_{12}(\tau) = |\gamma_{12}(\tau)| \exp(i\omega\tau - \tau/\tau_c)$, the intensity is modulated with τ in accordance with $I = I_0(1 + ve^{-\tau/\tau_c} \cos \omega\tau)$, where $v = 2\,|\gamma_{12}(\tau)|\,\sqrt{I_1 I_2}/(I_1 + I_2)$ is the *fringe visibility* of the resultant interference pattern. Note that as $\tau \to \tau_c$, the fringes begin to wash out due to the finite coherence time of the source. For $\tau \ll \tau_c$ and $I_1 = I_2$, the visibility is equal to the degree of coherence [i.e., $v = |\gamma_{12}(\tau)|$]. When $\tau \to 0$, the visibility reaches a maximum given by $v = 1$, in which case the intensity becomes

$I_{max} = 2I_1(1 + v) \to 4I_1$. Note that this last result is consistent with the definition of the intensity as the square of the sum of the coherent fields (i.e., $I = |E_1 + E_2|^2 = 4|E_1|^2 \equiv 4I_1$). ■

3.2.2. Van Cittert–Zernike Theorem

Probably the most important theorem in classical optics is the van Cittert–Zernike theorem. The theorem effectively provides a Fourier transform relationship between the correlation properties of the intensity at an observation plane and the spatial intensity distribution at the source aperture. As a consequence, it is very much analogous to the Fraunhofer theory of diffraction, which, as we have seen in Chapter 2, provides a Fourier transform relationship between the far-field complex amplitude distribution and the complex field distribution at the aperture. However, it should be remembered that the analogy does not extend to the underlying physical processes, which are quite different. In a sense the van Cittert–Zernike theorem is somewhat more general in that the correlation properties can be calculated for any point along the propagation path, whereas the Fraunhofer theory is restricted to the far field (or the focal plane of a lens).

A quasimonochromatic optical source of finite extent can always be considered as being composed of a large number of small coherent radiators that are randomly phased over the aperture distribution. If these radiators are unresolvable from the observation plane, we can make the simplifying assumption that the source distribution can be treated as continuous, with each radiator having an infinitesimal spatial extent. The differential contribution to the mutual intensity at the points P_1 and P_2 in Figure 3-2*a* may then be attributed to a differential element of disturbance area dS within the total source area S, that is,

$$dJ_U(\vec{z}_1, \vec{z}_2) = \langle U^*(\vec{z}_1)U(\vec{z}_2)\rangle\, dS \tag{3-16}$$

The total mutual intensity at P_1 and P_2 is then the integral of all such contributions over the surface area of the source; that is,

$$J_U(\vec{z}_1, \vec{z}_2) = \iint_S \langle U^*(\vec{z}_1)U(\vec{z}_2)\rangle\, dS \tag{3-17}$$

If we now assume that each elemental disturbance on S generates a spherical wave, the complex amplitudes at P_1 and P_2 due to an elemental source at (ξ, η) in S become

$$U(\vec{z}_1) = \frac{u(\xi, \eta)e^{ik|z_1|}}{|z_1|} \quad \text{and} \quad U(\vec{z}_2) = \frac{u(\xi, \eta)e^{ik|z_2|}}{|z_2|} \tag{3-18}$$

Substituting Eqs. (3-18) into Eq. (3-17) yields

$$J_U(\vec{z}_1, \vec{z}_2) = \iint_S \frac{\langle u(\xi, \eta)u^*(\xi, \eta)\rangle e^{ik(z_1 - z_2)}}{z_1 z_2} d\xi\, d\eta \tag{3-19}$$

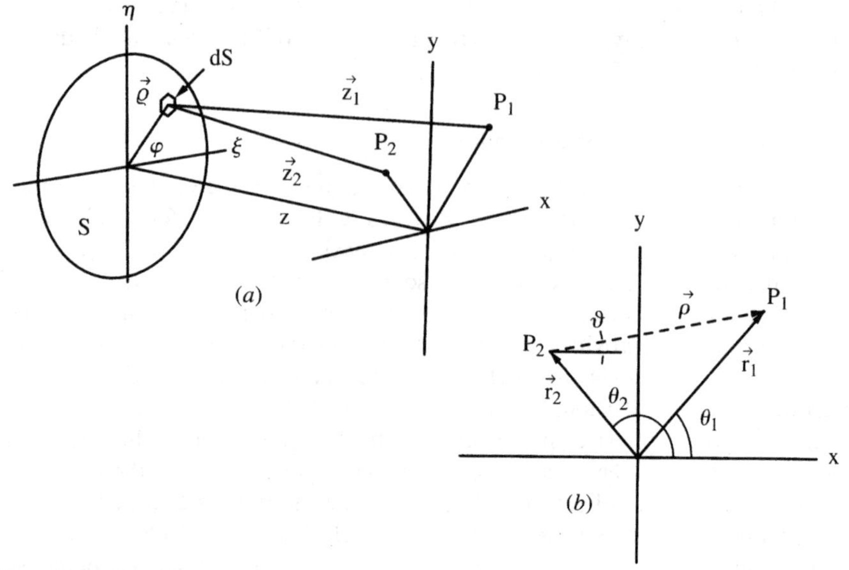

Figure 3-2. Coordinate systems used in describe the van Cittert–Zerniki theorem: (a) illumination of points P_1 and P_2 by a source element dS in the $\xi-\eta$ plane; (b) expanded view of the observation plane defined by (x,y).

where the mean frequencies at P_1 and P_2 are assumed equal. Equation (3-19) can be simplified by applying similar paraxial approximations as were used in the development of the Fraunhofer diffraction formula, namely, Eqs. (2-4) and (2-5). Since in this case the geometry involves coordinate differences, we will write the corresponding equations explicitly. Thus, beginning with

$$\begin{aligned} z_1^2 &= (x_1 - \xi)^2 + (y_1 - \eta)^2 + z^2 \\ z_2^2 &= (x_2 - \xi)^2 + (y_2 - \eta)^2 + z^2 \end{aligned} \tag{3-20}$$

we have in the large z limit,

$$\begin{aligned} z_1 &\approx z + \frac{(x_1 - \xi)^2 + (y_1 - \eta)^2}{2z} \\ z_2 &\approx z + \frac{(x_2 - \xi)^2 + (y_2 - \eta)^2}{2z} \end{aligned} \tag{3-21}$$

Hence

$$z_1 - z_2 \approx \frac{z_1^2 - z_2^2}{2z} - \frac{\Delta x \xi + \Delta y \eta}{z} \tag{3-22}$$

where $\Delta x = x_1 - x_2$ and $\Delta y = y_1 - y_2$. Thus, letting $z_1 \approx z_2 \approx z$ in the denominator of Eq. (3-19) and $\langle u(\xi, \eta) u^*(\xi, \eta) \rangle \equiv I(\xi, \eta)$, we have

$$J_U(\vec{z}_1, \vec{z}_2) \approx \iint_S \frac{I(\xi, \eta)}{z^2} \exp\left[\frac{ik(\Delta x\xi + \Delta y\eta)}{z} + \frac{ik(z_1^2 - z_2^2)}{2z}\right] d\xi\, d\eta \quad (3\text{-}23)$$

This expression can be written in terms of the complex degree of coherence by normalizing to the mutual intensity at a point; that is, we make the identification

$$J_U(\vec{z}_1, \vec{z}_1) = J_U(\vec{z}_2, \vec{z}_2) \equiv \iint_S \frac{I(\xi, \eta)}{z^2}\, d\xi\, d\eta \quad (3\text{-}24)$$

Dividing Eq. (3-23) by Eq. (3-24) results in

$$\mu_{12}(\Delta x, \Delta y) = \frac{e^{ik\psi} \displaystyle\iint_S I(\xi, \eta) e^{ik[(\Delta x\xi + \Delta y\eta)/z]}\, d\xi\, d\eta}{\displaystyle\iint_S I(\xi, \eta)\, d\xi\, d\eta} \quad (3\text{-}25)$$

where $\mu_{12} \equiv J_U(\vec{z}_1, \vec{z}_2)/[J_U(\vec{z}_1, \vec{z}_1) J_U(\vec{z}_2, \vec{z}_2)]^{1/2}$ is the complex degree of coherence, and $\psi = (z_1^2 - z_2^2)/2z$ is a phase term relating the phase of the wave at the points P_1 and P_2.

Equation (3-25) can be written in terms of polar coordinates by letting $\xi = \varrho\cos\varphi$, $\eta = \varrho\sin\varphi$, and $\Delta x = \rho\cos\vartheta$, $\Delta y = \rho\sin\vartheta$, where $\rho = \sqrt{\Delta x^2 + \Delta y^2}$ and $\vartheta = \tan^{-1}(\Delta y/\Delta x)$, as shown in Figure 3-2*b*. This results in

$$\mu_{12}(\rho, \vartheta) = \frac{e^{ik\psi} \displaystyle\iint_S I(\varrho, \varphi) e^{-i(k\varrho\rho/z)\cos(\vartheta - \varphi)} \varrho\, d\varrho\, d\varphi}{\displaystyle\iint_S I(\varrho, \varphi)\varrho\, d\varrho\, d\varphi} \quad (3\text{-}26)$$

The absolute magnitude of Eq. (3-26) corresponds to the degree of coherence for the two observation points and is given by

$$|\mu_{12}(\rho, \vartheta)| = \left| \frac{\displaystyle\iint_S I(\varrho, \varphi) e^{-i(k\varrho\rho/z)\cos(\vartheta - \varphi)} \varrho\, d\varrho\, d\varphi}{\displaystyle\iint_S I(\varrho, \varphi)\varrho\, d\varrho\, d\varphi} \right| \quad (3\text{-}27)$$

The van Cittert–Zernike theorem is seen to equate the degree of coherence to the absolute magnitude of the normalized Fourier transform of the intensity distribution at the aperture. Using Eq. (2-8) for the definition of the Bessel function of order zero, Eq. (3-27) can be written

$$|\mu_{12}(\rho, \vartheta)| = \left| \frac{2\pi \displaystyle\iint_S I(\varrho, \varphi) J_0(k\varrho\rho/z)\varrho\, d\varrho}{\displaystyle\iint_S I(\varrho, \varphi)\varrho\, d\varrho\, d\varphi} \right| \quad (3\text{-}28)$$

In the case of azimuthal symmetry, this reduces to

$$|\mu_{12}(\rho, \vartheta)| = \left| \frac{\displaystyle\int_0^\infty I(\varrho) J_0(k\varrho\rho/z)\varrho\, d\varrho}{\displaystyle\int_0^\infty I(\varrho)\varrho\, d\varrho} \right| \tag{3-29}$$

The van Cittert–Zernike theorem provides a framework within which numerous optical processes can be understood. The most important of these is the gain in coherence of incoherent radiation as a result of propagation through free space. In this case the resulting coherence properties of propagated stellar radiation allow for measurement of the angular width of stars that are unresolvable with conventional telescopes. Devices that perform this function are the classical Michelson stellar interferometer and the quantum mechanical Hanbury Brown and Twiss interferometer. They are designed to measure the spatial coherence of starlight as a function of the separation of two points on the surface of the earth and are direct demonstrations of the van Cittert–Zernike theorem. Since laser light becomes spatially incoherent through scattering and refractive turbulence effects, it is reasonable to assume that the van Cittert–Zernike theorem will also be of use in describing these optical phenomena as well. Indeed, this will be shown in the following sections.

Comment 3-2. The van Cittert–Zernike theorem is even more powerful than that described above in that it also allows for determination of the internal intensity profile of the source. Consider a nonuniform incoherent circular disk of radius a with a quadratic radial intensity profile given by $I(\varrho) = u_0^* u_0(1 - \kappa\varrho^2/a^2)$, where κ is a constant that determines the rate of roll-off of the intensity near the edge of the disk. (In stellar interferometry this is known as *limb darkening* and is a consequence of the different temperatures that can be observed when viewing the center or edge of a star.) Since the problem has azimuthal symmetry, we use Eq. (3-29), obtaining

$$|\mu_{12}| = \frac{2}{a^2(1-\kappa/2)} \left| \int_0^a \left(1 - \kappa\frac{\varrho^2}{a^2}\right) J_0\left(\frac{k\varrho\rho}{z}\right) \varrho\, d\varrho \right| \tag{3-30}$$

Using the Bessel identity given by Eq. 2-16 for the cases of $\kappa = 0$ and $\kappa = 1$, this integrates to

$$|\mu_{12}| = \begin{cases} \left|\dfrac{2J_1(av)}{av}\right| & \text{uniform } (\kappa = 0) \\[2ex] \left|\dfrac{8J_2(av)}{(av)^2}\right| & \text{nonuniform } (\kappa = 1) \end{cases} \tag{3-31}$$

where $v = k\rho/z = 2\pi\rho/\lambda z$ and J_1 and J_2 are Bessel functions of the first kind, first and second order, respectively. These functions are plotted in Figure 3-3 as

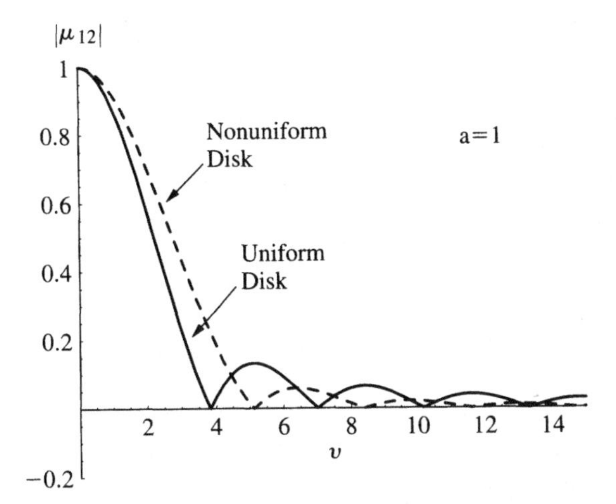

Figure 3-3. Degree of coherence versus $\upsilon = k\rho/z$ for uniform and nonuniform circular disks.

a function of υ. For the uniform disk, the Bessel function first goes to zero at $\upsilon = 3.83/a$ or $\alpha = 0.61\lambda/a$, where $\alpha = \Delta r/z$. Functionally, the uniform case is identical to that which occurred for the Fraunhofer diffraction pattern of a uniformly illuminated circular aperture. The difference here is that it is describing the degree of coherence of the propagating radiation, not the intensity. Thus the coherence falls off from full coherence to complete incoherence between $\upsilon = 0$ and $\upsilon = 3.83$ and then increases slightly again, as shown in the Figure 3-3. In the nonuniform case, the coherence first goes to zero at $\upsilon = 5.14/a$ or $\alpha = 0.82\lambda/a$. Since the aperture distribution is more concentrated near the center in this case, the result is analogous to the situation of a smaller aperture function producing a larger far-field central disk. Thus, by comparing the periods of oscillation of measured data to theory, one can estimate the degree of radial nonuniformity of the source. ■

3.3. SURFACE SCATTERING

Laser systems that interrogate solid targets must contend with the fact that most surfaces are rough or at least partially rough compared to the wavelength of the light being used. This results in the reflected energy not being contained within a narrow solid angle as in *specular reflection* but is, rather, spread over the full 2π steradian hemisphere bounding the scattering surface. In addition, the coherent nature of the illuminating laser light results in a random *granular* or *speckle pattern* in the scattered field. Coherent speckle effects play a fundamental role in optical detection theory and are explored extensively in the sections and chapters to follow. However, for the moment, we restrict our study to the surface itself and attempt to characterize the angular distribution of the scattered field and its dependence on surface structure.

One intuitively expects that a good first-order theory should be able to predict the dependence of the scatter distribution on the wavelength and angle of the incident field. Theories attempting to describe these phenomena have met with varying degrees of success, depending on the validity and generality of the underlying assumptions. In all such theories, the degree of roughness is usually restricted either in the maximum slope or the maximum radii of curvature relative to a wavelength that the spatial fluctuations can assume. The former restriction is characteristic of the Rayleigh–Rice[2,3] perturbation theory approach, and the latter is employed in the Kirchhoff[4] or physical optics approach, which we investigate in the next section.

For simplicity, the discussion is restricted to scalar fields with the incident polarization parallel to the surface. A scalar analysis implies that the theory will be unable to describe polarization effects in the scattered field, but this is not considered an undue restriction. Also, for the sake of brevity, the statistical aspects of the theory will be restricted to the limiting case of very rough normally distributed surfaces in which the specular component is essentially nonexistent. A more complete analysis that includes slightly rough and moderately rough surfaces with coexisting specular and diffuse components can be obtained from the theory[5] but is not pursued here. The scatter distribution of a very rough surface will be shown to compare well with Lambert's cosine law for the bistatic diffuse radiation emitted from an arbitrary surface and yields a relatively simple expression for the backscattered radiation associated with monostatic systems.

3.3.1. Scattering from a Rough Surface

Consider a rough surface S lying in the $x-y$ plane and with height variations given by $\zeta(x, y)$, as shown in Figure 3-4. The mean height is assumed to be $\langle\zeta(x, y)\rangle = 0$, corresponding to the $z = 0$ plane. The scattered field U_s at a point P in the far field of an illuminated spot on the surface can be found using the Helmholtz equation, given by

$$U_s(P) = \frac{1}{4\pi}\iint\limits_S \left(U\frac{\partial\psi}{\partial n} - \psi\frac{\partial U}{\partial n}\right)dS \tag{3-32}$$

where U is the total field on the surface, ψ is the scattered field at a distant point P, and $\partial/\partial n$ is the derivative along the unit local normal to the surface. The Kirchhoff approximation assumes that the surface in the vicinity of the point x, y is an infinite plane with slope equal to the tangent plane at x, y. This assumption requires that the local surface radii of curvature be large compared to a wavelength; that is, no sharp edges exist on the surface profile.

Consider then a unit amplitude plane wave that is incident on a surface S and propagating in the x, z plane with wavevector $\vec{k}_i$. The scalar field on the surface is described by

$$U_i = e^{i\vec{k}_i\cdot\vec{r}} \tag{3-33}$$

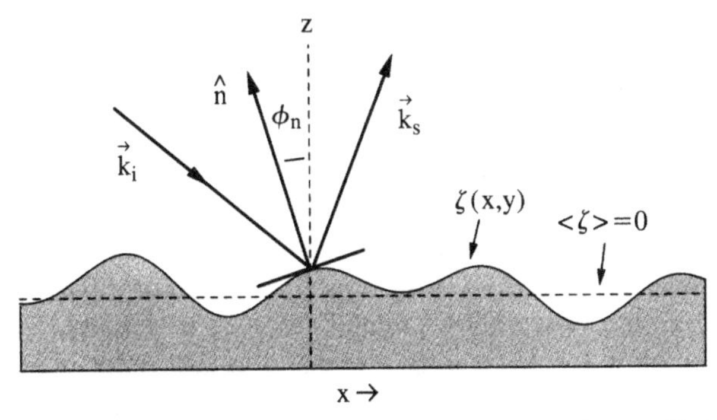

Figure 3-4. Two-dimensional view of a rough surface having a surface profile $\zeta(z)$ lying in the $x-y$ plane with a mean surface normal corresponding to the z-axis. A ray $\vec{k}_i$ is incident at the point x, y with local surface normal n and scattered ray $\vec{k}_s$.

where $\vec{r} = x\hat{i} + y\hat{j} + \zeta(x, y)\hat{k}$ is the vector from the origin to an arbitrary point on the surface as shown in Figure 3-5. The reflected wave on the surface in the Kirchhoff approximation is then

$$U_r = Re^{i\vec{k}_i \cdot \vec{r}} \tag{3-34}$$

where R is the local reflection coefficient at the point x, y. It may also be set equal to the mean reflectivity of the surface under the restriction that the surface gradients are small. The total field at the surface is therefore given by

$$U = U_i + U_r = (1 + R)U_i \tag{3-35}$$

and its normal derivative by

$$\frac{\partial U}{\partial n} = \frac{\partial(U_i + U_r)}{\partial n} = i(1 - R)\vec{k}_i \cdot \hat{n}U_i \tag{3-36}$$

Here $\partial\vec{r}/\partial n = \hat{n}$ is the unit normal of the local surface at the point x, y.

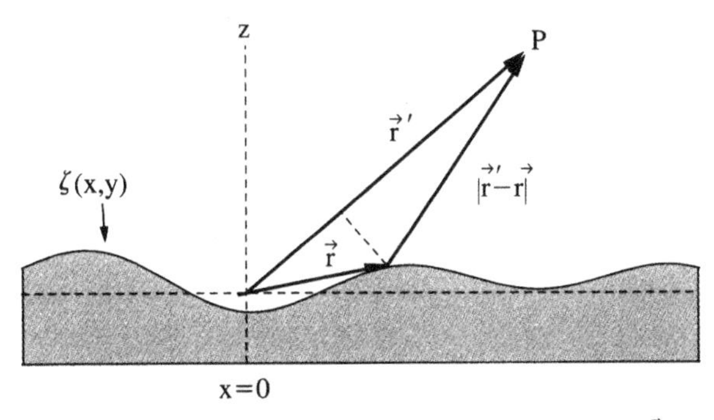

Figure 3-5. Local surface geometry in obtaining the approximate form of $|\vec{r}' - \vec{r}|$ as P extends to infinity.

Now the field at P may be assumed to be the result of scattering from an arbitrary point on the surface S and therefore given by a spherical wave of the form

$$\psi = \frac{e^{ik_s|\vec{r}' - \vec{r}|}}{|\vec{r}' - \vec{r}|} \tag{3-37}$$

where $k_s = 2\pi/\lambda$. (With minor rearrangements, ψ can be shown to represent the Green's function, with Eq. (3-32) representing Green's theorem.) Referring to Figure 3-5, as the observation point P recedes to infinity, $r' \gg r$, so that $|\vec{r}' - \vec{r}| = (r'^2 - 2\vec{r}' \cdot \vec{r} + r^2)^{1/2}$ can be expanded to first order in $\vec{r}$ in Eq. (3-37), to yield

$$\psi \approx \frac{e^{ik_s r' - i\vec{k}_s \cdot \vec{r}}}{r'} \tag{3-38}$$

where $\vec{k}_s = k_s \vec{r}'/r'$. The normal derivative of ψ is then

$$\frac{\partial \psi}{\partial n} = -i(\vec{k}_s \cdot \hat{n})\psi \tag{3-39}$$

Since the local surface normal is a random variable that may assume any orientation with respect to the positive z-axis, the scattered wavevector $\vec{k}_s$ may assume any direction in the hemisphere defined by $z \geq 0$. We therefore have, from Figure 3-6,

$$\vec{k}_i = k \sin\phi_i\, \hat{i} - k \cos\phi_i\, \hat{k} \tag{3-40}$$

$$\vec{k}_s = k \sin\phi_s \cos\theta_s\, \hat{i} + k \sin\phi_s \sin\theta_s\, \hat{j} + k \cos\phi_s\, \hat{k} \tag{3-41}$$

$$\hat{n} = \sin\phi_n \cos\theta_n\, \hat{i} + \sin\phi_n \sin\theta_n\, \hat{j} + \cos\phi_n\, \hat{k} \tag{3-42}$$

where $\hat{i}$, $\hat{j}$, $\hat{k}$ are unit vectors along the x, y, z directions, respectively. From Eqs. (3-40) and (3-41), a pair of sum and difference vectors can be defined as $\vec{p} = \vec{k}_i + \vec{k}_s$ and $\vec{v} = \vec{k}_i - \vec{k}_s$, where $\vec{p}$ and $\vec{v}$ are parallel and perpendicular to the local scattering surface, respectively.

With these definitions, Eq. (3-32) can be written

$$U_s(P) = \frac{ie^{ik_s r'}}{4\pi r'} \iint_S (R\vec{v} - \vec{p}) \cdot \hat{n} e^{i\vec{v}\cdot\vec{r}} dS \tag{3-43}$$

where

$$\vec{v} = k(\sin\phi_i - \sin\phi_s \cos\theta_s)\hat{i} - k \sin\phi_s \sin\theta_s \hat{j} - k(\cos\phi_i + \cos\phi_s)\hat{k} \tag{3-44}$$

$$\vec{p} = k(\sin\phi_i + \sin\phi_s \cos\theta_s)\hat{i} + k \sin\phi_s \sin\theta_s \hat{j} + k(\cos\phi_s - \cos\phi_i)\hat{k} \tag{3-45}$$

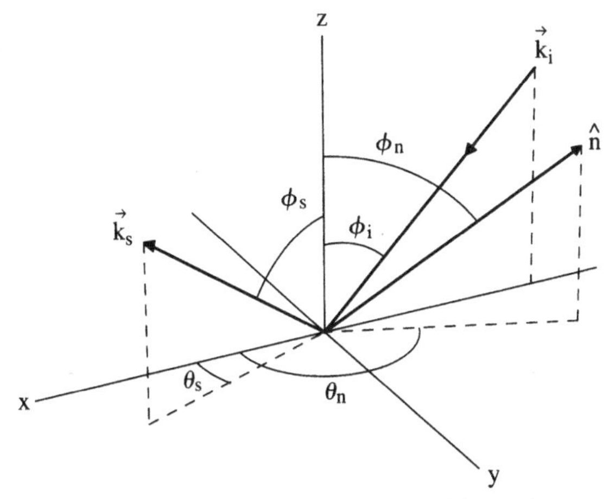

Figure 3-6. Spherical coordinate system used to define the incident and scattered rays for a rough surface.

Equation (3-43) can be put in scalar form by noting that $dS = dx\,dy/\cos\phi_n$ and $\tan\phi_n \cos\theta_n = d\zeta_x/dx = \zeta_x'$ and $\tan\phi_n \sin\theta_n = d\zeta_y/dy = \zeta_y'$. The first relationship is simply the projection of the local surface element dS onto the $x-y$ plane, while the second pair represents the slope of the surface feature at the point x, y. Thus

$$U_s(P) = K \int_{-L_x}^{L_x} \int_{-L_y}^{L_y} (a\zeta_x' + b\zeta_y' - c)e^{i\vec{v}\cdot\vec{r}}\,dx\,dy \tag{3-46}$$

where $K = ike^{ik_s r'}/4\pi r'$ and

$$\begin{aligned} a &= (1-R)\sin\phi_i + (1+R)\sin\phi_s\cos\theta_s \\ b &= (1+R)\sin\phi_s\sin\theta_s \\ c &= (1+R)\cos\phi_s - (1-R)\cos\phi_i \end{aligned} \tag{3-47}$$

Equation (3-46) is most easily solved by assuming that a, b, c are constants, which requires that the reflection coefficient R in Eqs. (3-47) be independent of position on the surface. Clearly, this is an approximation that is consistent with the assumption of small slopes and large radii of curvatures for the surface features such that R can be approximated by the mean reflection coefficient of the surface. With these assumptions, the simplest solution corresponds to that of a smooth surface where $\zeta_x = \zeta_y = \zeta_x' = \zeta_y' = 0$ and $\phi_i = \phi_s$, resulting in

$$\begin{aligned} U_s(P) &= -cK \int_{-L_x}^{L_x} \int_{-L_y}^{L_y} e^{i\vec{v}\cdot\vec{r}}\,dx\,dy \\ &= -2KAR\cos\phi_i \frac{\sin v_x L_x}{v_x L_x}\frac{\sin v_y L_y}{v_y L_y} \end{aligned} \tag{3-48}$$

where $A = 4L_x L_y$ is the area of the surface. In the case of a perfectly conducting surface, the reflection coefficient becomes $R = \pm 1$, where the plus sign corresponds to (vertical) polarization in the plane of incidence and the minus sign to (horizontal) polarization orthogonal to the plane of incidence. Equation (3-48) can then be written

$$U_{s0}^{\pm}(P) = \mp 2KA\cos\phi_i \frac{\sin v_x L_x}{v_x L_x} \frac{\sin v_y L_y}{v_y L_y} \tag{3-49}$$

where the plus sign in $U_{s0}^{\pm}(P)$ corresponds to vertical polarization. Thus we see that the solution to the surface scattering problem is bounded in the limit of a smooth surface by the diffraction pattern associated with a rectangular source of area $A = 2L_x \times 2L_y$. This is to be expected since the Kirchhoff theory becomes exact in this case. Later we will see that the solution is bounded at the other extreme by the radiation pattern of a diffuse Lambertian reflector.

Returning to Eq. (3-46), the integral can be evaluated by parts by letting

$$dV = a\zeta_x' e^{iv_z\zeta} \quad \text{and} \quad U = e^{iv_x x} \tag{3-50}$$

such that

$$V = \frac{ae^{iv_z\zeta}}{iv_z} \quad \text{and} \quad dU = iv_x e^{iv_x x} \tag{3-51}$$

which leads to

$$U_s(P) = -K\frac{av_x + bv_y + cv_z}{v_z}\int_{-L_x}^{L_x}\int_{-L_y}^{L_y} e^{i\vec{v}\cdot\vec{r}}\,dx\,dy + \frac{iK}{v_z}\left(a\int_{-L_y}^{L_y} e^{i\vec{v}\cdot\vec{r}}\,dy\Bigg|_{-L_x}^{L_x} + b\int_{-L_x}^{L_x} e^{i\vec{v}\cdot\vec{r}}\,dx\Bigg|_{-L_y}^{L_y}\right) \tag{3-52}$$

The second term in Eq. (3-52) is generally attributed to edge effects at the boundaries of the surface and may be considered small compared to the first term, especially near the specular direction. Indeed, numerous arguments have been put forth on the significance of the second term relative to the first. The most straightforward is the fact that the term ultimately leads to a factor $k^2 A \approx A/\lambda^2$ in the denominator after integration, which for practical-sized surfaces results in the second term being negligible compared to the first. For brevity, we assume the latter and substitute for v_x, v_y, and v_z from Eq. (3-44) in the first term of Eq. (3-52), which yields

$$U_s(P) = -2KRF\cos\phi_i \int_{-L_x}^{L_x}\int_{-L_y}^{L_y} e^{i\vec{v}\cdot\vec{r}}\,dx\,dy \tag{3-53}$$

where

$$F = \frac{1 - \sin\phi_i \sin\phi_s \cos\theta_s + \cos\phi_i \cos\phi_s}{\cos\phi_i(\cos\phi_i + \cos\phi_s)} \tag{3-54}$$

Equation (3-53) may be applied to any surface profile, including periodic surfaces. In that case, the solutions yield the well-known angular spectra of optical gratings. However, here we are concerned with random rough surfaces, which require statistical methods for a complete description. A first step in doing this is to explore the first moment or average value of U_s, that is,

$$\langle U_s\rangle = -2KRF\cos\phi_i \left\langle \int_{-L_x}^{L_x}\int_{-L_y}^{L_y} e^{i\vec{v}\cdot\vec{r}}\,dx\,dy \right\rangle \tag{3-55}$$

This can be rewritten as

$$\langle U_s\rangle = -2KRF\cos\phi_i G(iv_z) \left\langle \int_{-L_x}^{L_x}\int_{-L_y}^{L_y} e^{i(v_x x + v_y y)}\,dx\,dy \right\rangle \tag{3-56}$$

where

$$G(iv_z) = \int_{-\infty}^{\infty} p_s(z) e^{iv_z z}\,dz \tag{3-57}$$

is the characteristic function of the probability density function $p_s(z)$, which describes the surface statistics, and $z = \zeta(x, y)$. It therefore follows that the mean value of the field at P can be written

$$\langle U_s\rangle = G(iv_z)U_s \tag{3-58}$$

where U_s is given by Eq. (3-48). Note that $\langle U_s\rangle$ is in general complex since $G(iv_z)$ is complex. However, as we have seen in Chapter 1, if $p_s(z)$ is symmetric, then $G(iv_z)$ is real.

The quantity of ultimate interest is the mean power reflectivity in a given direction. This may be obtained by evaluating $\langle U_s U_s^*\rangle$ at the point P, that is,

$$\langle U_s U_s^*\rangle = 4|K|^2|R|^2F^2\cos^2\phi_i \left\langle \iint_S e^{i\vec{v}\cdot\vec{r}_1}dS_1 \iint_S e^{i\vec{v}\cdot\vec{r}_2}dS_2 \right\rangle \tag{3-59}$$

This leads to

$$\langle U_s U_s^*\rangle = 4|K|^2|R|^2F^2\cos^2\phi_i \times \iint_S \iint_S G(iv_z, -iv_z) e^{iv_x(x_1-x_2)+iv_y(y_1-y_2)} dS_1\,dS_2 \tag{3-60}$$

where $dS_1 \equiv dx_1\, dy_1$, $dS_2 \equiv dx_2\, dy_2$. Here

$$G(iv_z, -iv_z) = \int_{-\infty}^{\infty} \int_{-\infty}^{\infty} p_s(z_1, z_2) e^{iv_z(z_1 - z_2)}\, dz_1\, dz_2 \tag{3-61}$$

is the two-point characteristic function for the surface with $z_1 \equiv \zeta(x_1, y_1)$ and $z_2 \equiv \zeta(x_2, y_2)$.

To solve Eq. (3-61), we need to specify the statistics of the surface features. The most common assumption is to assume a homogeneous surface with height fluctuations described by a zero-mean Gaussian probability density function given by

$$p_s(z) = \frac{1}{\sqrt{2\pi\sigma_h^2}} e^{-z^2/2\sigma_h^2} \tag{3-62}$$

where σ_h^2 is the variance of the surface height fluctuations. The characteristic function associated with Eq. (3-62) is

$$G(iv_z) = e^{(-1/2)\sigma_h^2 v_z^2} \tag{3-63}$$

As will be seen repeatedly in the sections to follow, a random process requires at least first- and second-order statistics if a "complete" description of the process is to be obtained. In this case it is necessary to characterize not only the variations in height with position but also the mean distance between peaks or valleys. The latter is given by the normalized correlation coefficient $\sigma_s^2 = \langle z_1 z_2 \rangle / \langle z_1^2 \rangle$, which will be assumed to be given by the general expression

$$\sigma_s^2 = e^{-\rho^2/\rho_d^2} \tag{3-64}$$

where $\rho = [(x_1 - x_2)^2 + (y_1 - y_2)^2]^{1/2}$ and ρ_d is the *surface height correlation distance*. A two-point probability density function can then be written as

$$p_s(z_1, z_2) = \frac{1}{2\pi\sigma_h^2 \sqrt{1 - \sigma_s^2}} e^{-(z_1^2 - 2\sigma_s z_1 z_2 + z_2^2)/2\sigma_h^2(1 - \sigma_s^2)} \tag{3-65}$$

This leads to the characteristic function

$$G(iv_z, -iv_z) = e^{-\sigma_h^2 v_z^2 (1 - \sigma_s^2)} \tag{3-66}$$

Changing variables to $x_1 - x_2 = \rho \cos\vartheta$ and $y_1 - y_2 = \rho \sin\vartheta$, Eq. (3-60) becomes

$$\langle U_s U_s^* \rangle = 4\,|K|^2\,|R|^2\,F^2 \cos^2\phi_i \int_0^{\infty} \int_0^{2\pi} G(iv_z, -iv_z) e^{iv_x \rho \cos\vartheta + iv_y \rho \sin\vartheta} \rho\, d\rho\, d\vartheta \tag{3-67}$$

The integral over ϑ may be identified with the Bessel function of the first kind, zeroth order. Thus

$$\langle U_s U_s^* \rangle = 8\pi\, |K|^2\, |R|^2\, F^2 \cos^2\phi_i \int_0^\infty J_0(v_{xy}\rho) G(iv_z, -iv_z)\rho\, d\rho \tag{3-68}$$

where

$$J_0(v_{xy}\rho) = \frac{1}{2\pi} \int_0^\infty e^{iv_x\rho\cos\vartheta + iv_y\rho\sin\vartheta}\, d\vartheta \tag{3-69}$$

and $v_{xy} = \sqrt{v_x^2 + v_y^2}$. Substituting Eqs. (3-64) and (3-66) into Eq. (3-68) yields

$$\langle U_s U_s^* \rangle = 8\pi\, |K|^2\, |R|^2\, F^2 \cos^2\phi_i \int_0^\infty J_0(v_{xy}\rho) e^{-v_z^2\sigma_h^2(1-e^{-\rho^2/\rho_d^2})}\rho\, d\rho \tag{3-70}$$

For a very rough surface the quantity $v_z^2\sigma_h^2$ in the exponential is large since $v_z \propto k$ and $\sigma_h^2/\lambda^2 \gg 1$. Hence the integral will approach zero everywhere except near $\rho = 0$. It is therefore customary to let $\exp(-\rho^2/\rho_d^2) \approx 1 - \rho^2/\rho_d^2$, resulting in

$$\langle U_s U_s^* \rangle = 8\pi\, |K|^2\, |R|^2\, F^2 \cos^2\phi_i \int_0^\infty J_0(v_{xy}\rho) e^{-v_z^2\sigma_h^2\rho^2/\rho_d^2}\rho\, d\rho \tag{3-71}$$

Equation (3-71) can be integrated exactly using the identity

$$\int_0^\infty J_0(ax)e^{-bx^2} x\, dx = \frac{e^{-a^2/4b}}{2b} \tag{3-72}$$

yielding

$$\langle U_s U_s^* \rangle = \frac{4\pi\, |K|^2\, |R|^2\, F^2 \rho_d^2 \cos^2\phi_i}{v_z^2\sigma_h^2} e^{-v_{xy}^2\rho_d^2/4v_z^2\sigma_h^2} \tag{3-73}$$

This can be further simplified by defining a mean slope for the surface roughness as $\tan\phi_0 = 2\sigma_h/\rho_d$ such that $\tan\phi_n = v_{xy}/v_z$. Substituting for K from Eq. (3-46), we obtain

$$\langle U_s U_s^* \rangle = \frac{|R|^2\, F^2 \cos^2\phi_i \cot^2\phi_0 e^{-\tan^2\phi_n/\tan^2\phi_0}}{\pi r'^2(\cos\phi_i + \cos\phi_s)^2} \tag{3-74}$$

A final rearrangement results in a relatively simple expression for the mean power at P:

$$\langle U_s U_s^* \rangle = \frac{|R|^2 \cot^2\phi_0}{4\pi r'^2} \frac{v^4}{v_z^4} e^{-\tan^2\phi_n/\tan^2\phi_0} \tag{3-75}$$

where

$$\begin{aligned} v^2 &= 2k^2(1 - \sin\phi_i \sin\phi_s \cos\theta_s + \cos\phi_i \cos\phi_s) \\ v_z &= -k(\cos\phi_i + \cos\phi_s) \end{aligned} \tag{3-76}$$

To ensure energy conservation between the incident and scattered fields, the integral of the scattered field over the positive-z hemisphere must equate to unity when $|R| = 1$. This follows from our initial definition of a unit amplitude incident field given by Eq. (3-33), where $\langle U_i U_i^* \rangle \equiv 1$. We therefore define a normalization factor for Eq. (3-75) to correspond to the case of normal incidence when $|R| = 1$. Thus, using $\phi_i = 0$ in Eqs. (3-76), we obtain

$$\begin{aligned}\int_0^{2\pi}\int_0^{\pi/2} \langle U_s U_s^* \rangle \, d\Omega &= \frac{\cot^2\phi_0}{4\pi r'^2}\int_0^{2\pi}\int_0^{\pi/2} \frac{v^4}{v_z^4} e^{-\tan^2\phi_n/\tan^2\phi_0} \sin\phi_s \, d\phi_s \, d\theta_s \\ &= \frac{\cot^2\phi_0 e^{\cot^2\phi_0}}{\pi r'^2}\int_0^{2\pi}\int_0^{\pi/2} \frac{1}{(1+\cos\phi_s)^2} \\ &\quad \times e^{-2\cot^2\phi_0/1+\cos\phi_s} \sin\phi_s \, d\phi_s \, d\theta_s \\ &= \frac{1}{r'^2}(1 - e^{-\cot^2\phi_0}) \end{aligned} \tag{3-77}$$

where $v^2/v_z^2 = 2/(1+\cos\phi_s)$ and $\tan^2\beta = v_{xy}^2/v_z^2 = v^2/v_z^2 - 1$ were used. Dividing Eq. (3-75) by Eq. (3-77) results in an *inverse steradian power reflectivity* for the scattered field in the Kirchhoff theory, which we denote as $\rho_K(\pi)$, given by

$$\rho_K(\pi) = \frac{\rho_s v^4 \cot^2\phi_0}{4\pi v_z^4 N(\phi_0)} e^{-\tan^2\phi_n/\tan^2\phi_0} \tag{3-78}$$

where $N(\phi_0) = 1 - \exp(-\cot^2\phi_0)$ and $\rho_s = |R|^2$ is the mean power reflectivity of the surface. If the surface is not a perfect conductor, $\rho_s = 1 - \alpha_s$, where α_s is the absorption coefficient of the surface.

Equation (3-78) corresponds to the general *bistatic reflectance* of a rough surface. Three-dimensional renderings are shown in Figure 3-7 for an incidence angle of $\phi_i = 45^\circ$ and two values of ϕ_0 corresponding to $\rho_d/\sigma_h = 3.96$ and 7. The distribution can be seen to become more specular as ρ_d/σ_h increases. However, better insight into the quantitative behavior of the distributions may be obtained from two-dimensional linear-polar plots. Figure 3-8 shows the distributions resulting from incidence-plane scatter ($\theta_s = 0$) and perfect reflectivity ($\rho_s = 1$). Most presentations of scatter distributions are given on log-polar plots, but this is considered unnecessary here, for two reasons. First, reflectivity losses of several orders of magnitude typically cannot be tolerated by real systems, and second, the theory is not particularly accurate for large angles away from the specular direction.

An estimate of the lower bound on ρ_d/σ_h may be obtained by comparing Eq. (3-78) with the $\rho_L(\pi) = \cos\phi_s/\pi$ distribution of a diffuse Lambertian surface, where ϕ_s is the angle relative to the mean surface normal (cf. Comment 3-3 below). Such a distribution is also shown in Figure 3-8 as a

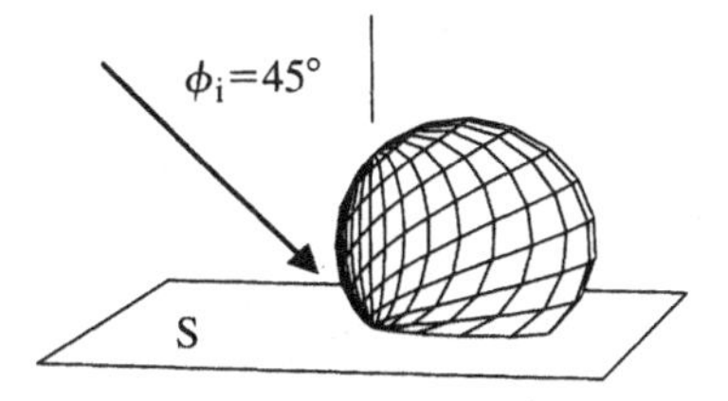

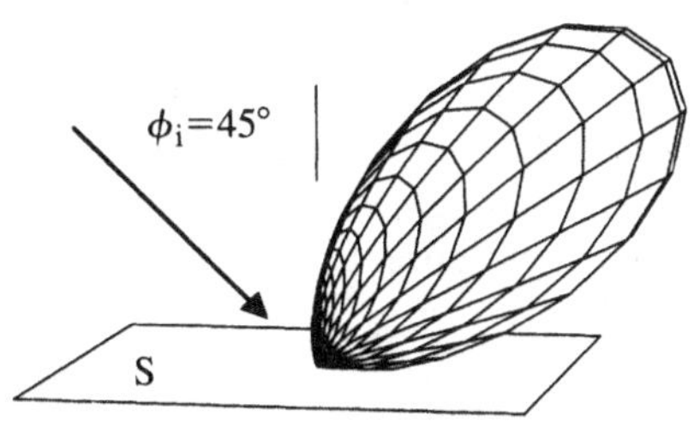

Figure 3-7. Bistatic scatter distributions resulting from the Kirchhoff theory of a very rough surface for an incidence angle $\phi_i = 45°$. The top figure corresponds to $1.5R_s$ for $\rho_d/\sigma_h = 3.96$ and the bottom figure R_s for $\rho_d/\sigma_h = 7$.

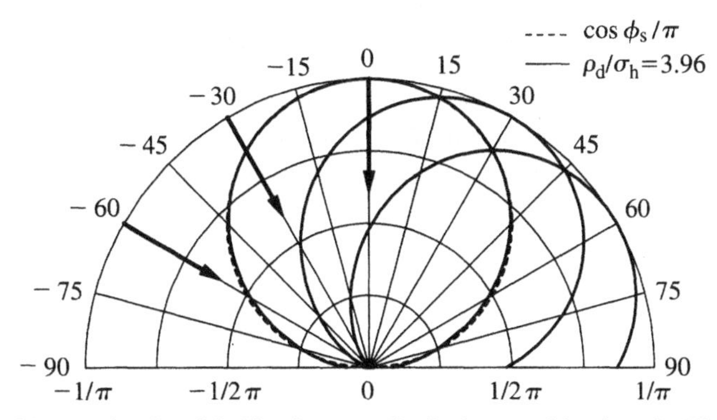

Figure 3-8. Linear-polar plot of the bistatic scatter distributions resulting from the Kirchhoff theory of a very rough surface with $\rho_d/\sigma_h = 3.96$ and incidence angles $\phi_i = 0°, 30°$, and $60°$. Also shown as a dashed curve is the distribution corresponding to a Lambertian reflector.

dashed curve. The Kirchhoff theory is seen to agree exceedingly well with Lambert's cosine law when $\rho_d/\sigma_h \approx 3.96$. Although this is a very pleasing result and somewhat remarkable considering the dissimilar nature of the two theories, it should not be taken too seriously, especially at the larger angles off the specular direction. However, it is certainly a strong indication that the Kirchhoff theory correctly represents the first-order features of surface scatter phenomena. Since by definition a Lambertian surface constitutes the maximum "diffuseness" that is physically realizable, corresponding to spatial white noise, the value $\rho_d/\sigma_h = 3.96$ may be considered a lower bound for the Kirchhoff theory. Also shown in Figure 3-8 for comparison are the distributions corresponding to $\phi_i = 30°$ and $60°$. It can be seen that the distribution rotates toward the specular direction for nonzero angles of incidence.

Figure 3-9 shows the scatter distributions obtained with $\rho_d/\sigma_h = 7$. There it can be seen that as ρ_s/σ_h increases, the distribution narrows about the specular direction while increasing in magnitude in order to conserve energy. An increasing ρ_d/σ_h ratio can be viewed as a reduction in the mean surface slopes, where ρ_d is a measure of the distance between peaks and σ_h is a measure of the height of the peaks. Such a reduction may be attributed to an increase in ρ_d or a decrease in σ_h, although the latter cannot be assumed to continue indefinitely; otherwise, the assumption that $\sigma_h/\lambda \gg 1$ made earlier in defining a very rough surface could be violated.

Two limiting cases of Eq. (3-78) may be considered. The first case involves the bistatic geometry of scatter in the specular direction, which can be obtained by letting $\phi_s = \phi_i$ and $\theta_s = 0$, resulting in

$$\rho_{KS} = \frac{\rho_s \cot^2 \phi_0}{4\pi N(\phi_0)} \tag{3-79}$$

This is the direction in which the Kirchhoff theory is most accurate but is generally too restricted in angle to be useful in remote sensing applications. The second case involves the monostatic geometry of backscatter along the incident direction,

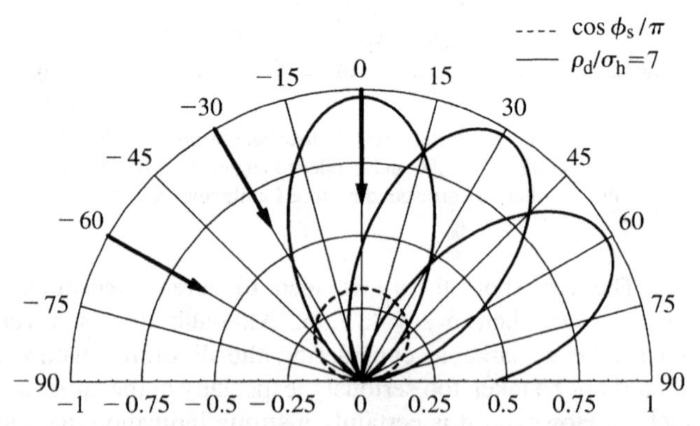

Figure 3-9. Linear-polar plot of the bistatic scatter distributions resulting from the Kirchhoff theory of a very rough surface with $\rho_d/\sigma_h = 7$ and incidence angles $\phi_i = 0°, 30°$, and $60°$. Also shown as a dashed curve is the distribution corresponding to a Lambertian reflector.

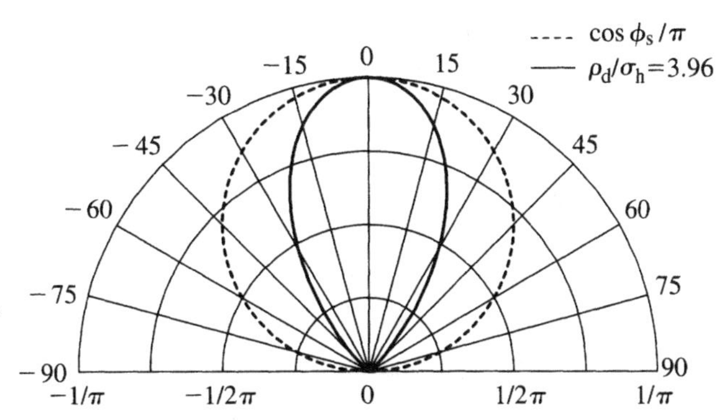

Figure 3-10. Linear-polar plot of the monostatic scatter distribution resulting from the Kirchhoff theory of a very rough surface with $\rho_d/\sigma_h = 3.96$ as a function of incidence angle ϕ_i. Also shown as a dashed curve is the distribution corresponding to a Lambertian reflector.

which may be obtained by letting $\phi_s = \phi_i$ and $\theta_s = \pi$, resulting in

$$\rho_{KB} = \frac{\rho_s \sec^4 \phi_i \cot^2 \phi_0}{4\pi(1 - e^{-\cot^2 \phi_0})} e^{-\tan^2 \phi_i / \tan^2 \phi_0} \tag{3-80}$$

where $v^2/v_z^2 \to \sec^2 \phi_i$ and $v_{xy}^2/v_z^2 \to v^2/v_z^2 - 1 = \tan^2 \phi_i$.

Equation (3-80) is plotted in Figure 3-10 for the case $\rho_d/\sigma_h = 3.96$. For comparison, a Lambertian distribution is also shown since this distribution is frequently used as a simple approximation to the diffuse target reflectivity in signal-to-noise ratio expressions of laser systems. It can be seen that even for the near-Lambertian case of $\rho_d/\sigma_h = 3.96$ (cf. Figure 3-8) there is a considerably faster roll-off with aspect angle ϕ_i than would be predicted using a Lambertian reflector. The reason for this is that the Lambertian model assumes that the scatter distribution is independent of incidence angle, while the Kirchhoff theory as presented here predicts a rotation of the scatter distribution toward the specular direction. This difference might be mitigated by inclusion of higher-order effects, such as multiple scattering and shadowing, but such a study is beyond our present goals.

Comment 3-3. Lambert's cosine law applies to the diffuse emission of radiation from a surface independent of the source of that radiation. Hence it applies to blackbody radiation as well as scatter from a diffusely reflecting surface. The law states that the power per unit solid angle emitted by a differential element of surface area dS depends on the angle ϕ_s measured with respect to the surface normal through the relationship

$$I(\phi) = I_0 \cos \phi_s \tag{3-81}$$

where I_0 is the power per unit solid angle in the direction of the normal and has units of watts per steradian (W/sr). Note that Eq. (3-81) says nothing about the

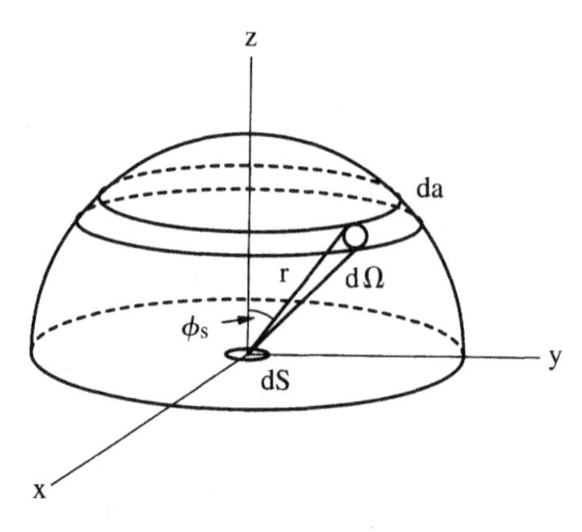

Figure 3-11. Definition of solid angles used in the derivation of the Lambertian distribution.

incidence angle ϕ_i and is independent of the azimuth angle θ_s. Therefore, we can define a differential element of solid angle as the annular area da centered on the surface normal and making an angle ϕ_s with the normal to the mean surface, as shown in Figure 3-11. The area of this annulus on the surface of a sphere located at a distance r from the source is given by

$$da = \int_0^{2\pi} r^2 \sin\phi_s \, d\phi_s \, d\theta_s = 2\pi r^2 \sin\phi_s \, d\phi_s \tag{3-82}$$

The corresponding solid angle is therefore $d\Omega = da/r^2$, or

$$d\Omega = 2\pi \sin\phi_s \, d\phi_s \tag{3-83}$$

The amount of power P contained in the differential solid angle $d\Omega$ is then $dP = I(\phi_s)d\Omega$, and the *radiance*, defined as the power emitted or scattered per unit solid angle per unit area of surface, is

$$L = \frac{dP}{dS\cos\phi_s} = \frac{I_0 \cos\phi_s \, d\Omega}{dS \cos\phi_s \, d\Omega} = \frac{I_0}{dS} \tag{3-84}$$

Hence the radiance of a Lambertian surface is independent of viewing angle.

The total power emitted into the hemisphere is

$$P = \int_0^{\pi/2} \int_0^{2\pi} I(\phi_s) \, d\Omega = 2\pi I_0 \int_0^{\pi/2} \cos\phi_s \sin\phi_s \, d\phi_s = \pi I_0 \qquad \text{(W/sr)} \tag{3-85}$$

If we now define the *reflectivity per steradian* as the ratio of reflected-to-incident power per unit solid angle in the direction of ϕ_s relative to the normal, we have

$$\rho_L(\pi) = \frac{I(\phi_s)}{P} = \frac{I_0 \cos\phi_s}{P} = \frac{\cos\phi_s}{\pi} \qquad (\mathrm{sr}^{-1}) \tag{3-86}$$

In the case of a surface that is not a perfect reflector, Eq. (3-86) can be multiplied by the mean power reflectivity of the surface, ρ_s, giving

$$\rho_L(\pi) = \frac{\rho_s \cos\phi_s}{\pi} \qquad (\mathrm{sr}^{-1}) \tag{3-87}$$

■

The Kirchhoff formulation of the surface scattering problem produces some useful results that provide insight into the general relationship between surface roughness and scatter distribution that may be observed as first-order effects in most surface materials. However, it must be recognized that its range of validity is restricted by the approximations assumed in its formulation. The assumption of large radii of curvature for the surface features can be shown to result in inequalities of the form $kr_c \cos\phi_i \gg 1$, where r_c is the maximum radius of curvature of the local surface feature. From this expression we see that the Kirchhoff approximations restrict the theory to nongrazing angles of incidence. However, this should not be too disturbing, since the theory does not account for multiple scattering and shadowing effects either, which can be significant at grazing incidence. Since the Kirchhoff theory is valid only for large radii of curvature, or equivalently, short wavelengths, the theory is occasionally referred to as a *'high-frequency' theory.*

Having gained some insight into the scatter distributions of a rough surface, the physics associated with the scattered radiation itself now becomes of interest to establish the fundamentals for optical detection theory. This is the subject of the next section.

3.3.2. Integrated Speckle Intensity

A laser system typically consists of a transmitter that places a certain amount of energy on target and a receiver that includes a set of collection optics that focuses the received radiation onto a photodetector surface. Assuming a stationary surface and a constant illumination profile, the speckle pattern generated by a rough surface is time invariant, with a mean granularity size that may or may not be small compared to the receiver aperture diameter. Hence it is important to develop a parameter that characterizes this granularity based on the correlation properties of the scattered field. Goodman[6] was the first to accomplish this using classical coherence theory, which is the primary focus of the sections to follow.

The simplest problem in detection statistics is that of a *point receiver*, that is, one whose aperture diameter is small compared to the smallest granularity of the pattern. However, of a more practical nature, especially for direct-detection receivers, are the statistics associated with an *area receiver*, that is, one in which the aperture diameter is equal to or larger than the mean granularity size of the pattern. In this case, the intensity fluctuations due to speckle are smoothed in a process known as *aperture averaging*. Such detection statistics are investigated in detail in Chapter 4. Here we wish to investigate the physical concepts upon which those statistics will be based. We therefore begin by defining the *correlation diameter* of the field scattered from a rough surface.

The correlation diameter of a speckle field may be defined by the number of correlation cells that fall within a given aperture size. The energy collected by the aperture is referred to as the *integrated energy*, or in the case of intensity, the *integrated intensity*. Consider a receive aperture of finite (i.e., nonzero) dimensions that collects a portion of a randomized beam pattern and passes it through a telescope assembly, which then focuses the radiation onto a detector, also of finite dimensions. (In actuality, the acceptance aperture of a telescope is more properly defined by the *pupil function*. However, for most telescope assemblies, the difference is small, so that we will use these terms synonymously.) Let I_0 be the integrated intensity at the aperture. Since the spatial distribution of the intensity is random, I_0 must be a random variable. Thus, for a circularly symmetric aperture,

$$I_0 = \frac{1}{S} \int_0^{\infty} \int_0^{2\pi} \zeta(r, \theta) I(r, \theta) r \, dr \, d\theta \tag{3-88}$$

where S is the total area of the aperture and ζ is a weighting factor that takes into account spatial nonuniformities in optical transmission through the system or in detector response. We will assume a uniform weighting such that

$$\zeta = \begin{cases} 1 & \text{within the aperture} \\ 0 & \text{outside the aperture} \end{cases} \tag{3-89}$$

By taking the average of Eq. (3-88) and interchanging the order of integration and averaging, the mean integrated intensity becomes

$$\langle I_0 \rangle = \frac{1}{S} \int_0^{\infty} \int_0^{2\pi} \zeta(r, \theta) \langle I \rangle r \, dr \, d\theta \tag{3-90}$$

where $\langle I \rangle$ is the mean intensity of the total pattern. Since $\langle I \rangle$ is independent of r and θ, it may be pulled out of the integral and we find that $\langle I_0 \rangle = \langle I \rangle$. That is, the mean of the integrated energy is equal to the mean energy of the total field. (This is meant in a local sense, of course, since any real and therefore finite speckle pattern will have a spatially dependent mean. See, for example, Figure (2-22.)

Consider now the covariance of the integrated intensity, a statistical quantity that will also prove useful later in the development of turbulence theory. In vector notation, we have

$$\sigma_{I_0}^2 = \frac{1}{S^2} \left\langle \int_0^{\infty} \int_0^{2\pi} \zeta(\vec{r}_1)[I(\vec{r}_1) - \langle I \rangle] \, d\vec{r}_1 \int_0^{\infty} \int_0^{2\pi} \zeta(\vec{r}_2)[I(\vec{r}_2) - \langle I \rangle] \, d\vec{r}_2 \right\rangle \tag{3-91}$$

where the vectors $\vec{r}_1$ and $\vec{r}_2$ lie in the observation plane as shown in Figure 3-2*b* and the averaging brackets indicate an ensemble average over many realizations of the speckle field. Interchanging the order of integration and averaging results in

$$\sigma_{I_0}^2 = \frac{1}{S^2}\int_0^\infty \int_0^{2\pi} \int_0^\infty \int_0^{2\pi} \zeta(\vec{r}_1)\zeta(\vec{r}_2)\langle[I(\vec{r}_1) - \langle I\rangle][I(\vec{r}_2) - \langle I\rangle]\rangle d\vec{r}_1\, d\vec{r}_2 \tag{3-92}$$

But the averaging process under the integral is simply the covariance of intensity, $\sigma_I^2(\vec{r}_1, \vec{r}_2)$, taken at two points in the observation plane. It can be rewritten as

$$\sigma_I^2(\vec{r}_1, \vec{r}_2) = \langle I(\vec{r}_1)I(\vec{r}_2)\rangle - \langle I\rangle^2 \tag{3-93}$$

where $\langle I(\vec{r}_1)\rangle = \langle I(\vec{r}_2)\rangle = \langle I\rangle$ and $\langle I(\vec{r}_1)I(\vec{r}_2)\rangle$ is the autocorrelation function of the intensity at the two points $\vec{r}_1$ and $\vec{r}_2$. Hence Eq. (3-92) becomes

$$\sigma_{I_0}^2 = \frac{1}{S^2}\int_0^\infty \int_0^{2\pi} \int_0^\infty \int_0^{2\pi} \zeta(\vec{r}_1)\zeta(\vec{r}_2)\sigma_I^2(\vec{r}_1, \vec{r}_2)\, d\vec{r}_1\, d\vec{r}_2 \tag{3-94}$$

For the moment, let us revert back to the difference coordinates defined by $\vec{\rho} = \vec{r}_1 - \vec{r}_2$ where $\rho = |\vec{r}_1 - \vec{r}_2| = (\Delta x^2 + \Delta y^2)^{1/2}$ and $\vartheta = \tan^{-1}(\Delta y/\Delta x)$ (cf. Figure 3-2*b*). From the assumption of spatial stationarity, $\sigma_I^2(\vec{r}_1, \vec{r}_2)$ must be a function only of these difference coordinates; thus

$$\sigma_{I_0}^2 = \frac{1}{S^2}\int_0^\infty \int_0^{2\pi} \psi(\vec{\rho})\sigma_I^2(\vec{\rho})\, d\vec{\rho} \tag{3-95}$$

where

$$\psi(\vec{\rho}) = \int_0^\infty \int_0^{2\pi} \zeta(\vec{r}_1)\zeta(\vec{r}_1 - \vec{\rho})\, d\vec{r}_1 \tag{3-96}$$

Here ψ may be identified as the autocorrelation function of the aperture or pupil function. Mathematically, it equates to the overlap integral of the aperture function with itself, with centers displaced by $\vec{\rho}$. Some specific examples of ψ are calculated in the following section.

Since S has units of area and $\sigma_{I_0}^2$ has units of intensity squared, the integral in Eq. (3-95) must have units of intensity squared as well. A normalized integrated intensity $\sigma_{N_0}^2 = \sigma_{I_0}^2/\langle I\rangle^2$ can therefore be defined as

$$\sigma_{N_0}^2 = \frac{1}{S^2}\int_0^\infty \int_0^{2\pi} \psi(\vec{\rho})\sigma_N^2(\vec{\rho})\, d\vec{\rho} \tag{3-97}$$

where

$$\sigma_N^2(\vec{\rho}) = \frac{\langle I(\vec{r}_1)I(\vec{r}_1 - \vec{\rho})\rangle}{\langle I\rangle^2} - 1 \tag{3-98}$$

is the *correlation coefficient* for the two points. Since we have said nothing about the statistics governing the intensity, Eqs. (3-95) and (3-97) may be applied to any stationary random field. Indeed, we now apply the concept of a normalized integrated intensity to the particular problem of a rough surface and later apply this concept to the problem of a turbulent field.

3.3.3. Speckle Correlation Diameter

The numerator on the right side of Eq. (3-98) may be recognized as the *autocorrelation function* of the intensity distribution, which may be written

$$R_I(\vec{r}_1, \vec{r}_2) = \langle I(\vec{r}_1) I(\vec{r}_2) \rangle \tag{3-99}$$

This, in turn, may be related to the autocorrelation function of the fields using the moment theorem for circular complex Gaussian fields.

Comment 3-4. The *moment theorem* for a circular complex Gaussian process is similar to that of multivariate normal distributions for real variables, namely, that the nth-order central product moment is zero if n is odd and is equal to a sum of products of covariances if n is even. Given a set of sample functions from a wide-sense stationary circular Gaussian process in either the spatial or temporal domains [i.e., $u_n \equiv u(t_n)$ or $u_n \equiv u(x_n)$], the moment theorem states that:

(a) If $s \neq t$, then

$$\langle u_{m_1}^* u_{m_2}^* \cdots u_{m_s}^* u_{n_1} u_{n_2} \cdots u_{n_t} \rangle = 0 \tag{3-100}$$

where m_k and n_j are a finite set of integers,

(b) If $s = t$, then

$$\langle u_{m_1}^* u_{m_2}^* \cdots u_{m_t}^* u_{n_1} u_{n_2} \cdots u_{n_t} \rangle = \sum_{\pi} \langle u_{m_{\pi(1)}}^* u_{n_1} \rangle \langle u_{m_{\pi(2)}}^* u_{n_2} \rangle \cdots \langle u_{m_{\pi(t)}}^* u_{n_t} \rangle \tag{3-101}$$

where π is a permutation of the set of integers $1, 2, 3, \ldots, t$. Proof of the theorem above is given by Reed.[7] This theorem leads to several useful identities, such as

$$\langle |u|^{2n} \rangle = n!\, \langle |u|^2 \rangle^n \tag{3-102}$$

$$\langle (u_1^* u_2)^n \rangle = n!\, \langle u_1^* u_2 \rangle^n \tag{3-103}$$

$$\langle u_1^* u_2^* u_3 u_4 \rangle = \langle u_1^* u_3 \rangle \langle u_2^* u_4 \rangle + \langle u_2^* u_3 \rangle \langle u_1^* u_4 \rangle \tag{3-104}$$

An infinity of such identities can be generated with appropriate permutations of the elements; for example, $\langle u_1^* u_2^* u_3^* u_4 u_5 u_6 \rangle$ results in the sum of $3! = 6$ products of three covariances. ■

Using the third identity given above, Eq. (3-99) becomes

$$\begin{aligned} R_I(\vec{r}_1, \vec{r}_2) &= \langle U^*(\vec{r}_1) U^*(r_2) U(\vec{r}_1) U(\vec{r}_2) \rangle \\ &= \langle U^*(\vec{r}_1) U(\vec{r}_1) \rangle \langle U^*(\vec{r}_2) U(\vec{r}_2) \rangle + \langle U^*(\vec{r}_2) U(\vec{r}_1) \rangle \langle U^*(\vec{r}_1) U(\vec{r}_2) \rangle \\ &= \langle I(\vec{r}_1) \rangle \langle I(\vec{r}_2) \rangle + |J_U(\vec{r}_1, \vec{r}_2)|^2 \end{aligned} \tag{3-105}$$

where $J_U(\vec{r}_1, \vec{r}_2) = \langle U^*(\vec{r}_1)U(\vec{r}_2)\rangle$ is the mutual intensity of the complex field amplitude U at points $\vec{r}_1$ and $\vec{r}_2$. Using Eq. (3-99) in Eq. (3-105), we obtain

$$|J_U(\vec{r}_1, \vec{r}_2)|^2 = \langle I(\vec{r}_1)I(\vec{r}_2)\rangle - \langle I(\vec{r}_1)\rangle\langle I(\vec{r}_2)\rangle \tag{3-106}$$

Equation (3-106) can be used to express the square of the absolute magnitude of the degree of coherence μ_{12} as

$$|\mu_{12}|^2 = \frac{|J_U(\vec{r}_1, \vec{r}_2)|^2}{\langle I(\vec{r}_1)\rangle\langle I(\vec{r}_2)\rangle} = \frac{\langle I(\vec{r}_1)I(\vec{r}_2)\rangle}{\langle I(\vec{r}_1)\rangle\langle I(\vec{r}_2)\rangle} - 1 \tag{3-107}$$

Comment 3-5. In Chapter 4 it is shown that for a fully developed speckle field, the standard deviation of the intensity at a point is equal to the mean. Thus, letting $\vec{r}_1 = \vec{r}_2$ in Eq. (3-107) results in $|\mu_{12}|^2 = \text{var}(I)/\langle I\rangle^2 = 1$; that is, the point $\vec{r}_1$ is fully *coherent* with itself. On the other hand, if $\vec{r}_1$ and $\vec{r}_2$ are widely separated, $I(\vec{r}_1)$ and $I(\vec{r}_2)$ become uncorrelated, with the result that $\langle I(\vec{r}_1)I(\vec{r}_2)\rangle \to \langle I(\vec{r}_1)\rangle\langle I(\vec{r}_2)\rangle$ so that $|\mu_{12}|^2 = 0$. The two points defined by $\vec{r}_1$ and $\vec{r}_2$ are then referred to as being *incoherent* with respect to each other. Therefore, the degree of coherence satisfies the requirement $0 \leq |\mu_{12}|^2 \leq 1$. ■

Using Eq. (3-99) in Eq. (3-107) yields for the autocorrelation function of the intensity

$$R_I(\vec{r}_1, \vec{r}_2) = \langle I(\vec{r}_1)\rangle\langle I(\vec{r}_2)\rangle[1 + |\mu_{12}|^2] \tag{3-108}$$

Now it was shown in Section 3.2.2 that as a consequence of the van Cittert–Zernike theorem, the degree of coherence $|\mu_{12}|$ is given by the absolute value of the normalized Fourier transform of the intensity distribution at the aperture. The relationship involves only difference coordinates such as Δx, Δy or equivalently ρ, ϑ, as defined in Figure 3-2*b*, for the case of circular symmetry. Thus Eq. (3-108) can be written

$$R_I(\rho, \vartheta) = \langle I\rangle^2[1 + |\mu_{12}(\rho, \vartheta)|^2] \tag{3-109}$$

where it is assumed that the mean intensities are equal when $\vec{r}_1 = \vec{r}_2$. Inserting Eq. (3-109) into Eq. (3-98) yields

$$\sigma_N^2(\rho, \vartheta) = \frac{R_I(\rho, \vartheta)}{\langle I\rangle^2} - 1 = |\mu_{12}(\rho, \vartheta)|^2 \tag{3-110}$$

Thus the degree of coherence is equal to the correlation coefficient for the two points in question. From the van Cittert–Zernike theorem we know that both of these functions may be obtained from the Fourier transform of the aperture distribution, or in the case of scattering from a rough surface, the field distribution just after reflection. Equation (3-97) therefore becomes

$$\sigma_{N_0}^2 = \frac{1}{S^2}\int_0^\infty \int_0^{2\pi} \psi(\rho, \vartheta)\,|\mu_{12}(\rho, \vartheta)|^2 \rho\, d\rho\, d\vartheta \tag{3-111}$$

Since $\sigma_{I_0}^2$ is a measure of the size of an intensity correlation cell, $\langle I \rangle^2/\sigma_{I_0}^2$ would be a reasonable estimate of the number of correlation cells M within the aperture. But this is simply the reciprocal of $\sigma_{N_0}^2$; hence

$$M = \left[\frac{1}{S^2} \int_0^\infty \int_0^{2\pi} \psi(\rho, \vartheta) \, |\mu_{12}(\rho, \vartheta)|^2 \rho \, d\rho \, d\vartheta \right]^{-1} \tag{3-112}$$

This definition seems intuitively correct since if ρ and ϑ increase, then $\sigma_{N_0}^2$, a measure of the correlation of intensity at two points, must necessarily decrease thereby increasing the speckle count M.

Additional insight into Eq. (3-112) may be obtained by considering its limiting forms. In the limit of large apertures (i.e., large compared to the correlation size of the speckle), $\psi(\rho, \vartheta) \to$ constant and can be taken outside the integral. It is also reasonable to assume that in this limit, $\mu_{12}(\rho, \vartheta) \to 0$. This yields

$$M = \lim_{S \to \infty} \left[\frac{\psi(\rho, \vartheta)}{S^2} \int_0^\infty \int_0^{2\pi} |\mu_{12}(\rho, \vartheta)|^2 \, \rho \, d\rho \, d\vartheta \right]^{-1} = \infty \tag{3-113}$$

In the case of $\psi(\rho, \theta) \to$ constant, Eq. (3-112) can be rearranged to read $M = S_m/S_c$, where $S_m = S^2/\psi(\rho, \theta)$ may be identified as the effective *measurement area* of the receiver aperture and

$$S_c = \int_0^\infty \int_0^{2\pi} |\mu_{12}(\rho, \vartheta)|^2 \, \rho \, d\rho \, d\vartheta \tag{3-114}$$

as the *correlation area* of the speckle distribution alluded to in Section 3.2.1. It should be noted that $S_m \to S$ for a simple aperture function as given by Eq. (3-89).

In the limit of small apertures, $\mu_{12}(\rho, \vartheta) = \mu_{12}(0, 0) \to 1$; hence

$$M = \lim_{S \to 0} \left[\frac{1}{S^2} \int_0^\infty \int_0^{2\pi} \psi(\rho, \vartheta) \rho \, d\rho \, d\vartheta \right]^{-1} = 1 \tag{3-115}$$

and only one correlation cell is intercepted by the aperture. Thus the interpretation of M as the number of correlation cells within the aperture is consistent with intuitive expectations.

We are now in a position to calculate the number of speckle cells for a given aperture function ψ and complex coherence factor μ_{12} using Eq. (3-112). Consider the case of a circular aperture of diameter D intercepting the radiation reflected from a rough surface that is illuminated by a Gaussian beam profile. The autocorrelation function of the aperture corresponds to the overlap area of two circles separated by ρ, ϑ, which is given by

$$\psi(\rho, D) = \begin{cases} \dfrac{D^2}{2} \cos^{-1} \dfrac{\rho}{D} - \dfrac{\rho}{D} \left[\left(1 - \dfrac{\rho^2}{D^2} \right)^{1/2} \right] & \rho \le D \\ 0 & \rho > D \end{cases} \tag{3-116}$$

where ρ is the radial separation of the centers.

Comment 3-6. Since the overlap integral above occurs repeatedly throughout this chapter, it is worthwhile to outline its derivation. Consider two overlapping circles lying on the x-axis as shown in Figure 3-12. The equation relating the radius a and the distance of the center from the origin b is given by

$$a^2 = b^2 + r^2 - 2br\cos\theta \tag{3-117}$$

This provides two solutions for r, namely,

$$r = b\cos\theta \pm \sqrt{a^2 - b^2\sin^2\theta} \tag{3-118}$$

Letting $b = a$ in Eq. (3-118) shows that the plus sign is the correct solution. Thus the total overlap area becomes

$$\begin{aligned} A_0 &= 2\pi a^2 - \int_{-\pi/2}^{\pi/2} r^2\, d\theta \\ &= \pi a^2 - 2b\sqrt{a^2 - b^2} - 2a^2 \tan^{-1}\frac{b}{\sqrt{a^2 - b^2}} \end{aligned} \tag{3-119}$$

Notice that for $b \to 0$, $A_0 \to \pi a^2$ and for $b \to a$, $A_0 \to 0$ as required. Letting $a = D/2$ and $b = \rho/2$, where D is the diameter of the circles and ρ is the separation of the centers, we have

$$\begin{aligned} A_0 &= \frac{\pi D^2}{4} - \frac{D^2}{2}\left(\tan^{-1}\frac{\rho}{\sqrt{1 - \rho^2/D^2}}\right) + \frac{\rho}{D}\sqrt{1 - \frac{\rho^2}{D^2}} \\ &= \frac{D^2}{2}\left[\left(\frac{\pi}{2} - \sin^{-1}\frac{\rho}{D}\right) - \frac{\rho}{D}\sqrt{1 - \frac{\rho^2}{D^2}}\right] \end{aligned} \tag{3-120}$$

Finally, using the identity $\cos^{-1}x + \sin^{-1}x = \pi/2$ in Eq. (3-120) leads to Eq. (3-116).

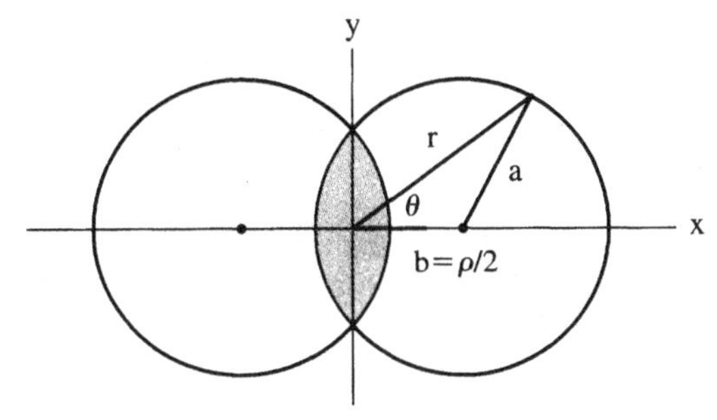

Figure 3-12. Overlap area of integration used in calculating the autocorrelation function of a circular aperture.

■

The complex spatial coherence factor $|\mu_{12}|$ for a circularly symmetric Gaussian illumination spot at normal incidence and with beam waist radius ω_0 may be obtained from Eq. (3-29) using

$$I(\varrho) = \frac{2}{\pi \omega_0^2} e^{-2\varrho^2/\omega_0^2} \tag{3-121}$$

where ϱ is once again a radial coordinate in the source or scattering plane (ξ, η) and $I(\varrho)$ is normalized to the total power P in the beam. We obtain

$$|\mu_{12}| = e^{-k^2 \rho^2 \omega_0^2/8z^2} \tag{3-122}$$

where $\rho = (\Delta x^2 + \Delta y^2)^{1/2}$. Using Eq. (3-122) in the defining expression for the coherence area given by Eq. (3-114) results in

$$S_c = \frac{\lambda^2 z^2}{\pi \omega_0^2} \tag{3-123}$$

A mean *correlation radius* ρ_c may be defined through the relationship $S_c = \pi \rho_c^2$. Using the latter equality in Eq. (3-123), we find that

$$\rho_c = \frac{\lambda z}{\pi \omega_0} \tag{3-124}$$

The right-hand side of Eq. (3-124) also happens to correspond to the far-field Gaussian beam radius that would have occurred at a distance z from the illumination plane if the beam had propagated without scattering. Thus ρ_c can quickly be estimated given knowledge of the illuminating spot size. With the definitions above, Eq. (3-122) becomes, after squaring,

$$|\mu_{12}|^2 = e^{-\rho^2/\rho_c^2} \tag{3-125}$$

Equation (3-112) can now be written

$$M = \left\{ \frac{1}{S^2} \int_0^{\infty} \int_0^{2\pi} \frac{D^2}{2} \left[\cos^{-1} \frac{\rho}{D} - \frac{\rho}{D} \left(1 - \frac{\rho^2}{D^2} \right)^{1/2} \right] e^{-\rho^2/\rho_c^2} \rho \, d\rho \, d\vartheta \right\}^{-1} \tag{3-126}$$

Letting $\rho' = \rho/D$ and $\rho'_c = \rho_c/D$, where D is the aperture diameter, and integrating over ϑ we obtain an expression for the number of speckles M in terms of the normalized variable ρ':

$$M = \frac{\pi}{16} \left\{ \int_0^1 \rho' [\cos^{-1} \rho' - \rho'(1 - \rho'^2)^{1/2}] e^{-\rho'^2/\rho'^2_c} d\rho' \right\}^{-1} \tag{3-127}$$

This can be integrated analytically resulting in

$$M = \beta^2\{1 - e^{-2\beta^2}[I_0(2\beta^2) + I_1(2\beta^2)]\}^{-1} \tag{3-128}$$

where $\beta = 1/2\rho_c' = D/2\rho_c$ is the reciprocal of the normalized correlation diameter and I_0 and I_1 are the modified Bessel functions of the first kind orders 0 and 1, respectively. Equation (3-128) is plotted in Figure 3-13, where it can be seen that the number of speckles $M = 1$ for an aperture that is small compared to the correlation radius, (i.e., a point receiver) and approaches $M = D^2/4\rho_c^2 = S_m/S_c$ for large apertures. Similar results can be obtained with a square aperture[8].

To determine the speckle count M as a function of range z from a scattering surface illuminated normally with a Gaussian beam of radius ω_0 and a given aperture diameter D, Eq. (3-127) can be written

$$M = \frac{\pi}{16}\left\{\int_0^1 \rho'[\cos^{-1}\rho' - \rho'(1-\rho')^{1/2}]e^{-(\pi^2\omega_0^2 D^2/\lambda^2 z^2)\rho'^2}\,d\rho'\right\}^{-1} \tag{3-129}$$

where Eq. (3-124) was used in the definition of ρ_c'. This also integrates to Eq. (3-128) but with $\beta \to 1/2z'$, where $z' = \lambda z/\pi\omega_0 D$ is a normalized range. This is plotted in Figure 3-14, where it can be seen that as z' extends beyond unity, the number of correlation cells begins to approach unity. Similar results can be derived for a square spot and a square receiving aperture (cf. Ref. 6).

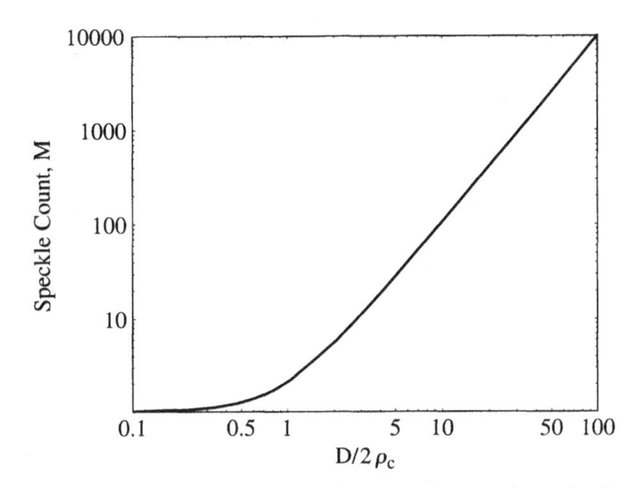

Figure 3-13. Speckle count M versus normalized aperture diameter for a circular aperture and a Gaussian illumination spot. D is the aperture diameter and ρ_c is the correlation radius of the speckle field.

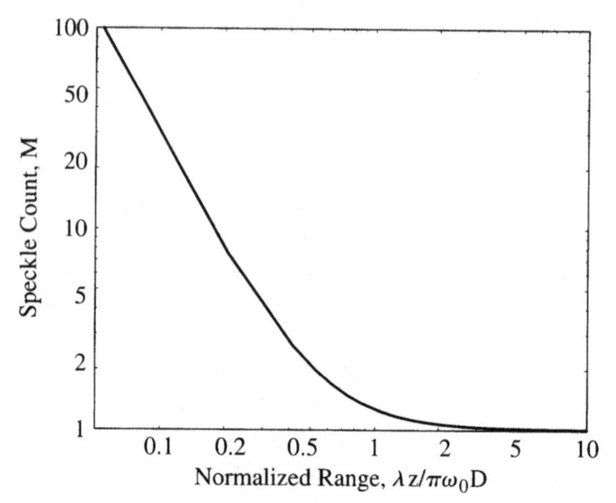

Figure 3-14. Speckle count M versus normalized range for a circular Gaussian illumination spot on a rough surface and a circular receiving aperture of diameter D. λ is the wavelength and ω_0 is the beam spot radius on the illuminated surface.

3.4. PROPAGATION THROUGH TURBULENT MEDIA

In Section 3.3 we considered the correlation properties of radiation scattered from the surface of a target that is rough compared to a wavelength. Consider now the correlation properties of a wave passing through a turbulent atmosphere. The turbulent atmosphere can produce significant perturbations on a monochromatic beam. The nonlinear mixing process of the turbulent flow results in interference between scattered wavelets in the down-range beam profile. As will be shown, this phenomenon is a multiplicative process for the electric field which can significantly affect the detection statistics of both coherent and noncoherent detection systems. Coherent systems are the most susceptible to these effects, due to their sensitivity to phase perturbations across the detector. Indeed, the mixing theorem developed in Chapter 2 states that spatially out-of-phase signal components mixing on a photodetector surface result in reduced signal output, hence the need for diffraction-limited optics to achieve a uniform phase distribution across the detector, even in the absence of turbulence.

The speckle distribution generated by a rough target is an example of a random-phase distribution and would produce similar effects if it was not for the fact that the minimum speckle diameter can be matched to the receive aperture diameter for all ranges. Although this can severely limit the aperture diameter of a short-wavelength system, it at least allows for a deterministic approach to the system design problem. Such is not the case for turbulence-induced speckle, however. In this case the ratio of aperture diameter to the beam transverse correlation diameter increases with range, resulting in a decreased effective aperture for the system. This, together with the degradation of detection statistics due to signal fluctuations, can result in severe range limitations for the coherent system. Furthermore, these effects are not just dependent on wavelength and range as in the target-induced speckle case, but are also dependent on local weather conditions that are necessarily unpredictable and uncontrollable.

The following subsections attempt to familiarize the reader with the basic phenomenology and mathematical foundations of turbulence theory as applied to the propagation of optical beams. It is not intended to be a comprehensive treatment of the subject since this would be well beyond the scope of the book. Topics have been selected based on their importance in the development of the associated detection statistics in Chapters 4 and 5. For those interested in greater detail, there are several excellent review articles, such as those by Lawrence and Strohbehn,[9] Brookner,[10] Clifford,[11] Strohbehn,[12] and Ishimaru[13].

3.4.1. Atmospheric Model

When a laser beam propagates through the atmosphere, it can experience random phase and amplitude fluctuations due to atmospheric turbulence. Turbulence effects on the propagation of laser beams arise from index of refraction fluctuations due to temperature gradients induced in the atmosphere by solar heating. The induced thermal gradients may be viewed as close-packed spherical regions or eddies having statistically varying diameters and indexes of refraction. These weak lenslike eddies, shown graphically in Figure 3-15, result in a randomized interference effect between different regions of the propagating beam such that the down-range beam profile exhibits an intensity pattern similar to, but statistically quite different from, the speckle patterns produced by diffusely reflecting targets. In addition, any component of the wind that is perpendicular to the beam propagation direction will cause temporal variations in the randomized beam pattern. Hence the turbulent eddies can be visualized as being fixed in the atmosphere and moving with the wind. This model is usually referred to as the *'frozen-in' model of turbulence*, implying that the temporal variations of the statistical parameters of interest are due to movement of the airmass, not to variation of the parameters themselves.

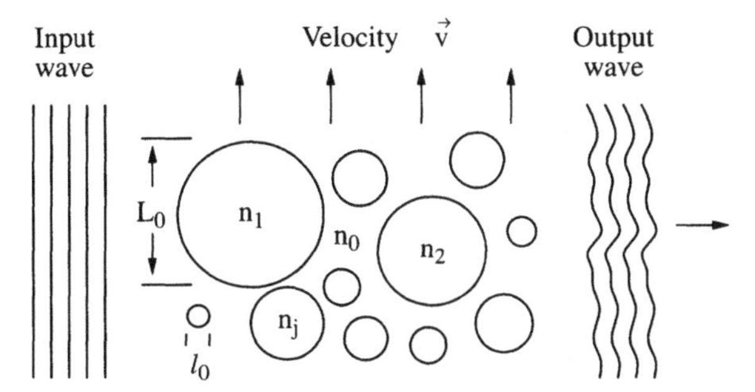

Figure 3-15. Spherical eddies with statistically different indices of refraction used for modeling the index of refraction variations in a turbulent atmosphere. The eddies are assumed to be moving transverse to the beam propagation direction with velocity $\vec{v}$.

Since the granularity of the scintillating beam is similar qualitatively to the speckle induced by target roughness, the coherence theory formalism discussed above can be used to describe some of the salient features of turbulence-induced speckle as well. Before presenting this, however, a brief review of the statistics associated with turbulence theory is warranted.

Turbulence theory was first developed successfully by Kolmogorov[14,15] for the case of incompressible fluids. Tatarskii[16,17] then applied the Kolmogorov theory to propagation of electromagnetic fields and sound waves in the turbulent atmosphere. By their very nature, turbulent media are extremely difficult to describe mathematically, even from a statistical point of view. This is due primarily to the presence of nonlinear mixing of observable quantities, which is fundamental to the process. As a result, most theories rely on several simplifying assumptions in order to reduce the complexity of the mathematics to manageable proportions. Beam propagation in a turbulent atmosphere is no different in this regard, so that some of these assumptions need to be discussed.

In almost all cases it is assumed that the atmosphere is a nondissipative medium for the propagating wave. This has two aspects. First, it does not mean that atmospheric absorption cannot take place, but rather, that if it does, the heat so generated is insignificant compared to diurnal contributions. (Such is not the case for very high powered lasers, however, where thermal blooming can occur. Thermal blooming arises due to local heating of the atmosphere through absorption of radiation from the propagating beam, which in turn perturbs the index of refraction of the medium.) Second, the process of scattering by the turbulent eddies does not result in loss of energy from the beam. Hence the mean energy in the presence of turbulence is assumed equal to the mean energy in the absence of turbulence. This assumption is generally valid for plane and spherical waves but can be violated for finite beams where energy can be scattered out of the beam. As a consequence, finite-beam or *beam-wave* theories are considerably more complex than those for plane or spherical waves and will therefore not be discussed. Plane-wave theory is generally applicable to starlight or laser beams propagated over great distances. On the other hand, spherical waves are good approximations to point sources of light, such as laser beams in the far field.

The concept of *stationarity* for the random variables of the field is usually assumed in order that the mean values of the observable quantities may be considered invariant with respect to time. Given this assumption, the *ergodic hypothesis* may then be invoked, which allows time averages taken at a point in space to be replaced by ensemble averages over many spatial realizations of the turbulent field. Indeed, as will be seen in the following pages, most of the statistical descriptions of optical turbulence effects are cast in the form of ensemble averages. Thus, with the ergodic principle in hand, the concept of temporal stationarity is easily generalized to the concept of *spatial homogeneity*. Homogeneity of the turbulent field implies that the mean values of the random variables are constant throughout the region of interest, and the mixed moments, such as the correlation function, depend only on the difference between two or more points of interest, not where those points are located. A final concept, that of statistical *isotropy*, removes any restrictions as to the directions of the vectors connecting the measurement points, thereby allowing vector equations to assume a simple scalar form.

Unfortunately, most of the meteorological observables encountered in real applications are not stationary or homogeneous (but usually isotropic), but exhibit slowly varying spatial and temporal characteristics. This would be the case, for example, for propagation paths between two different altitudes, in which case the medium is said to be *inhomogeneous* and slowly varying. Observables such as wind speed, temperature, and the refractive index, are driven by diurnal variations, local weather patterns, terrain features, and so on, and can therefore be inhomogeneous even for horizontal propagation paths. As long as these variations are slowly varying, they can be taken into account by integrating the variables over the propagation path. However, statistical theories, especially perturbation theories, may generate large-scale spatial and temporal results that may or may not have significance in understanding small-scale phenomena (i.e., phenomena within the bounds of the measurement sensor). Thus it would be highly desirable to incorporate a statistical measure that functions as a high-pass filter in the sense of removing uninteresting large-scale effects while retaining those with scales of importance to the sensor.

Kolmogorov introduced just such a measure, called the *structure function* $D(\rho)$, which for any random variable α takes the form

$$D_\alpha(\rho) = \langle[\alpha(x_1, t_1) - \alpha(x_2, t_2)]^2\rangle \tag{3-130}$$

The interesting feature about the structure function is that the filtering process mentioned above allows for the description of the field in terms of stationary random functions when the fundamental parameters (e.g., α) may not be stationary. Kolmogorov showed that the structure function of turbulent media conforms to a universal "two-thirds law" for any *conservative passive additive variable*, namely

$$D_\alpha(\rho) = C_\alpha^2 \rho^{2/3} \tag{3-131}$$

where C_α^2 is called the *structure constant* for the random variable α and ρ is the scalar distance between two transverse points in the medium. The concept of a conservative passive additive variable implies that the random variable will not change its value as it is displaced throughout the medium (i.e., conservative), and that the addition of such an observable to the medium does not change its statistics (i.e., passive additive).

Equation (3-131) is only valid for values of ρ within the *inertial subrange* (i.e., $\ell_0 \leq \rho \leq L_0$, where ℓ_0 and L_0 are defined as the inner and outer scales of turbulence, respectively. ℓ_0 is typically on the order of a few millimeters and is ultimately limited on the low end by viscous damping in the medium. A fundamental assumption in all of weak turbulence theory is that $\lambda \ll \ell_0$, where λ is the optical wavelength. L_0 is usually determined by the largest eddy that may be considered isotropic and is typically on the order of several meters in diameter. Both parameters are a function of the height above ground. Phenomena represented by descriptors that lie outside the inertial subrange violate the basic assumptions of the Kolmogorov theory of turbulence and must be treated with care. For the sake of brevity, we consider here only those phenomena that lie within the inertial subrange.

There are several conservative passive additives relevant to optical propagation, the most fundamental being temperature, more specifically, *potential temperature*, which is primarily responsible for driving the index of refraction fluctuations of the medium. It should be noted that absolute temperature T is not a conservative passive additive since it will change its value with altitude as per an adiabatic process. Hence a potential temperature $T_p = T - \gamma_h h$ is defined, where $\gamma_h = 9.8\,^\circ\text{C/km}$ and h is the altitude, such that T_p is conserved. The structure function for T_p is then

$$D_{T_P}(\rho) = C_{T_P}^2 \rho^{2/3} \tag{3-132}$$

where C_T^2 is the temperature structure constant for the medium. The relationship between the temperature of the atmosphere and its index of refraction is given by

$$n = 1 + 77.6(1 + 7.52 \times 10^{-3}\lambda^{-2})\frac{P}{T} \times 10^{-6} \tag{3-133}$$

where P is the pressure of the atmosphere in millibars, $T = T_P + \gamma_h h$ is the absolute temperature in kelvin, and λ is the wavelength of light in micrometers. Contributions to Eq. (3-133) due to humidity have been neglected since they are small at optical wavelengths. Taking the derivative of Eq. (3-133) for a constant wavelength while taking note of the fact that pressure changes are negligible compared to temperature changes leads to

$$dn = -\frac{7.9 \times 10^{-5} P dT_p}{\lambda T^2} \tag{3-134}$$

Equation (3-134) states that for a fixed height, since dT_p is a conservative passive additive, dn must also be a conservative passive additive. Hence the *refraction index structure function* can be written in the form of a two-thirds law:

$$D_n(\rho) = C_n^2 \rho^{2/3} \tag{3-135}$$

where C_n^2 is the *refractive index structure constant* that characterizes the strength of the refractive index fluctuations in the medium. Since $D_n(\rho)$ is dimensionless, C_n^2 has units of $m^{-2/3}$ and is a function of altitude. Equation (3-135) is a fundamental result of the Kolmogorov theory of turbulence and therefore plays a fundamental role in optical turbulence theory.

3.4.2. Weak Turbulence Theory

In this section we show that the solution to the wave equation describing optical propagation in a turbulent atmosphere leads directly to log-normal statistics. We begin with Maxwell's equations for the case of a spatially variant dielectric medium:

$$\vec{\nabla} \cdot \vec{\mathrm{H}} = 0 \tag{3-136}$$

$$\vec{\nabla} \times \vec{\mathrm{E}} = ik\vec{\mathrm{H}} \tag{3-137}$$

$$\vec{\nabla} \times \vec{\mathrm{H}} = -ikn^2\vec{\mathrm{E}} \tag{3-138}$$

$$\vec{\nabla} \cdot (n^2\vec{\mathrm{E}}) = 0 \tag{3-139}$$

where $k = 2\pi/\lambda, n$ is the refractive index of the medium which is a function of position [i.e., $n \equiv n(x, y, z)$], and $\vec{\nabla}$ is the vector gradient operator $(\hat{\mathrm{i}}\partial/\partial x, \hat{\mathrm{j}}\partial/\partial y, \hat{\mathrm{k}}\partial/\partial z)$ and $(\hat{\mathrm{i}}, \hat{\mathrm{j}}, \hat{\mathrm{k}})$ are unit vectors in the x, y, z directions. Taking the curl of Eq. (3-137) and substituting into Eq. (3-138) yields

$$\nabla^2\vec{\mathrm{E}} - \vec{\nabla}(\vec{\nabla} \cdot \vec{\mathrm{E}}) = -n^2k^2\vec{\mathrm{E}} \tag{3-140}$$

Expanding Eq. (3-139) and solving for $\vec{\nabla} \cdot \vec{\mathrm{E}}$ yields

$$\vec{\nabla} \cdot \vec{\mathrm{E}} = -\frac{1}{n^2}\vec{\mathrm{E}}(\vec{\nabla} \cdot n^2) = -\vec{\mathrm{E}} \cdot \vec{\nabla} \ln(n^2) \tag{3-141}$$

where the identity $\vec{\nabla} \ln f(x) = [\vec{\nabla} \cdot f(x)]f'(x)/f(x)$ was used. Substituting Eq. (3-141) into Eq. (3-140) results in

$$\nabla^2\vec{\mathrm{E}} + k^2n^2\vec{\mathrm{E}} + 2\nabla[\vec{\mathrm{E}} \cdot \vec{\nabla} \ln(n)] = 0 \tag{3-142}$$

The last term on the left-hand side represents a mixing term that characterizes the depolarization of the wave. Assuming single scattering events which are characteristic of the weak turbulence approximation being developed here, one might assume that this term may be set equal to zero. In point of fact, it has been shown both theoretically[18] and experimentally[19] that depolarization is insignificant in even for the strongest turbulence conditions. Thus Eq. (3-142) becomes

$$\nabla^2\vec{\mathrm{E}} + k^2n^2\vec{\mathrm{E}} = 0 \tag{3-143}$$

Since there is no coupling between coordinates in Eq. (3-143) independent scalar wave equations can be written for each of the three coordinates. Letting $U \equiv (E_x, E_y, E_z)$, we have

$$\nabla^2U + k^2n^2U = 0 \tag{3-144}$$

This, of course, can be recognized as the Helmholtz wave equation [Eq. (2-1)] with a spatially varying index of refraction.

There are several techniques that one may employ in solving Eq. (3-144). These include the geometric optics approach, the Born approximation, and the Rytov method. Of these, Tatarski has shown that the Rytov method has the wider range of validity, being dependent on the smallness of the gradients of the field rather than the smallness of the field itself. Following the methods of Tatarski, we now introduce the Rytov method by changing variables to $\psi = \ln U$. This transformation results in Eq. (3-144) being transformed to the *Riccati* equation:

$$\nabla^2 \psi + (\nabla \psi) \cdot (\nabla \psi) + k^2 n^2 = 0 \tag{3-145}$$

Letting $n = n_0 + n_1(\vec{r})$, where $n_0 \equiv 1$ is the mean index of refraction of the medium and $n_1(\vec{r})$ is the fluctuating part, which is assumed small relative to the mean, we introduce the method of smooth perturbations by letting $\psi = \psi_0 + \psi_1 + \cdots$. Substituting ψ into Eq. (3-145) and keeping only terms to first order yields the set of equations

$$\nabla^2 \psi_0 + (\nabla \psi_0) + k^2 = 0 \tag{3-146}$$

$$\nabla^2 \psi_1 + 2(\nabla \psi_0) \cdot (\nabla \psi_1) + 2k^2 n_1(\vec{r}) = 0 \tag{3-147}$$

where it was assumed that $n_1^2 \ll 1$ and $\nabla \psi_1 \ll \nabla \psi_0$, and Eq. (3-146) was used in Eq. (3-147). Now since $|\nabla \psi| \approx \left|\nabla e^{ikz}\right| \approx k$, the second of the assumptions above leads to the condition

$$\lambda \left|\nabla \psi_1\right| \ll 2\pi \tag{3-148}$$

which expresses the smallness of the gradient of ψ_1 over distances on the order of a wavelength. Let us also consider the electric field to consist of a series expansion in which the zeroth-order term corresponds to the field in the absence of the medium and the first-order term corresponds to the spatially dependent field perturbed by the medium, that is,

$$U = U_0 + U_1 \tag{3-149}$$

Then, with the Rytov change of variables, we have

$$\begin{aligned} U &= ue^{i\phi} = e^{\psi} \\ U_0 &= u_0 e^{i\phi_0} = e^{\psi_0} \end{aligned} \tag{3-150}$$

or

$$\begin{aligned} \psi &= \ln U = \ln u + i\phi \\ \psi_0 &= \ln U_0 = \ln u_0 + i\phi_0 \end{aligned} \tag{3-151}$$

Thus, since

$$\begin{aligned} \psi_1 &= \psi - \psi_0 = \ln \frac{U}{U_0} = \ln \frac{U_0 + U_1}{U_0} \\ &= \ln \left(1 + \frac{U_1}{U_0}\right) \approx \frac{U_1}{U_0} \end{aligned} \tag{3-152}$$

where $U_1/U_0 \ll 1$, we can write $\psi_1 = U_1/U_0 = U_1 e^{-\psi_0}$, which, when substituted into Eq. (3-147), yields, with the help of Eq. (3-146),

$$\nabla^2 U_1 + k^2 U_1 + 2k^2 n_1(\vec{r}) U_0 = 0 \tag{3-153}$$

The last term on the left side of Eq. (3-153) may be considered a source term in the solution for U_1. The solution is well known in electromagnetic theory and is given by the convolution of this source term with the time-independent Green's function, integrated over all space. Thus

$$U_1(\vec{r}) = 2k^2 \iiint_{V'} n_1(\vec{r}') G(\vec{r} - \vec{r}') U_0(\vec{r}')\, d^3\vec{r}' \tag{3-154}$$

where

$$G(\vec{r} - \vec{r}') = \frac{1}{4\pi} \frac{e^{ik|\vec{r}-\vec{r}'|}}{\left|\vec{r} - \vec{r}'\right|} \tag{3-155}$$

The Green's function represents spherical waves emitted from the source points r' which, if convolved with the free space solution U_0, yield the scattered fields at the points r. If the small angle approximation is used in a manner similar to that employed in the derivation of the Fraunhofer formula of Section 2.2.1, the argument of the exponential can be expanded to yield

$$k|\vec{r} - \vec{r}'| = k(z - z') \left[1 + \frac{|\vec{\varrho} - \vec{\varrho}'|^2}{2(z - z')^2} - \frac{|\vec{\varrho} - \vec{\varrho}'|^4}{8(z - z')^4} + \cdots \right] \tag{3-156}$$

where $\vec{\varrho} = x\hat{i} + y\hat{j}$ and $\vec{\varrho}' = x'\hat{i} + y'\hat{j}$ are the transverse displacements of $\vec{r}$ and $\vec{r}'$ from the z-axis and the vector difference $|\vec{r} - \vec{r}'|$ can be replaced by the scalar difference $z - z'$. The geometry is shown in Figure 3-16.

Now the maximum lateral displacement due to diffraction by the smallest eddy in the field may be estimated to be $|\vec{\varrho} - \vec{\varrho}'| \approx \lambda R/\ell_0$, where R is the distance from the source to the measurement point. Assuming that $z - z' \approx R$, the first term becomes $\pi \lambda R/\ell_0^2$. Hence, if $\lambda \ll \ell_0$ and $\ell_0 \ll \sqrt{\lambda R}$, where $\sqrt{\lambda R}$ is the first Fresnel zone, it is easy to show by direct substitution that the first term in the expansion is much greater than 1, while the second term is much less than 1. Thus, keeping only the first two terms, Eq. (3-154) becomes

$$U_1(\vec{r}) = \frac{k^2}{2\pi} \iiint_{V'} n_1(\vec{r}') U_0(\vec{r}') \frac{e^{ik[(z-z') + |\vec{\varrho} - \vec{\varrho}'|^2/2(z-z')]}}{z - z'} d^3\vec{r}' \tag{3-157}$$

and from Eq. (3-152), the general Rytov solution becomes

$$\psi_1(\vec{r}) = \frac{k^2}{2\pi} \iiint_{V'} n_1(\vec{r}') \frac{U_0(\vec{r}')}{U_0(\vec{r})} \frac{e^{ik[(z-z') + |\vec{\varrho} - \vec{\varrho}'|^2/2(z-z')]}}{z - z'} d^3\vec{r}' \tag{3-158}$$

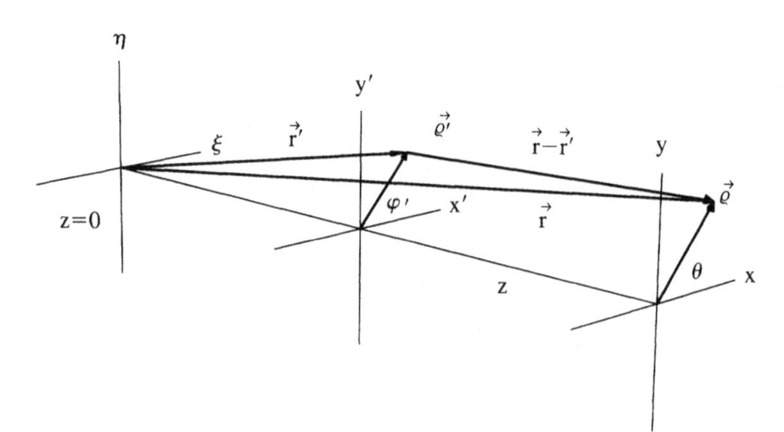

Figure 3-16. Geometry used to describe Rytov's theory of weak turbulence. Beam scintillation occurs at P_2 due to interference between the scattered radiation generated at P_1 and arriving along $\vec{r} - \vec{r}'$ and the primary beam arriving along $\vec{r}'$. Here $\vec{r}$ and $\vec{r}'$ may be considered much larger than $\vec{\varrho}$ and $\vec{\varrho}'$.

As mentioned earlier, we will consider two cases for the free-space propagating beam U_0 given by the plane- and spherical-wave solutions of Eq. (3-144). Thus

$$U_0 = \begin{cases} e^{ikz} & \text{plane wave} \\ \dfrac{e^{i\vec{k}\cdot\vec{r}}}{4\pi\vec{r}} & \text{spherical wave} \end{cases} \tag{3-159}$$

Consider the spherical-wave case first. If we make the same small-angle approximations in $U_0(\vec{r}')$ and $U_0(\vec{r})$ as in the Green's function, that is, $|\vec{r}| \approx z + (x^2 + y^2)/2z$ and $|\vec{r}'| \approx z' + (x'^2 + y'^2)/2z'$, Eq. (3-158) becomes

$$\psi_1(\vec{r}) = \frac{k^2}{2\pi} \iiint_V n_1(\vec{r}') \frac{e^{ik[|\vec{\varrho}' - \gamma\vec{\varrho}|^2/2\gamma(z-z')]}}{\gamma(z - z')} d^3\vec{r}' \tag{3-160}$$

where $\gamma = z'/z$. It is left to the reader to show that when $\gamma = 1$, the plane-wave solution is obtained. This is a very fortuitous occurrence since it necessarily follows that we can multiply $z - z'$ and ϱ by γ in subsequent solutions for the plane-wave case and obtain the corresponding spherical-wave results.

Now, from Eqs. (3-152), we have

$$\begin{aligned} \psi_1 &= \ln \frac{U}{U_0} = \ln\left(\frac{u}{u_0} e^{i(\phi - \phi_0)}\right) = \ln \frac{u}{u_0} + i(\phi - \phi_0) \\ &= \chi + iS \end{aligned} \tag{3-161}$$

where $\chi = \ln(u/u_0)$ and $S = \phi - \phi_0$ are called the *log amplitude* and *phase* of the turbulent medium. Thus we can identify the real and imaginary parts of $\psi_1(\vec{r})$

with χ and S to obtain

$$\chi(\vec{r}) = \frac{k^2}{2\pi}\int_{V'} \frac{n_1(\vec{r}')}{\gamma(z-z')} \cos\frac{k|\vec{\varrho}' - \gamma\vec{\varrho}|^2}{2\gamma(z-z')} d^3\vec{r}' \tag{3-162}$$

and

$$S(\vec{r}) = \frac{k^2}{2\pi}\int_{V'} \frac{n_1(\vec{r}')}{\gamma(z-z')} \sin\frac{k|\vec{\varrho}' - \gamma\vec{\varrho}|^2}{2\gamma(z-z')} d^3\vec{r}' \tag{3-163}$$

Equations (3-162) and (3-163) may be applied to any propagation problem within the constraints imposed by the assumptions outlined above.

The advantage of the Rytov method is that it assumes at the outset that the log-amplitude variable ψ must be additive, which implies that the cumulative effects on the field itself are multiplicative. This can be understood physically by imagining the turbulent medium to consist of a series of thin slabs through which the beam propagates. As it does so, each slab modulates the field from the previous slab's perturbation by some incremental amount. If we represent the jth incremental field by $U_j = e^{\psi_j}$, the total multiplicative field is $U = e^{(\psi_1+\psi_2+\cdots)}$, where the ψ_j are additive and therefore obey Gaussian statistics. In addition, the fact that ψ is a log-normal random variable is consistent with the central limit theorem, which states that the sum of a large number of random variables approaches a Gaussian or normal distribution as the sum approaches infinity.

3.4.2.1. Power Spectral Density. At this point it is worth considering the spectral nature of the refractive index fluctuations of the turbulent field since it is frequently the case that the spatial frequency domain is more amenable to analysis than the spatial domain. To do this we employ some mathematical procedures similar to the spatial Fourier transforms discussed in Chapter 2 but modified somewhat to account for the statistical nature of the problem. In particular, we seek the power spectrum $\Phi_n(\vec{K})$ of the refractive index fluctuations, where $\vec{K} = K_x\hat{i} + K_y\hat{j} + K_z\hat{k}$ is a spatial wave number defined by $K = 2\pi/\ell$, where ℓ is the scale size. We can find Φ_n by developing an expression relating the covariance of the refractive index to the refractive index structure function. Consider, then, the index of refraction as consisting of a mean and a fluctuating part:

$$n(\vec{r}) = \langle n(\vec{r})\rangle + n_1(\vec{r}) \tag{3-164}$$

where $\langle n_1(\vec{r})\rangle = 0$. The spatial spectrum of the fluctuating term can be obtained via a three-dimensional Fourier–Stieltjes transform yielding

$$n_1(\vec{r}) = \int e^{i\vec{K}\cdot\vec{r}}\, dN(\vec{K}) \tag{3-165}$$

where $dN(\vec{K})$ is the random spectral amplitude. The *Fourier–Stieltjes transform* is the statistical analog of the Fourier transform for deterministic functions. Here, both n_1 and N are random variables in their respective spaces (cf. Section 1.2.11).

Now since $\langle n_1(\vec{r})\rangle = 0$, the covariance of the refractive index for two points separated by $\vec{\rho} = \vec{r}_1 - \vec{r}_2$ is equal to the refractive index correlation function; that is,

$$\sigma_n^2(\vec{\rho}) = \langle [n_1(\vec{r}_1) - \langle n_1(\vec{r}_1)\rangle][n_1(\vec{r}_2) - \langle n_1(\vec{r}_2)\rangle]\rangle$$
$$= \langle n_1(\vec{r}_1) n_1(\vec{r}_2)\rangle \equiv B_n(\vec{\rho}) \tag{3-166}$$

where, to be consistent with Tatarski's work, we use the letter B to denote a correlation function. Substituting Eq. (3-165) into Eq. (3-166) yields

$$B_n(\vec{\rho}) = \iint e^{i\vec{K}\cdot\vec{r}_1 - i\vec{K}'\cdot\vec{r}_2} \langle dN(\vec{K})\, dN^*(\vec{K}')\rangle \tag{3-167}$$

Based on the assumption of homogeneity, we know that B_n must depend only on $\vec{\rho} = \vec{r}_1 - \vec{r}_2$ which, in turn, requires that

$$\langle dN(\vec{K}) dN^*(\vec{K}')\rangle = \delta(\vec{K} - \vec{K}')\Phi(\vec{K})\, d^3\vec{K}\, d^3\vec{K}' \tag{3-168}$$

Here δ is the three-dimensional Dirac delta function and $\Phi(\vec{K})$ is the power spectral density of the refractive index fluctuations. Substituting Eq. (3-168) into (3-167), we find that the spatial correlation function and the power spectral density are Fourier transform pairs, a manifestation of the Wiener–Khintchine theorem (cf. Chapter 1). Thus

$$B_n(\vec{\rho}) = \int \Phi(\vec{K}) e^{i\vec{K}\cdot\vec{\rho}} d^3\vec{K} \tag{3-169}$$

and

$$\Phi_n(\vec{K}) = \frac{1}{(2\pi)^3} \int B_n(\vec{\rho}) e^{-i\vec{k}\cdot\vec{\rho}} d^3\vec{\rho} \tag{3-170}$$

If we now invoke the assumption of statistical isotropy, then $B_n(\vec{r}_1, \vec{r}_2) = B_n(\vec{r}_2, \vec{r}_1) = B_n(\rho)$ and only the even (cosine) part of the exponential satisfies the foregoing condition. With these changes we can convert to spherical coordinates, where $d^3\vec{K} = K^2 \sin\theta\, d\theta\, d\phi\, dK$, and integrate Eq. (3-169) over the angular coordinates of the forward hemisphere to obtain

$$B_n(\rho) = \frac{4\pi}{\rho} \int_0^\infty dK\, K\Phi_n(K) \sin K\rho \tag{3-171}$$

It also follows from the assumption of homogeneity that

$$\langle n_1^2(r_1)\rangle = \langle n_1^2(r_2)\rangle = \langle n_1^2\rangle = B_n(0) \tag{3-172}$$

Thus, we can write for the refractive index structure function

$$D_n(\rho) = \langle [n_1(r_1) - n_1(r_2)]^2 \rangle$$
$$= 2B_n(0) - 2B_n(\rho) \qquad \text{(3-173)}$$

where $B_n(\rho) \equiv \langle n_1(r_1) n_1(r_2) \rangle$. Taking the limit of Eq. (3-171) as $\rho \to 0$, while recognizing that $\lim_{\rho \to 0} \sin K\rho / K\rho = 1$, and substituting for $B_n(0)$ in Eq. (3-173) gives

$$D_n(\rho) = 8\pi \int_0^\infty K^2 \Phi_n(K) \left(1 - \frac{\sin K\rho}{K\rho}\right) dK \qquad \text{(3-174)}$$

Notice that $(1 - \sin K\rho / K\rho) \to 0$ for $K < \rho^{-1}$, supporting the notion that $D_n(\rho)$ acts as a high-pass filter that rejects spatial periods that are large compared to the separation of measurement points.

It can be shown through some lengthy arguments that Eq. (3-174) can be inverted to yield[20]

$$\Phi_n(K) = \frac{1}{4\pi^2 K^2} \int_0^\infty \frac{\sin K\rho}{K\rho} \frac{d}{d\rho} \left[r^2 \frac{dD_n(\rho)}{d\rho} \right] d\rho \qquad \text{(3-175)}$$

If the Kolmogorov inertial subrange model given by Eq. (3-135) is substituted into Eq. (3-175), we obtain

$$\Phi_n(K) = \frac{5}{18\pi} C_n^2 K \int_{\ell_0}^{L_0} \rho^{-1/3} \sin K\rho \, d\rho \qquad \text{(3-176)}$$

Assuming that $2\pi/L_0 \ll K \ll 2\pi/\ell_0$ and letting the limits go to 0 and ∞ yields for the power spectral density of the refractive index fluctuations

$$\Phi_n(K) = 0.033 C_n^2 K^{-11/3} \qquad \text{(3-177)}$$

Equation (3-177) was first derived by Kolmogorov and plays a fundamental role in the descriptions of turbulent phenomena. Unfortunately, the Kolmogorov spectrum as written allows for infinitely large spatial wavenumbers K that extend well beyond those given by $2\pi/\ell_0$. Tatarski resolved this deficiency by requiring that the spectrum be exponentially limited for scale sizes less than ℓ_0. This modification, usually referred to as the *Tatarski spectrum*, reads

$$\Phi_n(K) = 0.033 C_n^2 K^{-11/3} e^{-K^2/K_m^2} \qquad \text{(3-178)}$$

where $K_m = 5.92/\ell_0$. As the eddy sizes become very large (very small wavenumbers), Eq. (3-178) suffers from an infinity at $K = 0$. Others[21] have used a modified form of the von Karman spectrum to further improve the range of usefulness of the spectrum, resulting in

$$\Phi_n(K) = \frac{0.033C_n^2}{(K^2 + K_0^2)^{11/6}} e^{-K^2/K_m^2} \tag{3-179}$$

where $K_0 \approx L_0^{-1}$ is a constant that limits the behavior of the expression to scale sizes no larger than L_0 as $K \to 0$. Equations (3-177) through (3-179) are key to the development of many of the useful properties of electromagnetic waves propagating in a turbulent medium, as shown in the following sections.

3.4.2.2. Correlation Function. Of fundamental importance to the statistical description of any random phenomena is the correlation function. In the case of optical propagation in a turbulent atmosphere, it is generally used as a starting point in the derivation of the higher-order moments of the turbulent field. In this section it is used to calculate the variance of the log-amplitude, σ_χ^2, which plays a key role in defining the strength of the turbulence in the log-normal probability density function. Although the derivation of the log-amplitude (and phase) correlation function is somewhat lengthy, it is included here for completeness and for the insights that it provides in the description of several key statistical parameters to follow.

Consider the first Rytov solution given by Eq. (3-160):

$$\psi_1(\vec{r}) = \frac{k^2}{2\pi} \iiint_V n_1(\vec{r}') \frac{e^{ik[|\vec{\varrho}' - \gamma\vec{\varrho}|^2/2\gamma(z-z')]}}{\gamma(z - z')} d^3\vec{r}' \tag{3-180}$$

where $\gamma = z'/z$ or 1 for the spherical- or plane-wave solutions, respectively, and $|\vec{\varrho}' - \gamma\vec{\varrho}|^2 = (x' - \gamma x)^2 + (y' - \gamma y)^2$. We wish to calculate at the position R along the propagation path the following correlation function:

$$B_\chi(\rho) = \langle \chi(R, \vec{\varrho}_1)\chi(R, \vec{\varrho}_2) \rangle \tag{3-181}$$

where χ is again given by Eq. (3-162) and $\rho = |\vec{\varrho} - \vec{\varrho}'|$. Experience has shown that it is mathematically more convenient to use the spectral representation in evaluating Eq. (3-181) rather than substituting directly from Eqs. (3-162) and (3-163). We therefore expand the fluctuating part of the refractive index in terms of a two-dimensional Fourier–Stieltjes transform at the point z' obtaining,

$$n_1(z', \vec{\varrho}') = \int e^{i\vec{\kappa}\cdot\vec{\varrho}'} dN(z', \vec{\kappa}) \tag{3-182}$$

where $\vec{\varrho}' = x'\hat{\imath} + y'\hat{\jmath}$ and $\vec{\kappa} = K_x\hat{\imath} + K_y\hat{\jmath}$. Equation (3-180) therefore becomes after substituting for $n_1(z', \vec{\varrho}')$ and rearranging,

$$\psi_1(\vec{r}) = k^2 \int_0^R dz' \int dN(z', \vec{\kappa}) \iint d\vec{\varrho}' e^{i\vec{\kappa}\cdot\vec{\varrho}'} \frac{e^{-k|\vec{\varrho}'-\gamma\vec{\varrho}|^2/2i\gamma(R-z')}}{2\pi\gamma(R-z')} \tag{3-183}$$

where $d\varrho' = dx'dy'$. Letting $\sigma = \sqrt{i\gamma(R-z')/k}$, the double integral over ϱ' in Eq. (3-183) can be written

$$\iint d\varrho' f(\varrho') = \frac{i}{k} \int_{-\infty}^{\infty} dx' e^{iK_x x'} \frac{e^{-(x'-\gamma x)^2/2\sigma^2}}{\sqrt{2\pi\sigma^2}} \int_{-\infty}^{\infty} dy' e^{iK_y y'} \frac{e^{-(y'-\gamma y)^2/2\sigma^2}}{\sqrt{2\pi\sigma^2}} \tag{3-184}$$

But these integrals are recognizable as Fourier transforms of Gaussian functions the solutions of which are also Gaussian functions (cf. Chapter 1). Thus Eq. (3-183) reduces to

$$\psi_1(\vec{\varrho}) = ik \int_0^R dz' \int dN(z', \vec{\kappa}) e^{i\gamma\vec{\kappa}\cdot\vec{\varrho}} e^{-\sigma^2\kappa^2/2} \tag{3-185}$$

Now we have shown in Eq. (3-161) that χ and S correspond to the real and imaginary parts of ψ_1. Hence

$$\begin{aligned} \chi(\vec{\varrho}) &= \frac{1}{2}[\psi_1(\vec{\varrho}) + \psi_1^*(\vec{\varrho})] \\ S(\vec{\varrho}) &= \frac{1}{2i}[\psi_1(\vec{\varrho}) - \psi_1^*(\vec{\varrho})] \end{aligned} \tag{3-186}$$

Letting $\kappa \to -\kappa$ in Eq. (3-185) and noting that $dN(z', \vec{\kappa}) = dN^*(z', -\vec{\kappa})$, the log-amplitude $\chi(\vec{\varrho})$ becomes, after substituting for ψ_1 and ψ_1^* in Eq. (3-186),

$$\chi(\vec{\varrho}) = k \int_0^R dz' \iint dN(z', \vec{\kappa}) e^{i\gamma\vec{\kappa}\cdot\vec{\varrho}} \sin \frac{-i\sigma^2\kappa^2}{2} \tag{3-187}$$

The log-amplitude correlation function given by Eq. (3-181) can now be written

$$B_\chi(\rho) = k^2 \int_0^R \int_0^R dz'dz'' \iint d\vec{\kappa} e^{i\gamma\vec{\kappa}\cdot\vec{\rho}} \sin^2 \frac{-i\sigma^2\kappa^2}{2} F_n(|z'-z''|, \vec{\kappa}) \tag{3-188}$$

where $\vec{\rho} = \vec{\varrho}_1 - \vec{\varrho}_2$ and use has been made of the two-dimensional analog of Eq. (3-168), namely,

$$\langle dN(z', \vec{\kappa}) dN^*(z'', \vec{\kappa}') \rangle = F_n(|z'-z''|, \vec{\kappa}) \delta(\vec{\kappa} - \vec{\kappa}')\, d\vec{\kappa}\, d\vec{\kappa}' \tag{3-189}$$

Here, $F_n(|z'-z''|, \vec{\kappa})$ is the two-dimensional spectral density of the refractive index fluctuations. We can put Eq. (3-188) into a more useful form by changing variables to $z_d = z' - z''$ and $\eta = (z' + z'')/2$, and by noting that since

$\sin(-i\sigma^2\kappa^2/2) = \sin[\gamma(R - z')\kappa^2/2k]$ is a slowly varying function of z', we can let $z' \to \eta$. Finally, by noting that

$$\int_{-\infty}^{\infty} F_n(|z_d|, \vec{\kappa})\, dz_d = 2\pi\, \Phi_n(\kappa) \tag{3-190}$$

the log-amplitude correlation function becomes

$$B_\chi(\rho) = 2\pi k^2 \int_0^R d\eta \iint d\vec{\kappa} e^{i\vec{\kappa}\cdot\vec{\rho}} \sin^2 \frac{\gamma(R-\eta)\kappa^2}{2k} \Phi_n(\kappa) \tag{3-191}$$

where $\gamma = \eta/R$. We can further reduce the number of integrals in Eq. (3-191) by transforming to cylindrical coordinates and by integrating over the azimuth coordinate ϕ. Thus with $K_x = K\cos\phi$, $K_y = K\sin\phi$, $x_d = \rho\cos\phi'$, $y_d = \rho\sin\phi'$, we have

$$B_\chi(\rho) = (2\pi)^2 k^2 \int_0^R d\eta \int_0^\infty \kappa d\kappa\, J_0(\gamma\kappa\rho) \sin^2 \frac{\gamma(R-\eta)\kappa^2}{2k} \Phi_n(\kappa) \tag{3-192}$$

where $J_0(\gamma\kappa\rho) = (1/2\pi)\int_0^{2\pi} d\phi\, e^{i\gamma\kappa\rho\cos(\phi-\phi')}$ is the zeroth-order Bessel function of the first kind. A similar derivation yields for the phase correlation function

$$B_S(\rho) = (2\pi)^2 k^2 \int_0^R d\eta \int_0^\infty d\kappa\kappa\, J_0(\gamma\kappa\rho) \cos^2 \frac{\gamma(R-\eta)\kappa^2}{2k} \Phi_n(\kappa) \tag{3-193}$$

Tatarski has shown via a relatively complex procedure that the normalized function $b_\chi(\rho) = B_\chi(\rho)/B_\chi(0)$ can be expanded in an infinite power series in the normalized variable $\rho/\sqrt{\lambda R}$, yielding

$$b_\chi(\rho) = 1 - 2.36\left(\frac{k\rho}{R}\right)^{5/6} + 1.71\left(\frac{k\rho}{R}\right) - 0.024\left(\frac{k\rho}{R}\right)^2 + 0.00043\left(\frac{k\rho}{R}\right)^4 + \cdots \tag{3-194}$$

Equation (3-194) is shown in Figure 3-17, where it can be seen that the correlation function oscillates around zero, first going negative in the vicinity of $\rho/\sqrt{\lambda R} \approx 1$. A negative correlation function implies that for two points separated by $\rho \approx \sqrt{\lambda R}$ in the transverse plane, one of the points, on the average, will be dark when the other is bright, and so on. We will see later that this has interesting consequences for the aperture-averaging problem.

A measure of the strength of the amplitude fluctuations, which plays a key role in the detection statistics of later chapters, is the variance of log amplitude commonly referred to as the *Rytov parameter*. It is given by Eq. (3-192) with $\rho = 0$ and the Kolmogorov spectrum, Eq. (3-177), substituted for $\Phi_n(\kappa)$. The latter implies that the scale sizes of interest, as defined by the Fresnel length,

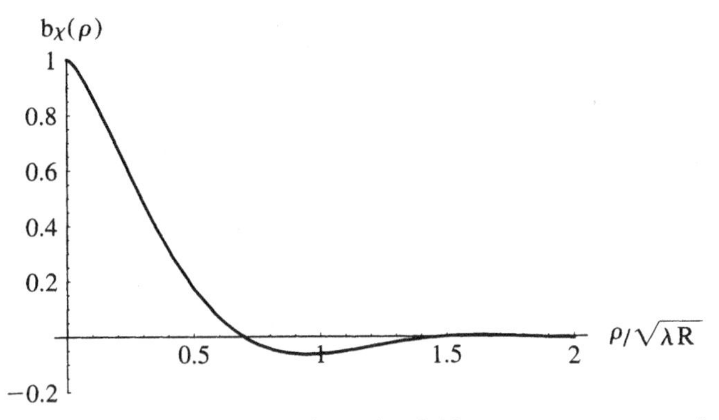

Figure 3-17. Correlation function for a weak turbulent field versus transverse separation ρ normalized to the Fresnel length $\sqrt{\lambda R}$.

are within the inertial subrange (i.e., $\ell_0 \ll \sqrt{\lambda R} \ll L_0$). Since $B_\chi(0) \equiv \sigma_\chi^2$, we obtain for the case of plane waves ($\gamma = 1$),

$$\sigma_\chi^2 = 0.033(2\pi)^2 k^2 \int_0^R d\eta C_n^2(\eta) \int_0^\infty d\kappa \kappa^{-8/3} \sin^2 \frac{(R-\eta)\kappa^2}{2k} \tag{3-195}$$

The inner integral can be solved exactly, yielding

$$\begin{aligned}\sigma_\chi^2 &= 0.033(2\pi)^2 \frac{(\sqrt{3}-1)\Gamma(-5/6)}{8\sqrt{2}} k^2 \int_0^R d\eta C_n^2(\eta) \left(\frac{R-\eta}{k}\right)^{5/6} \\ &= 0.563 k^{7/6} \int_0^R d\eta C_n^2(\eta)(R-\eta)^{5/6}\end{aligned} \tag{3-196}$$

For the case of $C_n^2(\eta) = C_n^2 \equiv$ constant, this integrates to

$$\sigma_\chi^2 = 0.307 C_n^2 k^{7/6} R^{11/6} \tag{3-197}$$

The corresponding spherical-wave case is obtained by inserting $\gamma = \eta/R$ in Eq. (3-192), which leads to

$$\sigma_\chi^2 = 0.563 k^{7/6} \int_0^R d\eta C_n^2(\eta) \left[\frac{\eta(R-\eta)}{R}\right]^{5/6} \tag{3-198}$$

Assuming again that $C_n^2(\eta) = C_n^2 \equiv$ constant, we obtain

$$\begin{aligned}\sigma_\chi^2 &= 0.563 C_n^2 k^{7/6} \left[\frac{3\Gamma(5/6)\Gamma(11/6)}{16\Gamma(5/3)} R^{11/6}\right] \\ &= 0.124 C_n^2 k^{7/6} R^{11/6}\end{aligned} \tag{3-199}$$

3.4.2.3. Mutual Coherence Function. As in the theory of scattered radiation from rough surfaces, mixed moments such as the mutual coherence function play an important role in turbulence theory as well. In this case it is useful in defining the correlation length of the propagating field, which, in turn, will be useful in defining the performance limitations of diffraction-limited systems. Consider two points in the transverse plane of the propagating field defined by the complex scalar functions $U = ue^{\chi(x,y)+i\phi(x,y)}$ and $U' = u'e^{\chi(x',y')+i\phi(x',y')}$. From Section 3.2.1, the mutual coherence function is defined as

$$\Gamma_{12} = \frac{\langle U^* U' \rangle}{U_0^* U_0'} \tag{3-200}$$

where $U_0 = u$ and $U_0' = u'$ are the fields in the absence of turbulence. Letting $\chi \equiv \chi(x, y)$, $\chi' \equiv \chi(x', y')$, $\phi \equiv \phi(x, y)$, and $\phi' \equiv \phi(x', y')$, and substituting for U yields

$$\Gamma_{12} = \langle e^{\chi+\chi'+i(\phi-\phi')} \rangle \tag{3-201}$$

Comment 3-7. It is not too difficult to show that $\chi + \chi'$ and $\phi - \phi'$ are statistically independent, despite the fact that both variables are driven by the same variations in the refractive index. Consider the joint statistic

$$\sigma_{\chi S}^2 = \langle (\chi + \chi')(\phi - \phi') \rangle = \langle \chi\phi \rangle - \langle \chi'\phi' \rangle + \langle \chi'\phi \rangle - \langle \chi\phi' \rangle \tag{3-202}$$

If $\chi + \chi'$ and $\phi - \phi'$ are correlated, this quantity will be nonzero. For a homogeneous medium, $\langle \chi\phi \rangle = \langle \chi'\phi' \rangle$, and for an isotropic medium $\langle \chi'\phi \rangle = \langle \chi\phi' \rangle$ so that $\sigma_{\chi S}^2 = 0$. Hence, $\chi + \chi'$ and $\phi - \phi'$ must be statistically independent Gaussian random variables. ■

Now it can be shown that any pair of statistically independent Gaussian random variables α and β satisfy the relationship

$$\langle e^{a\alpha+b\beta} \rangle = e^{\{(1/2)[a^2\langle(\alpha-\overline{\alpha})^2\rangle+b^2\langle(\beta-\overline{\beta})^2\rangle]+a\overline{\alpha}+b\overline{\beta}\}} \tag{3-203}$$

where a and b are constants. Thus

$$\Gamma_{12} = e^{\{(1/2)[\langle(\chi+\chi'-2\overline{\chi})^2\rangle-\langle(\phi-\phi')^2\rangle]+2\overline{\chi}\}} \tag{3-204}$$

Expanding the first averaging bracket yields

$$\begin{aligned} \langle(\chi + \chi' - 2\overline{\chi})^2\rangle &= \langle[(\chi - \overline{\chi}) + (\chi' - \overline{\chi'})]^2\rangle \\ &= \langle(\chi - \overline{\chi})^2 + (\chi' - \overline{\chi'})^2 + 2(\chi - \overline{\chi})(\chi' - \overline{\chi'})\rangle \end{aligned} \tag{3-205}$$

For a homogeneous and isotropic field, the first and second central moments are independent of position. Hence Eq. (3-205) reduces to

$$\langle(\chi + \chi' - 2\overline{\chi})^2\rangle = 2\sigma_\chi^2 + 2\langle(\chi - \overline{\chi})(\chi' - \overline{\chi'})\rangle \tag{3-206}$$

where $\sigma_\chi^2 = \langle(\chi - \overline{\chi})^2\rangle$ is the variance of log amplitude. Noting that $D_\phi(\rho) \equiv \langle(\phi_s - \phi'_s)^2\rangle$ from Eq. (3-116), where $D_\phi(\rho)$ is the phase structure function, Eq. (3-204) becomes

$$\Gamma_{12} = e^{\sigma_\chi^2 + \langle(\chi - \overline{\chi})(\chi' - \overline{\chi'})\rangle + 2\overline{\chi} - (1/2)D_\phi(\rho)} \tag{3-207}$$

Now the covariance of log amplitude is given by $\sigma_\chi^2(\rho) = \langle(\chi - \overline{\chi})(\chi' - \overline{\chi'})\rangle = \langle\chi\chi'\rangle - \overline{\chi}^2$. Also, it will be shown in the next section that the mean and variance of the log amplitude are related by $\langle\chi\rangle = -\sigma_\chi^2(0) \equiv -\sigma_\chi^2$, a well-known relationship for any log-normal random variable. With these substitutions, Eq. (3-207) becomes

$$\Gamma_{12} = e^{\sigma_\chi^2(\rho) - \sigma_\chi^2 - (1/2)D_\phi(\rho)} \tag{3-208}$$

In a similar manner, it is straightforward to show from the definition of the variance of a random variable that

$$\begin{aligned} D_\chi(\rho) &= \langle[\chi - \chi']^2\rangle = 2\langle\chi^2\rangle - 2\langle\chi\chi'\rangle \\ &= 2\sigma_\chi^2 - 2\sigma_\chi^2(\rho) \end{aligned} \tag{3-209}$$

We therefore obtain

$$\Gamma_{12}(\rho) = e^{-(1/2)D_\chi(\rho) - (1/2)D_\phi(\rho)} \tag{3-210}$$

The sum of the log amplitude and phase structure functions as derived by the Rytov method is defined as the *wave structure function, $D(\rho)$*. Thus a simple relationship is found between the mutual coherence function and the wave structure function, namely,

$$\Gamma_{12}(\rho) = e^{-(1/2)D(\rho)} \tag{3-211}$$

Equation (3-211) will prove useful in Section 3.4.2.6 in the study of the sensitivity of coherent optical systems operating in turbulent atmospheres.

The form of the wave structure function can be obtained with the help of Eqs. (3-192) and (3-193). With $B_\chi(\rho) = \langle\chi\chi'\rangle$ and $B_\chi(0) = \langle\chi^2\rangle$, the second form of the log amplitude structure function given in Eqs. (3-209) can be written

$$D_\chi(\rho) = 2B_\chi(0) - 2B_\chi(\rho) \tag{3-212}$$

and similarly for the phase structure function,

$$D_S(\rho) = 2B_S(0) - 2B_S(\rho) \tag{3-213}$$

Thus

$$D(\rho) = 2[B_\chi(0) - B_\chi(\rho)] + 2[B_S(0) - B_S(\rho)] \tag{3-214}$$

Performing these sums and using Eqs. (3-192) and (3-193) leads to

$$D(\rho) = 8\pi^2 k^2 \int_0^R d\eta \int_0^\infty d\kappa\kappa\,\Phi_n(\kappa)[1 - J_0(\kappa\rho\gamma)] \tag{3-215}$$

Inserting the Kolmogorov spectrum given by Eq. (3-177) for the case of homogeneous turbulence [i.e., $C_n^2(\eta) = C_n^2$], gives

$$D(\rho) = (0.033)8\pi^2 k^2 C_n^2 \int_0^R d\eta \int_0^\infty d\kappa\kappa^{-8/3}[1 - J_0(\kappa\rho\gamma)] \tag{3-216}$$

The integral is of the form[22]

$$\int_0^\infty dx x^{-p}[1 - J_0(ax)] = -\frac{2^{-p}\Gamma[(1-p)/2]}{\Gamma[(1+p)/2]} a^{p-1} \qquad 1 < p < 3 \tag{3-217}$$

which for plane waves ($\gamma = 1$) leads to

$$D(\rho) = 2.91 k^2 C_n^2 R \rho^{5/3} \tag{3-218}$$

and for spherical waves ($\gamma = \eta/R$),

$$D(\rho) = 1.09 k^2 C_n^2 R \rho^{5/3} \tag{3-219}$$

Equations (3-218) and (3-219) are valid within the inertial subrange (i.e., for $\ell_0 \ll \rho \ll L_0$). Tatarski has shown that for both the plane- and spherical-wave cases, the primary contribution to the wave structure function in the Rytov approximation is from the phase structure function, the contribution from the log-amplitude structure function being negligible by comparison.

Lutomirski and Yura[23] point out that the Kolmogorov spectrum can lead to unphysical results when extrapolated to wavenumbers that are large compared to L_0, despite the convergence of the integrals above. They suggest the use of the modified von Kármán spectrum given by Eq. (3-179) in Eq. (3-215). This leads to $\rho^{1/3}$ correction terms in Eqs. (3-218) and (3-219) and a $\rho^{3/2}$ dependence for the wave structure functions. For the sake of simplicity of discussion, we do not develop these expressions here but refer the interested reader to the article by Lutomirski and Yura.

Inserting Eqs. (3-218) and (3-219) into Eq. (3-211) and rearranging leads to

$$\Gamma_{12}(\rho) = e^{-(\rho/\rho_0)^{5/3}} \tag{3-220}$$

where we have let $D(\rho) = 2(\rho/\rho_0)^{5/3}$ with

$$\rho_0 \approx \begin{cases} (1.455 k^2 C_n^2 R)^{-3/5} & \text{plane waves} \\ (0.545 k^2 C_n^2 R)^{-3/5} & \text{spherical waves} \end{cases} \tag{3-221}$$

Here $\rho = \rho_0$ is the separation in measurement points where the mutual coherence function falls to the e^{-1} value. It is usually referred to as the *transverse coherence length* of the propagating field. It will be shown in Section 3.4.2.6 that ρ_0 places fundamental limitations on the performance of a coherent optical system.

3.4.2.4. Statistics of the Turbulent Field. Although theoretical treatments of optical propagation involving diffraction are necessarily discussed in terms of the amplitude and phase, it is the optical intensity that is ultimately measured and therefore might be considered the more fundamental descriptor. As stated earlier, numerous experiments have confirmed that the log intensity obeys a normal distribution, that is,

$$p(\ell) = \frac{1}{\sqrt{2\pi\sigma_\ell^2}} e^{-(\ell-\langle\ell\rangle)^2/2\sigma_\ell^2} \tag{3-222}$$

Here $\ell = \ln(I/\langle I\rangle)$ is the log intensity, I the optical intensity at a point, and σ_ℓ^2 the variance of log intensity. It is σ_ℓ^2 that is usually measured in characterizing the strength of the optical turbulence, and for the case of a uniform horizontal propagation path, are related to the variance of log amplitude derived earlier by [cf. Eqs. (3-197) and (3-199)]

$$\sigma_\ell^2 = 4\sigma_\chi^2 = \begin{cases} 1.23C_n^2 k^{7/6} R^{11/6} & \text{plane wave} \\ 0.496C_n^2 k^{7/6} R^{11/6} & \text{spherical wave} \end{cases} \tag{3-223}$$

where $k = 2\pi/\lambda$. Here C_n^2 is assumed to be uniform over the propagation path and typically ranges from 10^{-15} m$^{-2/3}$ (weak turbulence) to 10^{-12} m$^{-2/3}$ (strong turbulence). The transition from weak to strong turbulence has been found to occur in the range $1 < \sigma_\ell^2 < 2$.

By noting that $\langle e^\ell\rangle = \langle I/\langle I\rangle\rangle = 1$, it is easily shown using Eq. (3-222) that $\langle\ell\rangle = -\sigma_\ell^2/2$ such that Eq. (3-222) assumes the well-known form

$$p(\ell) = \frac{1}{\sqrt{2\pi\sigma_\ell^2}} e^{-(\ell+\sigma_\ell^2/2)^2/2\sigma_\ell^2} \tag{3-224}$$

The normalized variance of intensity may be related to the variance of log intensity as follows. Letting the variance of intensity be given by σ_I^2 and the mean intensity by $\langle I\rangle$, we have

$$\sigma_I^2 = \langle I^2\rangle - \langle I\rangle^2 = \langle I\rangle(\langle e^{2\ell}\rangle - \langle e^\ell\rangle^2) = \langle I\rangle(\langle e^{2\ell}\rangle - 1) \tag{3-225}$$

Using Eq. (3-203) for the case of a single variable, that is,

$$\langle e^{ag}\rangle = e^{a\langle g\rangle+(1/2)a^2\langle[g-\langle g\rangle]^2\rangle} \tag{3-226}$$

with $a = 1$, we find that $\langle e^{2\ell} \rangle = e^{\sigma_\ell^2}$, such that the normalized variance of intensity becomes

$$\sigma_N^2 = \frac{\sigma_I^2}{\langle I \rangle^2} = e^{\sigma_\ell^2} - 1 \tag{3-227}$$

It is frequently the case in calculating detection statistics that it is more convenient to work with the density function for the intensity rather than the log intensity. Following the usual procedures for transformation of random variables, Eq. (3-224) becomes with $\ell \to I/\langle I \rangle$,

$$p\left(\frac{I}{\langle I \rangle}\right) = \frac{1}{\sqrt{2\pi\sigma_\ell^2}} \frac{\langle I \rangle}{I} e^{-[\ln(I/\langle I \rangle) + (1/2)\sigma_\ell^2]^2/2\sigma_\ell^2} \tag{3-228}$$

By convention, the log amplitude χ is defined in terms of the log intensity ℓ via the relationship

$$\frac{I}{\langle I \rangle} = e^{\ell} \equiv e^{2\chi} \tag{3-229}$$

such that $\chi = \ell/2$. The probability density function for χ is therefore obtained by transforming Eq. (3-222) as follows:

$$p(\chi) = p(\ell)\frac{d\ell}{d\chi}\bigg|_{\ell \to 2\chi} = \frac{2}{\sqrt{2\pi\sigma_\ell^2}} e^{-4(\chi - \langle \chi \rangle)^2/2\sigma_\ell^2} \tag{3-230}$$

If in Eq. (3-230) we let $\sigma_\ell^2 = 4\sigma_\chi^2$ and calculate $\langle e^{2\chi} \rangle \equiv 1$, we find that $\langle \chi \rangle = -\sigma_\chi^2$ and the log-amplitude probability density function becomes

$$p(\chi) = \frac{1}{\sqrt{2\pi\sigma_\chi^2}} e^{-(\chi + \sigma_\chi^2)^2/2\sigma_\chi^2} \tag{3-231}$$

Using methods similar to those used to develop Eq. (3-227), it is straightforward to show that the normalized variance of amplitude is given by

$$\sigma_n^2 = \frac{\sigma_u^2}{\langle u \rangle^2} = 1 - e^{-\sigma_\chi^2} \tag{3-232}$$

where $I = u^2$. The probability density function for the normalized amplitude follows from Eq. (3-231):

$$p\left(\frac{u}{\langle u \rangle}\right) = \frac{1}{\sqrt{2\pi\sigma_\chi^2}} \frac{\langle u \rangle}{u} e^{-[\ln(u/\langle u \rangle) + (1/2)\sigma_\chi^2]^2/2\sigma_\chi^2} \tag{3-233}$$

where the transformation $\chi \to \ln(u/\langle u \rangle)$ was used.

3.4.2.5. Aperture Averaging in Direct-Detection Systems. We now return to our general equation for the normalized integrated intensity, Eq. (3-97), and redefine the normalized covariance of intensity σ_N^2 (frequently referred to as *covariance of irradiance* in turbulence literature) to account for the transverse correlations of a beam propagating in a turbulent atmosphere. This will then be used to define an aperture-averaging factor that takes into account the reduced signal variances associated with apertures that are large compared to the transverse correlation size of the propagating beam. This factor can easily be included in the point-detection statistics of the log-intensity field, a relatively straightforward process that is discussed in Chapter 4. Historically, the aperture-averaging factor was originally introduced by Fried[24] in lieu of a more complete theory describing atmospherically induced speckle analogous to that for surface-induced speckle and mathematical limitations associated with using log-normal statistics.

For the case of a turbulent atmosphere, the covariance of irradiance $\sigma_I^2(\rho)$ in Eq. (3-97) can be calculated as follows. We have seen that

$$\sigma_I^2(\rho) = \langle [I(x,y) - \langle I \rangle][I(x',y') - \langle I \rangle] \rangle \tag{3-234}$$

where $\rho = x - x'$. In terms of the log intensity ℓ, we have

$$\sigma_I^2(\rho) = \langle I \rangle^2 \langle [e^{\ell(x,y)} - 1][e^{\ell(x',y')} - 1] \rangle \tag{3-235}$$

Thus, by adopting the notation $\ell(x,y) \equiv \ell$ and $\ell(x',y') \equiv \ell'$, and using Eq. (3-203), we obtain

$$\begin{aligned} \sigma_I^2(\rho) &= \langle I \rangle^2 (\langle e^{\ell+\ell'} \rangle - \langle e^{\ell} \rangle - \langle e^{\ell'} \rangle + 1) \\ &= \langle I \rangle^2 \{ e^{2\langle \ell \rangle + (1/2)\langle [(\ell+\ell') - \langle \ell+\ell' \rangle]^2 \rangle} - 1 \} \end{aligned} \tag{3-236}$$

where the energy relations $\langle e^{\ell} \rangle = \langle e^{\ell'} \rangle = 1$ and $\langle \ell \rangle = \langle \ell' \rangle$ were used. Expanding the exponent in the exponential and collecting terms reduces Eq. (3-236) to

$$\sigma_I^2(\rho) = \langle I \rangle^2 \{ e^{2\langle \ell \rangle + (1/2)[\langle \ell^2 \rangle - \langle \ell \rangle^2 + \langle \ell'^2 \rangle - \langle \ell' \rangle^2 + 2\langle \ell\ell' \rangle - 2\langle \ell \rangle \langle \ell' \rangle]} - 1 \} \tag{3-237}$$

But from the definition of the variance of log intensity (i.e., $\sigma_\ell^2 = \langle \ell^2 \rangle - \langle \ell \rangle^2$) and with the assumption of the homogeneity of the field, the normalized covariance of irradiance becomes, with $\langle \ell \rangle = -\sigma_\ell^2/2$,

$$\sigma_N^2(\rho) = \frac{\sigma_I^2(\rho)}{\langle I \rangle^2} = e^{\sigma_\ell^2(\rho)} - 1 \tag{3-238}$$

where $\sigma_\ell^2(\rho) = \langle \ell\ell' \rangle - \langle \ell \rangle \langle \ell' \rangle$. Note that when $\rho \to 0$, Eq. (3-238) reduces to the normalized *variance* of irradiance given by Eq. (3-227) with $\sigma_\ell^2(0) \equiv \sigma_\ell^2$.

Consider now the normalized integrated intensity given by Eq. (3-97) for the case of a circular aperture. We have

$$\sigma_{N_0}^2 = \frac{1}{S^2} \int_0^{2\pi} \int_0^1 \psi(\rho')\sigma_N^2(\rho')\rho' d\rho' \, d\vartheta \tag{3-239}$$

where $\rho' = \rho/D$, ρ and ϑ are difference coordinates, and $S = \pi D^2/4$ is the area of the aperture. (It is sometimes the case that $\sigma_{N_0}^2$ is identified as the variance of the signal, but this is true only if we define the signal as the optical intensity at the aperture, not the electrical signal out of the detector. The latter requires that we include the statistics of the photoemmissive process as well (cf. Section 4.4).

Using Eq. (3-116) for the aperture function while dividing by the normalized variance of irradiance given by Eq. (3-227), we obtain, after integrating over ϑ,

$$\Theta = \frac{\sigma_{N_0}^2}{\sigma_N^2} = \frac{16}{\pi} \int_0^1 \rho'[\cos^{-1}\rho' - \rho'(1-\rho')^{1/2}] \frac{\sigma_I^2(\rho')}{\sigma_I^2(0)} d\rho' \tag{3-240}$$

Θ, referred to as an *aperture averaging factor*, assumes values in the range $0 \leq \Theta \leq 1$. Thus a spatially averaged variance of log intensity σ_Θ^2 may be defined as

$$\sigma_N^2 = \Theta(e^{\sigma_\ell^2} - 1) \equiv e^{\sigma_\Theta^2} - 1 \tag{3-241}$$

or

$$\sigma_\Theta^2 = \ln[\Theta(e^{\sigma_\ell^2} - 1) + 1] \tag{3-242}$$

where σ_N^2 is the normalized variance of intensity and σ_ℓ^2 is the single-point variance of log intensity.

Comment 3-8. The reasons for developing an aperture-averaging factor Θ rather than a turbulence-induced speckle count M_{atm} analogous to the target-induced speckle count M are twofold. First, even though the atmosphere is modeled as a set of closely packed independent spherical eddies, the field does not conform fully to this model. The very fact that the correlation function of the field is negative for certain values of ρ implies that such a simple model is only an approximation to the real situation. This will become more apparent later in this section. Second, even if M_{atm} could be calculated by using the reciprocal of Eq. (3-239), for example, the use of such a parameter in the detection statistics would require knowledge of the characteristic function for the log intensity distribution, which has yet to be developed mathematically. ■

Calculation of Θ is accomplished by substituting Eqs. (3-227) and (3-238) into Eq. (3-240), which yields

$$\Theta = \frac{16}{\pi} \int_0^1 \rho'[\cos^{-1}\rho' - \rho'(1-\rho')^{1/2}] \frac{e^{\sigma_\ell^2(D\rho')} - 1}{e^{\sigma_\ell^2(0)} - 1} d\rho' \tag{3-243}$$

Evaluation of the integral in Eq. (3-243) requires calculation of the covariance of the log intensity, $\sigma_\ell^2(\rho)$. Intuitively, one might expect a D^{-2} dependence[25] for Θ. However, we will now show[26] that by using Eqs. (3-227) and (3-238) in the limit of small fluctuations, in which case $\sigma_N^2(0) \approx \sigma_\ell^2(0)$ and $\sigma_N^2(\rho) \approx \sigma_\ell^2(\rho)$, respectively, the dependence is more like $D^{-7/3}$. When substituted directly into Eq. (3-240), these approximations yield

$$\Theta = \frac{16}{\pi} \int_0^1 \rho' [\cos^{-1} \rho' - \rho'(1 - \rho')^{1/2}] \frac{\sigma_\ell^2(\rho')}{\sigma_\ell^2(0)} d\rho' \tag{3-244}$$

where $\sigma_\ell^2(0)$ is given by either of Eqs. (3-223).

Under the assumption of $\langle \chi \rangle = 0$, the covariance of the log amplitude is equal to the log-amplitude correlation function. The latter was derived earlier and is given by Eq. (3-193). Thus for the case of a smoothly varying turbulent medium, with $\sigma_\ell^2(\rho) = 4\sigma_\chi^2(\rho) = 4B_\chi(\rho)$ and interchanging the order of integration, we have

$$\sigma_\ell^2(\rho) = 16\pi^2 k^2 \int_0^\infty d\kappa \kappa \Phi_n(\kappa) \int_0^R d\eta J_0(\kappa\rho\gamma) \sin^2 \frac{\gamma(R - \eta)\kappa^2}{2k} \tag{3-245}$$

where J_0 is the zero-order Bessel function of the first kind, R is the path length, γ is a geometric factor that is unity for a plane wave and η/R for a spherical wave, $k = 2\pi/\lambda$ is the optical wavenumber, and $\Phi_n(\kappa)$ is the spectrum of refractive index fluctuations, which for a small inner scale of turbulence (i.e., $\ell_0 \ll \sqrt{R/k}$), is given by Eq. (3-177). Inserting the latter into Eq. (3-245) leads to

$$\sigma_\ell^2(\rho) = 5.21 k^2 C_n^2 \int_0^\infty d\kappa \kappa^{-8/3} \int_0^R d\eta J_0(\kappa\rho\gamma) \sin^2 \frac{\gamma(R - \eta)\kappa^2}{2k} \tag{3-246}$$

[Additional cases may be considered for various ratios of coherence diameter to inner scale size ℓ_0 and Fresnel zone $\sqrt{R/k}$, in which case Tatarski's power spectral density, Eq. (3-178), is used in Eq. (3-245) instead of Eq. (3-177).]

Changing variables to $\rho' = \rho/D$ in Eq. (3-246) and inserting into Eq. (3-244) yields for the plane-wave case,

$$\Theta = 21.6 k^{5/6} R^{-11/6} \int_0^\infty d\kappa \kappa^{-8/3} \int_0^R d\eta \sin^2 \frac{(R - \eta)\kappa^2}{2k}$$
$$\times \int_0^1 d\rho' \rho' J_0(\kappa D \rho') [\cos^{-1} \rho' - \rho'(1 - \rho')^{1/2}] \tag{3-247}$$

where the plane-wave version of Eqs. (3-223) was used. The integrals over ρ' and η can be done exactly, the former evaluating to $\pi J_1^2(\kappa D/2)/\kappa^2 D^2$. Letting $u = \kappa^2 R/k$ in the resulting expression leads to

$$\Theta_{\text{pw}} = 8.47 \left(\frac{kD^2}{4R} \right)^{5/6} \int_0^\infty du\, u^{-14/3} J_1^2(u) \left(1 - \frac{kD^2}{4Ru^2} \sin \frac{4Ru^2}{kD^2} \right) \tag{3-248}$$

For small aperture, $J_1(u) \approx 1/2u$, which results in $\Theta = 1$ (i.e., no spatial averaging). For large apertures, $kD^2/4R \gg 1$, and u will be small. Expanding the sine function in powers of u and keeping only the first two terms leads to

$$\Theta_{\text{pw}} \approx 1.41 \left(\frac{kD^2}{4R}\right)^{-7/6} \int_0^\infty du\, u^{-2/3} J_1^2(u) = 0.932 \left(\frac{kD^2}{4R}\right)^{-7/6} \tag{3-249}$$

Equation (3-249) shows a slight departure from the D^{-2} behavior that one might have expected on the basis of intuitive arguments. The reasons for this are that if the true model of the irradiance fluctuations consisted of independent, nonoverlapping spherical regions in the spatial domain, the argument for a D^{-2} law would be quite valid. However, the effects of small-angle scattering into adjacent regions of the beam causes the correlation function to go negative for some separation ρ (cf. Figure 3-17). It has been suggested that this negative correlation coefficient is responsible for the departure from the expected D^{-2} behavior.

Given the limiting behavior of Eq. (3-248), a simple interpolation formula suitable for engineering applications can be developed. Using the results given by Eq. (3-249), we have for the case of infinite plane waves in homogeneous, isotropic media,

$$\Theta_{\text{pw}} \approx \frac{1}{1 + 1.07(kD^2/4R)^{7/6}} \tag{3-250}$$

A similar analysis may be performed for the case of spherical waves by letting $\gamma = \eta/R$ in Eq. (3-246). In that case, using the spherical-wave version of Eq. (3-223), we obtain

$$\Theta_{\text{sw}} = 41.9 \left(\frac{kD^2}{4R}\right)^{5/6} \int_0^\infty du\, u^{-14/3} J_1^2(u) \int_0^1 d\gamma\, \gamma^{5/3} \times \sin^2\left[\frac{1}{2}\frac{4Ru^2}{kD^2}\left(\frac{1}{\gamma} - 1\right)u^2\right] \tag{3-251}$$

where $u = \eta\kappa D/2R$. Once again, for large D the sine function reduces to its argument and the integrals can be evaluated exactly, leading to the spherical-wave interpolation formula

$$\Theta_{\text{sw}} \approx \frac{1}{1 + 0.214(kD^2/4R)^{7/6}} \tag{3-252}$$

The interpolation formulas above are plotted in Figures 3-18 and 3-19 along with the exact results given by Eqs. (3-248) and (3-251). It can be seen that the plane-wave approximation compares well with the exact equation, while the spherical-wave case is a little less accurate. However, from a systems point of view, these formulas are perfectly adequate since turbulence conditions are inherently unpredictable and only gross trends should be expected for use in systems modeling. Equations (3-250) and (3-252) are used in Chapter 4 to

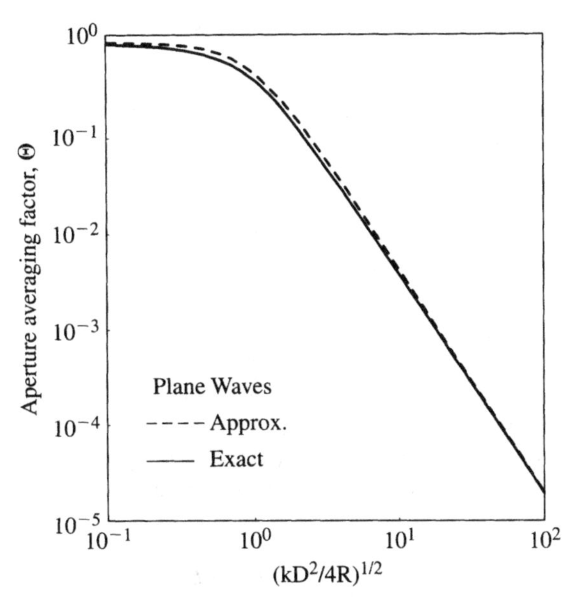

Figure 3-18. Comparison of exact and approximate formulas for the aperture-averaging function for plane waves. (From J. H. Churnside, *Appl. Opt.*, Vol. 30, No. 15, May 20, 1991.)

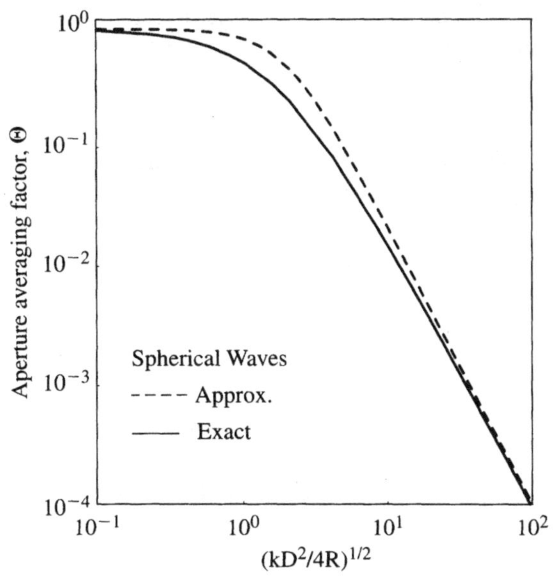

Figure 3-19. Comparison of exact and approximate formulas for the aperture-averaging function for spherical waves. (From J. H. Churnside, *Appl. Opt.*, Vol. 30, No. 15, May 20, 1991.)

calculate the detection probability of a direct-detection system operating in the presence of turbulence.

3.4.2.6. Turbulence-Limited Performance of Coherent Systems. The impact of atmospheric turbulence on coherent detection systems is quite different from that of direct-detection systems. The fact that out-of-phase signal components mixing on a detector surface of a heterodyne system can cancel suggests that the random spatial phase fluctuations of a scintillating beam could also limit the sensitivity of such a system. Indeed, we will now show, by rewriting Eq. (3-97) in terms of the fields, that there is an upper limit to the size of the receiving aperture, or equivalently, the range performance, of a coherent system operating in a turbulent atmosphere. In this section we follow Fried's[27] original analysis of the problem that relates the coherence diameter of the field to the aperture diameter of the system.

Consider an infinitesimal element of current on a photodetector surface. We can write for the case of a spatially nonuniform signal field mixing with a uniform local oscillator field,

$$di = \tfrac{1}{2}\Re\xi(x, y)[E_o + E_s(x, y)]^*[E_o + E_s(x, y)]\, dx\, dy \tag{3-253}$$

where $E_o = u_o e^{i\omega_o t + \phi_o}$ is the local oscillator field, $E_s = u_s e^{i\omega_s t + \phi_s}$ is the signal field, $\Re = \eta_q q/h\nu$ is the detector responsivity expressed in units of amperes per watt, and $\xi(x, y)$ is a weighting function taking into account any spatial nonuniformities in the aperture function or the detector response.

For the case of signals propagating through weak turbulence, we can write

$$E_s = u_s e^{i(\omega_s t + \phi_s)} e^{\chi(x,y)} \tag{3-254}$$

where χ is the log amplitude random variable defined by $\chi = \ln(u/\overline{u})$ (cf. Section 3.4.2). The total current therefore becomes

$$i = \frac{\Re}{2} \int_{-\infty}^{\infty} \int_{-\infty}^{\infty} \xi(x, y)[u_o^2 + 2u_o \overline{u}_s e^{\chi(x,y)} \cos(\Delta\omega t + \Delta\phi)]\, dx\, dy \tag{3-255}$$

where $\Delta\omega = \omega_o - \omega_s$ and $\Delta\phi = \phi_o(x, y) - \phi_s(x, y)$. The signal current is therefore

$$i_s = u_o \overline{u}_s \Re \int_{-\infty}^{\infty} \int_{-\infty}^{\infty} \xi(x, y)[e^{\chi(x,y)} \cos(\Delta\omega t + \Delta\phi)]\, dx\, dy \tag{3-256}$$

The mean signal power out of the detector is therefore

$$\begin{aligned} P_s &= \langle i_s^2 \rangle R_L \\ &= (\Re u_o \overline{u}_s)^2 R_L \left\langle \left[\int_{-\infty}^{\infty} \int_{-\infty}^{\infty} \xi(x, y)\{e^{\chi(x,y)} \cos[\Delta\omega t + \Delta\phi(x, y)]\}\, dx\, dy \right] \right. \\ &\quad \left. \times \left[\int_{-\infty}^{\infty} \int_{-\infty}^{\infty} \xi(x', y')\{e^{\chi(x',y')} \cos[\Delta\omega t + \Delta\phi(x', y')]\}\, dx'\, dy' \right] \right\rangle \end{aligned} \tag{3-257}$$

where R_L is the load resistor across the detector. Bringing the averaging process inside the integrals yields

$$P_s = K \int_{-\infty}^{\infty} \int_{-\infty}^{\infty} \int_{-\infty}^{\infty} \int_{-\infty}^{\infty} dx\, dy\, dx'\, dy' \xi(x, y)\xi(x', y')$$
$$\times \langle e^{\chi(x,y)+\chi(x',y')} \cos[\Delta\omega t + \Delta\phi(x, y)] \cos[\Delta\omega t + \Delta\phi(x', y')]\rangle \tag{3-258}$$

where $K = (\Re u_o \overline{u}_s)^2 R_L$. Using the identity $\cos(a + b)\cos(c + d) = \frac{1}{2}[\cos(a + b + c + d) + \frac{1}{2}\cos(a + b - c - d)]$, the averaging term in brackets reduces to

$$\tfrac{1}{2}\langle e^{\chi(x,y)+\chi(x',y')} \cos[2\Delta\omega t + \Delta\phi(x, y) + \Delta\phi(x', y')]$$
$$+ \cos[\phi_s(x, y) - \phi_s(x', y')]\rangle \tag{3-259}$$

But the first cosine term averages to zero due to the presence of $2\Delta\omega t$ in the argument, leaving

$$P_s = \frac{K}{2} \int_{-\infty}^{\infty} \int_{-\infty}^{\infty} \int_{-\infty}^{\infty} \int_{-\infty}^{\infty} dx\, dy\, dx' dy' \xi(x, y)\xi(x', y')$$
$$\times \langle e^{\chi(x,y)+\chi(x',y')} \cos[\phi_s(x, y) - \phi_s(x', y')]\rangle \tag{3-260}$$

Now the term in averaging brackets can be shown to be related to the mutual coherence function $\Gamma(\rho)$, where $\rho = |\vec{x} - \vec{x}'|$, as follows. For simplicity of notation, let $\chi \equiv \chi(x, y)$, $\chi' \equiv \chi(x', y')$, $\phi \equiv \phi(x, y)$, and $\phi' \equiv \phi(x', y')$. Then using the exponential form for the cosine function, we have

$$\langle e^{\chi+\chi'} \cos(\phi_s - \phi'_s)\rangle = \tfrac{1}{2}[\langle e^{\chi+\chi'+i(\phi_s-\phi'_s)}\rangle + \langle e^{\chi+\chi'-i(\phi_s-\phi'_s)}\rangle] \tag{3-261}$$

Using Eq. (3-203) while noting that $\langle \phi_s - \phi'_s \rangle = 0$, we find that the two bracketed terms in Eq. (3-261) are identical. Thus

$$\langle e^{\chi+\chi'} \cos(\phi_s - \phi'_s)\rangle = e^{\{(1/2)[\langle(\chi+\chi'-2\overline{\chi})^2\rangle - \langle(\phi_s-\phi'_s)^2\rangle]+2\overline{\chi}\}} \tag{3-262}$$

But the right-hand side of Eq. (3-262) is identical in form to right-hand side of Eq. (3-204). We therefore find our sought-after result, namely,

$$\langle e^{\chi+\chi'} \cos(\phi_s - \phi'_s)\rangle = \langle e^{\chi+\chi'} e^{i(\phi-\phi')}\rangle \equiv \Gamma_{12}(\rho) \tag{3-263}$$

where $\Gamma_{12}(\rho)$ is the mutual coherence function given by Eq. (3-211). Returning to Eq. (3-260), we can now write for the mean heterodyne signal,

$$P_s = \frac{K}{2} \int_{-\infty}^{\infty} \int_{-\infty}^{\infty} \int_{-\infty}^{\infty} \int_{-\infty}^{\infty} dx\, dy\, dx'\, dy' \xi(x, y)\xi(x', y')\Gamma_{12}(\rho) \tag{3-264}$$

Converting to the difference coordinates $\Delta x = x - x'$ and $\Delta y = y - y'$ followed by conversion to polar coordinates ρ and ϑ, we obtain, after integration over ϑ,

$$P_s = \pi(\Re u_o \bar{u}_s)^2 R_L \int_0^D \psi(\rho, D)\Gamma_{12}(\rho)\rho\, d\rho \tag{3-265}$$

Now the mean noise power is obtained by integrating Eq. (3-265) for the case of $\bar{u}_s = 0$, yielding

$$N = q\Re u_o^2 R_L \frac{\pi D^2}{4} \tag{3-266}$$

The mean signal-to-noise power ratio then becomes

$$\mathrm{SNR} = \frac{4\Re\bar{u}_s^2}{qD^2} \int_0^D \psi(\rho, D)\Gamma_{12}(\rho)\rho\, d\rho \tag{3-267}$$

For purposes of normalization, we divide Eq. (3-267) by the SNR that would be achieved in the absence of turbulence with an aperture size equal to the "coherence diameter" of the field. Calling this diameter ρ_0, with the understanding that it has yet to be defined, the signal power becomes $P_s = (\Re u_o \bar{u}_s \pi\rho_0^2/4)^2 R_L/2$ and the noise power $N = q\Re u_o^2 R_L \pi\rho_0^2/4$, such that

$$\mathrm{SNR}_o = \frac{\Re}{2q}\bar{u}_s^2 \frac{\pi\rho_0^2}{4} \tag{3-268}$$

Dividing Eq. (3-267) by Eq. (3-268) results in a normalized signal-to-noise ratio,

$$\mathrm{SNR}_N = \frac{32}{\pi\rho_0^2 D^2} \int_0^D \psi(\rho, D)\Gamma_{12}(\rho)\rho\, d\rho \tag{3-269}$$

For a circular aperture, $\psi(\rho, D)$ is given once again by Eq. (3-116). Thus

$$\mathrm{SNR}_N = \frac{16}{\pi\rho_0^2} \int_0^D \left[\cos^{-1}\frac{\rho}{D} - \frac{\rho}{D}\left(1 - \frac{\rho^2}{D^2}\right)^{1/2}\right]\Gamma_{12}(\rho)\rho\, d\rho \tag{3-270}$$

From Section 3.4.2.3 we know that

$$\Gamma_{12}(\rho) = e^{-(\rho/\rho_0)^{5/3}} \tag{3-271}$$

Letting $U = D/\rho_0$ and $\rho' = \rho/D$ in Eq. (3-260), we obtain

$$\mathrm{SNR}_N = \frac{16U^2}{\pi} \int_0^D [\cos^{-1}\rho' - \rho'(1 - \rho'^2)^{1/2}]e^{-(U\rho')^{5/3}}\rho' d\rho' \tag{3-272}$$

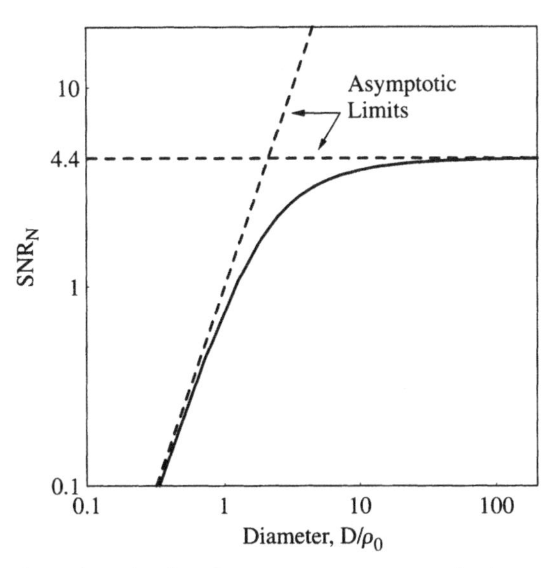

Figure 3-20. Signal-to-noise ratio of a coherent system versus normalized aperture diameter D/ρ_0, where ρ_0 is the transverse correlation length of the medium.

To show a logical progression to our final result, Eq. (3-272) is plotted in Figure 3-20. There it can be seen that the maximum SNR that is achievable by a coherent detection system operating in a turbulent atmosphere is approximately 7 dB greater than that for a system operating in a nonturbulent atmosphere with an aperture diameter of $D = \rho_0$. It can also be seen that the knee of the curve occurs at $D \approx 2.1\rho_0$, at which point the mutual coherence function assumes the value $\Gamma_{12}(\rho) = e^{-(2.1)^{5/3}} = e^{-3.44} = 0.032$. This is a significant loss in spatial coherence for the signal, which is reflected in the sensitivity loss experienced by the system.

Physical significance can be placed on the value 2.1 by defining a *saturation length* $r_0 = 2.1\rho_0$, characterizing the sensitivity of a coherent detection system to the coherence properties of the medium. In this case, Eq. (3-272) becomes by direct substitution into the exponential and normalizing with $\text{SNR}_0(r_0)$ rather than $\text{SNR}_0(\rho_0)$,

$$\text{SNR}_N = \frac{16U^2}{\pi} \int_0^D [\cos^{-1}\rho' - \rho'(1 - \rho'^2)^{1/2}] e^{-3.44(U\rho')^{5/3}} \rho' d\rho' \qquad (3\text{-}273)$$

where $U = D/r_0$. This last form of the normalized signal-to-noise ratio is plotted in Figure 3-21. It can be seen that the knee of the curve now occurs at $D/r_0 = 1$ and $\text{SNR}_N \to 1$ as $U \to \infty$. It should also be noted that the vertical asymptotes in Figures 3-20 and 3-21 express the D^2 dependence of the SNR at small values of D/r_0. Once again we may conclude that there is little to be gained in sensitivity from further increases in aperture diameter beyond $D = r_0$. This can, of course, also be interpreted as a limitation in range, where from Eqs. (3-273)

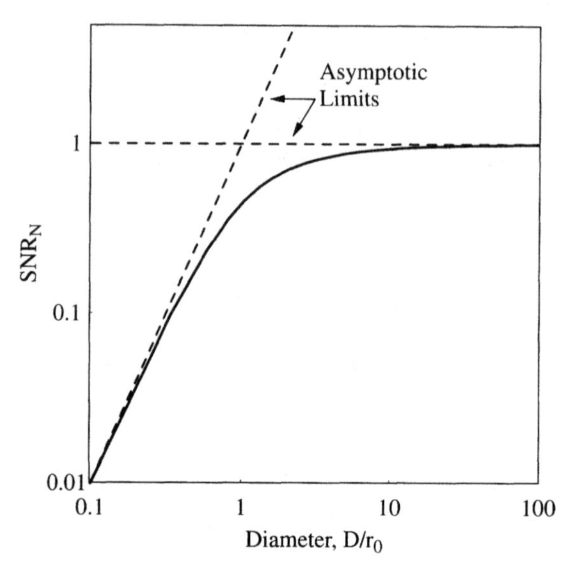

Figure 3-21. Signal-to-noise ratio of a coherent system versus normalized aperture diameter D/r_0, where r_0' is Fried's saturation length of the medium given by $r_0 = 2.1\rho_0$. (From D. L. Fried, *Proc. IEEE*, Vol. 55, No. 1, Jan. 1967.)

with D equal to a constant, $U = D/(1.5k^2C_n^2R)^{-3/5}$ or $U = D/(0.5k^2C_n^2R)^{-3/5}$. Such behavior has been verified experimentally by varying either the aperture diameter[28] or range.

Also, since the receiver is responding only to an effective aperture diameter of $D = r_0$, Eq. (3-273) can also be viewed as an expression for the limiting angular resolution of the system in a turbulent atmosphere. Indeed, it can be shown that when recast in the formalism of the optical transfer function (OTF)[29], r_0 constitutes the limiting resolution of any diffraction-limited optical system, not just heterodyne detection systems.

It should be noted that the coherence diameter ρ_0, and therefore r_0, are a function of wavelength. This is shown in Figure 3-22, where ρ_0 for the plane-wave case in Eq. (3-221) is plotted as a function of range for two widely separated wavelengths, 1.5 and 10.6 μm, and three values of C_n^2, corresponding to strong, medium, and weak turbulence levels. (Care must be exercised in using the data at high range values and strong turbulence levels where the Rytov approximation breaks down. This occurs for $\sigma_\ell^2 \approx 1$. Strong turbulence theory is discussed in Section 3.4.3.) It can be seen that near-IR and optical coherent systems are significantly limited in range performance for a given aperture size compared to the far-IR.

3.4.2.7. Beam Wander. It was shown earlier that turbulent eddies whose size are within the inertial subrange can significantly affect the coherence properties and point density functions of a beam propagating in a turbulent medium.

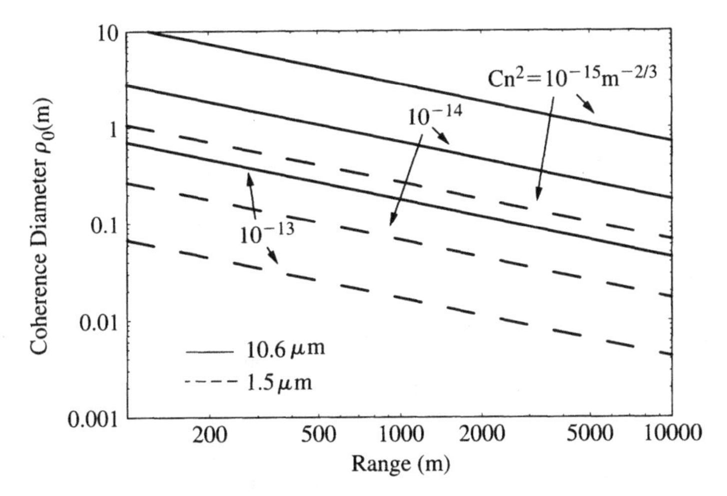

Figure 3-22. Transverse coherence length ρ_0 versus range for 1.5- and 10.6-μm wavelengths.

However, eddies that are equal to or larger than the beam can cause random fluctuations in the overall propagation direction of the beam. This effect, commonly referred to as *beam wander,* can result in the beam missing the target when operating against point targets or the generation of Rayleigh fluctuations when operating against large rough targets.

In this section we wish to develop a simple set of analytic equations that describe the variance in beam displacement resulting from beam wander for various beam-focusing geometries. The treatment follows that of Churnside and Lataitis,[30] using a relatively simple geometric optics formulation that produces results that are within a few percent of more exact calculations. The formulation is valid under the restriction that $\lambda R/\rho_0 \leq D \ll L_0$, where λ is the wavelength of the propagating beam, R the path length, ρ_0 the spherical-wave coherence length, D the aperture diameter, and L_0 the outer scale dimension of the turbulence.

For generality, consider a focused optical beam propagating through an elemental optical wedge as shown in Figure 3-23. Here the wedge is comprised of an index gradient along the x-direction perpendicular to the z-axis and has a constant thickness dz. A beam of diameter $D = 2\omega(z)$ that traverses the slab therefore sees a greater optical thickness at the bottom of the slab than at the top. The slab therefore functions as an optical prism, producing an elemental deflection angle $d\theta$ given by

$$d\theta = \frac{n(\omega, z) - n(-\omega, z)}{2\omega(z)}\, dz = \frac{\Delta n(z)}{2\omega(z)}\, dz \tag{3-274}$$

The lateral displacement of the beam in the observation plane located a distance R from the aperture is

$$d\rho = (R - z)\, d\theta \tag{3-275}$$

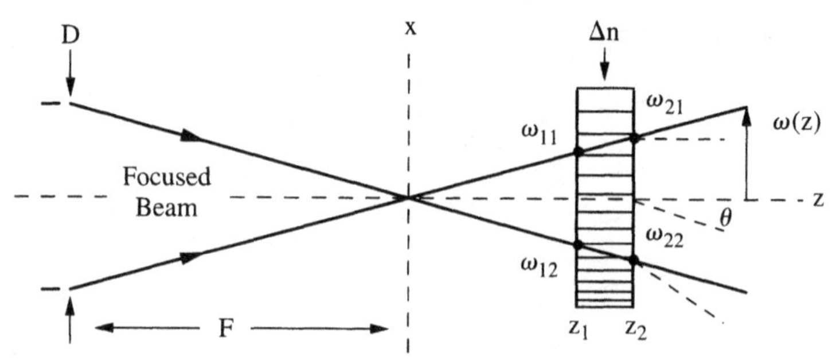

Figure 3-23. Beam parameters for a focused propagating beam in the geometric optics approximation. The region between z_1 and z_2 represents an infinitesimal slab containing a refractive index gradient transverse to the propagation direction.

The single-axis integrated effect from source to observation plane is then given by

$$\rho = \int_0^R \frac{\Delta n(z)}{2\omega(z)}(R - z)\, dz \tag{3-276}$$

Assuming that $\Delta n(z)$ is a zero-mean random variable determined by the degree of turbulence, we can write for the variance of the transverse displacement,

$$\sigma_w^2 = \langle \rho_1 \rho_2 \rangle = \int_0^R \int_0^R \langle \Delta n(z_1) \Delta n(z_2) \rangle \frac{R - z_1}{2\omega(z_1)} \frac{R - z_2}{2\omega(z_2)}\, dz_1\, dz_2 \tag{3-277}$$

where the order of integration and averaging has been reversed. The refractive index gradients $\Delta n(z_1)$ and $\Delta n(z_2)$ in Eq. (3-277) involve the four beam positions ω_{11}, ω_{12}, ω_{21}, and ω_{22} as depicted in Figure 3-23, thus yielding

$$\begin{aligned} \langle \Delta n(z_1) \Delta n(z_2) \rangle &= \langle [n(\omega_{11}) - n(\omega_{12})][n(\omega_{21}) - n(\omega_{22})] \rangle \\ &= \langle n(\omega_{11}) n(\omega_{21}) \rangle - \langle n(\omega_{11}) n(\omega_{22}) \rangle \\ &\quad - \langle n(\omega_{12}) n(\omega_{21}) \rangle + \langle n(\omega_{12}) n(\omega_{22}) \rangle \end{aligned} \tag{3-278}$$

Now each of the four terms above corresponds to a correlation function. Hence with the help of Eq. (3-173), Eq. (3-278) can be written in terms of index structure functions, that is,

$$\begin{aligned} \langle \Delta n(z_1) \Delta n(z_2) \rangle &= \tfrac{1}{2}[D_n(\omega_{11}, \omega_{22}) + D_n(\omega_{12}, \omega_{21}) \\ &\quad - D_n(\omega_{11}, \omega_{21}) - D_n(\omega_{12}, \omega_{22})] \end{aligned} \tag{3-279}$$

where the order of the terms has been rearranged and the $B(0)$ terms cancel out.

We have also seen in Section 3.3.1 that the Kolmogorov theory of turbulence predicts a two-thirds power law for the index structure function when separations are within the inertial subrange. Thus using Eq. (3-135) together with the geometry of Figure 3-23, Eq. (3-279) becomes

$$\langle \Delta n(z_1)\Delta n(z_2)\rangle = C_n^2 \left\{(z_2 - z_1)^2 + [\omega(z_1) + \omega(z_2)]^2\right\}^{1/3} + C_n^2 \left\{(z_2 - z_1)^2 + [\omega(z_1) - \omega(z_2)]^2\right\}^{1/3} \quad (3\text{-}280)$$

Now for large separations of $z_2 - z_1$ the correlation functions in Eq. (3-278) necessarily go to zero. If we also assume a beam radius that is slowly varying with propagation distance such that $\omega(z_1) \approx \omega(z_2)$, Eq. (3-280) reduces to

$$\langle \Delta n(z_1)\Delta n(z_2)\rangle = C_n^2[(z_2 - z_1)^2 + 4\omega^2(z_1)]^{1/3} + C_n^2(z_2 - z_1)^{2/3} \quad (3\text{-}281)$$

The beam radius $\omega(z)$ can be written in the geometric optics approximation as

$$\omega(z) = \frac{D}{2}\left|1 - \frac{z}{F}\right| \quad (3\text{-}282)$$

where F is the focal range of the optics. Thus for homogeneous media,

$$\sigma_w^2 = C_n^2 \int_0^R \int_0^R \frac{(R - z_1)^2}{b^2}\{[(z_2 - z_1)^2 + b^2]^{1/3} - (z_2 - z_1)^{2/3}\}\, dz_1\, dz_2 \quad (3\text{-}283)$$

where $b = D\,|1 - z/F|$. Performing the z_2-integral first while factoring out $b^{2/3}$ and letting $y = (z_2 - z_1)/b$ results in

$$\sigma_w^2 = C_n^2 \int_0^{(R-z_1)/b} \frac{(R - z_1)^2}{b^{1/3}} \int_0^R [(1 + y^2)^{1/3} - y^{2/3}]\, dy\, dz_1 \quad (3\text{-}284)$$

Since we have assumed that $l_0 \ll z_2 - z_1 \ll L_0$, the integration limits of the z_2-integral may be extended to $\pm\infty$ without introducing much error. We obtain

$$\int_{-\infty}^{\infty} f(y)\, dy = \frac{\sqrt{\pi}\,\Gamma(\frac{-5}{6})}{\Gamma(\frac{-1}{3})} \approx 2.91 \quad (3\text{-}285)$$

The single-axis variance of beam displacement therefore becomes, after substituting for b and letting $x = z_1/R$,

$$\sigma_w^2 = 2.91 C_n^2 D^{-1/3} R^3 \int_0^1 \frac{(1 - x)^2}{|1 - (R/F)x|^{1/3}}\, dx \quad (3\text{-}286)$$

For $R/F \leq 1$, the integral reduces to

$$\sigma_w^2 = 0.97 C_n^2 D^{-1/3} R^3 {}_2F_1(\tfrac{1}{3}, 1; 4, R/F) \quad (3\text{-}287)$$

where ${}_2F_1$ is the hypergeometric function defined by the series expansion ${}_2F_1(\delta, \varepsilon; \mu; z) = \sum_{n=0}^{\infty} (\delta)_n(\varepsilon)_n/(\mu)_n z^n/n!$. For a collimated beam, $F = \infty$ and ${}_2F_1 = 1$, resulting in

$$\sigma_{wc}^2 \approx 0.97 C_n^2 D^{-1/3} R^3 \tag{3-288}$$

For a focused beam $R = F$, we have

$$\sigma_{wf}^2 \approx 1.09 C_n^2 D^{-1/3} R^3 \tag{3-289}$$

We can use Eq. (3-288) to normalize Eq. (3-286), obtaining

$$\frac{\sigma_w^2}{\sigma_{wc}^2} = 3 \int_0^1 \frac{(1-x)^2}{|1-(R/F)x|^{1/3}} dx \tag{3-290}$$

Equation (3-290) is plotted in Figure 3-24 as a function of the ratio R/F. A maximum beam deflection is seen to occur for a propagation range R that is about 2.6 times the focal range F. Negative values of R/F correspond to an expanded or diverging beam (a virtual focus), which results in zero variance as $R/F \to -\infty$.

It is instructive to consider the linear beam displacement for a collimated beam relative to the diameter of a Gaussian beam at the same range. Such a comparison should be allowed for the analysis above given the assumptions of a slowly varying beam expansion, as long as the Rytov variance is maintained within the weak turbulence limit ($\sigma_\ell^2 \leq 1.2$). First, it should be noted that σ_c is independent of wavelength and inversely proportional to aperture diameter ($D^{-1/6}$). From Eq. (2-27) the e^{-2} spot diameter of a Gaussian beam can be

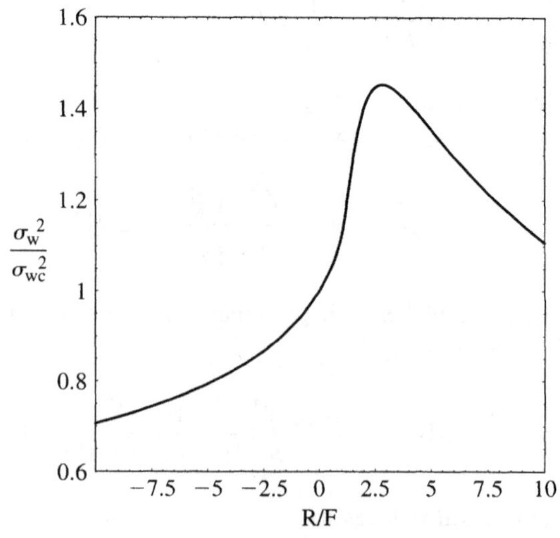

Figure 3-24. Single-axis beam wander variance normalized to the collimated variance versus the range-to-focus ratio R/F. (From J. H. Churnside, *Appl. Opt.*, Vol. 29, No. 7, Mar. 1, 1990.)

written

$$D_b = \sqrt{D^2 + \left(\frac{4\lambda R}{\pi D}\right)^2} \tag{3-291}$$

Thus

$$\frac{\sigma_{wc}}{D_b} = \sqrt{\frac{0.97 C_n^2 R^3}{D^{7/3}[1 + (4\lambda R/\pi D^2)^2]}} \qquad \sigma_\ell^2 \le 1.2 \tag{3-292}$$

where R must be constrained to stay within the weak turbulence limit. In the near field, $D \gg \sqrt{4\lambda R/\pi}$, we find that for all practical values of D and λ,

$$\frac{\sigma_{wc}}{D_b} \approx \sqrt{C_n^2 D^{-7/6} R^{3/2}} < 1 \tag{3-293}$$

Similarly, in the far field where $D \ll \sqrt{4\lambda R/\pi}$, we find that for modest ranges,

$$\frac{\sigma_{wc}}{D_b} \approx \frac{\pi}{4\lambda} D^{5/6} \sqrt{C_n^2 R} < 1 \tag{3-294}$$

Thus in the weak turbulence limit, collimated beam displacements are generally less than a beamwidth. These results would have to be modified at extended ranges where cumulative turbulence effects begin to violate the limits of the Rytov variance.

3.4.3. Strong Turbulence Theory

As described earlier, weak turbulence theory holds only for small values of the Rytov parameter σ_χ^2. As σ_χ^2 increases beyond about $\sigma_\chi^2 = 0.3$, *saturation* effects must be taken into account. Saturation occurs when the coherence diameter of the beam approaches the inner scale of turbulence. Since σ_χ^2 is a function of C_n^2, range, and wavelength, there are a variety of ways that turbulence-induced scintillation can enter the saturated regime. Thus propagation of short wavelengths over long ranges can result in saturation even for small-to-moderate values of C_n^2. Experiments in Russia and the United States have shown that the variance of log intensity not only saturates at a particular value of the Rytov parameter but also exhibits the unusual property of reaching a maximum value and then decreasing as the Rytov parameter continues to increase. This is shown in Figure 3-25. The region beyond the peak of the curve is usually referred to as the *supersaturated regime*, where fluctuations of intensity actually decrease with increasing range. This behavior can be modeled empirically with a two-parameter function of the form

$$\sigma_{\chi s}^2 = \frac{\sigma_\chi^2}{1 + A\sigma_\chi^{2B}} \tag{3-295}$$

where A and B are, in general, wavelength dependent.

From a detection theory point of view, obtaining a probability density function that accurately models the statistics of beam propagation in strong turbulence would be of great use. Early measurements suggested that the statistics remained log-normal well into the saturated regime, but this was later shown theoretically

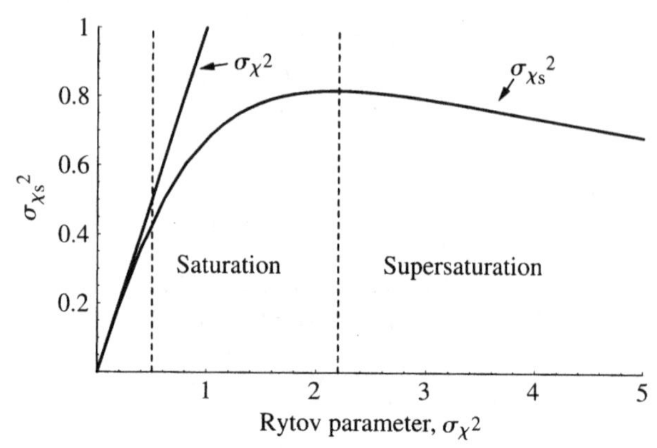

Figure 3-25. Variance of log amplitude versus weak-turbulence Rytov parameter for the case of saturated optical scintillation.

and experimentally to be only partially true. Indeed, if the resulting field is due to many scattering events, one would expect that by virtue of the central limit theorem, circular complex Gaussian statistics should play a fundamental role, with the Rayleigh and negative exponential statistics representing the amplitude and intensity, respectively. However, as pointed out by Churnside, this must be tempered by the fact that if the many scattering amplitudes are log-normal variates, the slow convergence of the central limit theorem[31,32] for the log amplitude distribution could account for the observed component of log amplitude statistics in the moderately strong turbulence regime. However, in the limit of very large values of the Rytov parameter, it is generally accepted that the irradiance statistics approach a negative exponential distribution with the normalized variance of irradiance $\sigma_I^2/\langle I\rangle \rightarrow 1$.

Ideally, one would like to have a *universal theory* that transitions smoothly from the weak turbulence model to the strong turbulence model and asymptotically approaches the negative exponential limit as the turbulence level increases indefinitely. In contrast to the weak turbulence theory, most of the strong turbulence theories developed thus far account only for the saturated and supersaturated regimes and are essentially empirical in nature. Indeed, most theories are constructed to account for a limited set of observed effects, primarily the fluctuation statistics, and are chosen based on best fits to the data. This is understandable since the strong turbulence regime involves strong nonlinear mixing phenomena, which remains a difficult area of mathematical physics.

However, a reasonable degree of success in constructing universal theories capable of accounting for weak as well as strong turbulence effects on beam propagation has been achieved. Two competing heuristic models of the irradiance distribution yield good agreement with experiment while providing plausible physical arguments for their selection: the *log-normal-Rician distribution*[33] and the *I − K distribution*[34]. Here the I and K refer to the fact that the probability density function contains Bessel functions of the first and second kinds, respectively. Both distribution functions have well-behaved limiting forms. In the limit of moderately strong turbulence, the log-normal Rician distribution reduces to the *log-normally modulated negative exponential distribution*[35] and the I–K distribution reduces to the *K distribution*.[36,37] In the limit of extremely strong turbulence, both distributions reduce to the negative exponential distribution, which is the commonly accepted distribution for extremely strong turbulence. The advantage of the I–K distribution is the fact that it assumes a closed form that is amenable to the calculation of detection probabilities and related effects due to turbulence. Due to the complexity of these theories, especially the many limiting forms of the probability density functions, only a brief outline of the essential features is given.

The heuristic argument for the I–K model is based on the nonstationary nature of the statistics of real meteorological fields that are subjected to diurnal and wind-speed variations. The I–K distribution is based on the assumption that the propagating field consists of the vector sum of m scattered waves superimposed on a coherent background, the coherent component dominating for weak turbulent fields and the random or diffuse component dominating for strong turbulent fields. The statistics of the intensity is then given by the expected value of the conditional probability for this composite field. Assuming that the mean of the random component obeys negative exponential statistics, the probability density function for the intensity is obtained by averaging over the mean value of the intensity I_s of the random component. Thus

$$p(I) = \int_0^\infty p(I \mid I_s) p(I_s)\, dI_s \tag{3-296}$$

where

$$p(I_s) = \frac{1}{\langle I_s \rangle} e^{-I_s/\langle I_s \rangle} \tag{3-297}$$

is the exponentially distributed mean of the random component with mean value $\langle I_s \rangle$ and

$$p(I \mid I_s) = \frac{m}{I_s}\left(\frac{I}{I_c}\right)^{(m-1)/2} e^{-m(I+I_c)/I_s} I_{m-1}\left(\frac{2m}{I_s}\sqrt{I I_c}\right) \tag{3-298}$$

is the conditional single-point probability density function for the summed coherent and random components. Here I_c is the intensity of the coherent component; I_{m-1} the modified Bessel function of the first kind, order $m-1$; and m the effective number of scatterers in the volume of interest. Substituting Eqs. (3-297) and (3-298) into Eq. (3-296) while assuming m to be continuous results in a known tabulated integral[38] that yields

$$p(I) = \begin{cases} \frac{2m}{\langle I_s \rangle} \left(\frac{I}{I_c} \right)^{(m-1)/2} K_{m-1}\left(2\sqrt{\frac{mI_c}{\langle I_s \rangle}} \right) I_{m-1}\left(2\sqrt{\frac{mI}{\langle I_s \rangle}} \right) & I < I_c \\ \frac{2m}{\langle I_s \rangle} \left(\frac{I}{I_c} \right)^{(m-1)/2} I_{m-1}\left(2\sqrt{\frac{mI_c}{\langle I_s \rangle}} \right) K_{m-1}\left(2\sqrt{\frac{mI}{\langle I_s \rangle}} \right) & I > I_c \end{cases} \tag{3-299}$$

where K_{m-1} is the modified Bessel function of the second kind, order $m-1$. It is easy to show using the series expansion

$$I_{m-1}(y) = \frac{(y/2)^{m-1}}{j!\,\Gamma(m+j)} \sum_{j=0}^{\infty} \left(\frac{y^2}{4} \right)^j \tag{3-300}$$

that in the limit of $I_c \to 0$, Eq. (3-299) reduces to the *K distribution*:

$$p(I) = \frac{2}{\Gamma(m)} \frac{m^{(m+1)/2}}{\langle I_s \rangle} \left(\frac{I}{\langle I_s \rangle} \right)^{(m-1)/2} K_{m-1}\left(2\sqrt{\frac{mI}{\langle I_s \rangle}} \right) \tag{3-301}$$

where $\Gamma(m)$ is the gamma function, defined by the integral $\Gamma(z) = \int_0^\infty e^{-z} t^{z-1} dt$, $\mathrm{Re}\, z > 0$ (cf. Appendix A).

Equation (3-299) can be written in terms of normalized variables by letting $y = I/\langle I \rangle = I/(\langle I_s \rangle + I_c)$ and $\kappa = I_c/\langle I_s \rangle$, where κ is referred to as the *coherence parameter* for the propagating beam. κ ranges from 0 when the beam is totally randomized with $I_c = 0$ to infinity for a perfectly coherent beam when $\langle I_s \rangle = 0$ Thus

$$p(y) = \begin{cases} 2m(1+\kappa) \left[\frac{(1+\kappa)y}{\kappa} \right]^{(m-1)/2} \\ \quad \times K_{m-1}(2\sqrt{m\kappa}) I_{m-1}\{2[my(1+\kappa)]^{1/2}\} & y < \frac{\kappa}{1+\kappa} \\ 2m(1+\kappa) \left[\frac{(1+\kappa)y}{\kappa} \right]^{(m-1)/2} \\ \quad \times I_{m-1}(2\sqrt{m\kappa}) K_{m-1}\{2[my(1+\kappa)]^{1/2}\} & y > \frac{\kappa}{1+\kappa} \end{cases} \tag{3-302}$$

Clearly, the weak turbulence limit occurs for large κ where the coherent component is dominant and the strong turbulence limit occurs for small κ where the diffuse component is dominant. Note that the I–K and K distributions are determined by the parameters m, κ, and m, respectively. These are chosen based on best fits of the first two normalized moments to experimental data. Given the closed-form nature of the I–K distribution, the moments become

$$\langle y^n \rangle = \int_0^\infty y^n p(y)\, dy = \frac{n!}{m^n(1+\kappa)^n} \sum_{k=0}^{\infty} \frac{\Gamma(m+n)}{\Gamma(m+k)} \frac{(m\kappa)^k}{k!} \qquad \text{(3-303)}$$

The heuristic argument for the log-normally modulated exponential distribution is that the exponential fluctuations are associated with small-scale eddies, while the log-normal fluctuations are associated with large-scale eddies. Since most practical measurements are made with relatively large receiving apertures to maximize signal-to-noise ratio (i.e., apertures on the order of $\sqrt{\lambda R}$ or larger), the small-scale effects tend to be averaged out, and the resulting aperture-averaged statistics appear log-normal, as frequently observed. However, measurements made using small apertures and photon-counting techniques have resulted in data that closely fit theoretical predictions. The probability density function for the normalized intensity is therefore written as a convolution:

$$p(y) = \int_0^\infty p(y \mid z) p(z)\, dz \qquad \text{(3-304)}$$

where

$$p(y \mid z) = \frac{1+\kappa}{z} e^{-\{\kappa+[(1+\kappa)/z]y\}} I_0 \left\{ 2 \left[\frac{(1+\kappa)\kappa y}{z} \right]^{1/2} \right\} \qquad \text{(3-305)}$$

is the *Rice–Nakagami distribution* and

$$p(z) = \frac{1}{\sqrt{2\pi\sigma_z^2}} \frac{1}{z} e^{-(\ln z + \sigma_z^2/2)^2/2\sigma_z^2} \qquad \text{(3-306)}$$

is the log intensity distribution. Here y and κ are as given above, $z = \exp(2\chi)$, I_0 is the modified Bessel function of the first kind, order zero, and σ_z^2 is the variance of the irradiance modulation factor z. Inserting Eqs. (3-305) and (3-306) into Eq. (3-304) results in an integral that cannot be evaluated analytically but is so constructed as to have useful limiting forms. For example, in the weak turbulence limit when $\kappa \to \infty$, it can be shown that $p(y)$ reduces to the one-parameter log-intensity distribution given earlier by Eq. (3-228). In the strong turbulence limit when $\kappa \to 0$, Eq. (3-305) reduces to the negative exponential distribution, which results in Eq. (3-304) reducing to the one-parameter *log-normally modulated exponential distribution* given by

$$p(y) = \frac{1}{\sqrt{2\pi\sigma_z^2}} \int_0^\infty z^{-2} e^{-y/z-(\ln z+\sigma_z^2/2)^2/2\sigma_z^2}\, dz \qquad \text{(3-307)}$$

An analytic expression for the moments of Eq. (3-304) can be obtained and is given by

$$\langle y^n \rangle = \int_0^\infty y^n p(y)\,dy = \frac{(n!)^2}{(1+\kappa)^n} e^{n(n-1)\sigma_z^2/2} \sum_{m=0}^{n} \frac{\kappa^m}{(m!)^2(n-m)!} \tag{3-308}$$

Both the I–K and log-normal Rician models are two-parameter models that require fitting to at least the first three moments of measured normalized intensity data to obtain a valid solution. Expressions can be derived in both cases that relate the second and third moments with the model parameters. Estimates of the moments from measured data then allow for calculation of the various model parameters, κ and m in the case of the I–K model and κ and σ_z^2 in the case of the log-normal Rician model. In the former case it is easy to show using the second and third moments as given by Eq. (3-303) that κ and m may be obtained from the polynomial expressions

$$\left[\langle y^2\rangle^2 - \langle y^2\rangle - \frac{1}{3}\langle y^3\rangle\right](\kappa^3+1) + [3\langle y^2\rangle^2 - 2\langle y^2\rangle - \langle y^3\rangle](\kappa^2+\kappa) + \frac{\kappa^3}{3} = 0 \tag{3-309}$$

and

$$m = \left[\frac{1}{2}(1+\kappa)\langle \mathrm{I}^2\rangle - \frac{1}{2}\frac{\kappa^2}{1+\kappa} - 1\right]^{-1} \tag{3-310}$$

In the case of the log-normal Rician distribution, the second and third moments given by Eq. (3-308) lead to

$$\frac{(\kappa^2+4\kappa+2)^3}{(1+\kappa)^3(\kappa^3+9\kappa^2+18\kappa+6)} = \frac{\langle y^2\rangle^3}{\langle y^3\rangle} \tag{3-311}$$

and

$$\sigma_z^2 = \ln\left[\frac{(1+\kappa)^2}{\kappa^2+4\kappa+2}\langle y^2\rangle\right] \tag{3-312}$$

Thus, given measured values of the second and third moments of irradiance, parameter values may be obtained from Eqs. (3-309) through (3-312), which may then be used in the corresponding probability density functions for comparison with measured data. Example density functions are shown in Figures 3-26 through 3-28 for a broad range of log intensity variances.[39] Here I ($\equiv y$) represents the normalized intensity. The major objection to the I–K model is the presence of a cusp, which can be seen at the peak of the distributions in Figures 3-26 and 3-27, that tends to overestimate the probability density function in that region. At high turbulence levels, the cusp disappears and the fit to the data improves. On the other hand, the log-normal Rician distribution has been shown to underestimate the data in the vicinity of the peak, while both models show excellent

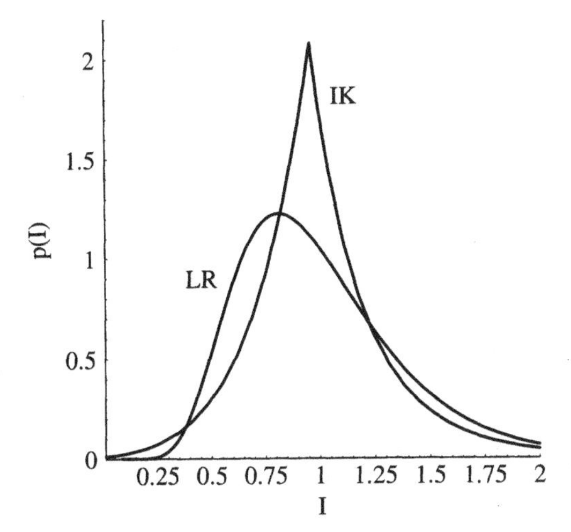

Figure 3-26. Linear scale comparisons of the I–K ($\kappa = 23$, $m = 0.69$) and log-normal Rician (LR) ($\kappa = \infty, \sigma_z^2 = 0.12$) distributions for the case of weak turbulence having a measured log-irradiance variance $\sigma_z^2 = 0.12$ and a third moment of 1.43. (After J. H. Churnside, *J. Opt. Soc. Am.*, Vol. 6, No. 11, Nov. 1, 1989.)

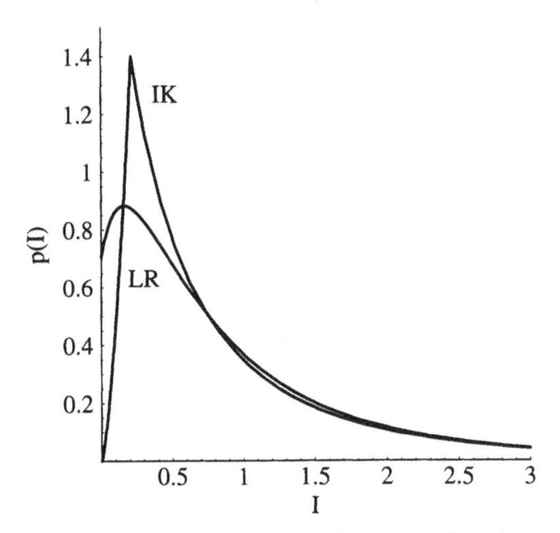

Figure 3-27. Linear comparisons of the I–K ($\kappa = 0.27$, $m = 2.3$) and log-normal Rician (LR) ($\kappa = 1.7, \sigma_z^2 = 0.36$) distributions for the case of strong turbulence having a measured log-irradiance variance $\sigma_z^2 = 1.3$ and a third moment of 10.2. (After J. H. Churnside, *J. Opt. Soc. Am.*, Vol. 6, No. 11, Nov. 1, 1989.)

fits in the tails of the distributions where turbulence levels are high. However, the cusp remains in the I–K case. Figure 3-28 shows the density functions for very strong turbulence, although the process of parameter selection becomes somewhat arbitrary at this point. Note that the cusp is not present in this case.

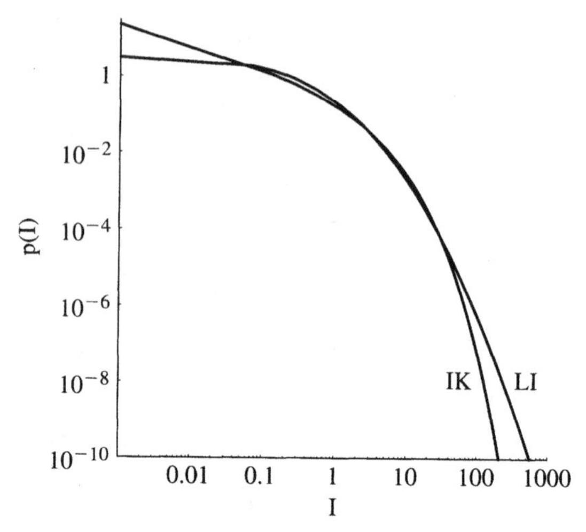

Figure 3-28. Log scale comparison of the I–K ($\kappa = 0$, $m = 0.48$) and log-normal Rician (LR) ($\kappa = 0$, $\sigma_z^2 = 1.13$) distributions for the case of very strong turbulence having a measured log-irradiance variance $\sigma_z^2 = 5.2$ and a third moment of 122. The third moment was arbitrarily reduced by 30% for the LR model and increased by 28% for the I–K model to fit the data. (After J. H. Churnside, *J. Opt. Soc. Am.*, Vol. 6, No. 11, Nov. 1, 1989.)

Despite the partial successes of the theories above, they remain two-parameter models (that reduce to one-parameter models in the strong turbulence limit) that must be fit to the data before the parameters can be determined. This makes their general use in engineering applications not much better than the empirical formula given by Eq. (3-295). Also, it is not clear from the theories above how the parameters vary with wavelength or how to select such parameters without extensive curve fitting to data sets. Indeed, the process of parameter selection in Figure 3-28 was not obvious, corresponding to more of a trial-and-error process than one would like to see in a rigorous theory. However, these deficiencies are partially compensated for by the physical insights provided by the heuristic arguments, which may lead to more general theories in the future.

REFERENCES

[1]M. Born and E. Wolf, *Principles of Optics*, 5th ed., Pergamon Press, Elmsford, 1975, p. 508.

[2]Lord Rayleigh, "On the dynamical theory of gratings," *Proc. R. Soc.*, Vol. A79, 1907, pp. 399–416.

[3]S. O. Rice, "Reflection of electromagnetic waves from slightly rough surfaces," *Commun. Pure Appl. Math.*, Vol. 4, 1951, pp. 351–378.

[4]P. Beckmann and A. Spizzichino, *The Scattering of Electromagnetic Waves from Rough Surfaces*, Artech House, Norwood, MA, 1987.

[5]See the work of Beckmann and Spizzichino (ref. 4) and J. A. Ogilvy, *Theory of Wave Scattering from Random Rough Surfaces*, Adam Hilger, New York, 1991.

[6]J. W. Goodman, "Some effects of target-induced scintillation on optical radar performance," *Proc. IEEE*, Vol. 53, No. 11, Nov. 1963, p.1688.

[7]I. S. Reed, "On a moment theorem for complex Gaussian processes," *IRE Trans. Inf. Theory*, Vol. IT-8, Apr. 1962, p. 194.

[8]J. W. Goodman, "Laser speckle and related phenomena," in *Topics in Applied Physics*, 2nd ed., Vol. 9, ed. J. C. Dainty, Springer-Verlag, New York, 1984, p. 50.

[9]R. S. Lawrence and J. W. Strohbehn, "A survey of clear-air propagation relevant to optical communications," *Proc. IEEE*, Vol. 58, No. 10, Oct. 1970.

[10]E. Brookner, "Atmospheric propagation and communication channel model for laser wavelengths," *IEEE Trans. Commun. Technol.*, Vol. 18, No. 4, Aug. 1970, p. 396.

[11]S. F. Clifford, "The classical theory of wave propagation in a turbulent medium," in "Laser Beam Propagation in the Atmosphere," Topics in Applied Physics, Vol. 25, Springer-Verlag, New York, NY, Editor J. W. Strohbehn, 1978, p. 9–43.

[12]J. W. Strohbehn, "Line-of-sight propagation through the turbulent atmosphere," *Proc. IEEE*, Vol. 56, No. 8, Aug. 1968, p. 1301.

[13]A. Ishimaru, *Wave Propagation and Scattering in Random Media*, Vol. 2, Academic Press, San Diego, CA, 1978.

[14]A. N. Kolmogorov, "The local structure of turbulence in incompressible viscous fluid for very large Reynolds numbers," *Dokl. Akad. Nauk. USSR*, Vol. 30, 1941.

[15]A. Kolmogorov, "Turbulence", in *Classic Papers on Statistical Theory*, ed. S. K. Friedlander and L. Topper, Interscience, New York, 1961, p. 151.

[16]V. I. Tatarskii, *Wave Propagation in a Turbulent Medium* (translated by R. A Silverman), McGraw-Hill Book Company, New York, 1961.

[17]V. I. Tatarskii, *The Effects of the Turbulent Atmosphere on Wave Propagation*, (translated by Israel Program for Scientific Translations; originally published in 1967), U.S. Department of Commerce, National Technical Information Service, Springfield, VA, 1971.

[18]J. W. Strohbehn and S. F. Clifford, "Polarization and angle of arrival fluctuations for a plane wave propagated through a turbulent medium," *IEEE Trans. Antennas Propag.*, Vol. AP-15, 1967, p. 416.

[19]D. H. Hohn, "Depolarization of a laser beam at 6328 Å due to atmospheric transmission," *Appl. Opt.*, Vol. 8, No. 2, Feb. 1969, p. 367.

[20]Ibid., ref. 17, p. 27.

[21]A. Ishimaru, "The beam wave case and remote sensing," in *Tcpics in Applied Physics: Laser Beam Propagation in the Atmosphere*, Vol. 25, Springer-Verlag, New York, 1978 p. 134.

[22]I. S. Gradshteyn and I. M. Ryzhik, *Table of Integrals, Series, and Products*, 4th ed., Academic Press, San Deigo, CA, 1965, p. 684, Eq. 6.561.14.

[23]R. F. Lutomirski and H. T. Yura, "Wave structure function and mutual coherence function of an optical wave in a turbulent atmosphere," *J. Opt. Soc. Am.*, Vol. 61, No. 4, Apr. 1971, p. 482.

[24]D. L. Fried, "Aperture-averaging of scintillation," *J. Opt. Soc. Am.*, Vol. 57, No. 2, Feb. 1967, p. 169.

[25]D. L. Fried and J. D. Cloud, "Propagation of an infinite plane wave in a randomly inhomogeneous medium," *J. Opt. Soc. Am.*, Vol. 56, No. 12, Dec. 1966, p. 1667.

[26]J. H. Churnside, "Aperture-averaging of optical scintillations in the turbulent atmosphere," *Appl. Opt.*, Vol. 30, No. 15, May 20, 1991, p. 1982.

[27]D. L. Fried, "Optical heterodyne detection of an atmospherically distorted signal wave front," *Proc. IEEE*, Vol. 55, No. 1, Jan. 1967, p. 57.

[28]I. Goldstein, P. A. Miles, and A. Chabot, "Heterodyne measurements of light propagation through atmospheric turbulence," *Proc. IEEE*, Vol. 53, Sept. 1965, p. 1172.

[29]J. W. Goodman, *Statistical Optics*, John Wiley & Sons, New York, 1985, p. 429.

[30]J. H. Churnside and R. J. Lataitis, "Wander of an optical beam in the turbulent atmosphere," *Appl. Opt.*, Vol. 29, No. 7, Mar. 1, 1990, p. 926.

[31]R. Barakat, "Sums of independent log-normally distributed random variables," *J. Opt. Soc. Am.*, Vol. 66, 1976, pp. 211–216.

[32]A. Consortini and L. Ronchi, "Probability distribution of the sum of N complex random variables," *J. Opt. Soc. Am.*, Vol. 67, 1977, pp. 181–185.

[33]J. H. Churnside and S. F. Clifford, "Log-normal Rician probability-density function of optical scintillations in the turbulent atmosphere," *J. Opt. Soc. Am.*, Vol. 4, No. 10, Oct. 1987, pp. 1923–1930.

[34]L. C. Andrews and R. L. Phillips, "I–K distribution as a universal propagation model of laser beams in atmospheric turbulence," *J. Opt. Soc. Am.*, Vol. 2, No. 2, Feb. 1985, p. 160.

[35]J. H. Churnside and R. J. Hill, "Probability density of irradiance scintillations for strong path-integrated refractive turbulence," *J. Opt. Soc. Am.*, Vol. 4, No. 4, Apr. 1987, p. 727.

[36]G. Parry and P. N. Pusey, "K distributions in atmospheric propagation of laser light," *J. Opt. Soc. Am.*, Vol. 69, No. 5, May 1979, p. 796.

[37]S. F. Clifford and R. J. Hill, "Relation between irradiance and log-amplitude variance for optical scintillation described by the K distribution," *J. Opt. Soc. Am. Lett.*, Vol. 71, No. 1, Jan. 1981, p. 113.

[38]Ibid., ref. 22, p.719, Eq. 6.635-3

[39]J. H. Churnside and R. G. Frehlich, "Experimental evaluation of log-normally modulated Rician and IK models of optical scintillation in the atmosphere," *J. Opt. Soc. Am.*, Vol. 6, No. 11, Nov. 1989, p. 1760.

4

Single-Pulse Direct-Detection Statistics

4.1. INTRODUCTION

In this chapter the probability density functions and associated detection statistics are derived for several classes of direct-detection receivers operating in the presence of receiver, target, and atmospheric-induced noise. The *receiver class* is herein defined by its dominant noise characteristic, that is, whether it is signal, dark current, amplifier, background, or backscatter noise limited. Figure 4-1 shows a generic *direct-detection receiver*. The receiver includes energy collecting optics, a square-law detector, a bandlimiting filter, a preamplifier/amplifier chain, and a signal processor. The *signal processor* performs constant false alarm rate (CFAR) processing, thresholding, and detection. The detected signal is then passed to a data processor, where *parameter estimates* can be performed, such as range or amplitude measurements. The energy collection optics usually consists of a telescope for increasing the effective aperture of the receiver and some optical filtering to reduce undesired background radiation at the detector.

In Chapter 3 we saw that the power spectral densities of the various noise contributions, each multiplied by the bandwidth of the receiver, are summed to yield an overall *noise equivalent power* (NEP) for the receiver. This, together with the received power P_R is then sufficient to determine the optical power signal-to-noise ratio for the system. The detection problem for each receiver class then reduces to calculating the detection and false alarm probabilities using the Neyman–Pearson criterion and the probability density functions developed for the signal-plus-noise and noise alone.

Many of the target and receiver classes discussed in this and later chapters achieve significant speckle noise reduction through various types of *diversity* techniques. *Speckle diversity* arises when multiple correlation cells are available for summing during the measurement interval. *Aperture-averaging*, *polarization diversity*, and *noncoherent illumination* are methods for achieving speckle diversity in direct-detection receivers on a single-pulse basis. As will be seen, much of the groundwork in describing these phenomenologies may be attributed to the pioneering work of Goodman[1] in the field of statistical optics. Later works have extended these results or have introduced new theories that describe more modern detection mechanisms, such as the avalanche photodiode receiver.

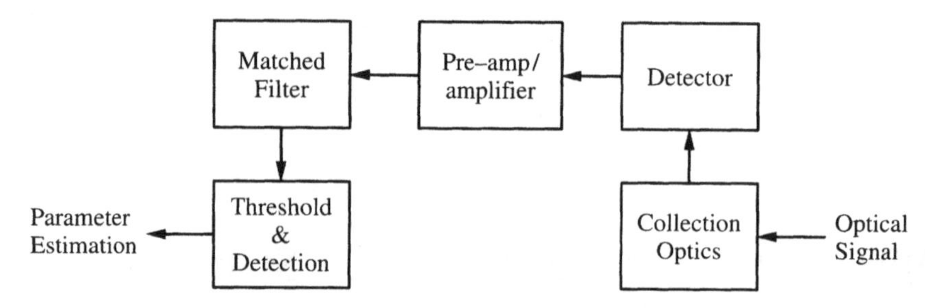

Figure 4-1. Generic block diagram of a direct-detection receiver and associated signal processing.

4.2. SINGLE-POINT STATISTICS OF FULLY DEVELOPED SPECKLE

The physics of speckle generated by the scattering of optical and infrared radiation from a rough surface is identical to that which generates target fading in conventional microwave radar systems. In both cases the incident radiation is scattered from a random distribution of a large number of scatterers of similar magnitude such that the far-field spatial distribution is random in both amplitude and phase. A computer generated example of speckle is shown in Figure 4-2. This image corresponds to the central portion of the pattern shown in Figure 2-23 magnified several times. Despite the common physical basis, there are some interesting differences between microwave and optical speckle that result primarily from the large difference in wavelengths of the two spectral regions.

First, the minimum diameter of a speckle cell in the far field of an illuminated target is given approximately by $d_s = \lambda R/d$, where d is the spot size on the target, R is the range, and λ is the wavelength. Since the wavelength is four

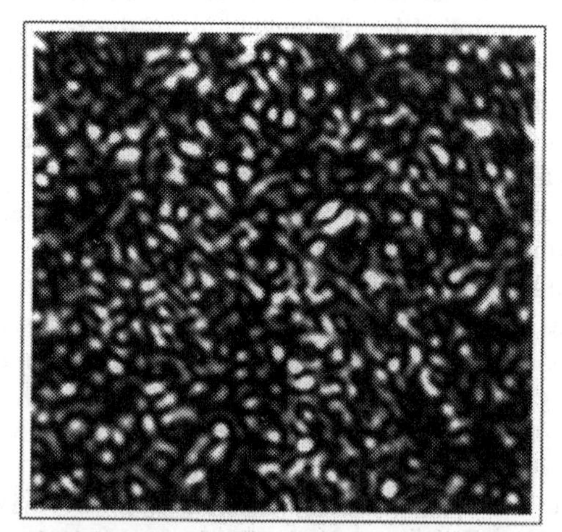

Figure 4-2. Simulated far-field image of a fully developed speckle pattern using the Gaussian beam model of Section 2.4 with the same spatial and phase distributions used in Figures 2-22 and 2-23 but with smaller beam radii ($\times \frac{1}{2}$). The image is generated using a 128×128 FFT.

orders of magnitude smaller in the optical spectrum than in the microwave spectrum (1 μm versus 3 cm, for example), the resulting speckle is proportionately smaller. Consequently, speckle averaging can be accomplished on a single-pulse basis with an optical energy receiver that is only a few centimeters in diameter. This condition usually cannot be satisfied with conventional radar antennas, due to the large size of the speckle cells or lobes compared to the antenna. (However, radar systems can achieve speckle averaging in the time domain when operating against moving targets by using multiple-pulse integration.) Second, the surface roughness of most man-made materials is essentially zero at conventional radar wavelengths, implying that target fading is essentially a consequence of the geometric structure of the target, not the surface finish. At optical wavelengths, the surface finish can dominate the target signature since few surfaces have a roughness of less than a wavelength.

Consider the field amplitude at a point x, y in the plane of a receiver aperture at a distance z from a rough target[2] that is illuminated by coherent radiation. Neglecting frequency-dependent terms, the resultant field is described by a linear superposition of scattered waves given by

$$U(x, y, z) = \frac{1}{\sqrt{n}} \sum_{k=1}^{n} u_k e^{j\varphi_k} \tag{4-1}$$

where n is the number of scattering elements on the surface, $u_k \equiv |u_k|$ is the absolute magnitude of the field amplitude of the kth scattered wave, and ϕ_k is the phase of the kth wave. Rewriting Eq. (4-1) in terms of its real and imaginary parts yields

$$\begin{aligned} U_R &= \frac{1}{\sqrt{n}} \sum_{k=1}^{n} u_k \cos\phi_k \\ U_I &= \frac{1}{\sqrt{n}} \sum_{k=1}^{n} u_k \sin\phi_k \end{aligned} \tag{4-2}$$

where the resultant complex field amplitude is $U = U_R + iU_I$ and $U^*U = I$, where I is the intensity of the resultant field. Taking an ensemble average of Eqs. (4-2) over many independent realizations of the scattered field leads to

$$\begin{aligned} \langle U_R \rangle &= \frac{1}{\sqrt{n}} \sum_{k=1}^{n} \langle u_k \cos\phi_k \rangle = \frac{1}{\sqrt{n}} \sum_{k=1}^{n} \langle u_k \rangle \langle \cos\phi_k \rangle = 0 \\ \langle U_I \rangle &= \frac{1}{\sqrt{n}} \sum_{k=1}^{n} \langle u_k \sin\phi_k \rangle = \frac{1}{\sqrt{n}} \sum_{k=1}^{n} \langle u_k \rangle \langle \sin\phi_k \rangle = 0 \end{aligned} \tag{4-3}$$

Here it is assumed that u_k and ϕ_k are statistically independent and can therefore be averaged separately, while letting $\langle \cos\phi_k \rangle = \langle \sin\phi_k \rangle = 0$ implies a uniformly distributed phase over the interval $(0, 2\pi)$.

For a large number of scatterers, the central limit theorem can be invoked to obtain zero-mean Gaussian probability density functions for the real and imaginary components (cf. Section 1.2.6). Thus

$$p(U_R) = \frac{e^{-U_R^2/2\sigma_R^2}}{\sqrt{2\pi\sigma_R^2}} \quad \text{and} \quad p(U_I) = \frac{e^{-U_I^2/2\sigma_I^2}}{\sqrt{2\pi\sigma_I^2}} \tag{4-4}$$

where

$$\begin{aligned}
\sigma_R^2 &= \operatorname{var}(U_R) = \langle U_R^2\rangle - \langle U_R\rangle^2 = \langle U_R^2\rangle \\
&= \frac{1}{n}\sum_{k,m}\langle u_k u_m\rangle\langle\cos\phi_k\cos\varphi_m\rangle = \frac{1}{n}\sum_{k=1}^{n}\frac{\langle u_k^2\rangle}{2} \\
\sigma_I^2 &= \operatorname{var}(U_I) = \langle U_I^2\rangle - \langle U_I\rangle^2 = \langle U_I^2\rangle \\
&= \frac{1}{n}\sum_{k,m}\langle u_k u_m\rangle\langle\sin\phi_k\sin\varphi_m\rangle = \frac{1}{n}\sum_{k=1}^{n}\frac{\langle u_k^2\rangle}{2}
\end{aligned} \tag{4-5}$$

Hence

$$\sigma_R^2 = \sigma_I^2 \equiv \sigma^2 \tag{4-6}$$

We can also calculate the mixed moment $\sigma_{RI} = \langle U_R U_I\rangle$; that is,

$$\langle U_R U_I\rangle = \frac{1}{n}\sum_{k,m}\langle u_k u_m\rangle\langle\cos\phi_k\sin\varphi_m\rangle = 0 \tag{4-7}$$

To summarize, the real and imaginary parts of the complex field have zero mean, identical variances, and are uncorrelated.

Since U_R and U_I are uncorrelated, the joint probability density function can be written as the product of the individual density functions:

$$p(U_R, U_I) = \frac{e^{-(U_R^2+U_I^2)/2\sigma^2}}{2\pi\sigma^2} \tag{4-8}$$

where Eq. (4-6) was used.

We now wish to express Eq. (4-8) in terms of the resultant amplitude U. To do so requires a transformation of variables as outlined in Section 1.2.4; that is,

$$p(U,\theta) = p(U_R, U_I)\,|J| \tag{4-9}$$

where $|J|$ is the absolute value of the Jacobian of the transformation given by

$$J = \begin{vmatrix} \dfrac{\partial U_R}{\partial U} & \dfrac{\partial U_R}{\partial\theta} \\[2ex] \dfrac{\partial U_I}{\partial U} & \dfrac{\partial U_I}{\partial\theta} \end{vmatrix} \tag{4-10}$$

With $U_R = U\cos\theta$, $U_I = U\sin\theta$, $|J| = U$ and

$$p(U,\theta) = \frac{Ue^{-U^2/2\sigma^2}}{2\pi\sigma^2} \tag{4-11}$$

The marginal probability density function for the amplitude alone is obtained by integrating Eq. (4-11) over all possible phase angles, which yields

$$\begin{aligned} p(U) &= \int_0^{2\pi} \frac{Ue^{-U^2/2\sigma^2}}{2\pi\sigma^2}\, d\theta \\ &= \frac{Ue^{-U^2/2\sigma^2}}{\sigma^2} \end{aligned} \tag{4-12}$$

Equation (4-12) is the well-known *Rayleigh* probability density function that is found to govern the statistics of a wide variety of physical phenomena (cf. Comment 1-9 for more details).

The probability density function for phase can be obtained in a similar manner, that is, by integrating over the amplitude U in Eq. (4-11), yielding

$$p(\theta) = \frac{1}{2\pi} \tag{4-13}$$

Hence the phase is uniformly distributed as in our original assumption.

The amplitude and phase are not of particular interest in direct-detection systems because these parameters cannot be measured without a phase reference such as exists in coherent or heterodyne detection systems. Direct-detection systems measure intensity $I = U^2$ only so that a transformation of the variable U to I in Eq. (4-12) is warranted. Since this is a single-variable transformation, we write

$$\begin{aligned} p(I) &= p(U)\left.\frac{dU}{dI}\right|_{U\to\sqrt{I}} \\ &= \frac{e^{-I/2\sigma^2}}{2\sigma^2} \end{aligned} \tag{4-14}$$

The first and second moments become

$$\begin{aligned} \langle I\rangle &= \int_0^\infty I p(I)\, dI = 2\sigma^2 \\ \langle I^2\rangle &= \int_0^\infty I^2 p(I)\, dI = 8\sigma^4 = 2\langle I\rangle^2 \end{aligned} \tag{4-15}$$

Therefore,

$$\sigma_I^2 = \langle I^2\rangle - \langle I\rangle^2 = \langle I^2\rangle \tag{4-16}$$

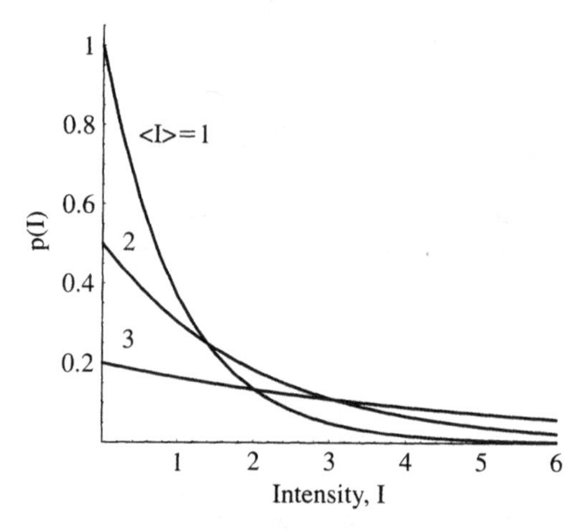

Figure 4-3. Negative exponential probability density function.

or

$$\sigma_I = \langle I \rangle \tag{4-17}$$

and

$$p(I) = \frac{e^{-I/\langle I \rangle}}{\langle I \rangle} \tag{4-18}$$

Equation (4-18) is known as the *negative exponential probability density function* and represents the statistics of the intensity of a speckle field at a point. It is determined by the single parameter $\langle I \rangle$ and is shown in Figure 4-3.

From Eq. (4-17) we see that the standard deviation of the intensity is equal to the mean intensity. From a system point of view this is an important conclusion. It implies that for a point receiver, increasing the mean signal strength of the system will not reduce the relative strength of signal fading due to target speckle. In coherent radar and laser radar terminology this effect is known as *Rayleigh fading*, since coherent systems respond to the amplitude of the field, not the intensity. In direct-detection systems, which respond to the intensity of the field, the effect might be better characterized as *exponential fading*.

The cumulative distribution function for the exponential probability density function is given by

$$F(I) = \int_0^I p(I')\,dI' = 1 - e^{-I/\langle I \rangle} \tag{4-19}$$

and represents the probability of the intensity being less than I.

Comment 4-1. As alluded to above, the negative exponential statistics of the speckle field is the result of a series of measurements at a single point on an ensemble of microscopically different surfaces having identical statistics.

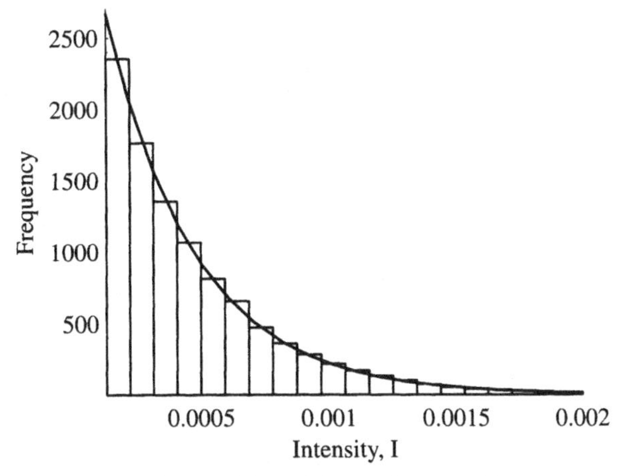

Figure 4-4. Histogram of the simulated data of Figure 4-2 normalized such that the area of the histogram sums to 1. The data closely approximate the negative exponential probability density function shown as a solid line which has the same mean value as the data.

However, in the spirit of the ergodic hypothesis, it is just as valid to consider the statistics associated with n single-point measurements within a single realization of the ensemble, which should lead to the same distribution. Indeed, the latter case is easily demonstrated using the simulated image data shown in Figure 4-2. Figure 4-4 shows a histogram plot of the intensity data from Figure 4-2 overlaid with a negative exponential density curve having the *same mean value*. The agreement is clearly evident. It is left to the reader to show that the former case may also be demonstrated through simulation by histograming the peak intensities of n realizations of m summed sine waves having uniformly distributed phases. ■

4.3. SUMMED STATISTICS OF FULLY DEVELOPED SPECKLE

We have seen that the amplitude and intensity of polarized speckle patterns measured at a single point obey Rayleigh and exponential probability density functions, respectively. Also of interest for later calculations of the statistical distributions of photoemmissive processes in the presence of noise is the calculation of the density function for the *summed* or *integrated energy* intercepted by a receiving aperture. The integrated intensity during the measurement interval τ is given by

$$I_0 = \frac{1}{S}\int_{-\infty}^{\infty}\int_{-\infty}^{\infty} g(x, y)I(x, y)\,dx\,dy \tag{4-20}$$

where $g(x, y)$ is the transfer function of the aperture at position x, y, $I(x, y)$ is the intensity at the point x, y within the aperture, and the integral is taken over the surface area S of the pupil function of the aperture.

From Eq. (4-20) we see that I_0 represents the intensity contribution from all speckle cells falling within the receiver aperture dimensions and is therefore a random variable when taken over an ensemble of rough targets. Figure 4-2 is an example of such windowed data where the actual size of the speckle pattern is much larger than the pattern displayed. An approximate density function for I_0 has been derived by Mandel[3] and Goodman[4] based on the time-domain formulation of Rice.[5] In effect, Goodman's "quasiphysical" approach models the continuous spatial distribution of speckle intensity with an idealized series of piecewise continuous functions. However, the summing process need not be solely spatial in origin but can also be temporal so that a similar piecewise continuous function may be developed in the temporal domain, as indicated in Figure 4-5. This equivalence of formulations may be viewed as an expression of the ergodic hypothesis.

When applied to the *summed statistics* of a speckle field, the ergodic hypothesis may be interpreted as stating that the *area statistics* of an ensemble of statistically identical *stationary* rough surfaces are equivalent to the *summed statistics* of a single *rotating* rough surface evaluated at a point in space. In the former case, the summing process is achieved by using a receiver aperture that is large compared to the size of a correlation cell. In the latter case, summing is achieved by rotating the surface about an axis normal to the illuminating beam by an amount $\theta \geq \lambda/D$, where D is the diameter of the illuminating spot, and electronically integrating the output from a point detector. This equivalence is shown schematically in Figure 4-6.

A necessary condition for the ergodic hypothesis to be true is that the mean intensity of each speckle realization be given by $\langle I_m \rangle = \langle I \rangle / M$, where $\langle I \rangle$ is the mean intensity of the total field. Since the random variable I_0 is the sum of m independent random variables, I_m, we can use the method of characteristic functions to obtain the density function for the integrated intensity, $p(I_0)$ (cf.

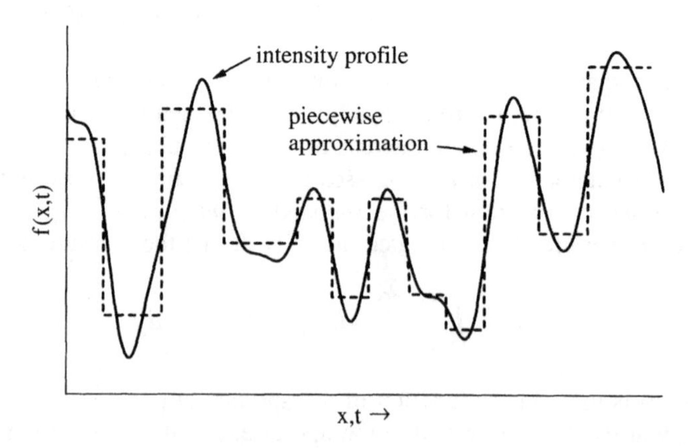

Figure 4-5. Piecewise continuous approximation to the intensity profile of a speckle pattern.

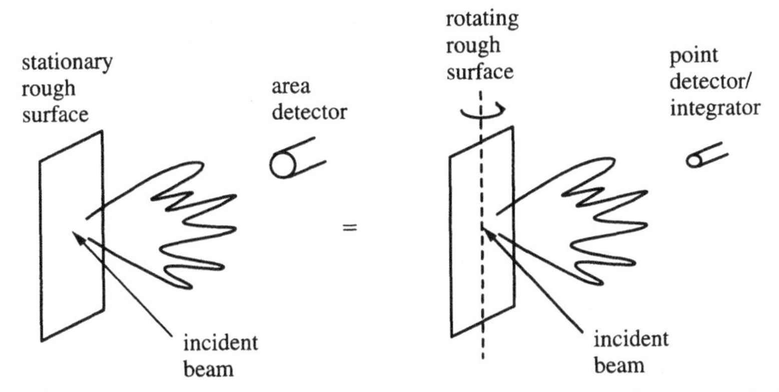

Figure 4-6. Statistical equivalence of a stationary and a rotating rough surface measured with an area detector and a point detector–integrator, respectively.

Section 1.2.5). Thus we write for the mth speckle field,

$$G_{I_0}(i\upsilon) = \prod_{m=1}^{M} G_{I_m}(i\upsilon) = \prod_{m=1}^{M} \langle e^{i\upsilon I_m} \rangle \tag{4-21}$$

where, using Eq. (4-18),

$$\begin{aligned} G_{I_m}(i\upsilon) &= \int_0^\infty p(I_m) e^{i\upsilon I_m} dI_m \\ &= \frac{1}{\langle I_m \rangle} \int_0^\infty e^{-I_m/\langle I_m \rangle} e^{i\upsilon I_m} dI_m \\ &= \frac{1}{1 - i\upsilon \langle I_m \rangle} \end{aligned} \tag{4-22}$$

Here the fact that $I_m \geq 0$ restricts the integration to positive values of I_m. Since $\langle I_m \rangle = \langle I \rangle / M$, the characteristic function of the integrated intensity becomes, using Eq. (4-21),

$$G_{I_0}(i\upsilon) = \left(1 - \frac{i\upsilon \langle I \rangle}{M}\right)^{-M} \tag{4-23}$$

Performing the inverse (Fourier) transform,

$$\begin{aligned} p(I_0) &= \frac{1}{2\pi} \int_{-\infty}^\infty G_{I_0}(i\upsilon) e^{-i\upsilon I_0} d\upsilon \\ &= \frac{1}{2\pi} \int_{-\infty}^\infty \frac{e^{-i\upsilon I_0}}{(1 - i\upsilon \langle I \rangle / M)^M} d\upsilon \end{aligned} \tag{4-24}$$

Equation (4-24) has a pole at $\upsilon = -iM/\langle I \rangle$ and can easily be evaluated using a contour integral over the lower half of the complex plane (cf. Appendix B). This yields

$$p(I_0) = \begin{cases} \left(\dfrac{M}{\langle I \rangle}\right)^M \dfrac{(I_0)^{M-1} e^{-MI_0/\langle I \rangle}}{\Gamma(M)} & I_0 \geq 0 \\ 0 & I_0 < 0 \end{cases} \tag{4-25}$$

Here $\Gamma(M)$ is the well-known *gamma function*, defined by

$$\Gamma(a) = \begin{cases} \displaystyle\int_0^\infty x^{a-1} e^{-x} dx & a > 0 \\ (a-1)! & a = \text{positive integer} \end{cases} \tag{4-26}$$

Equation (4-25) is known as the *gamma distribution* for the integrated intensity of a fully developed speckle field. Note that $p(I_0)$ represents a continuous probability density function despite the fact that M is frequently assumed to be an integer, in which case one can let $\Gamma(M) \to (M-1)!$. But even in this case, $p(I_0)$ is still a continuous density function, although with discrete parameters. Note also that the gamma distribution corresponds to a χ^2 distribution with $2n$ degrees of freedom. The latter distribution, introduced in Chapter 1, will be seen frequently in the sections to follow as a generator of various target models.

The gamma distribution can be written in generalized notation as

$$p(x) = \frac{\lambda^\eta x^{\eta-1} e^{-\lambda x}}{\Gamma(\eta)} \qquad x \geq 0 \tag{4-27}$$

Its mean and variance are

$$\langle x \rangle = \frac{\eta}{\lambda} \tag{4-28}$$

and

$$\sigma_x^2 = \frac{\eta}{\lambda^2} \tag{4-29}$$

respectively. Letting $x = I_0$, $\eta = M$, and $\lambda = M/\langle I \rangle$ in Eqs. (4-28) and (4-29), we find that

$$\langle I_0 \rangle = \langle I \rangle \tag{4-30}$$

and

$$\sigma_{I_0}^2 = \frac{\langle I \rangle^2}{M} \tag{4-31}$$

That is, the mean of the integrated intensity is equal to the mean of the total intensity of the speckle field, and the variance equals the square of the mean intensity divided by the speckle count M. These moments are consistent with those defined in Sections 3.3.2 and 3.3.3 and lead to an effective signal-to-noise ratio at the aperture of $\langle I \rangle / \sigma_{I_0} = \sqrt{M}$. As stated earlier, the parameter $\langle I \rangle / M$ appearing in the characteristic function of Eq. (4-23) can be viewed as the mean intensity of an individual speckle cell. Hence the introduction of M into the characteristic function has assured energy conservation, as given by Eq. (4-30).

At this point, in anticipation of later discussions, we change variables in Eq. (4-25) to the integrated energy $W_0 = I_0 A_d \tau$, where A_d is the detector area and τ is the measurement interval, obtaining

$$p(W_0) = \begin{cases} \left(\dfrac{M}{\overline{W}}\right)^M \dfrac{(W_0)^{M-1} e^{-M W_0/\overline{W}}}{\Gamma(M)} & W_0 \geq 0 \\ 0 & W_0 < 0 \end{cases} \tag{4-32}$$

Comparing Eqs. (4-27) and (4-32), we see that $\lambda = M/\overline{W}$ may be viewed as the reciprocal of the mean energy per speckle cell. In mathematical statistics, $\eta(= M)$ and $\lambda(= M/\overline{W})$ are referred to as the shape and scaling parameters of the probability density function, respectively. This follows from the fact that if λ is held constant and η is varied, the shape of the density function varies, while if η is held constant and λ varied, the density function scales in its parameters (mean, mode, etc.) but not its shape. These effects are shown in Figures 4-7 and 4-8, where $p(W_0)$ is plotted versus W_0 with $M/\overline{W}$ held constant and M held constant, respectively.

From Figure 4-7 we see that for $M \to 1$, which corresponds to an aperture that is small compared to the size of a correlation cell, $p(W_0)$ approaches a negative exponential density function, which was assumed at the outset. On the

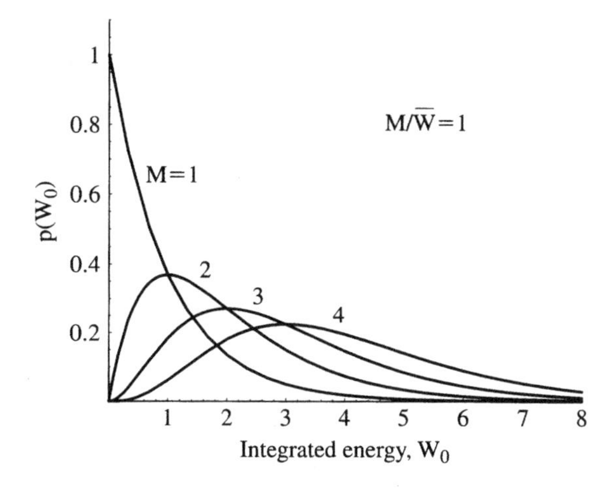

Figure 4-7. Gamma distributions with speckle count to energy ratio M/W held constant.

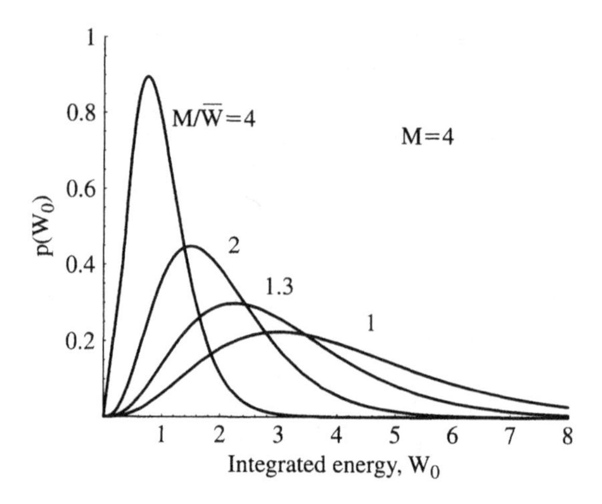

Figure 4-8. Gamma distributions with speckle count held M constant.

other hand, as $M \to \infty$, which corresponds to an aperture that is large compared to the size of a correlation cell, $p(W_0)$ approaches a Gaussian distribution as per the central limit theorem. Also, with $M \to \infty$, the variance $\sigma_{W_0}^2 \to 0$, since in that limit all the speckle cells are contained within the aperture, resulting in $W_0 \to W_T$ (i.e., the integrated energy is equal to the total scattered energy, which is necessarily constant).

Equation (4-32) represents an approximation to the exact density function for the integrated energy. Mathematically, the exact density function is based on a *Karhunan–Loeve expansion* and is relatively complex to derive[6–8] and somewhat cumbersome to use in systems analysis. We will therefore skip a detailed discussion of the exact theory and refer the reader to the works of Goodman,[9] where it is shown that the gamma distribution is a reasonable approximation to the exact theory, certainly suitable for engineering applications.

The cumulative distribution function for the gamma distribution is

$$F(W_0 \le W_0') = \int_0^{W_0'} p(W_0)\,dW_0$$

$$= \frac{a^M}{\Gamma(M)} \int_0^{W_0'} W_0^{M-1} e^{-aW_0}\,dW_0 \tag{4-33}$$

Changing integration variables to $x_0 = aW_0$ yields

$$F(W_0 \le W_0') = \frac{1}{\Gamma(M)} \int_0^{x_0'/a} x_0^{M-1} e^{-x_0}\,dx_0$$

$$= \frac{\gamma(M, W_0')}{\Gamma(M)} \tag{4-34}$$

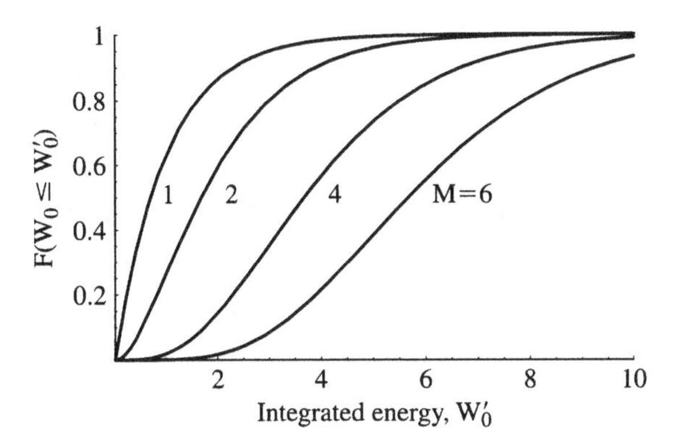

Figure 4-9. Cumulative gamma distribution with the speckle count M as a parameter.

where $W_0' = x_0'/a$ and $\gamma(M, W_0')$ is the *incomplete gamma function*, defined by

$$\gamma(a, t) = \int_0^t x^{a-1} e^{-x} dx \qquad a > 0 \tag{4-35}$$

Equation (4-34) represents the probability that W_0 will be *less than or equal to* W_0'. Alternatively, the probability that W_0 will be *greater than* W_0' is given by

$$P(W_0 > W_0') = \frac{\Gamma(M, W_0')}{\Gamma(M)} \tag{4-36}$$

where $\Gamma(M, W_0')$ is defined through the relationship

$$\Gamma(M) = \Gamma(M, W_0') + \gamma(M, W_0') \tag{4-37}$$

Figure 4-9 shows $F(W_0 \leq W_0')$ plotted as a function of integrated energy W_0'.

4.4. POISSON SIGNAL IN POISSON NOISE

Most laser systems detect radiation with photodetectors that are based on the photoelectric effect. Semiclassical radiation theory predicts, and is indeed based on the assumption, that the average rate of emission of photoelectrons by a photoemitting surface which is irradiated with a constant intensity source of optical radiation is given by

$$\langle r \rangle = \frac{\eta I A_d}{h\nu} \tag{4-38}$$

where η is the quantum efficiency of the detector, $h\nu$ the photon energy at frequency ν, I the optical intensity of the classical wave, and A_d the detector surface area. η is seen to be a semiclassical parameter relating the classical-wave concept of intensity to the quantum mechanical concept of particle emission. It is generally identified as the probability of photoelectron emission given incident energy $h\nu$.

The statistics associated with photoemissive processes were originally developed by Purcell[10] and Mandel[11] within a semiclassical framework. We attempt to provide a simplified derivation of the probability density function for photoemission based on the probability P_k of k photoemissive events occurring in a finite time interval $d\tau$. From Eq. (4-38), the probability of an event occurring in a small time interval $d\tau$ is $\langle r\rangle\, d\tau$, provided that $d\tau$ is small enough that the probability of two or more events occurring during $d\tau$ is negligible. The probability $P_k(\tau + d\tau)$ is then the sum of two mutually exclusive occurrences:

1. k emission events occur by time τ with probability $P_k(\tau)$ and no events occur during the interval $d\tau$, the latter probability being $(1 - \langle r\rangle\tau)$. Since these events are independent by assumption, their joint probability is $(1 - \langle r\rangle\tau)P_k(\tau)$.
2. $k - 1$ emission events occur in time τ with probability $P_{k-1}(\tau)$ and one emission occurs during the interval $d\tau$, the latter probability being $\langle r\rangle\, d\tau$. Once again, independence leads to a joint probability of $\langle r\rangle\, d\tau\, P_{k-1}(\tau)$.

Thus

$$P_k(\tau + d\tau) = (1 - \langle r\rangle\, d\tau)P_k(\tau) + \langle r\rangle\, d\tau\, P_{k-1}(\tau) \tag{4-39}$$

or

$$\frac{P_k(\tau + d\tau) - P_k(\tau)}{d\tau} = \langle r\rangle[P_{k-1}(\tau) - P_k(\tau)] \tag{4-40}$$

Letting $d\tau \to 0$ yields

$$\frac{dP_k}{d\tau} = \langle r\rangle[P_{k-1}(\tau) - P_k(\tau)] \tag{4-41}$$

Assuming that $k = 0$ emissions occur in the interval $d\tau$ and $P_{-1}(\tau) \equiv 0$, we have

$$\frac{dP_0}{d\tau} = -\langle r\rangle P_0(\tau) \tag{4-42}$$

Integrating with the boundary condition $P_0(0) = 1$ yields

$$P_0(\tau) = e^{-\langle r\rangle\tau} \tag{4-43}$$

Similarly, for $k = 1$ emissions in $d\tau$,

$$\begin{aligned}\frac{dP_1}{d\tau} &= \langle r\rangle[P_0(\tau) - P_1(\tau)] \\ &= \langle r\rangle[e^{-\langle r\rangle\tau} - P_1(\tau)]\end{aligned} \tag{4-44}$$

With the boundary condition $P_1(0) = 0$, we obtain

$$P_1(\tau) = \langle r \rangle \tau e^{-\langle r \rangle \tau} \tag{4-45}$$

Continuing in this manner, it can be concluded by induction that

$$P_k(\tau) = \frac{(\langle r \rangle \tau)^k}{k!} e^{-\langle r \rangle \tau} \tag{4-46}$$

Now the mean optical power $\overline{P}$ is related to the mean intensity $\overline{I}$ by $\overline{P} = \overline{I} A_d$ and the mean optical energy $\overline{W}$ to the power in the time interval τ by $\overline{W} = \overline{P}\tau$. We therefore have for the average number of photoelectrons emitted in time τ,

$$\overline{N} = \overline{r}\tau = \frac{\eta \overline{I} A_d \tau}{h\nu} = \frac{\eta \overline{W}}{h\nu} \tag{4-47}$$

where, for simplicity of notation, the averaging notation has been changed from brackets to overbars. Thus, using Eq. (4-47) in Eq. (4-46), while designating the function $q(\cdot)$ as a discrete probability density function, we obtain for the distribution of k photoelectrons,

$$q(k; \overline{N}) = \frac{(\overline{N})^k e^{-\overline{N}}}{k!} \tag{4-48}$$

Equation (4-47) is commonly referred to as the *discrete Poisson probability density function* with random count variable k. It is shown in Figure 4-10 for several values of $\overline{N}$.

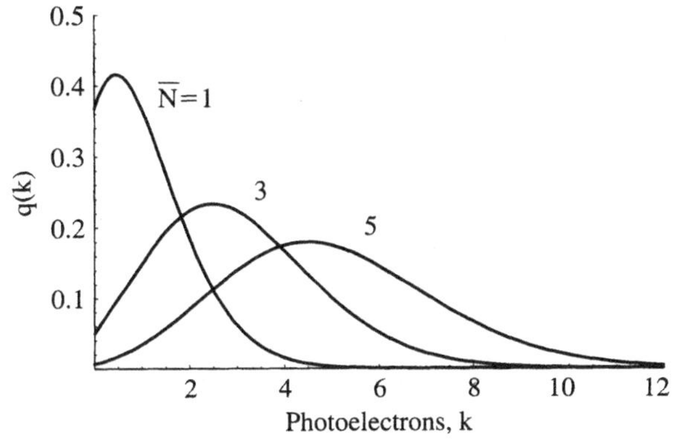

Figure 4-10. Envelopes of the discrete Poisson distributions with mean photoelectron count $\overline{N}$ as a parameter.

Comment 4-2. It can be shown that $\overline{N}$ is indeed the mean of the random count variable k, as well as develop an expression for the variance of k, by using the method of characteristic functions. Since $q(k; \overline{N})$ is a discrete probability density function, we write

$$G(i\upsilon) = \langle e^{i\upsilon k} \rangle = \sum_{k=0}^{n} q(k) e^{i\upsilon k} \tag{4-49}$$

or, using Eq. (4-48),

$$\begin{aligned} G(i\upsilon) &= e^{-\overline{N}} \sum_{k=0}^{\infty} \frac{\overline{N}^k}{k!} e^{i\upsilon k} \\ &= e^{-\overline{N}} e^{\overline{N} e^{i\upsilon}} \\ &= e^{\overline{N}(e^{i\upsilon}-1)} \end{aligned} \tag{4-50}$$

The first moment is therefore given by

$$\begin{aligned} \langle k \rangle &= -i \frac{dG(i\upsilon)}{d\upsilon}\bigg|_{\upsilon=0} \\ &= -i\overline{N}(-\sin\upsilon + i\cos\upsilon) e^{\overline{N}(\cos\upsilon + i\sin\upsilon - 1)}|_{\upsilon=0} \\ &= \overline{N} \end{aligned} \tag{4-51}$$

Similarly, the second moment is given by

$$\begin{aligned} \langle k^2 \rangle &= (-i)^2 \frac{d^2G(i\upsilon)}{d\upsilon^2}\bigg|_{\upsilon=0} \\ &= -\{\overline{N}(-\cos\upsilon - i\sin\upsilon) + \overline{N}^2(-\sin\upsilon + i\cos\upsilon)\} e^{\overline{N}(\cos\upsilon + i\sin\upsilon - 1)}|_{\upsilon=0} \\ &= \overline{N}^2 + \overline{N} \end{aligned} \tag{4-52}$$

Hence the variance in photoelectron count is given by

$$\begin{aligned} \mathrm{var}(k) &= \langle k^2 \rangle - \langle k \rangle^2 \\ &= \overline{N} \end{aligned} \tag{4-53}$$

Note that the ratio of the standard deviation to the mean is $\sigma/\langle k \rangle = 1/\sqrt{\overline{N}}$. From a systems point of view, this means that increasing the mean pulse energy $\overline{N}$ can reduce the fluctuations associated with the detector photoemissive process.

Finally, it will be useful in later analysis to derive the limiting form of the Poisson distribution for large $\overline{N}$. From Eq. (4-50) we have

$$\begin{aligned} G(i\upsilon) &= e^{\overline{N}(e^{i\upsilon}-1)} \\ &= e^{\overline{N}(\cos\upsilon + i\sin\upsilon - 1)} \\ &= e^{i\overline{N}\sin\upsilon - 2\overline{N}\sin^2\upsilon/2} \end{aligned} \tag{4-54}$$

For large $\overline{N}$, the only significant contributions occur when υ is small. Thus

$$G(i\upsilon) \approx e^{i\overline{N}\upsilon - \overline{N}\upsilon^2/2} \tag{4-55}$$

But this is recognizable as the characteristic function of the Gaussian probability density function; hence

$$p(k) \approx \frac{1}{\sqrt{2\pi\overline{N}^2}} e^{-(k-\overline{N})^2/2\overline{N}^2} \tag{4-56}$$

Here Eq. (4-56) is written as a continuous probability density function since for $\overline{N}$ very large, the integer nature of k becomes irrelevant for most calculations. Note also that $p(k)$ remains a single parameter function, as in the Poisson distribution. ■

The Poisson density function of Eq. (4-48) represents the distribution of emitted photoelectrons during the counting interval τ, where $\overline{N}$ is the mean number of photoelectrons during τ. In practice, τ is identified with the measurement interval of the detection process, which for an optimally matched receiver, is the signal pulse width.

In addition to the signal there are several noise contributions to the photodetection statistics. Consider first those noise sources that are Poisson distributed. These can include background radiation due to solar scatter off clouds, terrain, or targets and detector dark currents. The latter arise when photoelectrons are thermally or quantum mechanically generated within the detector material.

Let $\overline{N}_s$ and $\overline{N}_n$ be the count variables representing the mean signal and noise photoelectrons, respectively. Since these processes are statistically independent, we can write, using Eq. (4-50), the characteristic function for the signal plus noise as a simple product of the individual characteristic functions, that is,

$$\begin{aligned} G_{s+n}(i\upsilon) &= e^{\overline{N}_s(e^{i\upsilon}-1)} e^{\overline{N}_n(e^{i\upsilon}-1)} \\ &= e^{\overline{N}_{s+n}(e^{i\upsilon}-1)} \end{aligned} \tag{4-57}$$

where $\overline{N}_{s+n} = \overline{N}_s + \overline{N}_n =$ the mean signal plus noise photoelectron count due to photons arriving during the measurement interval τ. But $G_{s+n}(i\upsilon)$ has the same form as $G(i\upsilon)$; that is, it is the characteristic function of a

Poisson variate but with different parameters. The inverse transformation is therefore

$$q_{s+n}(k; \overline{N}_{s+n}) = \frac{(\overline{N}_s + \overline{N}_n)^k e^{-\overline{N}_s + \overline{N}_n}}{k!} \tag{4-58}$$

The probability of detecting a target in the presence of background or dark noise can be calculated using the Neyman–Pearson criterion of Chapter 1. In this case the a priori probability density function associated with the hypothesis H_0 that noise alone is present is given by Eq. (4-58) with $\overline{N}_s = 0$, and the hypothesis H_1 that signal plus noise is present is given by Eq. (4-58). Thus, for a specified false alarm probability P_{fa}, the threshold photoelectron count k_{th} is given by the largest integer satisfying the inequality

$$P_{\text{fa}} \geq \sum_{k=k_{\text{th}}}^{\infty} \frac{(\overline{N}_n)^k e^{-\overline{N}_n}}{k!} = 1 - \sum_{k=0}^{k_{\text{th}}-1} \frac{(\overline{N}_n)^k e^{-\overline{N}_n}}{k!} \tag{4-59}$$

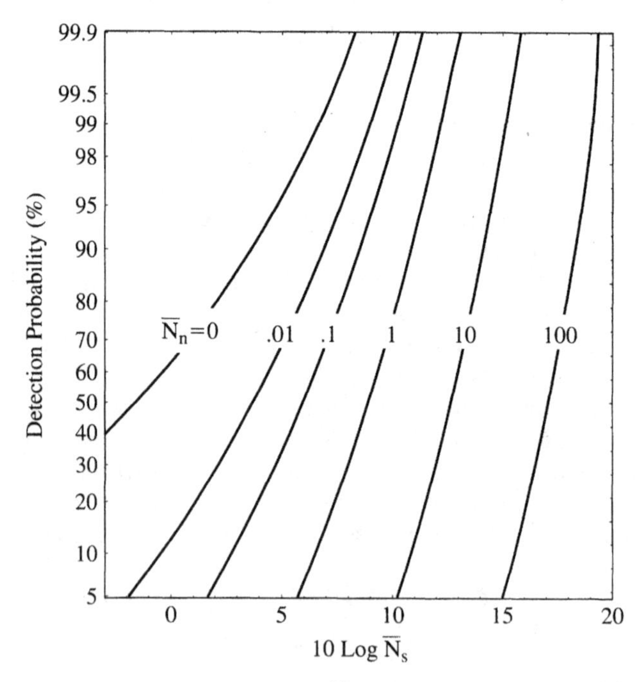

Figure 4-11. Detection probability versus $10 \log \overline{N}_s$ for a Poisson signal in Poisson noise, where $\overline{N}_s$ is the mean signal photoelectron count, several values of the mean noise photoelectron count $\overline{N}_n$, and a false alarm probability of $P_{\text{fa}} = 10^{-6}$. (After J. W. Goodman, Proc. IEEE, Vol. 53, No. 11, Nov. 1963.)

The corresponding detection probability is then

$$P_d = \sum_{k=k_{\text{th}}}^{\infty} q_{s+n}(k; \overline{N}_{s+n}) = 1 - \sum_{k=0}^{k_{\text{th}}-1} q_{s+n}(k; \overline{N}_{s+n}) \tag{4-60}$$

Note that the second forms of Eqs. (4-59) and (4-60) obviate the need for calculating infinite sums. Equation (4-60) is plotted in Figure 4-11 for a maximum $P_{\text{fa}} = 10^{-6}$. Threshold settings are $N_t = 1, 3, 5, 10, 29$, and 152 for $N_n = 0, 10^{-2}, 0.1, 1, 10$, and 10^2, respectively.

Comment 4-3. Since the statistical variable k is an integer, a quantization effect is introduced into the calculation of any arbitrary P_{fa}. For $\overline{N}_n$ small, this implies large differences in P_{fa} for each integer change in k. For example, for $\overline{N}_n = 0.01$, a threshold setting of $k_{\text{th}} = 3$ electrons in Eq. (4-60) yields $P_{\text{fa}} = 1.65 \times 10^{-7}$; for $k_{\text{th}} = 2$ and 4 electrons, $P_{\text{fa}} = 4.97 \times 10^{-5}$ and 4.13×10^{-10}, respectively, which are significantly different. The sensitivity of P_{fa} to integer changes in k_{th} becomes less at higher values of $\overline{N}_n$, where the discrete density function becomes broader. ■

4.5. NEGATIVE BINOMIAL SIGNAL IN POISSON NOISE

We have seen that the distribution of photoelectrons generated by a photodetector operating against a specular target is described by Poisson statistics. By definition the specular target is one that produces a constant intensity signal at the receiver over the measurement interval τ. However, we have also seen in Section 4.3 that a diffuse or rough target results in a gamma distribution for the integrated intensity or energy at the receiver. Hence, to calculate the probability of k photoelectrons being emitted, the conditional probability distribution $q_{s+n}(k \mid W)$, which represents the probability of k photoelectrons being emitted given an integrated energy W, must be averaged over all possible realizations of W. Here, the subscripts in the integrated variables W_0 and I_0 have been dropped for conciseness while the vertical bar in $q_{s+n}(k \mid W)$ is meant to indicate that W is a (continuous) random variable. The conditional probability $q_{s+n}(k \mid W)$ is frequently referred to as a *doubly stochastic Poisson process*[12] since the rate process W is also a random variable. We therefore define the *Poisson transform* of $p(W)$ as

$$q(k) = \int_0^{\infty} q_{s+n}(k \mid W) p(W)\, dW \tag{4-61}$$

where, from Eq. (4-58),

$$q_{s+n}(k \mid W) = \frac{[(\eta W/h\nu) + \overline{N}_n]^k e^{-[(\eta W/h\nu)+\overline{N}_n]}}{k!} \tag{4-62}$$

and from Eq. (4-32) with $a = M/\overline{W}$,

$$p(W) = \frac{a^M W^{M-1} e^{-aW}}{\Gamma(M)} \tag{4-63}$$

Substituting Eqs. (4-63) and (4-62) into Eq. (4-61) yields

$$q_{s+n}(k) = \frac{a^M e^{-\overline{N}_n}}{k!\,\Gamma(M)} \int_0^\infty (\overline{N}_n + bW)^k W^{M-1} e^{-(a+b)W}\, dW \tag{4-64}$$

where $b = \eta/h\nu$. Using the binomial expansion[13]

$$(\overline{N}_n + bW)^k = \sum_{j=0}^{k} \frac{k!}{j!\,(k-j)!} \overline{N}_n^j (bW)^{k-j} \tag{4-65}$$

Eq. (4-64) can be written

$$q_{s+n}(k) = \frac{a^M e^{-\overline{N}_n}}{k!\,\Gamma(M)} \sum_{j=0}^{k} \frac{k!\,\overline{N}_n^j b^{k-j}}{j!\,(k-j)!} \int_0^\infty W^{k-j+M-1} e^{-(a+b)W}\, dW \tag{4-66}$$

But the integral is of the form

$$\int_0^\infty x^{\alpha-1} e^{-\beta x}\, dx = \frac{(\alpha-1)!}{\beta^\alpha} \tag{4-67}$$

Therefore,

$$q_{s+n}(k) = \frac{a^M e^{-\overline{N}_n}}{k!\,\Gamma(M)} \sum_{j=0}^{k} \frac{(k+M-j-1)!\,\overline{N}_n^j}{j!(k-j)!} \frac{b^{k-j}}{(a+b)^{k-j+M}} \tag{4-68}$$

Substituting for a and b in the last fraction yields

$$\frac{b^{k-j}}{(a+b)^{k-j+M}} = \frac{(\eta/h\nu)^{k-j}}{(M/\overline{W} + \eta/h\nu)^{k-j+M}} = \frac{\overline{N}_s^{k-j}\,\overline{W}^M}{(M+N_s)^{k-j+M}}$$

$$= \frac{\overline{N}_s^{k-j}\,\overline{W}^M}{(M+N_s)^{k-j}(M+N_s)^M} \tag{4-69}$$

where $\overline{N}_s = \eta\overline{W}/h\nu$. We therefore obtain, after removing j independent terms from the summation,

$$q_{s+n}(k) = \left(\frac{M}{M+\overline{N}_s}\right)^M \left(\frac{\overline{N}_s}{M+\overline{N}_s}\right)^k \frac{e^{-\overline{N}_n}}{\Gamma(M)}$$

$$\times \sum_{j=0}^{k} \frac{(k+M-j-1)!}{j!\,(k-j)!} \left[\frac{\overline{N}_n(M+\overline{N}_s)}{\overline{N}_s}\right]^j \tag{4-70}$$

Equation (4-70) can be cast in a more useful form by rearranging the summation term. Letting $z = \overline{N}_n(M + \overline{N}_s)/\overline{N}_s$ and pulling the factor $(k + M - 1)!/k!$ out of the sum leads to

$$q_{s+n}(k) = K\frac{\Gamma(k+M)}{\Gamma(k+1)}\left[1 + \frac{k}{(k+M-1)}\frac{z}{1!} + \frac{k(k-1)}{(k+M-1)(k+M-2)}\frac{z^2}{2!} + \cdots + \frac{k!(M-1)!}{(k+M-1)!}\frac{z^k}{k!}\right] \quad (4\text{-}71)$$

where K represents all factors prior to the sum in Eq. (4-70) and the factorials have been generalized to gamma functions. Now Eq. (4-71) can also be written

$$q_{s+n}(k) = K\frac{\Gamma(k+M)}{\Gamma(k+1)} \times \left[1 + \frac{-k}{(-k-M+1)}\frac{z}{1!} + \frac{-k(-k+1)}{(-k-M+1)(-k-M+2)}\frac{z^2}{2!} + \cdots\right] \quad (4\text{-}72)$$

The summation term can be identified as *Kummer's confluent hypergeometric function* ${}_1F_1(\alpha, \beta; z)$, defined as

$${}_1F_1(\alpha, \beta; z) = 1 + \frac{(\alpha)_1 z}{(\beta)_1} + \frac{(\alpha)_2}{(\beta)_2}\frac{z^2}{2!} + \cdots + \frac{(\alpha)_n z^n}{(\beta)_n k!} \quad (4\text{-}73)$$

where

$$\alpha = -k \quad (4\text{-}74)$$

and

$$\beta = -k - M + 1 \quad (4\text{-}75)$$

and the notation

$$\begin{aligned}(\alpha)_n &= \alpha(\alpha+1)(\alpha+2)\cdots(\alpha+n-1)\\ (\beta)_n &= \beta(\beta+1)(\beta+2)\cdots(\beta+n-1)\end{aligned} \quad (4\text{-}76)$$

where $(\alpha)_0 = (\beta)_0 = 1$. Equation (4-70) can now be written as

$$q_{s+n}(k; \overline{N}_s, \overline{N}_n, M) = \left(\frac{M}{M+\overline{N}_s}\right)^M \left(\frac{\overline{N}_s}{M+\overline{N}_s}\right)^k \frac{e^{-\overline{N}_n}}{\Gamma(M)}\frac{\Gamma(k+M)}{\Gamma(k+1)}\,{}_1F_1(\alpha, \beta; z) \quad (4\text{-}77)$$

where $z = \overline{N}_n(M + \overline{N}_s)/\overline{N}_s$. The discrete density function given by Eq. (4-77) will be referred to as the *Kummer distribution*. It should be noted that the number of photoelectrons k is necessarily an integer, while the number of speckles M is not necessarily an integer. The Kummer function will be discussed shortly, but for the moment we let $\overline{N}_n = 0$, in which case ${}_1F_1(a, b; 0) = 1$ (from boundary conditions on Kummer's differential equation), obtaining

$$q_s(k; \overline{N}_s, 0, M) = \left(\frac{M}{M + \overline{N}_s}\right)^M \left(\frac{\overline{N}_s}{M + \overline{N}_s}\right)^k \frac{\Gamma(k + M)}{\Gamma(M)\Gamma(k + 1)} \tag{4-78}$$

Equation (4-78) may be recognized as a *negative binomial distribution*. Indeed, it is not too difficult to show that the negative binomial distribution can be obtained directly from Eq. (4-61) using Eq. (4-62) with $\overline{N}_n = 0$ together with Eq. (4-63).

Figure 4-12 shows the negative binomial distribution for several values of M with $\overline{N}_s = 10$ photoelectrons. Figure 4-13 shows the corresponding set of distribution functions for the case of $\overline{N}_n = 2$ photoelectrons and $\overline{N}_s = 10$. It can be seen that the effect of a nonzero noise contribution is to move the peaks of the curves to the right.

Comment 4-4. We have previously defined the Poisson distribution as one that gives the probability of exactly x events occurring during a given period of time if the events take place independently and at a constant rate. We can now define the negative binomial distribution as one that gives the probability of exactly x events occurring during a given period of time when the events *do not* occur at a constant rate and the occurrence rate is a *random variable* that is represented by a gamma distribution.

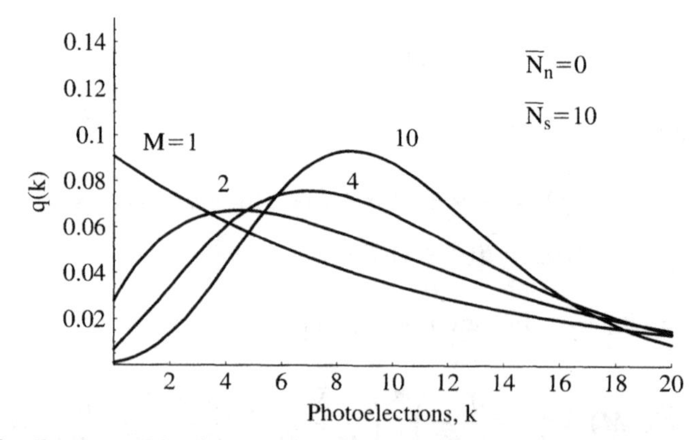

Figure 4-12. Envelopes of the discrete probability density function representing a negative binomial signal with mean photoelectron count $\overline{N}_s = 10$ and speckle count M as a parameter.

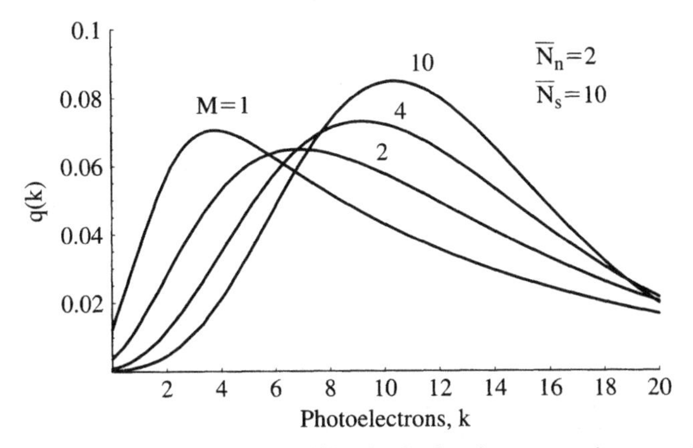

Figure 4-13. Envelopes of the discrete probability density function representing a negative binomial signal immersed in Poisson noise. The mean signal photoelectron count $\overline{N}_s = 10$ and the mean background count $\overline{N}_n = 2$, with the speckle count M as a parameter.

Recall from elementary statistics that the *binomial distribution* may be written

$$q_B(x; n) = \binom{n}{x} p^x q^{n-x} \tag{4-79}$$

where the notation

$$\binom{n}{x} = \frac{n!}{x!(n-x)!}$$

represents the binomial coefficients. Equation (4-79) expresses the probability that the number of successes obtained in an experiment of n Bernoulli trials, each having probability p, with $q = 1 - p$, will be x, where n is a fixed number. On the other hand, the negative binomial distribution expresses the probability that the number of Bernoulli trials required to achieve the rth success will be x. Here the number of trials x is not fixed but is a statistical variable. The corresponding density function is frequently written

$$q_{NB}(x; r) = \binom{x-1}{r-1} p^r q^{x-r} \tag{4-80}$$

A simple change of variables to $y = x - r$ leads to

$$q_{NB}(y; r) = \binom{y+r-1}{y} p^r q^y \tag{4-81}$$

This form of the distribution is interpreted as the probability that the number of Bernoulli trials in *excess* of the rth success will be $y(= x - r)$.

It should be noted that just as $q_B(x; n)$ is generated via an expansion of $(p+q)^n$, $q_{NB}(x; r)$ is generated by an expansion of $p^r(1-q)^{-r}$ —hence the name *negative binomial*. Indeed, it is not too difficult to show by expansion that

$$(1-q)^{-r} = \sum_{x=r}^{\infty} \binom{x-1}{r-1} q^{x-r} = \sum_{y=0}^{\infty} \binom{y+r-1}{y} q^y \tag{4-82}$$

such that $p^r(1-q)^{-r} = 1$.

The discussion thus far has been for the case of r being an integer, which is actually a special case of the negative binomial distribution known as the *Pascal distribution* (but is still frequently referred to as the *negative binomial distribution*). When r is not an integer, the factorials in Eq. (4-81) must be replaced with gamma functions. Thus, letting $y = k$ and $r = M$ in Eq. (4-81) and performing this replacement, we obtain the negative binomial distribution

$$q_{NB}(k; M) = \frac{\Gamma(k+M)}{\Gamma(k+1)\Gamma(M)} q^k p^M \tag{4-83}$$

which is identical to Eq. (4-78) under the substitutions

$$q = \frac{\overline{N}_s}{M + \overline{N}_s} \quad \text{and} \quad p = \frac{M}{M + \overline{N}_s} \tag{4-84}$$

It should be noted that the forms above for the binomial probabilities p and q appear frequently in the statistics associated with speckle fields and should not be confused with the continuous and discrete probability density functions $p(x)$ and $q(x)$, which will always be written as functions. ■

The limiting forms of the Kummer and negative binomial distributions are worth discussing. When $M = 1$ and $\overline{N}_n = 0$, Eqs. (4-77) and (4-78) reduce to the *Bose–Einstein distribution*, given by

$$q_{s+n}(k; \overline{N}_s, 0, 1) = \frac{1}{1+\overline{N}_s} \left(\frac{\overline{N}_s}{1+\overline{N}_s} \right)^k \tag{4-85}$$

where ${}_1F_1(-k, -k; 0) = 1$. Equation (4-85) corresponds to the probability distribution of the occupation number of a single mode of a quantum mechanical boson gas (i.e., a gas consisting of integral spin particles such as photons) at thermal equilibrium and is given by the $M = 1$ curve in Figure 4-12. However, it is instructive to show that the Bose–Einstein distribution may be obtained directly from a Poisson transform of the negative exponential distribution of a polarized speckle pattern measured at a single point. From Eq. (4-61) we have

$$q_s(k) = \int_0^{\infty} q(k \mid N_s) p(N_s)\, dN_s \tag{4-86}$$

where $q_s(k)$ is the discrete density function for the signal photoelectrons, $q(k \mid N_s)$ the conditional Poisson probability of obtaining k photoelectrons given N_s signal photoelectrons, and $p(N_s)$ the gamma distributed signal photoelectron count given by Eq. (4-25) with $I_0 \equiv I$. In the present case of $M = 1$ and $\overline{N}_n = 0$, $p(N_s)$ reduces to the negative exponential distribution given by Eq. (4-18). Thus, with $N_s = \eta W/h\nu = \eta I A_d \tau/h\nu$, Eq. (4-86) becomes

$$\begin{aligned} q_s(k) &= \int_0^\infty \frac{N_s^k e^{-N_s}}{k!} \frac{e^{-N_s/\overline{N}_s}}{\overline{N}_s} dN_s \\ &= \frac{1}{k!\overline{N}_s} \int_0^\infty N_s^k e^{-N_s(1+1/\overline{N}_s)} dN_s \end{aligned} \tag{4-87}$$

where Eq. (4-18) has been transformed to the statistical variable N_s. The integral in Eq. (4-87) is of the form

$$\int_0^\infty x^k e^{-x(1+\alpha)} dx = \frac{\Gamma(1+k)}{(1+\alpha)^{1+k}} = \frac{k!}{(1+\alpha)^{1+k}} \tag{4-88}$$

Therefore,

$$\begin{aligned} q_s(k) &= \frac{1}{\overline{N}_s} \left(\frac{\overline{N}_s}{1+\overline{N}_s} \right)^{k+1} \\ &= \frac{1}{1+\overline{N}_s} \left(\frac{\overline{N}_s}{1+\overline{N}_s} \right)^k \end{aligned} \tag{4-89}$$

which is the Bose–Einstein distribution.

When the number of photoelectrons is much smaller than the number of speckle cells (i.e., $\overline{N}_s/M \ll 1$), Eq. (4-78) reduces to a Poisson distribution. To show this, note that in the above limit

$$\left(\frac{M}{M+\overline{N}_s} \right)^M \to e^{-\overline{N}_s} \tag{4-90}$$

and

$$\left(\frac{\overline{N}_s}{M+\overline{N}_s} \right)^k \to \left(\frac{\overline{N}_s}{M} \right)^k \tag{4-91}$$

With $k \ll M$, the asymptotic form of $\Gamma(x)$ can be used to obtain (cf. Appendix A)

$$\frac{\Gamma(k+M)}{\Gamma(k+1)\Gamma(M)} \approx \frac{1}{k!} \frac{\sqrt{2\pi} e^{-M} M^{M+k-1/2}}{\sqrt{2\pi} e^{-M} M^{M-1/2}} \to \frac{M^k}{k!} \tag{4-92}$$

where k is an integer. Hence

$$q(k; \overline{N}_s) = \frac{\overline{N}_s^k}{k!} e^{-\overline{N}_s} \tag{4-93}$$

which is identical to Eq. (4-18) under the change of variable $N_s = \eta I A_d \tau / h\nu$. Thus a single speckle cell obeys the same statistics as that of a point detector operating against a specular target.

It will be useful for later comparisons to develop expressions for the mean and variance of the negative binomial distribution. Once again it is convenient to use the characteristic function for the distribution. Using Eq. (4-81) for $q(y; r)$, we have

$$\begin{aligned} G(iv) = \langle e^{ivy} \rangle &= \sum_{y=0}^{\infty} q(y; r) e^{ivy} \\ &= p^r \sum_{y=0}^{\infty} \frac{(y + r - 1)!}{(y)!(r - 1)!} (qe^{iv})^y \\ &= p^r (1 - qe^{iv})^{-r} \end{aligned} \tag{4-94}$$

The moments then follow from the derivatives of G, that is

$$\begin{aligned} \langle x \rangle &= -i \frac{dG}{dv} \bigg|_{v \to 0} \\ &= p^r \frac{d}{dv} (1 - qe^{iv})^{-r} |_{v \to 0} \\ &= \frac{rq}{p} \end{aligned} \tag{4-95}$$

Similarly,

$$\begin{aligned} \langle x^2 \rangle &= (-i)^2 \frac{d^2G}{dv^2} \bigg|_{v \to 0} \\ &= p^r \frac{d}{dv} [irqe^{iv} (1 - qe^{iv})^{-(r+1)}] |_{v \to 0} \\ &= \frac{r(r + 1)q^2}{p^2} + \frac{rq}{p} \end{aligned} \tag{4-96}$$

Using $q + p = 1$, the variance becomes

$$\text{var}(x) = \langle x^2 \rangle - \langle x \rangle^2 = \frac{rq}{p^2} \tag{4-97}$$

With $r = M$ and Eqs. (4-84) of Comment 4-4, we find that

$$\langle x \rangle = \frac{rq}{p} = \overline{N}_s$$
$$\text{var}(x) = \frac{rq}{p^2} = \overline{N}_s \frac{M + \overline{N}_s}{M} \tag{4-98}$$

Consider now the Kummer term ${}_1F_1$ in the M-dependent density function given by Eq. (4-77). It is a solution to Kummer's differential equation

$$zy'' + (b - z)y' - ay = 0 \tag{4-99}$$

with the boundary conditions ${}_1F_1(\alpha, \beta; 0) = 1$ and $\partial[{}_1F_1(\alpha, \beta; z)]/\partial z|_{z=0} = \alpha/\beta$. The confluent hypergeometric functions can be related to many other functions, depending on the parameters α, β, z and boundary conditions. One of these functions, the parabolic cylinder functions, is discussed later in the case of Gaussian speckle signals in Gaussian noise. A brief review of the confluent hypergeometric functions is given in Appendix A, but the interested reader who desires a deeper understanding of the confluent hypergeometric functions is referred to the many excellent textbooks on mathematical functions that are available.[14]

For the moment we wish to generalize our previous results for the mean and variance of the negative binomial distribution corresponding to $\overline{N}_n = 0$, to the more general case of Kummer's distribution wherein $\overline{N}_n \neq 0$. The characteristic function may be obtained using Eq. (4-81) and the expansion given by Eq. (4-71). With the variables $r = M$ and $y = k$, we have

$$\begin{aligned} G_k(iv) &= \sum_{y=0}^{\infty} \binom{y + r - 1}{y} p^r q^y e^{-qz} \\ &\times \left[1 + \frac{y}{y + r - 1}\frac{z}{1!} + \frac{y(y-1)}{(y + r - 1)(y + r - 2)}\frac{z^2}{2!} \right. \\ &\left. + \cdots + \frac{y!}{(y + r - 1)!}\frac{z^y}{y!}\right] e^{ivy} \end{aligned} \tag{4-100}$$

where $z = \overline{N}_n/q$. Using Eq. (4-82), the first term becomes

$$p^r(1 - qe^{iv})^{-r} e^{-qz} \tag{4-101}$$

The second term becomes

$$ze^{-qz} p^r \sum_{y=0}^{\infty} \binom{y + r - 1}{y} (qe^{iv})^y \frac{y}{y + r - 1} \tag{4-102}$$

Since the first term in the expansion of Eq. (4-102) is zero, we let $y = s + 1$, to obtain

$$
\begin{aligned}
ze^{-qz} p^r \sum_{s=0}^{\infty} \frac{(s+r)!}{(s+1)!(r-1)!} (qe^{iv})^{s+1} \frac{s+1}{s+r} \\
= qze^{-qz} e^{iv} p^r \sum_{s=0}^{\infty} \frac{(s+r-1)!}{s!(r-1)!} (qe^{iv})^s \\
= qze^{-qz} e^{iv} p^r (1 - qe^{iv})^{-r} \qquad \text{(4-103)}
\end{aligned}
$$

The procedure can be continued to higher order such that

$$
\begin{aligned}
G_y(iv) = p^r (1 - qe^{iv})^{-r} e^{-qz} \left(1 + \frac{qze^{iv}}{1!} + \frac{q^2 z^2 e^{2iv}}{2!} + \cdots \right) \\
= p^r (1 - qe^{iv})^{-r} e^{-qz(1-e^{iv})} \qquad \text{(4-104)}
\end{aligned}
$$

Thus, with $qz = \overline{N}_n$ and $r = M$, the characteristic function is simply the product of the characteristic functions of a negative binomial distributed signal [Eq. (4-94)] and a Poisson-distributed noise [Eq. (4-50)]. This is a very interesting but not too surprising result since the Kummer density function was originally formulated to represent a negative binomial signal immersed in Poisson noise.

The mean and variance of $q_{s+n}(k; \overline{N}_s, \overline{N}_n, M)$ can be obtained from Eq. (4-104) in the usual manner, which yields

$$
\langle k \rangle = \frac{rq}{p} + \overline{N}_n = \overline{N}_s + \overline{N}_n \qquad \text{(4-105)}
$$

and

$$
\operatorname{var}(k) = \frac{rq}{p^2} + \overline{N}_n = \overline{N}_s \left(1 + \frac{\overline{N}_s}{M}\right) + \overline{N}_n \qquad \text{(4-106)}
$$

The detection probability of a negative binomial signal immersed in Poisson noise can now be calculated in the same manner as was done for the specular target in Poisson noise [i.e., Eqs. (4-59) and (4-60), but with Eq. (4-60) now including an M dependence]. We have

$$
P_d = 1 - \sum_{k=0}^{k_{\text{th}}-1} q_{s+n}(k; \overline{N}_s, \overline{N}_n, M) \qquad \text{(4-107)}
$$

where $q_{s+n}(k; \overline{N}_s, \overline{N}_n, M)$ is given by Eq. (4-77). Figures 4-14 and 4-15 show Eq. (4-107) plotted as a function of $10 \log \overline{N}_s$ for a false alarm probability of $P_{\text{fa}} \approx 10^{-6}$, $\overline{N}_n = 0$ and $\overline{N}_n = 1$, respectively, and several values of M. The

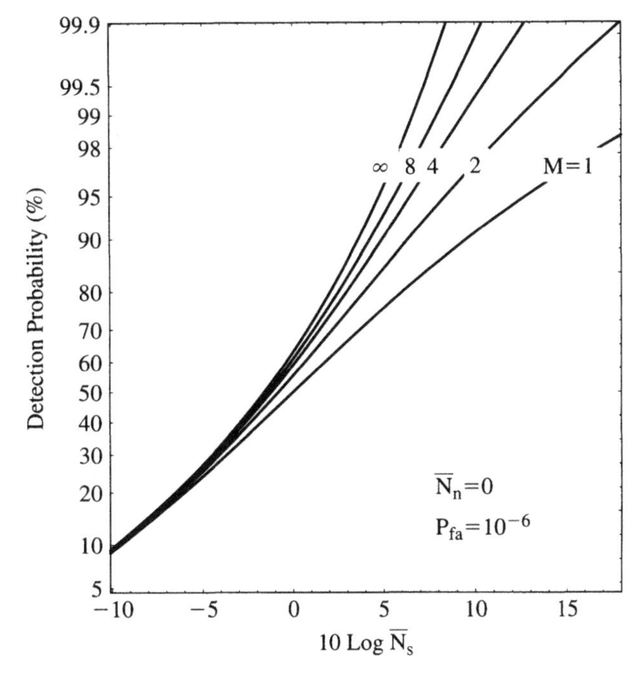

Figure 4-14. Detection probability versus $10 \log \overline{N}_s$ for a negative binomial signal in Poisson noise, where $\overline{N}_s$ is the mean signal photoelectron count, with a mean noise photoelectron count $\overline{N}_n = 0$, a false alarm probability of $P_{fa} = 10^{-6}$, and several values of speckle count M. Note that the $M = \infty$ curve corresponds to $\overline{N}_n = 0$ curve of Figure 4-11. (After J. W. Goodman in part, *Proc. IEEE*, Vol. 53, No. 11, Nov. 1963.)

curve labeled $\overline{N}_n = 0$ corresponds to the negative binomial distribution, while the curve labeled $M \to \infty$ corresponds to the specular target in Poisson noise discussed earlier.

It can be seen in Figure 4-15 that the case $M = 1$, which corresponds to a single speckle realization at the aperture, shows reduced performance compared to that of the specular target case due to the finite probability that the aperture can sit in a fade or null of the speckle pattern. This argument reverses at low probabilities, however, where the finite probability of sitting on a speckle peak increases the probability of detection over that of the specular target case.

4.6. NONCENTRAL NEGATIVE BINOMIAL SIGNAL IN POISSON NOISE

In Section 4.5 we dealt with the statistics associated with fully developed speckle, that is, speckle whose point statistics are described by the negative exponential distribution. There are other distributions, however, which are important in modeling laser radar and communications systems. One such distribution is

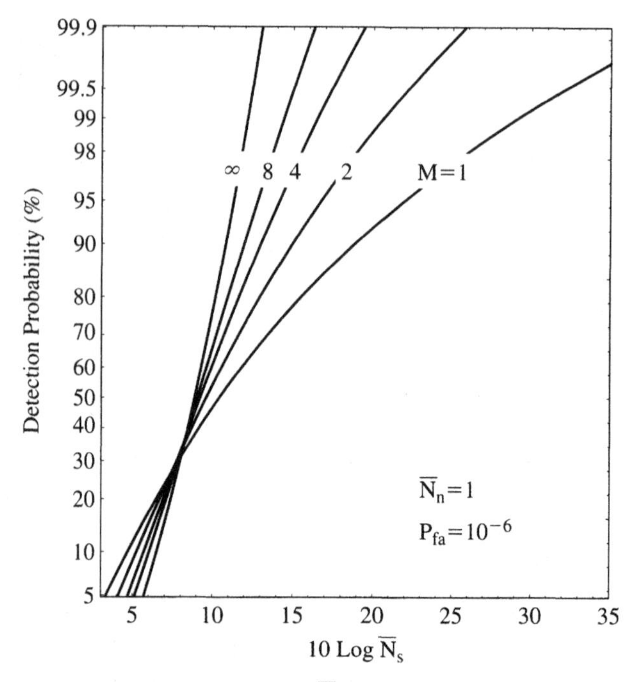

Figure 4-15. Detection probability versus $10 \log \overline{N}_s$ for a negative binomial signal in Poisson noise, where $\overline{N}_s$ is the mean signal photoelectron count, with a mean noise photoelectron count $\overline{N}_n = 1$, a false alarm probability of $P_{fa} = 10^{-6}$, and several values of speckle count M. Note that the $M = \infty$ curve corresponds to $\overline{N}_n = 1$ curve of Figure 4-11. (After J. W. Goodman, *Proc. IEEE*, Vol. 53, No. 11, Nov. 1963.)

that resulting from the addition of a coherent component to a fully developed speckle field. The probability density function in this case does not obey negative exponential statistics so that the resultant speckle field is referred to as being *partially developed.* Examples of such fields are many, including scatter from surfaces at near-normal incidence in which the RMS (root mean square) surface roughness is larger than a wavelength but the density of scatterers is low, or rough surfaces that contain small areas of specular reflection frequently referred to as *glints.* Another example is the receiver that employs an optical preamplifier. In this case, spontaneous emission noise from the amplifier is added coherently to the signal resulting in partially developed speckle at the amplifier output. Optical preamplification is becoming increasingly important as a means for achieving high sensitivity in both laser radar and communications systems.

Consider the case of a diffusely reflecting surface with a coherent specular component. By *coherent* is meant that there is interference between the specular and diffuse components. This, of course, implies that the linear dimensions of the scattering spot must be less than the coherence length and width of the transmitted beam. Additional assumptions are that the diffuse component is fully developed, which greatly simplifies the mathematical complexity, and that the scattered radiation is fully polarized. Indeed, many targets exhibit such features to varying degrees depending on their surface structure and the angle of incidence of the illuminating beam.[15] In these cases, both the specular and diffuse components contribute to the signal so that the detection statistics can differ markedly from that of fully developed speckle alone.

The analysis is accomplished by first generalizing the joint probability density function given by Eq. (4-8) to include a coherent specular component. This is done, without loss of generality, by adding a constant real vector to the real part of the random vector U, giving

$$p(U_R, U_I) = \frac{e^{-[(U_R - \sqrt{I_c})^2 + U_I^2]/2\sigma^2}}{2\pi\sigma^2} \tag{4-108}$$

where I_c is the point intensity of the *coherent* or specular component. With $U_R = \sqrt{I}\cos\theta$ and $U_I = \sqrt{I}\sin\theta$, the Jacobian for the transformation $|J| = \frac{1}{2}$, so that Eq. (4-108) can be transformed to read

$$p(I, \theta) = \begin{cases} \dfrac{1}{4\pi\sigma^2} e^{-(I + I_c - 2\sqrt{I I_c}\cos\theta)/2\sigma^2} & I \geq 0,\ 0 \leq \theta \leq 2\pi \\ 0 & I < 0 \end{cases} \tag{4-109}$$

where $\int_0^\infty p(I, \theta)\, dI\, d\theta = 1$. Now from Section 4.2 we know that $\langle I \rangle = 2\sigma^2$ is the mean intensity of the diffuse component alone. To make the differences between specular and diffuse components clear in the following discussions, we let $\langle I \rangle \to \langle I_s \rangle$, where the subscript s stands for the *scattered* (or diffuse) component. The marginal probability density function for the intensity alone of a partially developed speckle field is then found by integrating Eq. (109) over the limits of $0 < \theta < 2\pi$. This yields[16]

$$p(I) = \begin{cases} \dfrac{1}{\langle I_s \rangle} e^{-(I + I_c)/\langle I_s \rangle} I_0 \left(\dfrac{2\sqrt{I I_c}}{\langle I_s \rangle} \right) & I \geq 0 \\ 0 & I < 0 \end{cases} \tag{4-110}$$

where $I_0(x)$ is the modified Bessel function of the first kind, zeroth order, defined by

$$I_0(2x) = \frac{1}{2\pi} \int_0^{2\pi} e^{2x\cos\theta}\, d\theta \tag{4-111}$$

Equation (4-110) is known as the *modified Rician distribution* since it describes the distribution of intensities rather than the fields. The mean and variance are readily found to be

$$\langle I \rangle = \langle I_s \rangle + I_c \tag{4-112}$$

and

$$\text{var}(I) = \langle I_s \rangle (\langle I_s \rangle + 2I_c) \tag{4-113}$$

The modified Rician distribution is shown in Figure 4-16 for several values of the coherent intensity I_c. The case $I_c = 0$ corresponds to the negative exponential distribution.

A simulated image of partially developed speckle is shown in Figure 4-17. Here the same random beam pattern was used that generated Figure 4-2 but with

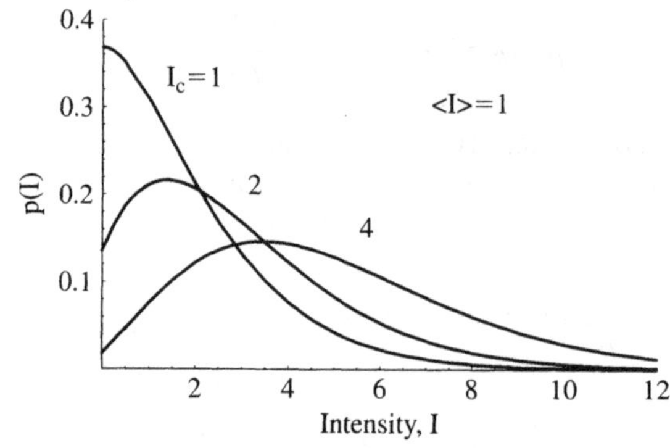

Figure 4-16. Probability density function representing the modified Rician distribution for several values of the coherent component I_c with the diffuse component $\langle I_s \rangle$ held constant. The case $I_c = 0$ corresponds to the negative exponential distribution.

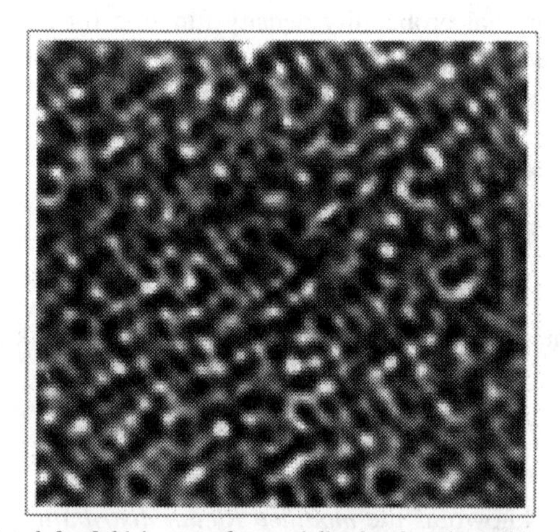

Figure 4-17. Simulated far-field image of a partially developed speckle pattern using the same parameters for the diffuse component as Figure 4-2 but with an additional beam located at the center of the aperture and having $10\times$ the mean amplitude of the diffuse component. Note the different intensity distribution compared to Figure 4-2, a consequence of interference with the coherent component. The image is again generated using a 128×128 FFT.

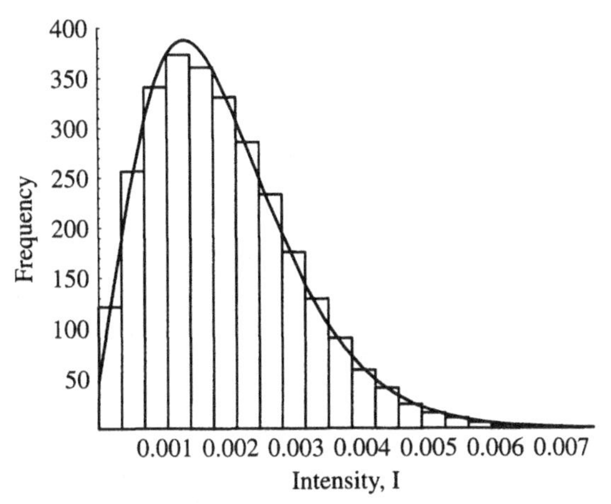

Figure 4-18. Histogram of the simulated data of Figure 4-17 normalized such that the area of the histogram sums to 1. The data closely approximate the modified Bessel distribution, shown as a solid line, corresponding to $\langle I_s \rangle \approx 0.0015$, $I_c \approx 0.006$, and a ratio $\kappa = I_c/\langle I_s \rangle \approx 4$.

an additional beam located at the center of the pattern and having 10 times the mean amplitude of the diffuse beams. Note that the distribution is quite different and of lower contrast than that of Figure 4-2 despite the fact that the random phase and spatial distributions of the Gaussian beams at the aperture are identical. The low contrast is a consequence of the presence of the strong coherent component, while the redistribution of the speckle pattern compared to that of Figure 4-2 is a consequence of interference between the diffuse component and specular components. Figure 4-18 shows the histogram corresponding to the FFT data of Figure 4-18. A curve fit of Eq. (4-110) to the data is also shown where $\langle I \rangle$ and $\langle I_s \rangle$, obtained from Figures 4-17 and 4-2, respectively, were used to calculate I_c. The resulting ratio $I_c/\langle I_s \rangle \approx 4$ corresponds to the $I_c = 4$ curve shown in Figure 4-16.

Comment 4-5. The contrast of an interference pattern is defined by $C = \sqrt{\text{var}(I)}/\langle I \rangle$. Thus, for a partially developed speckle field, $C = (1 + 2\kappa)^{1/2}/(1 + \kappa)$, where $\kappa = I_c/\langle I_s \rangle$ is called[17] the coherence parameter for the field. Thus, for a fully developed speckle field $\kappa = 0$, so that $C = 1$, while for a partially developed speckle field, $0 < \kappa \leq \infty$, so that $0 \leq C < 1$. ■

4.6.1. Summed Statistics of Partially Developed Speckle

As in the case of the gamma distribution, the solution for the integrated intensity of a partially developed speckle field may also be obtained using contour integration. However, we will follow the original derivation[18] and integrate term by term in a series expansion. The modified Rician distribution can be put in the form of a noncentral χ^2 distribution of order 2 (cf. Chapter 1) by letting $x = 2I/\langle I_s \rangle$ and $\lambda = I_c/\langle I_s \rangle$ in Eq. (4-110), which yields

$$p(x) = \tfrac{1}{2} e^{-(x/2)-\lambda} I_0(\sqrt{2\lambda x}) \tag{4-114}$$

where λ is the noncentrality factor. Notice that for $\lambda = 0$, Eq. (4-114) reduces to the χ^2 distribution of order 2, corresponding to a negative exponential distribution. The characteristic function corresponding to Eq. (4-114) is, from Eq. (1-118),

$$G(i\upsilon) = \frac{e^{2i\upsilon\lambda/(1-2i\upsilon)}}{1-2i\upsilon} \tag{4-115}$$

We now wish to find the probability density function for the integrated intensity of a *single* partially developed speckle field, assuming uncorrelated speckle *cells* within that field. As was pointed out in Section 4.3, the summing of M speckle cells in a single speckle field of mean intensity $\langle I_s \rangle$ is statistically equivalent to the summing of M uncorrelated speckle fields at a single point, each of mean intensity $\langle I_m \rangle = \langle I \rangle / M$. Both approaches yield the Gamma distribution for the probability density function. Thus we will require that the mth field and the integrated field satisfy the energy relationships $\langle I_m \rangle = \langle I_{sm} \rangle + I_{cm}$ and $\langle I \rangle = \langle I_s \rangle + I_c$, respectively, such that $\langle I_{sm} \rangle = \langle I_s \rangle / M$ and $I_{cm} = I_c/M$. With these definitions in mind, we will use the series form of Eq. (4-115) given by Eq. (1-121) with $n = 2$, to represent the mth speckle field, obtaining

$$G_m(i\upsilon) = \sum_{j=0}^{\infty} \frac{\lambda_m^j e^{-\lambda_m}}{j!}(1-2i\upsilon)^{-(1+j)} \tag{4-116}$$

where $\lambda_m = I_{cm}/\langle I_{sm} \rangle = I_c/\langle I_s \rangle$. Assuming M independent speckle fields, the total characteristic function becomes, after multiplying out the product and resuming,

$$\begin{aligned} G(i\upsilon) &= \prod_{m=1}^{M} G_m(i\upsilon) \\ &= \frac{e^{2i\upsilon\lambda M/(1-2i\upsilon)}}{(1-2i\upsilon)^M} \\ &= \sum_{j=0}^{\infty} \frac{(M\lambda_m)^j e^{-M\lambda_m}}{j!(1-2i\upsilon)^{M+j}} \end{aligned} \tag{4-117}$$

The final density function can now be found by performing an inverse (Fourier) transform on the individual terms of Eq. (4-117), which, after summing, leads to

$$p(x) = \sum_{j=0}^{\infty} \frac{(M\lambda_m)^j x^{M+j-1} e^{-(x/2+M\lambda_m)}}{2^{M+j} j!\Gamma(M+j)} \tag{4-118}$$

Recognizing that[19]

$$I_{M-1}(y) = \left(\frac{y}{2}\right)^{(M-1)} \sum_{j=0}^{\infty} \frac{(y^2/4)^j}{j!\Gamma(M+j)} \tag{4-119}$$

where $y = \sqrt{2Mx\lambda}$ and $I_{M-1}(y)$ represents the modified Bessel function of the first kind, order $M - 1$, the density function becomes a noncentral χ^2 distribution with $2M$ degrees of freedom, that is,

$$p(x) = \left(\frac{x}{M\lambda_m}\right)^{(1/2)(M-1)} \frac{e^{-(x/2+M\lambda_m)}}{2^{(M+1)/2}} I_{M-1}(\sqrt{2Mx\lambda_m}) \tag{4-120}$$

Returning to intensity variables with $x = 2I/\langle I_{sm}\rangle = 2MI/\langle I_s\rangle$ and $\lambda_m = I_c/\langle I_s\rangle$, Eq. (4-120) becomes

$$p(I) = \begin{cases} \dfrac{M}{\langle I_s\rangle}\left(\dfrac{I}{I_c}\right)^{(1/2)(M-1)} e^{-M(I+I_c)/\langle I_s\rangle} I_{M-1}\left(\dfrac{2M\sqrt{II_c}}{\langle I_s\rangle}\right) & I \geq 0 \\ 0 & I < 0 \end{cases} \tag{4-121}$$

Equation (4-121) is the direct analog of the gamma distribution for fully developed speckle and leads to moments given by

$$\langle I\rangle = \langle I_s\rangle + I_c \tag{4-122}$$

and

$$\mathrm{var}(I) = \frac{\langle I_s\rangle}{M}(\langle I_s\rangle + 2I_c) \tag{4-123}$$

Note that Eq. (4-122) conserves energy, being independent of the speckle count M, while the variance of the integrated intensity is inversely proportional to the speckle count M, as expected.

Equation (4-121) reduces to the gamma distribution given by Eq. (4-32). To show this, we rewrite Eq. (4-121) in terms of energy variables, obtaining

$$p(W) = a\left(\frac{W}{W_c}\right)^{(M-1)/2} e^{-a(W+W_c)} I_{M-1}(2a\sqrt{W_c W}) \tag{4-124}$$

where W is the integrated energy and $a = M/\overline{W}_s$. Using Eq. 9.6.7 of Ref. 19,

$$\lim_{z\to 0}[I_{M-1}(z)/z^{M-1}] = [2^{M-1}(M-1)!]^{-1} \tag{4-125}$$

Eq. (4-124) can easily be shown to reduce to the gamma distribution of Eq. (4-32).

Equation (4-121) is plotted in Figures 4-19, 4-20, and 4-21 for speckle counts $M = 1, 3$, and 10 and several values of I_c and $\langle I_s\rangle$. It can be seen that when $\langle I_s\rangle \gg I_c$, the curves approach the gamma distribution of a fully developed speckle pattern, and when $I_c \gg \langle I_s\rangle$, the distribution function approaches the delta function of a specular target. From these limiting cases it may be concluded that Eq. (4-121) can be no less accurate than the gamma distribution since, as I_c increases, the distribution approaches the specular target distribution, which is exact.

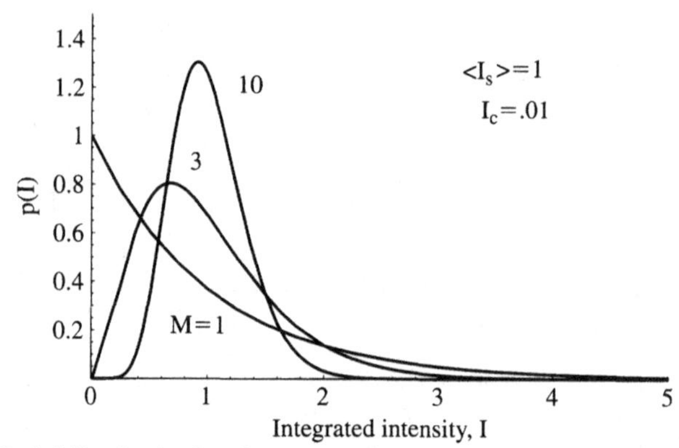

Figure 4-19. Probability density function representing the integrated intensity of a partially developed speckle field for the case $\langle I_s \rangle \gg I_c$. Here the distributions approach those of the gamma distribution. (From G. R. Osche, *Appl. Opt.*, Vol. 39, No. 24, Aug. 20, 2000.)

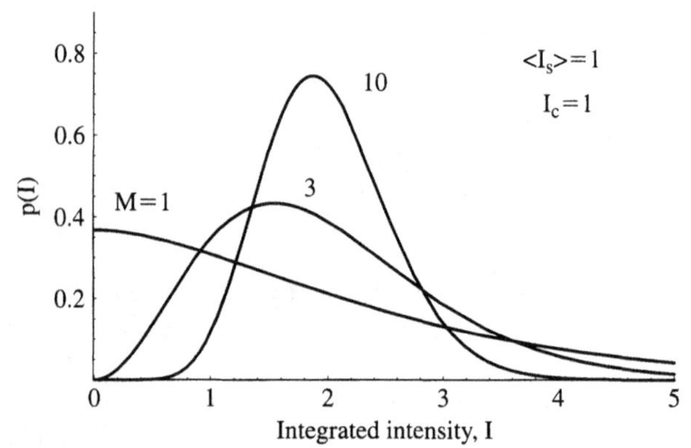

Figure 4-20. Probability density function representing the integrated intensity of a partially developed speckle field for the case $\langle I_s \rangle = I_c$. (From G. R. Osche, *Appl. Opt.*, Vol. 39, No. 24, Aug. 20, 2000.)

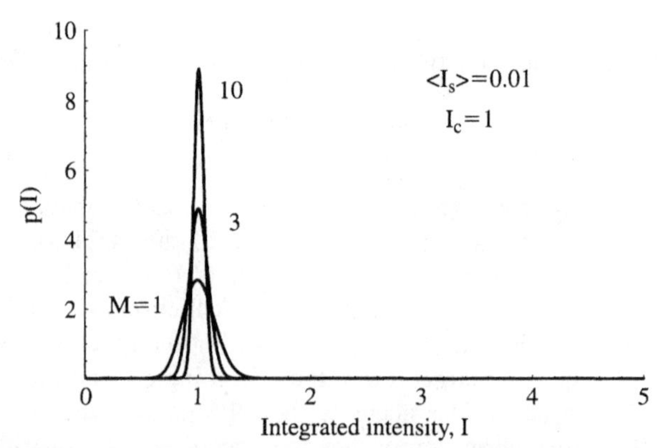

Figure 4-21. Probability density function representing the integrated intensity of a partially developed speckle field for the case $\langle I_s \rangle \ll I_c$. Here the distributions approach a delta function. (From G. R. Osche, *Appl. Opt.*, Vol. 39, No. 24, Aug. 20, 2000.)

Comment 4-6. At this point it is worthwhile pointing out some parallels between speckle theory and the heuristic theories of strong turbulence theory given in Section 3.4.3. Specifically, Eq. (3-298) was used in the I–K theory to model the coherent sum of m random scatters superimposed on a coherent background (with a mean intensity obeying exponential statistics). Therefore, it is not surprising that Eqs. (3-298) and (4-121) are identical under the replacements $m \leftrightarrow M$. Similarly, the Rice–Nakagami distribution given by Eq. (3-305) in the log-normal Rician model can be rewritten in terms of intensity variables by letting $y = I/\langle I \rangle = I/(\langle I_s \rangle + I_c)$ and $\kappa = I_c/\langle I_s \rangle$, to obtain

$$p(I) = \frac{1}{z\langle I_s \rangle} e^{-[I/z\langle I_s \rangle + I_c/\langle I_s \rangle]} I_0 \left(\frac{2}{\langle I_s \rangle} \sqrt{\frac{I I_c}{z}} \right)$$

where $z = \exp(2\chi)$ is a log-normal modulation on the mean random component of the diffuse field. In the absence of turbulence, $z = 1$, and the Rice–Nakagami distribution reduces to the modified Rician distribution given above. This close connection between the formalisms of surface scatter and turbulence theory is not too surprising since both scattering processes clearly involve, albeit from physically different processes, the coherent summation of coherent and random components. ■

4.6.2. Single-Pulse Detection Statistics

Development of the discrete density function for partially developed speckle in the presence of Poisson background noise follows the same procedures as those outlined in Section 4.5 for fully developed speckle. Thus, inserting the Poisson statistics given by Eq. (4-62) and the summed statistics given by Eq. (4-124) into the convolution integral given by Eq. (4-61), we obtain for the discrete density function of k signal plus noise photoelectrons

$$q_{s+n}(k) = a \frac{e^{-(\overline{N}_n + aW_c)}}{k!} \int_0^\infty (\overline{N}_n + bW)^k \left(\frac{W}{W_c} \right)^{(M-1)/2} \times e^{-(a+b)W} I_{M-1}(2a\sqrt{W W_c})\, dW \tag{4-126}$$

where $a = M/\langle I_s \rangle = (\eta/h\nu)M/\overline{N}_s$, $b = \eta/h\nu$, $p = M/(M + \overline{N}_s)$, $q = \overline{N}_s/(M + \overline{N}_s)$, $N_c = \eta W_c/h\nu$, and W represents the energy in both the specular and diffuse components. The integral in Eq. (4-126) may be solved using the series form for I_{M-1} given by Eq. (4-119) with $y = 2a(W W_c)^{1/2}$. This leads to

$$q_{s+n}(k) = \frac{a e^{-(\overline{N}_n + bW_c)}}{k!} \int_0^\infty (\overline{N}_n + bW)^k e^{-(a+b)W} \times \sum_{j=0}^\infty \frac{a^{M+2j-1} W_c^j M^{M+j-1}}{j!\Gamma(M+j)} dW \tag{4-127}$$

Exchanging the order of summation and integration yields

$$q_{s+n}(k) = \frac{e^{-(\overline{N}_n + aW_c)}}{k!} \sum_{j=0}^{\infty} \frac{(aW_c)^j a^{M+j}}{j!\Gamma(M+j)} \int_0^{\infty} (\overline{N}_n + bW)^k e^{-(a+b)W} W^{M+j-1} dW \tag{4-128}$$

The integral in Eq. (4-128) can be solved using the binomial expansion given by Eq. (4-65) and the integral identity of Eq. (4-67), yielding

$$q_{s+n}(k) = e^{-(\overline{N}_n + aW_c)} \sum_{j=0}^{\infty} \frac{(aW_c)^j a^{M+j}}{j!\Gamma(M+j)(a+b)^{M+j}} \times \sum_{r=0}^{k} \frac{(M+k+j-r-1)!}{r!(k-r)!} (\overline{N}_n)^r \left(\frac{\overline{N}_s}{M+\overline{N}_s} \right)^{k-r} \tag{4-129}$$

or

$$q_{s+n}(k) = p^M q^k e^{-(\overline{N}_n + pN_c/q)} \sum_{j=0}^{\infty} \frac{1}{j!} \left(\frac{p^2 N_c}{q} \right)^j \frac{\Gamma(M+k+j)}{\Gamma(M+j)\Gamma(k+1)} \times {}_1F_1\left(\alpha, \beta_j; \frac{\overline{N}_n}{q} \right) \tag{4-130}$$

where the parameters p and q have been defined in Eq. (4-84) and $\alpha = -k$ and $\beta_j = -M - k - j + 1$.

Consider the limiting forms of Eq. (4-130). Letting $j = 0$ and $N_c = 0$ causes all terms containing N_c to disappear except the first term, where $\lim_{N_c \to 0, j \to 0} (N_c)^j \to 0^0 = 1$. This results in the fully developed speckle discrete density function of Eq. (4-77). On the other hand, letting $\overline{N}_n \to 0$ leaves

$$q_s(k) = p^M q^k e^{-pN_c/q} \sum_{j=0}^{\infty} \left(\frac{p^2 N_c}{q} \right)^j \frac{\Gamma(M+k+j)}{\Gamma(M+j)\Gamma(k+1)} = p^M q^k \frac{\Gamma(M+k)}{\Gamma(M)\Gamma(k+1)} e^{-pN_c/q} {}_1F_1\left(k+M, M; \frac{p^2 N_c}{q} \right) \tag{4-131}$$

where ${}_1F_1(\alpha, \beta_j; 0) = 1$. Equation (4-131) represents the signal statistics for partially developed speckle in the absence of noise and is the direct analog of the negative binomial distribution for fully developed speckle. Since it is the Poisson transform of a noncentral χ^2 distribution of order $2M$, it will be referred to as a *noncentral negative binomial distribution* (also referred to as a *generalized negative binomial distribution* in ref. 18) and therefore the discrete density function of Eq. (4-130) as a *noncentral negative binomial signal in Poisson noise*.

Since the ratio of specular to diffuse scattered radiation is expected to be a constant for a given target aspect, we can express Eq. (4-130) in terms of the coherence parameter $\kappa = N_c/\overline{N}_s$ (cf. Comment 4-5), obtaining

$$q_{s+n}(k) = p^M q^k e^{-(\overline{N}_n+\kappa M)} \sum_{j=0}^{\infty} \frac{(\kappa p M)^j}{j!} \frac{\Gamma(M+k+j)}{\Gamma(M+j)\Gamma(k+1)} {}_1F_1\left(\alpha, \beta_j; \frac{\overline{N}_n}{q}\right) \tag{4-132}$$

The detection probability follows in the usual manner using Eqs. (4-61) and (4-62). Examples are plotted in Figures 4-22 and 4-23 as a function of the *total* signal photoelectron count $\overline{N}_s + N_c = \overline{N}_s(1+\kappa)$, expressed in decibels, for the two cases of $\kappa = 0.1$ and 5, and a false alarm probability of $P_{\text{fa}} = 10^{-6}$.

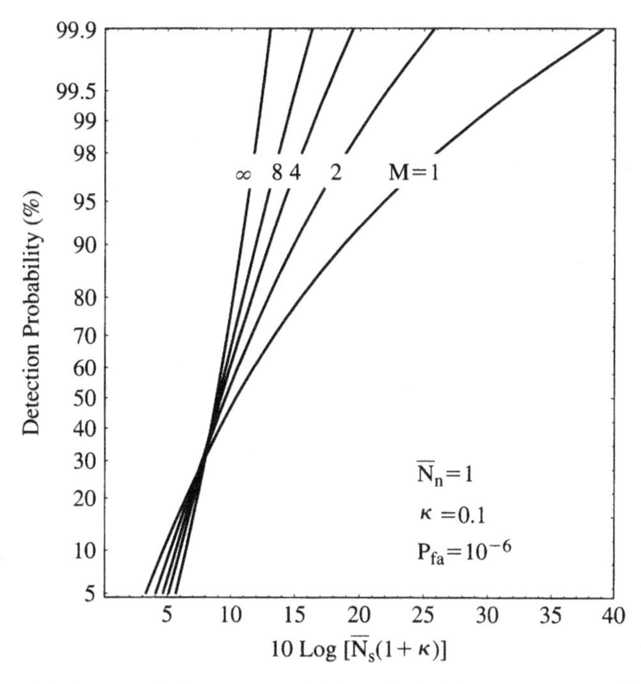

Figure 4-22. Detection probability versus the total received photoelectron count $10 \log \overline{N}_s(1+\kappa)$ for a partially developed speckle field with $\kappa = 0.1$ and a false alarm probability of $P_{\text{fa}} = 10^{-6}$. Here $\kappa = N_c/\overline{N}_s$ is a coherence parameter that represents the ratio of specular to diffuse photoelectron counts and $\overline{N}_n = 1$ represents the noise photoelectrons. (After G. R. Osche, *Appl. Opt.*, Vol. 39, No. 24, Aug. 20, 2000.)

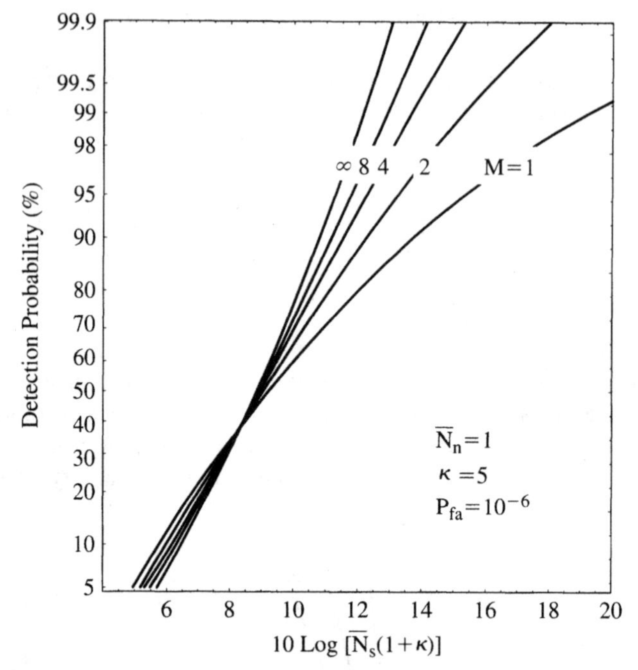

Figure 4-23. Detection probability versus the total received photoelectron count $10 \log \overline{N}_s(1+\kappa)$ for a partially developed speckle field with $\kappa = 5$ and a false alarm probability of $P_{\text{fa}} = 10^{-6}$. Here $\kappa = N_c/\overline{N}_s$ is a coherence parameter that represents the ratio of specular to diffuse photoelectron counts and $\overline{N}_n = 1$ represents the noise photoelectrons. (After G. R. Osche, *Appl. Opt.*, Vol. 39, No. 24, Aug. 20, 2000.)

Note that the $M = \infty$ curves in the two figures are identical while the remaining curves approach the $M = \infty$ case as κ increases. The former effect is a consequence of plotting the data as a function of the total signal (i.e., diffuse plus specular components), while the latter effect is a consequence of the statistics becoming more specular as κ increases. In the limit of $M \to \infty$, the discrete density function approaches that of a specular target given by

$$q_{s+n}(k) = \frac{(\overline{N}_n + \overline{N}_s + N_c)^k e^{-(\overline{N}_n+\overline{N}_s+N_c)}}{k!} \tag{4-133}$$

We again see the crossover of the curves at low detection probabilities as in the $\kappa = 0$ case shown in Figure 4-15.

Comment 4-7. Equation (4-131) can be shown to represent the radiation statistics at the output of an optical preamplifier, such as the *erbium doped fiber amplifier* (EDFA). These devices are used to provide high signal gain just prior to photodetection. Li and Teich[20] have derived the density function for such a receiver by using a statistical rate-analysis for the photon number n. They obtained $p(n) =$

$(\bar{n}_{th})^n(1+\bar{n}_{th})^{-(n+M_m)}\exp[-G\overline{N}_{in}/(1+\bar{n}_{th})]L_n^{(M_m-1)}(z)$, where $\bar{n}_{th}=\bar{n}_o/M_m$, $\bar{n}_o$ is the mean spontaneous-emission population, $\overline{N}_{in}$ is the mean number of photons at the amplifier input, M_m is the number of spontaneous-emission modes emitted by the amplifier, G is the gain of the medium, $L_n^{(\alpha)}(z)$ is the generalized Laguerre polynomial, and $z=G\overline{N}_{in}/[\overline{N}_{th}(1+\overline{N}_{th})]$. The two formulations may be shown to be equivalent by using the Kummer–Laguerre function identity in Table A-1 of Appendix A, Kummer's transformation, Eq. (A-30), and Pochhammer's symbol, Eq. (A-28). Noting that the photon statistics are preserved in the detection process, a change of variables $n\to k$, $M_m\to M$, $G\overline{N}_{in}\to\overline{N}_c$ then leads to the desired result. Note that under a similar change of variables, Eq. (4-130) represents the statistics of an optical preamplifier in the presence of Poisson (background) noise, an important consideration for free-space optical links. ■

4.7. PARABOLIC-CYLINDER SIGNAL IN GAUSSIAN NOISE

A receiver may not be Poisson noise limited as in previous section but may be amplifier noise limited, depending on detector/amplifier parameters. In such cases, the noise voltage is characterized by a continuous, zero-mean Gaussian random variable. In addition to amplifier noise, the signal itself may exhibit Gaussian statistics in the limit of large photoelectron counts k where the Poisson process of photoemission becomes Gaussian [cf. Comment 4-2, Eq. (4-55)]. The solution therefore establishes a large k performance bound for the general optical signal detection problem.

Consider then a zero-mean Gaussian noise process described by

$$p_n(k)=\frac{1}{\sqrt{2\pi\sigma_n^2}}e^{-k^2/2\sigma_n^2}\qquad k\geq 0\tag{4-134}$$

where k is treated as a continuous random variable and σ_n^2 is the dark noise variance. *Dark noise* is defined as that noise which remains after the receiver aperture is closed and usually consists of contributions from dark current and thermal noise sources only. Consider also a signal that obeys conditionally Gaussian statistics. We have

$$p_s(k\mid N_s)=\frac{1}{\sqrt{2\pi\sigma_s^2}}e^{-(k-N_s)^2/2\sigma_s^2}\tag{4-135}$$

where σ_s^2 is the signal variance and N_s is a continuous random variable representing the mean signal during the measurement interval. The characteristic functions for $p_n(k)$ and $p_s(k\mid N_s)$ are

$$G_n(iv)=e^{-\sigma_n^2v^2/2}\quad\text{and}\quad G_s(iv)=e^{iN_sv-\sigma_s^2v^2/2}\tag{4-136}$$

respectively. Hence,

$$G_{s+n}(iv)=e^{iN_sv-(\sigma_s^2+\sigma_n^2)v^2/2}\tag{4-137}$$

and by comparing to $G_s(iv)$, the inverse transform of Eq. (4-137) follows immediately as

$$p_{s+n}(k \mid N_s) = \frac{1}{\sqrt{2\pi\sigma^2}} e^{-(k-N_s)^2/2\sigma^2} \tag{4-138}$$

where $\sigma^2 = \sigma_s^2 + \sigma_n^2$.

We are now in a position to calculate the probability density function for the integrated speckle embedded in Gaussian noise. Since the detectors of interest here generally have unity gain and are therefore dark-noise limited, we let $\sigma_n^2 \gg \sigma_s^2$ such that $\sigma^2 \cong \sigma_n^2$. We again begin with Eq. (4-61) but with the conditional probability $p_{s+n}(k \mid W)$ being given by

$$p_{s+n}(k \mid W) = \frac{1}{\sqrt{2\pi\sigma^2}} e^{-[k-(\eta/h\nu)W]^2/2\sigma^2} \tag{4-139}$$

which follows from Eq. (4-138) with $N_s = \eta W/h\nu$. The integrated energy is given by the gamma distribution discussed earlier; Eq. (4-63), which is repeated here for convenience,

$$p(W) = \frac{(a)^M W^{M-1} e^{-aW}}{\Gamma(M)} \tag{4-140}$$

so that

$$p_{s+n}(k) = \frac{a^M}{\sqrt{2\pi\sigma^2}} \frac{e^{-k^2/2\sigma^2}}{\Gamma(M)} \int_0^\infty W^{M-1} e^{-b^2W^2/2\sigma^2-(a-bk/\sigma^2)W} dW \tag{4-141}$$

where $a = M/\overline{W}$ and $b = \eta/h\nu$. The integral in Eq. (4-141) is of the form

$$\int_0^\infty y^{\nu-1} e^{-\beta y^2-\gamma y} dy = (2\beta)^{-\nu/2}\Gamma(\nu) e^{\gamma^2/8\beta} D_{-\nu}\left(\frac{\gamma}{\sqrt{2\beta}}\right) \tag{4-142}$$

where $D_{-\nu}(z)$ is a *parabolic-cylinder function* of order $-\nu$ (cf. Appendix A for more details). Letting $y = W$, $\nu = M$, $\beta = b^2/2\sigma^2$, and $\gamma = a - bk/\sigma^2$, Eq. (4-141) becomes

$$\begin{aligned} p_{s+n}(k) &= \frac{a^M}{\sqrt{2\pi\sigma^2}} \left(\frac{b^2}{\sigma^2}\right)^{-M/2} e^{-(a-bk/\sigma^2)^2/8(b^2/2\sigma^2)-k^2/2\sigma^2} D_{-M}\left(\frac{a - bk/\sigma^2}{\sqrt{b^2/\sigma^2}}\right) \\ &= \frac{1}{\sqrt{2\pi\sigma^2}} \left(\frac{a\sigma}{b}\right)^M e^{(1/4)(a\sigma/b-k/\sigma)^2-k^2/2\sigma^2} D_{-M}\left(\frac{a\sigma}{b} - \frac{k}{\sigma}\right) \end{aligned} \tag{4-143}$$

Or, by letting $a' = a/b = M/\overline{N}_s$,

$$p_{s+n}(k; a', M) = \frac{(a'\sigma)^M}{\sqrt{2\pi\sigma^2}} e^{-[(1/4)(a'\sigma-k/\sigma)^2-k^2/2\sigma^2]} D_{-M}\left(a'\sigma - \frac{k}{\sigma}\right) \tag{4-144}$$

Equation (1-144) is referred to as a *parabolic-cylinder density function* and was first derived by Mecherle.[21] Note that for $\sigma_n = 0$, $\sigma \to \sigma_s$ in Eq. (4-144), in which case the expression represents the integrated statistics of a Gaussian signal in the absence of noise. Referred to as the *parabolic-cylinder signal*, this is analogous to the integrated statistics of a Poisson signal in the absence of noise (i.e., the negative binomial signal).

Consider the case $M = 1$. Expanding the exponent in Eq. (4-144) and rewriting in terms of the normalized variables $\alpha = \overline{N}_s/\sigma$ and $x = k/\overline{N}_s$ yields

$$p_{s+n}(x;\alpha,M) = \frac{\alpha}{\sqrt{2\pi}\,\overline{N}_s}\left(\frac{M}{\alpha}\right)^M e^{-(\alpha^2x^2/4+Mx/2-M^2/4\alpha^2)} D_{-M}\left(\frac{M}{\alpha}-\alpha x\right) \tag{4-145}$$

From Appendix A, Eq. (A-40), we have

$$D_{-1}(z) = \sqrt{\frac{\pi}{2}}e^{z^2/4}\operatorname{erfc}\left(\frac{z}{\sqrt{2}}\right) \tag{4-146}$$

where $\operatorname{erfc}(x)$ is the complementary error function, defined by

$$\operatorname{erfc}(z) = 1 - \operatorname{erf}(z) = \frac{2}{\sqrt{\pi}}\int_z^\infty e^{-t^2}dt \tag{4-147}$$

Thus,

$$p_{s+n}(x) = \frac{1}{2\overline{N}_s}e^{(1/2\alpha^2)-x}\operatorname{erfc}\left(\frac{1}{\sqrt{2}\alpha}-\frac{\alpha x}{\sqrt{2}}\right) \tag{4-148}$$

Equation (4-148) is plotted in Figures 4-24, 4-25, and 4-26 for several values of the signal-to-noise ratio parameter $\alpha = \overline{N}_s/\sigma$ and $M = 1$, 2, and 3, respectively. It can be seen in Figure 4-24 that the $M = 1$ case approaches a negative exponential distribution as expected at high α and a Gaussian distribution at low α. Figures 4-25 and 4-26 show that, in general, the parabolic density function approaches a Gaussian distribution as the speckle count M increases.

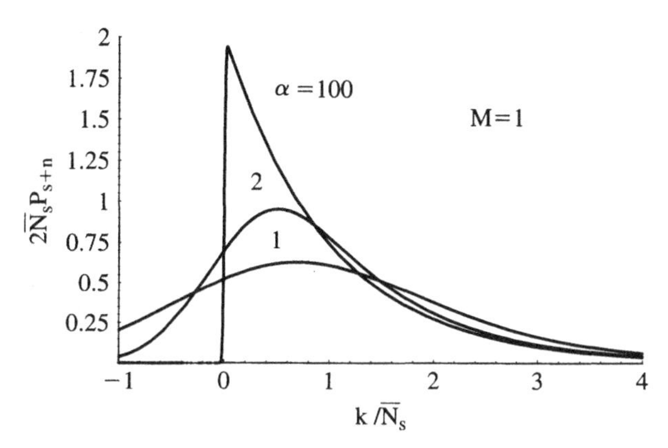

Figure 4-24. Parabolic cylinder density function for $M = 1$. (After G. S. Mecherle, *J. Opt. Soc. Am.*, Vol. 1, No. 1, Jan. 1984).

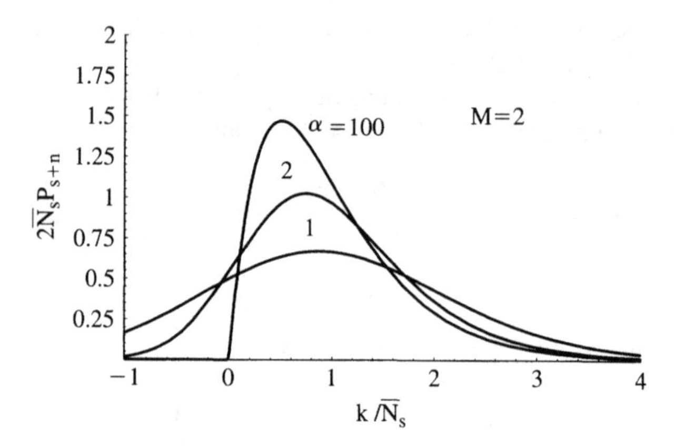

Figure 4-25. Parabolic cylinder probability density function for speckle count $M = 2$.

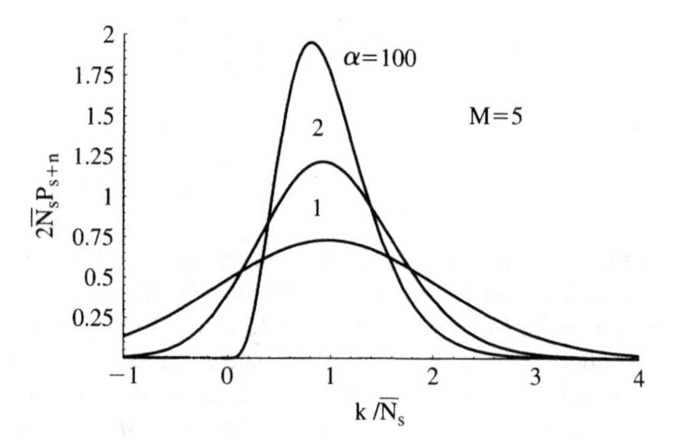

Figure 4-26. Parabolic cylinder probability density function for speckle count $M = 5$.

Returning to the general parabolic-cylinder distribution, Eq. (4-144), the first two moments of the distribution are useful in estimating signal-to-noise ratios. These can be calculated using the method of characteristic functions. We have

$$G_k(i\upsilon) = \langle e^{ik\upsilon} \rangle = \int_{-\infty}^{\infty} e^{ik\upsilon} P_{s+n}(k; a', M)\, dk \tag{4-149}$$

Using the integral form of $D_\nu(z)$ as given in Appendix A, Eq. (A-37), leads to

$$G_k(i\upsilon) = \frac{(a'\sigma)^M}{\sqrt{2\pi\sigma^2}} \frac{1}{\Gamma(M)} \int_{-\infty}^{\infty} e^{i\upsilon k - [(1/4)(a'\sigma - k/\sigma)^2 - k^2/2\sigma^2]} \times e^{-(1/4)(a'\sigma - k/\sigma)^2} \int_0^{\infty} u^{M-1} e^{-(u^2/2) - u(a'\sigma - k/\sigma)}\, du\, dk \tag{4-150}$$

or

$$G_k(i\upsilon) = \frac{(a'\sigma)^M}{\sqrt{2\pi\sigma^2}} \frac{1}{\Gamma(M)} \int_{-\infty}^{\infty} e^{i\upsilon k - k^2/2\sigma^2} \times \int_0^{\infty} u^{M-1} e^{-(u^2/2) - u(a'\sigma - k/\sigma)} du\, dk \tag{4-151}$$

Interchanging the order of integration yields

$$G_k(i\upsilon) = \frac{(a'\sigma)^M}{\sqrt{2\pi\sigma^2}} \frac{1}{\Gamma(M)} \int_0^{\infty} u^{M-1} e^{-(u^2/2) - ua'\sigma} \int_{-\infty}^{\infty} e^{-k^2/2\sigma^2 + [(u/\sigma) + i\upsilon]k} dk\, du \tag{4-152}$$

But the integral over k in Eq. (4-152) is of the form

$$\int_{-\infty}^{\infty} e^{-p^2x^2 \pm qx} dx = e^{q^2/4p^2} \frac{\sqrt{\pi}}{p} \qquad p > 0 \tag{4-153}$$

Thus, with $q = u/\sigma + i\upsilon$ and $p = 1/\sqrt{2}\sigma$, we have

$$G_k(i\upsilon) = \frac{(a'\sigma)^M}{\Gamma(M)} e^{-\upsilon^2\sigma^2/2} \int_0^{\infty} u^{M-1} e^{-u(a'\sigma - i\upsilon\sigma)} du \tag{4-154}$$

Once again the integral has a known solution, namely

$$\int_0^{\infty} x^{\nu-1} e^{-\mu x} dx = \frac{\Gamma(\nu)}{\mu^{\nu}} \qquad \operatorname{Re}\nu > 0, \quad \operatorname{Re}\mu > 0 \tag{4-155}$$

yielding

$$G_k(j\upsilon) = \left(\frac{a'}{a' - j\upsilon}\right)^M e^{-(1/2)\sigma^2\upsilon^2} \tag{4-156}$$

Using the moment generating function of Section 1.2.5, the mean photoelectron count becomes, after some straightforward algebra,

$$\langle k \rangle = -i \left. \frac{dG_k}{d\upsilon} \right|_{\upsilon=0} = \frac{M}{a} = \overline{N}_s \tag{4-157}$$

Similarly, the second moment becomes

$$\langle k^2 \rangle = (-i)^2 \left. \frac{d^2G_k}{d\upsilon^2} \right|_{\upsilon=0} = \overline{N}_s^2 \left(1 + \frac{1}{M}\right) + \sigma^2 \tag{4-158}$$

Thus the variance becomes

$$\begin{aligned}\text{var}(k) &= \langle k^2 \rangle - \langle k \rangle^2 \\ &= \frac{\overline{N}_s^2}{M} + \sigma^2\end{aligned} \tag{4-159}$$

We can now define an *output power signal-to-noise ratio* as

$$\text{SNR}_p = \frac{\langle k \rangle^2}{\text{var}(k)} = \frac{\overline{N}_s^2}{(\overline{N}_s^2/M) + \sigma^2} \tag{4-160}$$

In the limit of small $\overline{N}_s^2/M$, $\text{SNR}_p \approx \overline{N}_s^2/\sigma^2$ and the receiver becomes dark-noise limited, as expected, whereas in the limit of large $\overline{N}_s^2/M$, the receiver becomes speckle noise limited (i.e., $\text{SNR}_p \approx M$).

The detection statistics associated with the parabolic density function of Eq. (4-144) are obtained in the usual manner. Thus for zero-mean Gaussian noise, the false alarm probability becomes

$$\begin{aligned}P_{\text{fa}} &= \frac{1}{\sqrt{2\pi\sigma^2}} \int_{k_{\text{th}}}^{\infty} e^{-k^2/2\sigma^2}\, dk \\ &= \frac{1}{2} \text{erfc}\left(\frac{k_{\text{th}}}{\sqrt{2}\sigma}\right)\end{aligned} \tag{4-161}$$

and the detection probability

$$P_d = \int_{k_{\text{th}}}^{\infty} p_{s+n}(k)\, dk \tag{4-162}$$

Mecherle has shown that a closed-form expression for the detection probability may be obtained using $p_{s+n}(k)$ from Eq. (4-144) and the integral representation of the parabolic cylinder functions, Eq. (A-37) of Appendix A, in Eq. (4-162). Thus

$$P_d = \frac{(a'\sigma)^M}{\sqrt{2\pi\sigma^2}} \frac{1}{\Gamma(M)} \int_{k_{\text{th}}}^{\infty} e^{-k^2/2\sigma^2} \int_0^{\infty} u^{M-1} e^{-(u^2/2) - u(a'\sigma - k/\sigma)}\, du\, dk \tag{4-163}$$

where Eq. (4-150) was used. Interchanging the order of integration

$$P_d = \frac{(a'\sigma)^M}{\sqrt{2\pi\sigma^2}} \frac{1}{\Gamma(M)} \int_0^{\infty} u^{M-1} e^{-(u^2/2) - ua'\sigma} \int_{k_{\text{th}}}^{\infty} e^{-(k^2/2\sigma^2) - ku/\sigma}\, dk\, du \tag{4-164}$$

The integral over k is of the form

$$\int_{\mu}^{\infty} e^{-(x^2/4\beta) - \gamma x}\, dx = \sqrt{\pi\beta}\, e^{\beta\gamma^2} \text{erfc}\left(\gamma\sqrt{\beta} + \frac{\mu}{2\sqrt{\beta}}\right) \quad \text{Re}\,\beta > 0,\ \text{Re}\,\mu > 0 \tag{4-165}$$

Letting $\beta = \sigma^2/2$ and $\gamma = -u/\sigma$ in the integral over k, we have

$$\int_{k_{\text{th}}}^{\infty} e^{-(k^2/2\sigma^2)-ku/\sigma} dk = \sqrt{\frac{\pi}{2}} \sigma e^{u^2/2} \operatorname{erfc}\left(-\frac{u}{\sqrt{2}} + \frac{k_{\text{th}}}{\sqrt{2}\sigma}\right) \tag{4-166}$$

Therefore,

$$P_d = \frac{(a'\sigma)^M}{2\Gamma(M)} \int_0^{\infty} u^{M-1} e^{-ua'\sigma} \operatorname{erfc}\left(-\frac{u}{\sqrt{2}} + \frac{k_{\text{th}}}{\sqrt{2}\sigma}\right) du \tag{4-167}$$

Equation (4-167) can be integrated by parts by letting

$$U = \operatorname{erfc}\left(-\frac{u}{\sqrt{2}} + \frac{k_{\text{th}}}{\sqrt{2}\sigma}\right) \quad \text{and} \quad dU = \sqrt{\frac{2}{\pi}} e^{-u^2/2 + k_{\text{th}}u/\sigma - k_{\text{th}}^2/2\sigma^2} du \tag{4-168}$$

such that

$$dV = u^{M-1} e^{-ua'\sigma} du \quad \text{and} \quad V = \int_0^{\infty} u^{M-1} e^{-ua'\sigma} du$$

Using the series representation for the integral in V, namely

$$\int x^n e^{cx} dx = \frac{e^{cx}}{c} \left[x^n + \sum_{j=1}^{n} \frac{(-1)^j n!}{(n-j)!} \frac{x^{n-j}}{c^j} \right] \tag{4-169}$$

we obtain

$$V = e^{-a'\sigma u} \left[\frac{u^{M-1}}{-a'\sigma} + \sum_{j=1}^{n} \frac{(-1)^j (M-1)!}{(M-j-1)!} \frac{u^{M-j-1}}{(-a'\sigma)^{j+1}} \right] \tag{4-170}$$

Now

$$\int_0^{\infty} U dV = UV \Big|_0^{\infty} - \int_0^{\infty} V dU \tag{4-171}$$

and

$$\begin{aligned} UV \Big|_0^{\infty} &= \frac{a'\sigma^M}{-a'\sigma} \frac{e^{-a'\sigma u}}{2\Gamma(M)} \\ &\quad \times \left[u^{M-1} + \sum_{j=1}^{n} \frac{(-1)^j (M-1)!}{(M-j-1)!} \frac{u^{M-j-1}}{(-a'\sigma)^j} \right] \\ &\quad \times \operatorname{erfc}\left(-\frac{u}{\sqrt{2}} + \frac{k_{\text{th}}}{\sqrt{2}\sigma}\right) \Big|_0^{\infty} \end{aligned} \tag{4-172}$$

The upper limit in Eq. (4-172) forces the entire expression to zero, due to the exponential multiplier [where $\text{erfc}(-\infty) = 2$]. The lower limit needs to be inspected more closely. Clearly, the series in brackets contains polynomial terms in powers of u, all of which vanish except for the last one ($j = M - 1$), which becomes $\lim_{u\to 0} u^0$, which is indeterminate. However, by using L'Hôpital's rule,

$$\lim_{u\to 0} \log u^0 \to 0 \tag{4-173}$$

and therefore

$$\lim_{u\to 0} u^0 \to 1 \tag{4-174}$$

Thus, for each value of M there is one surviving term in the series, resulting in

$$\begin{aligned} UV\Big|_0^\infty &= 0 - \left[-\frac{1}{2\Gamma(M)}\frac{(M-1)!}{0!}\right]\text{erfc}\left(\frac{k_{\text{th}}}{\sqrt{2}\sigma}\right) \\ &= \frac{1}{2}\text{erfc}\left(\frac{k_{\text{th}}}{\sqrt{2}\sigma}\right) \end{aligned} \tag{4-175}$$

where the solution has been restricted to the case of M being an integer via the assumption $\Gamma(M) = (M - 1)!$. (Clearly, M is not, in general, an integer since a nonintegral number of speckles can be intercepted by the aperture. However, for engineering applications this restriction is quite acceptable.)

We now need to evaluate the second half of the integration by parts, that is

$$\begin{aligned} -\int_0^\infty V\,dU = \sqrt{\frac{2}{\pi}}\frac{(a'\sigma)^{M-1}}{2\Gamma(M)}\int_0^\infty \left[\mu^{M-1} + \sum_{j=1}^{M-1}\frac{(-1)^j(M-1)!}{(M-j-1)!}\frac{u^{M-j-1}}{(-a'\sigma)^j}\right] \\ \times e^{-u^2/2 + k_{\text{th}}u/\sigma - k_{\text{th}}^2/2\sigma^2 - a'\sigma u}\,du \end{aligned} \tag{4-176}$$

Interchanging the order of summation and integration yields

$$\begin{aligned} -\int_0^\infty V\,dU = \frac{1}{\sqrt{2\pi}}\frac{(a'\sigma)^{M-1}}{\Gamma(M)}e^{-k_{\text{th}}^2/2\sigma^2}\Bigg[\int_0^\infty u^{M-1}e^{-(u^2/2)-(a'\sigma - k_{\text{th}}/\sigma)u}\,du \\ + \sum_{j=1}^{M-1}\frac{(M-1)!}{(M-j-1)!}\frac{(-1)^j}{(-a'\sigma)^j}\int_0^\infty u^{M-j-1}e^{-(u^2/2)-(a'\sigma - k_{\text{th}}/\sigma)u}\,du\Bigg] \end{aligned} \tag{4-177}$$

But we can employ another integral expression,

$$\int_0^\infty x^{\nu-1} e^{-\beta x^2 - \gamma x} dx = (2\beta)^{-\nu/2}\Gamma(\nu) e^{-\gamma^2/8\beta} D_{-\nu}\left(\frac{\gamma}{\sqrt{2\beta}}\right) \tag{4-178}$$

to obtain

$$\begin{aligned} -\int_0^\infty V dU = {} & \frac{(a'\sigma)^{M-1}}{\sqrt{2\pi}} e^{a'^2\sigma^2/4} - a'k_{\text{th}}/2 - k_{\text{th}}^2/4\sigma^2 \\ & \times \left[D_{-M}\left(a'\sigma - \frac{k_{\text{th}}}{\sigma}\right) + \sum_{j=1}^{M-1} (a'\sigma)^{-j} D_{-(M-j)}\left(a'\sigma - \frac{k_{\text{th}}}{\sigma}\right)\right] \end{aligned} \tag{4-179}$$

The detection probability therefore becomes

$$\begin{aligned} P_d = {} & \frac{1}{2}\operatorname{erfc}\left(\frac{k_{\text{th}}}{\sqrt{2}\sigma}\right) + \frac{(a'\sigma)^{M-1}}{\sqrt{2\pi}} e^{a'^2\sigma^2/4 - a'k_{\text{th}}/2 - k_{\text{th}}^2/4\sigma^2} \\ & \times \left[D_{-M}\left(a'\sigma - \frac{k_{\text{th}}}{\sigma}\right) + \sum_{j=1}^{M-1} (a'\sigma)^{-j} D_{-(M-j)}\left(a'\sigma - \frac{k_{\text{th}}}{\sigma}\right)\right] \end{aligned} \tag{4-180}$$

The first term in Eq. (4-180) is identical to the false alarm probability given by Eq. (4-161) and is therefore negligible for most applications. The detection probability P_d is plotted in Figures 4-27 and 4-28. The curves are shown as a function of the (voltage) signal-to-noise ratio, $\text{SNR} = \overline{N}_s/\sigma$, for false alarm probabilities $P_{\text{fa}} = 10^{-4}$ and $P_{\text{fa}} = 10^{-8}$, respectively, with M as a parameter. Equation (4-138) was used in Eq. (4-162) to calculate the $M = \infty$ case. The corresponding power signal-to-noise ratio curves are obtained by multiplying the abscissa values by 2.

As in the Poisson case of Section 4.5, we again see a crossover of the curves at low values of SNR and essentially for the same reason. The advantage of aperture averaging is clearly evident at the high signal levels where significantly greater signal levels are required to achieve a given probability of detection for the $M = 1$ case compared to the $M = \infty$ case.

4.8. DETECTION OF SIGNALS IN APD EXCESS NOISE

This section considers the detection statistics associated with receivers that employ *avalanche photodiode* (APD) *detectors*. These devices are useful for

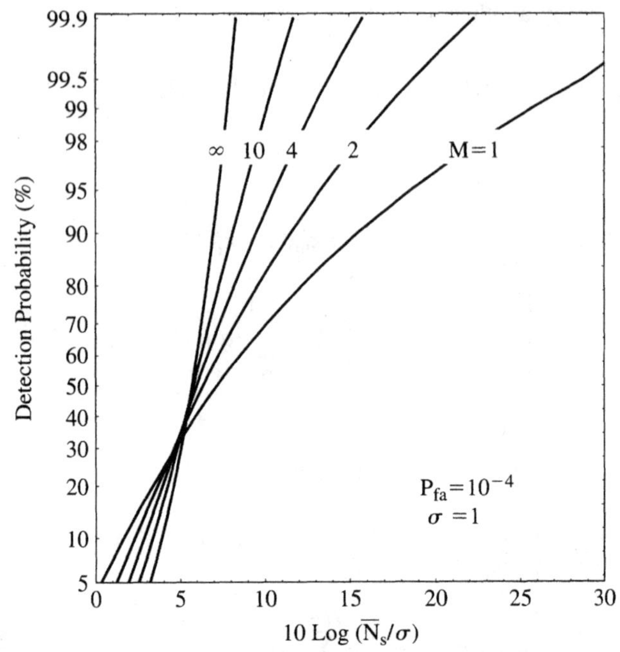

Figure 4-27. Detection probability for a parabolic-cylinder signal immersed in Gaussian noise with speckle count M as a parameter and a false alarm probability $P_{fa} = 10^{-4}$.

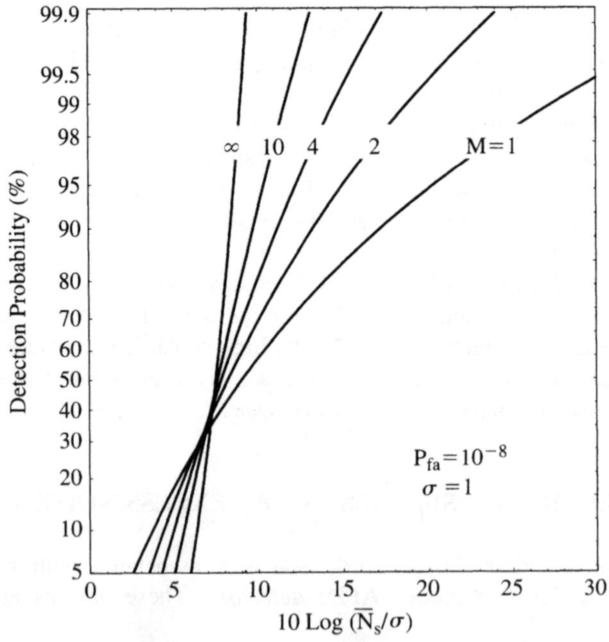

Figure 4-28. Detection probability for a parabolic-cylinder signal immersed in Gaussian noise with speckle count M as a parameter and a false alarm probability $P_{fa} = 10^{-8}$.

low-level photon detection because of their internal gain mechanisms. Detectors that exhibit gain offer the potential advantage of shot-noise-limited performance, whereby signal shot noise dominates internal detector and external circuit noise. Not all receivers achieve shot-noise-limited performance, however, especially if preamplifier noise is high. In those cases, Gaussian or other suitable representations of the receiver statistics must be used.

There are a variety of noise sources that can affect receiver performance. Since we are concerned with obtaining a discrete density function that may be used in generating the detection statistics of an APD receiver, we will be concerned primarily with two sources of noise: (1) randomness in the generation of primary hole–electron carriers, including those generated by photoinjection processes, such as signal and background radiation, and those generated by internal thermal and tunneling processes, such as dark current; and (2) randomness in the carrier multiplication process in the depletion region of the detector. The first process results in fluctuations in the generation of *primary* photoelectrons for injection into the multiplying region. We considered such processes in previous sections, where it was found that the specular target case was characterized by Poisson-distributed signal and noise photoelectrons, while the rough target case was characterized by a product of negative binomial and Kummer distributions.

The second process (i.e., gain variations within the multiplying region of the detector) results from the nonuniform fields associated with the finite geometries of the devices and, more fundamentally, the inherently stochastic nature of the impact ionization process of the gain mechanism. Characterization of the multiplication statistics and the associated excess noise properties of APD detectors has been the subject of continuing study since the devices first became available. McIntyre[22] was the first to develop a discrete density function representing the distribution of multiplied photoelectrons generated by an APD detector. This formulation was later verified experimentally by Conradi[23] using a silicon APD. McIntyre's somewhat complex formulation was later simplified by Webb et al.[24] (WMC). All of the works mentioned above resulted in approximate analytic expressions that prove useful in calculating detection statistics, as shown in this section. (These theories are currently being improved upon by more exact computer simulations that take into account some of the more subtle properties of APD devices, such as dead space[25,26] and device thickness.) In this section we consider the McIntyre and WMC formulations as well as a simple Gaussian approximation to the latter, which is useful in modeling performance in the high signal-to-noise ratio limit.

Most conventional APD devices are designed such that the photoinjection of either holes or electrons occurs on one side of the multiplying region in order to minimize or eliminate excess noise due to the other carrier type. For simplicity, we assume such a device in the following paragraphs. We also assume that the primary carriers are electrons and that the *photoinjected excess noise* is given by

$$F = k_e G + \left(2 - \frac{1}{G}\right)(1 - k_e) \tag{4-181}$$

where k_e is the effective hole/electron ionization rate ratio ($k_e \equiv k_{\text{eff}}$ in Chapter 2) and G is the average gain. As discussed in Section 2.9.3, excess noise generated by dark current can generally be neglected.

4.8.1. Poisson Signal

The desired discrete density function that represents the statistics of the multiplied photoelectrons m follows from

$$q(m) = \sum_{k=1}^{\infty} q(m \mid k) q(k) \tag{4-182}$$

where $q(m \mid k)$ is the conditional probability of obtaining m multiplied electrons given k primary photoelectrons and $q(k)$ is the probability of k primary photoelectrons. The derivation of $q(m \mid k)$ is quite complex, taking into account such issues as the nonuniform fields within the detector as well as the impact ionization processes of semiconductor devices, and is therefore outside the scope of this book. We therefore use the *McIntyre density function* (cf. Ref. 23, Eq. 16a), given by

$$q_M(m \mid k) = \frac{k(1-k_e)^{m-k}}{[k(1-k_e)+k_e m](m-k)!} \frac{\Gamma(m/(1-k_e))}{\Gamma(k+k_e m/(1-k_e))} \times \left[\frac{1+k_e(G-1)}{G}\right]^{k+k_e m/(1-k_e)} \left(\frac{G-1}{G}\right)^{m-k} \tag{4-183}$$

Here the random variable has been changed from McIntyre's r to $m = k + r$, where r is the number of ionizations in the multiplying region, m the number of multiplied electrons, $G(= \overline{m}/\overline{N})$ the average gain, $\overline{N}$ the mean primary photoelectron count in the measurement interval, k the photoelectron count number, k_e the effective hole-to-electron ionization rate ratio.

From Eq. (4-58) the discrete density function representing the distribution of signal plus noise photoelectrons is Poisson,

$$q(k) = \frac{(\overline{N}_s + \overline{N}_n)^k e^{-(\overline{N}_s + \overline{N}_n)}}{k!} \tag{4-184}$$

Equation (4-182), representing the distribution of multiplied electrons, therefore becomes

$$q_{SM}(m; \overline{N}_s, \overline{N}_n, G) = \sum_{k=1}^{\infty} q(m \mid k) q(k)$$

$$= \sum_{k=1}^{\infty} \frac{k(1-k_e)^{m-k}}{[k(1-k_e)+k_e m](m-k)!} \frac{\Gamma(m/(1-k_e))}{\Gamma(k+k_e m/(1-k_e))}$$
$$\times \left[\frac{1+k_e(G-1)}{G}\right]^{k+k_e m/(1-k_e)} \left(\frac{G-1}{G}\right)^{m-k}$$
$$\times \frac{(\overline{N}_s+\overline{N}_n)^k e^{-(\overline{N}_s+\overline{N}_n)}}{k!} \tag{4-185}$$

where the subscript S implies the specular target case. McIntyre has shown that the mean multiplied electron count given by Eq. (4-185) is $G(\overline{N}_s + \overline{N}_n)$.

Equation (4-185) is relatively complex and therefore not very convenient to use in engineering applications. Recognizing this, Webb, McIntyre, and Conradi proposed a much simpler expression that closely approximates Eq. (4-185). Denoting this density function as the *WMC approximation*, it is applicable only in the case of Poisson-distributed primary photoelectrons.[27] In principle, these photoelectrons can be generated by signal, background, and/or dark-current noise sources, as given by Eq. (4-184); however, we will follow conventional usage of the WMC density function and include only signal and background contributions. (Although dark current is also characterized by Poisson statistics, it is usually associated with a different excess noise factor, as mentioned earlier.) Thus, letting $\overline{N}_n \to \overline{N}_b$,

$$q_{SW}(x) = \frac{e^{-x^2/2(1+x/\lambda)}}{\sqrt{2\pi}(1+x/\lambda)^{3/2}} \qquad -\frac{F-1}{F}\frac{G-1}{G} < \frac{x}{\lambda} < \infty \tag{4-186}$$

where

$$x = \frac{m - G\overline{N}}{\sqrt{FG^2\overline{N}}} \qquad \lambda = \frac{\sqrt{\overline{N}F}}{F-1} \qquad \overline{N} = \overline{N}_s + \overline{N}_b \tag{4-187}$$

and the subscript W denotes WMC. [Note that although the WMC and related density functions to follow will be treated mathematically as continuous probability density functions, they are written explicitly as discrete density functions $q(x)$ as a reminder of the fact that x represents a discrete random variable.] In fact, Eq. (4-186) can be written explicitly in terms of the multiplied electrons, m, using standard methods for changing statistical variables; thus

$$q_{SW}(m) = q_{SW}(x)\frac{dx}{dm}\bigg|_{x \to m - \overline{N}G/\sqrt{\overline{N}G^2F}} \tag{4-188}$$

In this particular case, dx/dm is equivalent to a normalization factor $1/\sqrt{\overline{N}G^2F}$, which therefore yields

$$q_{SW}(m; \overline{N}, G, F) = \frac{\exp\left(-\frac{(m-\overline{N}G)^2}{2\overline{N}G^2F\{1+[(m-\overline{N}G)/\overline{N}GF](F-1)\}}\right)}{\sqrt{2\pi\overline{N}G^2F}\left[1+\frac{(m-\overline{N}G)}{\overline{N}GF}(F-1)\right]^{3/2}} \tag{4-189}$$

Note that the background noise $\overline{N}_b$ contained in $\overline{N}$ constitutes a bias term in $q_{SW}(m; \overline{N}, G, F)$, which shifts the peak or mode of the distribution and therefore can affect the false alarm statistics. This term can be set equal to zero for ac-coupled receivers depending on the low-frequency cutoff of the receiver passband.

It is not too difficult to verify that $q_{SW}(m \mid \overline{N})$ is normalized such that $\sum_{m=1}^{\infty} q_{SW}(m; \overline{N}, G, F) = 1$. One can also treat m as a continuous variable and, lacking a closed-form expression for the integral over m, numerically integrate $q_{SW}(m; \overline{N}, G, F)$ to obtain $\int_0^{\infty} q_{SW}(m; \overline{N}, G, F) = 1$. Unfortunately, the intractable nature of the integral expression prevents any analytic calculation of the moments. However, we do know from Eq. (4-185) that in regions where the approximation is good (i.e., $\overline{N}G \gg 1$), the first moment must approach $\overline{N}G$.

Figure 4-29 compares Eqs. (4-185) and (4-189) for several values of $\overline{N}$. It can be seen that the WMC approximation is quite good for values of $\overline{N}_s + \overline{N}_b$ greater than about 4. Because of the additive property of the Poisson distribution, $\overline{N}_s$ and $\overline{N}_b$ functionally affect the density function in the same way. Another way of viewing this is that signal and background photons are indistinguishable as far as the detector is concerned.

It can also be seen in Figure 4-29 that the McIntyre and WMC distributions approach a Gaussian distribution for large $\overline{N}$. Indeed, in the limit of $\overline{N}/F \gg 1$, Eq. (4-189) may be approximated by a continuous Gaussian density function given by

$$p_{SG}(m; \overline{N}) = \frac{1}{\sqrt{2\pi \overline{N} G^2 F}} e^{-(m-\overline{N}G)^2/2\overline{N}G^2 F} \tag{4-190}$$

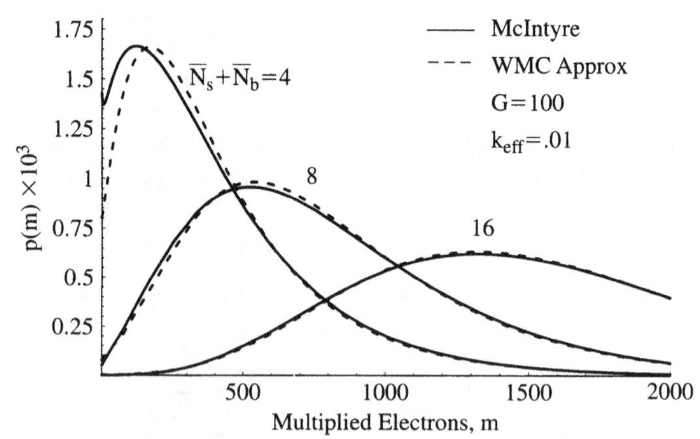

Figure 4-29. Comparison of probability density functions for McIntyre and WMC approximations with the total mean signal and mean noise photoelectron counts as a parameter.

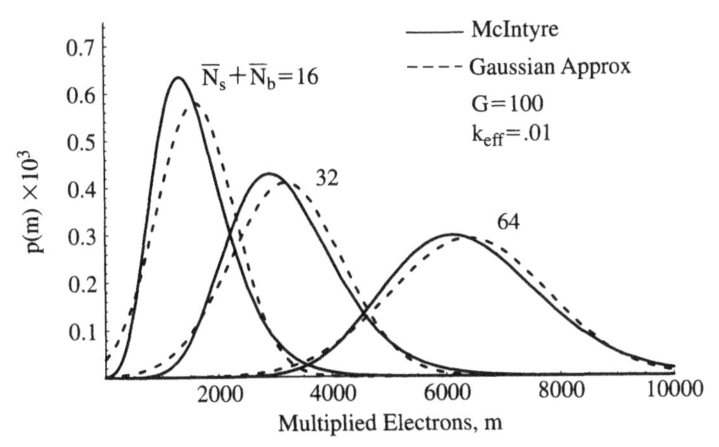

Figure 4-30. Comparison of probability density functions for McIntyre and Gaussian approximations with the total mean signal and mean noise photoelectron counts as a parameter.

where $p(\cdot)$ denotes a continuous probability distribution, the subscript G stands for Gaussian, and the parameters G and F have been dropped in the argument of p_{SG} for simplicity of notation. Figure 4-30 compares Eq. (4-190) with McIntyre's distribution for large values of $\overline{N}$. At high values of $\overline{N}$, the Gaussian approximation is a reasonable one and is frequently employed for quick performance estimates of high signal-to-noise ratio systems.

Ultimately, we wish to develop expressions for APD detection and false alarm probabilities for receivers operating at low photon counts. Before attempting this, however, it will be instructive to continue developing the Gaussian high signal-to-noise ratio limit of the theory in order to obtain some analytical insights. This limiting case is applicable not only to APD detectors operating at high photoelectron counts but also to unity-gain detectors with $G = 1$ and $F = 1$. As discussed in Chapter 3, the central limit theorem may be invoked to characterize the noise in a direct-detection receiver as a zero-mean Gaussian variate with a noise variance equal to the sum of the individual noise variances. The receiver noise characteristics will therefore be represented by a continuous Gaussian probability density function given by

$$p_n(m; \sigma) = \frac{1}{\sqrt{2\pi\sigma_n^2}} e^{-m^2/2\sigma_n^2} \tag{4-191}$$

where m represents the multiplied noise electrons, a continuous random variable, and $\sigma_n^2 = \sigma_a^2 + \sigma_t^2 + \sigma_d^2$ represents the *dark noise* of the receiver (not to be confused with the dark current noise), that is, the sum of all noise contributions that are still present when the receiver aperture is closed. It consists of the sum of amplifier, thermal, and dark current noise variances, respectively.

The total discrete density function for the signal plus noise can now be obtained by convolving the probability density function for the signal, Eq. (4-190), with that for the noise, Eq. (4-191). The corresponding characteristic functions representing the noise and the signal are

$$\begin{aligned} G_n(iv) &= \int_{-\infty}^{\infty} e^{ivm} p_n(m;\sigma_n)\,dm \\ &= e^{-(1/2)\sigma_n^2 v^2} \\ G_s(iv) &= \int_{-\infty}^{\infty} p_{SG}(m;\overline{N}) e^{ivm}\,dm \\ &= e^{iG\overline{N}v-(1/2)FG^2\overline{N}v^2} \end{aligned} \tag{4-192}$$

respectively, where $\overline{N} = \overline{N}_s + \overline{N}_b$. Thus

$$G_{s+n}(iv) = G_s(iv)G_n(iv) = e^{iG\overline{N}v-(1/2)(\sigma_n^2+FG^2\overline{N})v^2} \tag{4-193}$$

The signal-plus-noise probability density function is therefore Gaussian distribution, as expected:

$$\begin{aligned} p_{s+n}(m;\overline{N}_s,\overline{N}_b) &= \frac{1}{2\pi}\int_{-\infty}^{\infty} G_{s+n}(iv)e^{-imv}\,dv \\ &= \frac{1}{2\pi}\int_{-\infty}^{\infty} e^{-i(m-G\overline{N})v-(1/2)(\sigma_n^2+FG^2\overline{N})v^2}\,dv \\ &= \frac{1}{\sqrt{2\pi\sigma^2}} e^{-(m-G\overline{N})^2/2\sigma^2} \end{aligned} \tag{4-194}$$

where $\sigma^2 = \sigma_s^2 + \sigma_b^2 + \sigma_n^2$, $\sigma_s^2 = FG^2\overline{N}_s$, $\sigma_b^2 = FG^2\overline{N}_b$, and $\overline{N} = \overline{N}_s + \overline{N}_b$. Notice that the excess noise factor F multiplies both the multiplied signal and background photoelectrons. For the case of unity-gain detectors, $G = F = 1$, and $m = k$ in Eq. (4-194). From Section 4.7 we know that Eq. (4-194) should be applied to APD devices only when $\overline{N}/F \gg 1$, so that care must be exercised in its use if accurate results are to be obtained.

The false alarm probability may be obtained by integrating Eq. (4-194) over the range $m_t \leq m \leq \infty$ with $\overline{N}_s = 0$. Treating m as a continuous random variable (see Comment 4-9), this yields

$$\begin{aligned} P_{\text{fa}} &= \int_{m_t}^{\infty} p_{s+n}(m;0,\overline{N}_b)\,dm \\ &= \frac{1}{2}\,\text{erfc}\left(\frac{\text{TNR}_V}{\sqrt{2}}\right) \end{aligned} \tag{4-195}$$

where $\text{TNR}_V = (m_t - G\overline{N}_b)/\sqrt{\sigma_n^2 + \sigma_b^2}$ is the voltage threshold-to-noise ratio, m_t is the threshold level of the multiplied electrons, and $\text{erfc}(x) = 1 - \text{erf}(x)$. Here we see that the presence of background radiation requires an increase in the threshold level to achieve any given P_{fa}. A similar threshold level shift would occur for atmospheric backscatter. The corresponding detection probability becomes

$$P_d = \int_{m_t}^{\infty} p_{s+n}(m; \overline{N}_s, \overline{N}_b)\, dm$$
$$= \frac{1}{2}\,\text{erfc}\left[\frac{m_t - G\overline{N}_s - G\overline{N}_b}{\sqrt{2(\sigma_n^2 + FG^2\overline{N}_s + FG^2\overline{N}_b)}}\right] \tag{4-196}$$

Here we have written out the noise terms explicitly to emphasize the fact that the mean primary signal photoelectron count, $\overline{N}_s$, contributes not only to the signal in the numerator but also to the noise in the denominator. An effective voltage signal-to-noise ratio can be defined as the ratio of the mean electron count to the square root of the noise variance, that is

$$\text{SNR}_V = \frac{G\overline{N}_s}{\sqrt{\sigma_n^2 + FG^2\overline{N}_s + FG^2\overline{N}_b}} \tag{4-197}$$

The power signal-to-noise ratio, SNR_P, is the square of Eq. (4-197). Note that the signal or shot noise limit of Eq. (4-197) yields $\text{SNR}_P = \overline{N}_s/F$, and for the dark-noise limit, $\text{SNR}_P = G\overline{N}_s/\sigma_n$. Hence, a low excess noise factor is important in maintaining a high SNR in signal-limited systems, while high detector gain enhances the SNR in amplifier or Gaussian noise-limited systems.

Comment 4-8. Equations (4-194), (4-195), and (4-197) can be written in terms of primary photoelectrons as follows. Substituting $m = Gk$, where k = primary photoelectron count, we have

$$p_{s+n}(k; \overline{N}_s, \overline{N}_b) = \frac{1}{\sqrt{2\pi\sigma^2}} e^{-(m-G\overline{N})^2/2\sigma^2} \left.\frac{dm}{dk}\right|_{m=Gk}$$
$$= \frac{1}{\sqrt{2\pi\sigma'^2}} e^{-(k-\overline{N})^2/2\sigma'^2} \tag{4-198}$$

where $\sigma'^2 = \sigma_n^2/G^2 + F\overline{N}_s + F\overline{N}_b$. Similarly, Eqs. (4-195) and (4-197) become

$$P_{\text{fa}} = \frac{1}{2}\text{erfc}\left[\frac{k_t - \overline{N}_b}{\sqrt{2(\sigma_n^2 + \sigma_b^2)/G^2}}\right] \tag{4-199}$$

$$\mathrm{SNR}_V = \frac{\overline{N}_s}{\sqrt{\sigma_n^2/G^2 + F\overline{N}_s + F\overline{N}_b}} \tag{4-200}$$

■

Some representative examples of P_d versus SNR_V(dB) for nonavalanche detectors ($G = 1$, $F = 1$) and APD detectors ($G = 100$, $F = 3$) are shown in Figure 4-31 with P_{fa} as a parameter. Here $\mathrm{SNR}_V(\mathrm{dB}) = 10\log(\overline{N}_s/\sigma_n)$ with $\sigma_n = 10$ and $\overline{N}_b = 2$. Notice that 5 to 10 dB of added signal is required by the unity-gain detector compared to the APD detector for a given probability of detection.

Given the foregoing insights, the previous procedures can be applied to the convolution of the WMC density function and the Gaussian noise density function, Eq. (4-191), to find the detection statistics associated with low photoelectron counts. Also, the asymmetric shape of the WMC distribution compared to the Gaussian distribution at low electron counts can be rather important in optical communications systems[28] where precise threshold levels are required for good estimates of the bit-error rate of the link (cf. Comment 1-12). As mentioned

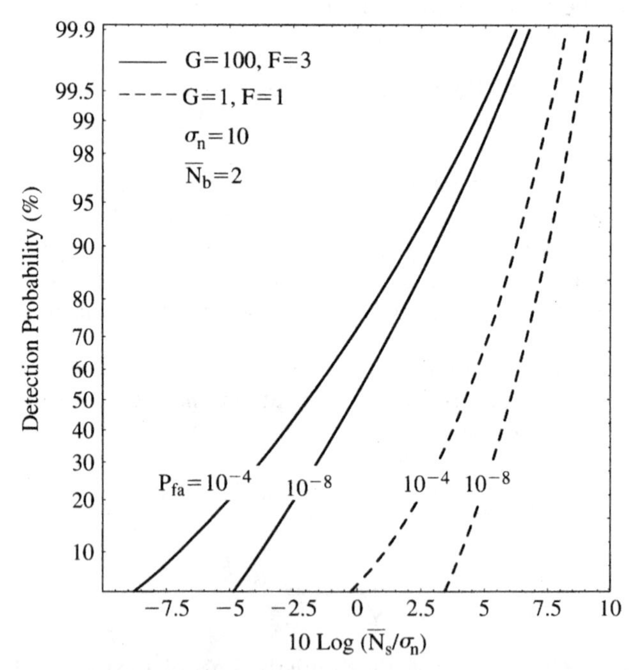

Figure 4-31. Comparative detection performance of unity-gain pin and $G = 100$ gain APD detectors for the case of a mean background photoelectron count $\overline{N}_b = 2$ and a dark noise photoelectron count $\sigma_n = 10$. Note that the excess noise, F, is gain dependent.

earlier, the integrals related to the WMC density function cannot be solved exactly, so a more numerical method is required.

One such method is the discrete Fourier transform (DFT). Since the characteristic function is basically a Fourier transform, the DFT is a convenient tool for generating these functions. Following the same procedures as with analytic functions, the inverse transform of the product of the forward transforms should yield the desired discrete density function. We therefore write

$$q_{s+n}^{\mathrm{wmc}}(m; \overline{N}_s, \overline{N}_b, \sigma_n) = F^{-1}\{F[q_{SW}(m; \overline{N}_s, \overline{N}_b)] \times F[q_n(m; \sigma_n)]\} \qquad (4\text{-}201)$$

where $q_{SW}(m; \overline{N}_s, \overline{N}_b)$ is given by Eq. (4-189), $q_n(m; \sigma_n)$ is given by Eq. (4-191), and F and F^{-1} represent the Fourier transform and its inverse, respectively. Many modern software packages contain DFT routines that can easily handle Eq. (4-201). A 256-point transform of $q_{s+n}^{\mathrm{wmc}}(m; \overline{N}, \sigma_n)$ is shown in Figure 4-32 for several values of $\overline{N}$ and $\sigma_n = 5$.

The false alarm probability is calculated using the smallest integer m_t that satisfies

$$P_{\mathrm{fa}} \geq \sum_{m=m_t}^{\infty} q_{s+n}^{\mathrm{wmc}}(m; 0, \overline{N}_b, \sigma_n) \qquad (4\text{-}202)$$

If all photoinjected carriers are zero (i.e., $\overline{N}_s = \overline{N}_b = 0$), $q_{s+n}^{\mathrm{wmc}}(m; 0, 0, \sigma_n)$ reduces to $q_n(m; \sigma_n)$ given by Eq. (4-201), resulting in

$$P_{\mathrm{fa}} \geq \sum_{m=m_t}^{\infty} q_n(m; \sigma_n) \qquad (4\text{-}203)$$

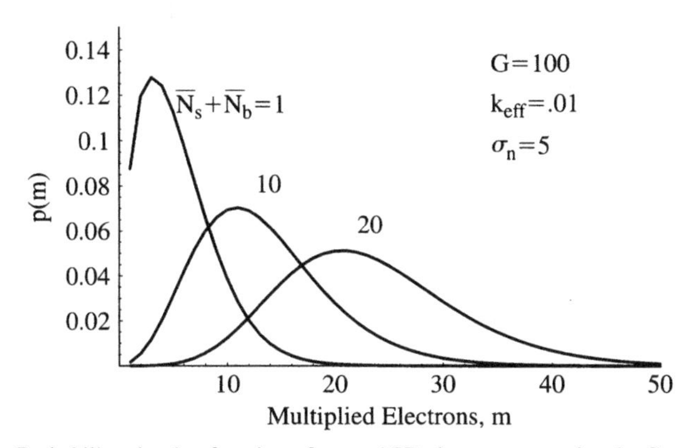

Figure 4-32. Probability density functions for an APD detector operating in Gaussian amplifier noise. The plots represent convolutions of the WMC and Gaussian density functions with the total mean signal and mean noise photoelectron counts as a parameter and a dark-noise photoelectron count of $\sigma_n = 5$.

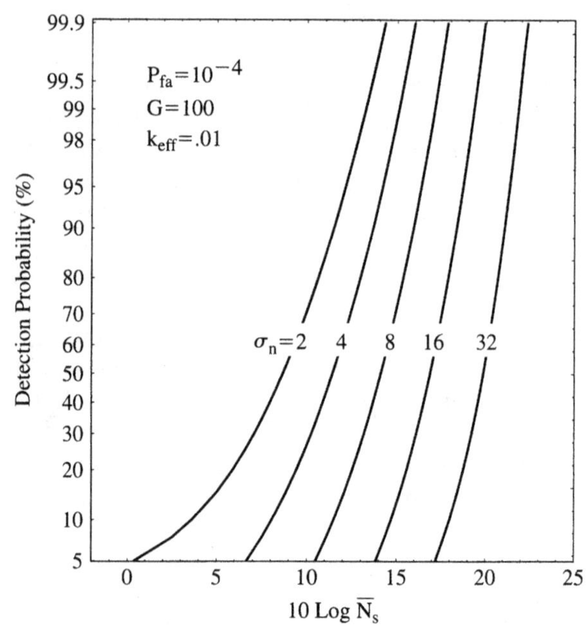

Figure 4-33. Detection probability versus mean signal photoelectron count for an APD detector having a gain $G = 100$, a hole-to-electron ionization rate ratio $k_{\text{eff}} = 0.01$, and a false alarm probability $P_{\text{fa}} = 10^{-4}$ with dark noise as a parameter.

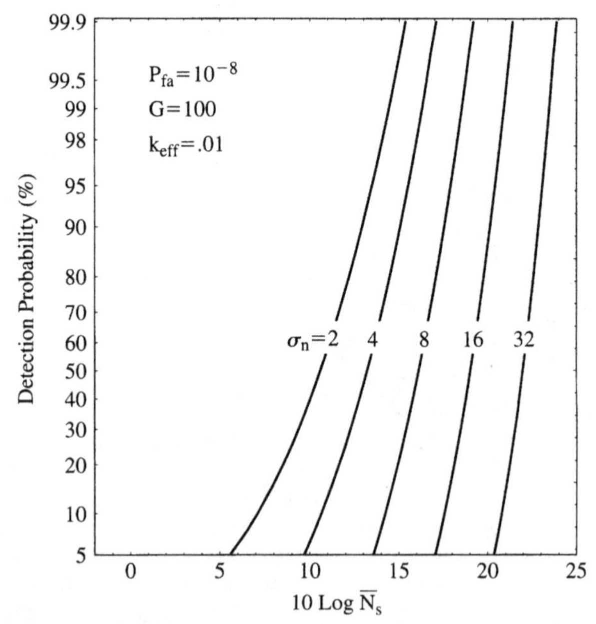

Figure 4-34. Detection probability versus mean signal photoelectron count for an APD detector having a gain $G = 100$, a hole-to-electron ionization rate ratio $k_{\text{eff}} = 0.01$, and a false alarm probability $P_{\text{fa}} = 10^{-8}$ with dark noise as a parameter.

The detection probability is given by

$$P_d = \sum_{m=m_t}^{\infty} p_{s+n}^{\text{wmc}}(m; \overline{N}_s, \overline{N}_b, \sigma) \tag{4-204}$$

Some representative detection probabilities are shown in Figures 4-33 and 4-34 for false alarm probabilities of $P_{\text{fa}} = 10^{-4}$ and 10^{-8}, respectively, using Eq. (4-195) to solve for m_t.

Comment 4-9. Use of the continuous form of the noise density function, Eq. (4-195), rather than the discrete sum given by Eq. (4-203) can lead to errors if not interpreted properly. Comment 4-3 of Section 4.4 considered the quantization errors resulting from calculation of P_{fa} from a discrete noise density function. The integral representation of Eq. (4-195) must be considered a continuous approximation to the discrete probability density function given by Eq. (4-203). Hence from the theory of calculus we see that each incremental rectangular area is equal to $p_n(m; \sigma)\Delta m$, as shown in Figure 4-35. Therefore, Eq. (4-195) could just as well have been written

$$P_{\text{fa}} \geq \sum_{m=m_t}^{\infty} p_n(m; \sigma_N)\Delta m \tag{4-205}$$

where $\Delta m \equiv 1$ for electrons. Inspection of Figure 4-35 also shows that the integral representation underestimates the threshold setting required to achieve a given P_{fa} due to the larger area of the discrete distribution (shaded region). For $\sigma_N \geq 2$, the difference is always less than unity, so that rounding up m_t obtained from Eq. (4-195) assures identical results. For example, to achieve a

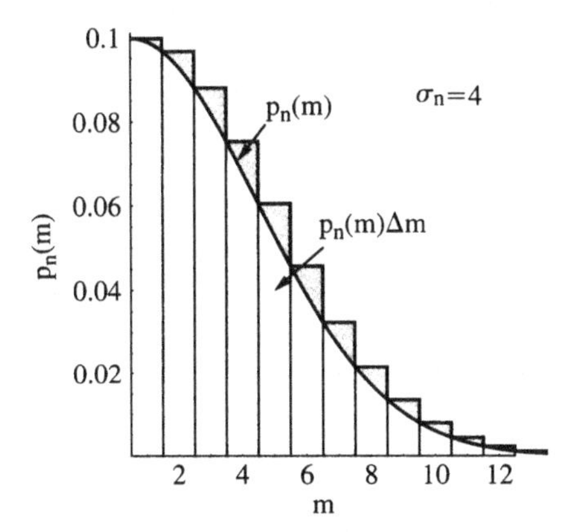

Figure 4-35. Graphical illustration of errors involved in approximating the discrete noise density function with a continuous density function in calculating the false alarm probability. Note that the continuous distribution always underestimates the required threshold setting to achieve a given P_{fa}.

false alarm probability less than or equal to $P_{fa} = 10^{-4}$ with $\sigma_N = 2$, the discrete case requires $m_t = 8$ with an associated $P_{fa} = 5.1 \times 10^{-5}$, whereas Eq. (4-195) predicts $m_t = 7.44$, which, when rounded up, equals 8. The difference between the two representations decreases at higher values of σ_N. ■

4.8.2. Negative Exponential Signal

In the case of a rough target, $q(k)$ in Eq. (4-182) assumes the form of the Kummer distribution given by Eq. (4-77). Thus, with $q(m \mid k)$ given by Eq. (4-183), we have the rather formidable expression for the multiplied photoelectron count,

$$
\begin{aligned}
q_{RM}(m) &= \sum_{k=1}^{\infty} q(m \mid k) q(k) \\
&= \sum_{k=1}^{\infty} \frac{k(1-k_e)^{m-k}}{[k(1-k_e)+k_e m](m-k_e)!} \frac{\Gamma(m/(1-k_e))}{\Gamma(k+k_e m/(1-k_e))} \\
&\quad \times \left[\frac{1+k_e(G-1)}{G}\right]^{k+k_e m/(1-k_e)} \left(\frac{G-1}{G}\right)^{m-k} \\
&\quad \times p^M q^k \frac{e^{-\overline{N}_n}}{\Gamma(M)} \frac{\Gamma(k+M)}{\Gamma(k+1)} {}_1F_1(-k, -k-M+1; z) \qquad (4\text{-}206)
\end{aligned}
$$

where the parameters p, q, and M are the same as those defined in Section 4.5. It was pointed out in Section 4.5 that for $\overline{N}_n = 0$, the confluent hypergeometric distribution ${}_1F_1(a, b; 0) = 1$ and $e^{-\overline{N}_n} = 1$ such that the last term in Eq. (4-206) can be simplified somewhat to the negative binomial distribution. Equation (4-206) is plotted in Figures 4-36 and 4-37 for several values of the mean noise count $\overline{N}_n$ with $M = 1$ and 4, respectively. Note once again the negative exponential-like distribution for $\overline{N}_n = 0$ and $M = 1$, as expected for a single speckle realization. As $\overline{N}_n$ increases, the mean of the distribution moves to the right while the shape approaches that of a Gaussian.

Use of Eq. (4-206) to calculate the detection statistics is computationally intensive and will not be explored further. However, a simpler approach is possible using a modified form of the WMC approximation discussed earlier. The modification allows the WMC density function to be generalized to the rough target case by introducing a *Fano factor* into the definition of the excess noise.[29]

Consider any point process representing the generation of k primary photoinjected carriers within a measurement interval τ. If the jth carrier is multiplied by a gain g_j, the total number of multiplied electrons is given by the sum $m = \sum_{j=1}^{k} g_j$. It then follows that if k and g represent stationary processes that are statistically independent, the mean and variance of the secondary

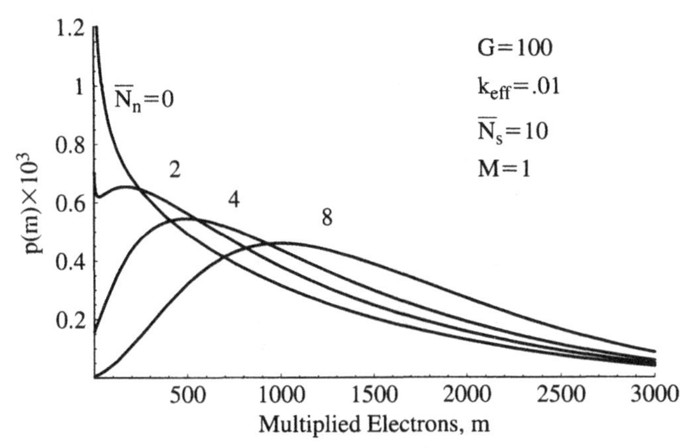

Figure 4-36. Probability density function of an APD detector operating against a rough target for the case of a speckle count of $M = 1$ with background noise count as a parameter and a mean signal count of $\overline{N}_s = 10$.

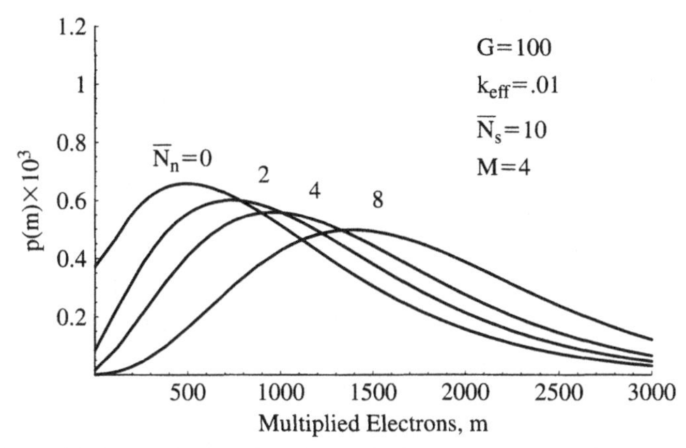

Figure 4-37. Probability density function of an APD detector operating against a rough target for the case of a speckle count of $M = 4$, a mean signal count of $\overline{N}_s = 10$, and the noise count $\overline{N}_n$ as a parameter.

(multiplied) electrons are given by

$$\langle m \rangle = \langle g \rangle \langle k \rangle \tag{4-207}$$

and

$$\text{var}(m) = \langle g \rangle^2 \text{var}(k) + \langle k \rangle \text{var}(g) \tag{4-208}$$

Equation (4-208) is known as the *Burgess variance theorem*.[30] A derivation[31] of Eqs. (4-207) and (4-208) is given in Appendix B. These relationships hold for any arbitrary distribution of g or k. If Eq. (4-208) is divided by Eq. (4-207), we obtain

$$f_m = \langle g \rangle f_k + f_g \tag{4-209}$$

Here f is commonly referred to as a *Fano factor*, defined as

$$f_x = \frac{\text{var}(x)}{\langle x \rangle} \tag{4-210}$$

In photon statistics, the Fano factor is a measure of the degree of *photon bunching* ($f > 1$) or *antibunching* ($f < 1$) of the distribution. For a Poisson distribution, $f = 1$.

Earlier, we defined the excess noise factor to be

$$F_e = \frac{\langle g^2 \rangle}{\langle g \rangle^2} \tag{4-211}$$

Using Eq. (4-211) in Eq. (4-209) yields

$$f_m = \langle g \rangle (f_k + F_e - 1) \tag{4-212}$$

or, with Eq. (4-207) while noting that $\langle g \rangle \equiv G$, we have

$$\text{var}(m) = \langle k \rangle G^2 (f_k + F_e - 1) \tag{4-213}$$

Here f_k represents the Fano factor for the primary or photoinjected carriers. For negative binomial processes,[32] the results of Eq. (4-98) can be used to obtain

$$f_k = \frac{\text{var}(k)}{\langle k \rangle} = 1 + \frac{\langle k \rangle}{M} \tag{4-214}$$

where M is the number of speckles intercepted by the receive aperture.

If we now replace F_e in the WMC expression for the multiplied electron distribution of an APD detector with $f_k + F_e - 1$, the resulting expression provides a reasonably good approximation to the rough target statistics of Eq. (4-206) for the case of $\overline{N}_n = 0$. Since $\langle k \rangle = \overline{N}_s$ for a negative binomial distribution, the modified excess noise factor becomes

$$F = k_e G + \left(2 - \frac{1}{G}\right)(1 - k_e) + \frac{\overline{N}_s}{M} \tag{4-215}$$

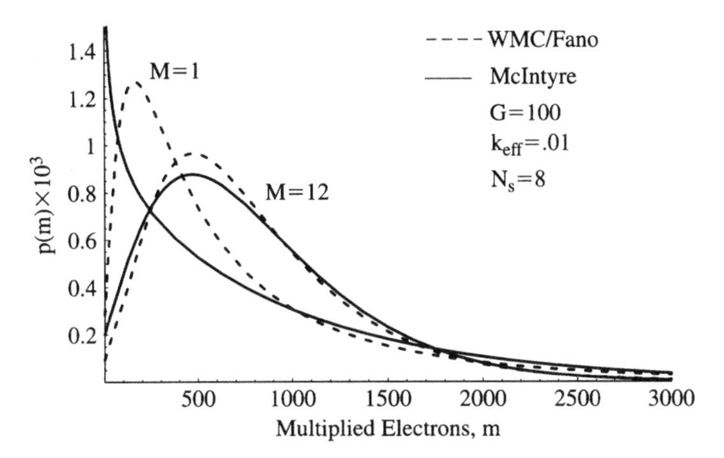

Figure 4-38. Comparison of McIntrye and WMC–Fano distributions with M as a parameter.

Figure 4-38 compares the McIntyre distribution with the WMC distribution using the Fano factor to account for speckle fluctuations. It can be seen that the mode of the distribution aligns well with the exact distribution as M increases, although the shape of the distributions are not precisely the same, even at large electron counts. The latter discrepancy is a minor one, however, compared to the computational advantages of the approach and affects the detection probability by only a couple of percent at the lower values of P_d.

The detection probability is calculated from Eq. (4-201) with F given by Eq. (4-215). Figures 4-39 and 4-40 show the detection probability plotted as a function of SNR_V, where $\mathrm{SNR}_V = G\overline{N}_s/\sigma_n$, $\sigma_n = 4$, and $P_{\mathrm{fa}} = 10^{-4}$, with M as a parameter. As in previous figures with M as a parameter, a crossover of the curves can be seen at the lower detection probabilities.

4.8.3. Geiger-Mode APD Statistics

The Geiger mode of the APD detector, also called the *single-photon avalanche diode* (SPAD), is achieved by raising the bias voltage on the detector above the breakdown voltage where the gain can reach values on the order of 10^5 to 10^7. At such high gain levels, a single bulk or photo-generated primary carrier can cause a self-sustaining avalanche that completely depletes the gain region of the device. As a consequence, the rise time or leading edge of the output pulse is very short ($\sim$100 ps), providing an excellent time reference for range or other measurements. The performance of these devices is usually characterized by their *photon detection efficiency*, which is the product of the probability of a photon

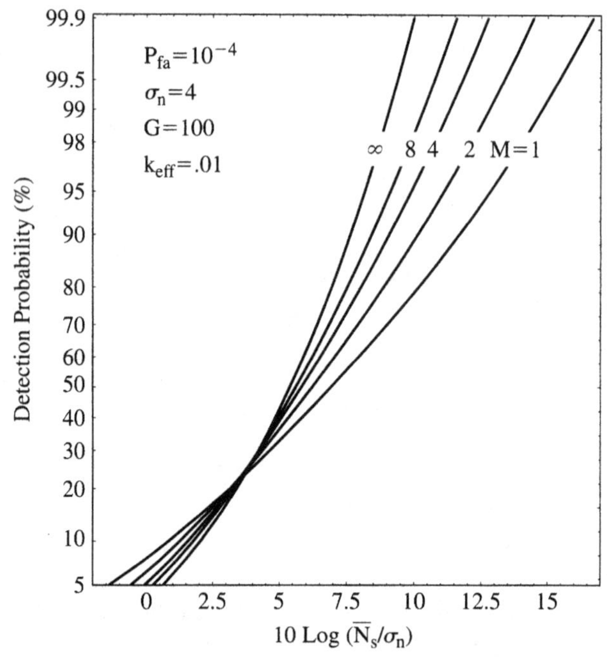

Figure 4-39. Detection probability versus signal-to-noise ratio, expressed in decibels, for an APD detector represented by the WMC–Fano density function with a false alarm probability $P_{fa} = 10^{-4}$.

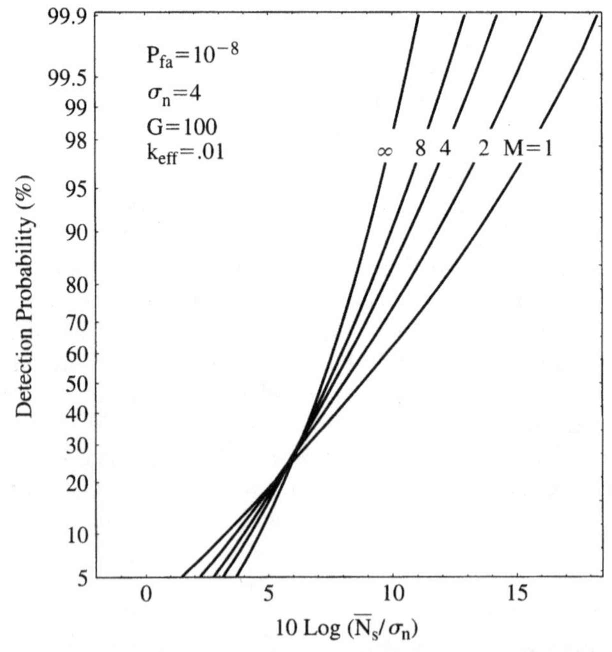

Figure 4-40. Detection probability versus signal-to-noise ratio, expressed in decibels, for an APD detector represented by the WMC–Fano density function with a false alarm probability $P_{fa} = 10^{-8}$.

being absorbed and creating a primary hole–electron pair (i.e., the quantum efficiency) and the probability that the pair will trigger an avalanche. The latter process is usually referred to as the *photoelectron detection probability* P_e and can attain values as high as 85% in silicon APD devices[33] operating in the visible region of the spectrum. Considerably lower values have been reported for APD materials such as indium gallium arsenide/indium phosphide[34] (InGaAs/InP) or germanium[35] (Ge) operating in the near-infrared. This high sensitivity to single-photon events is the primary advantage of Geiger-mode APD detectors over conventional linear-mode APD devices for low-light-level applications. However, it is accompanied by some other characteristics that need to be considered if a practical system is to be obtained.

One such characteristic is the fact that the Geiger-mode APD acts like a "one-shot" diode in the sense that the output is not proportional to the input. As a consequence, amplitude data is usually not useful. In addition, the avalanche pulse is followed by a relatively long recovery time, commonly referred to as *dead time*, during which the circuit capacitances must recharge to be ready for the next pulse. The recovery time depends on how the device circuit is designed (i.e., whether it employs a *passive* or *active quenching circuit*).[36] In a passive quenching circuit, a load resistor limits the APD current to a level at which continuous discharge cannot occur. In this case, dead times on the order of 300 ns are typical, with dark-count rates of approximately 10^6 counts/s. The latter generally limits their use to low-count-rate applications. In active quenching circuits, the voltage is dropped just below the breakdown voltage immediately upon initiation of an avalanche pulse. With this approach, the dead time can be reduced by up to an order of magnitude ($\sim$30 ns). Dark count rates of $<$1 count/s have been reported at $-20\,^\circ$C in silicon APD devices.

If we assume a specular target and a Si:APD device, in which electrons dominate as the primary carrier, the production rate of photoelectrons is governed by Poisson statistics, as discussed earlier. Since background and dark current statistics are also Poisson, the density functions are additive and Eqs. (4-59) and (4-60) may be used for the false alarm and detection probabilities, respectively. (In addition to dark current there are also occurrences of "after pulses" arising from trapped electrons in impurity sites of the detector material[37] that may be released randomly well after the initial avalanche, but these are not discussed here.) However, since the full output of the device can be obtained with as little as one noise photoelectron, thresholding is not very useful as a discriminant against noise. Hence we set $k^{\text{th}} = 1$ in Eqs. (4-59) and (4-60) and write for the probability of detecting at least one signal photoelectron in the presence of noise photoelectrons,

$$P_d = \sum_{k=1}^{\infty} \frac{(\overline{N}_{s+n})^k e^{-\overline{N}_{s+n}}}{k!} = 1 - e^{-\overline{N}_{s+n}} \tag{4-216}$$

and the false alarm probability when no signal is present,

$$P_{\text{fa}} = \sum_{k=1}^{\infty} \frac{\overline{N}_n^k e^{-\overline{N}_n}}{k!} = 1 - e^{-\overline{N}_n} \tag{4-217}$$

where $\overline{N}_{s+n} = \overline{N}_s + \overline{N}_n$ and $\overline{N}_s$ and $\overline{N}_n$ represent the mean signal and noise primary photoelectrons generated during the measurement interval, respectively. Note that the mean noise count $\overline{N}_n = \overline{N}_b + \overline{N}_d$ consists of both background and dark current photoelectrons.

As was implied earlier, avalanche processes initiated by signal or noise photoelectrons are indistinguishable, so that one must ensure that $\overline{N}_s \gg \overline{N}_n$ if only signal photons are to be counted. It is usually the case that in properly designed APD devices, $\overline{N}_b \gg \overline{N}_d$, while $\overline{N}_b$ can be limited through the use of narrowband optical filters or by limiting operation to low background environments. The photon detection probability is obtained by multiplying Eq. (4-216) by the product of the quantum efficiency and photoelectron detection probability (i.e., $\eta_q \times P_e$).

In the case of a rough target, Eq. (4-77) representing a negative binomial signal in Poisson noise must be used in place of the Poisson distribution given by Eq. (4-58). In that case the detection and false alarm probabilities become

$$P_d = \sum_{k=1}^{\infty} q_{s+n}(k) = 1 - \left(\frac{M}{M + \overline{N}_s}\right)^M e^{-\overline{N}_n} \tag{4-218}$$

and

$$P_{\text{fa}} = \sum_{k=1}^{\infty} q_n(k) = 1 - e^{-\overline{N}_n} \tag{4-219}$$

where ${}_1F_1(0, \beta; z) \equiv 1$. Note that in the limit of $\overline{Ns}/M \ll 1$, Eq. (4-218) approaches the Poisson detection probability given by Eq. (4-216) [cf. Eq. (4-90)]. Equation (4-218) is plotted in Figure 4-41 as a function of the mean signal count $\overline{N}_s$ for a noise count of $\overline{N}_n = 0.05$ and several values of speckle count M. These results are very similar to those of Figure 4-14, the difference being, of course, the noise count $\overline{N}_n$ and the fact that the threshold level has been assumed to be fixed at $k^{\text{th}} = 1$.

In the event that multiple sequential measurements are made over a given time period, usually referred to as a *processing window* (cf. Figure 4-42), the finite probability of detecting a noise photoelectron in measurement intervals other than that in which the signal resides will necessarily affect the detection statistics.[38] The most common example of sequential measurements made on a single-pulse basis are the *range or time bins* of a laser rangefinder. In such cases it is usually advantageous to use a passively quenched detector in order to limit the number of avalanches per pulse to one. With this restriction, the probabilities of avalanche production in different measurement intervals must be considered not only *statistically independent* but also *mutually exclusive*.

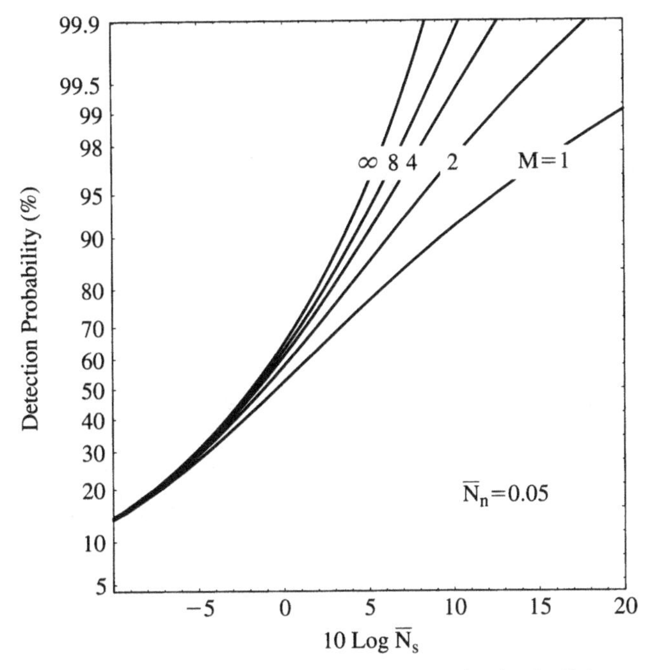

Figure 4-41. Detection probability for an APD detector operating in the Geiger mode against a rough target. The $M = \infty$ curve corresponds to the Poisson statistics of a specular target.

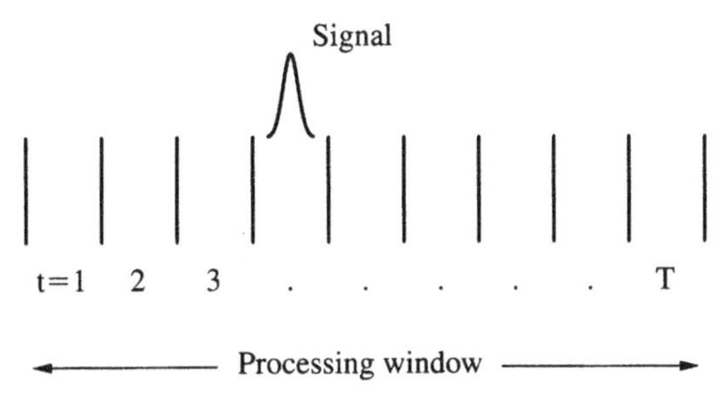

Figure 4-42. Processing window divided into T measurement intervals, where it is assumed that the signal resides in a single measurement interval.

To demonstrate the mutual exclusivity of the process, consider a signal from a specular target that resides in either the *first* or the *last* measurement interval of the processing window. Letting $P_d(t)$ represent the detection probability of a signal residing in interval t, the former case is given by

$$P_d(1) \equiv P_d \tag{4-220}$$

where P_d is the detection probability given by Eq. (4-216). When the signal resides in the *last* interval, assuming $T - 1$ statistically independent intervals prior to the signal interval,

$$P_d(T) = P_d \times (P_{0\mathrm{fa}})^{T-1}$$
$$= (1 - e^{-\overline{N}_{s+n}})e^{-\overline{N}_n(T-1)} \tag{4-221}$$

where $P_{0\mathrm{fa}} = 1 - P_{\mathrm{fa}}$ is the probability of no false alarms in a single measurement interval. It should be noted that a noise detection occurring in the signal interval cannot be distinguished from a true detection and therefore must be treated as a true detection.

The false alarm probability given by Eq. (4-217) must be modified similarly. When a signal is present, noise detections arriving in time intervals prior to the signal interval must be treated separately from those arriving after the signal. In general, this requires expressing the single-interval false alarm probability in terms of the interval number t and summing over all intervals other than the signal interval. However, in those cases where the signal resides in either the first or last interval of the window, the false alarm probability reduces to a pair of relatively simple expressions that are sufficient to demonstrate the desired effects. (The more general approach is discussed in Chapter 6.) Thus, in the case of the signal residing in the *first* interval, we have, assuming a probability P_m of a missed signal detection and no false alarms in the signal interval,

$$P_{\mathrm{fa}}(1) = P_m \times [1 - (P_{0\mathrm{fa}})^{T-1}]$$
$$= [1 - (1 - e^{-\overline{N}_{s+n}})] \times [1 - (e^{-\overline{N}_n})^{T-1}]$$
$$= e^{-\overline{N}_{s+n}}[1 - e^{-\overline{N}_n(T-1)}] \tag{4-222}$$

where the term in brackets in the first line represents the probability for a false alarm in the remaining $T - 1$ intervals. In the case of the signal residing in the *last* interval,

$$P_{\mathrm{fa}}(T) = 1 - (P_{0\mathrm{fa}})^{T-1}$$
$$= 1 - e^{-\overline{N}_n(T-1)} \tag{4-223}$$

Notice that Eq. (4-223) is a constant independent of the signal for any given $\overline{N}_n$ and T.

Comment 4-10. The form of the expressions containing $P_{0\mathrm{fa}}$ in Eqs. (4-221) through (4-223) can be understood by considering the cumulative binomial distribution for one or more successes in n Bernoulli trials. To see this, we write

$$P = \sum_{j=m}^{n} \binom{n}{j} p^j (1-p)^{n-j} = 1 - \sum_{j=0}^{m-1} \binom{n}{j} p^j (1-p)^{n-j}$$
$$= 1 - (1-p)^n \quad \text{for } m = 1 \tag{4-224}$$

■

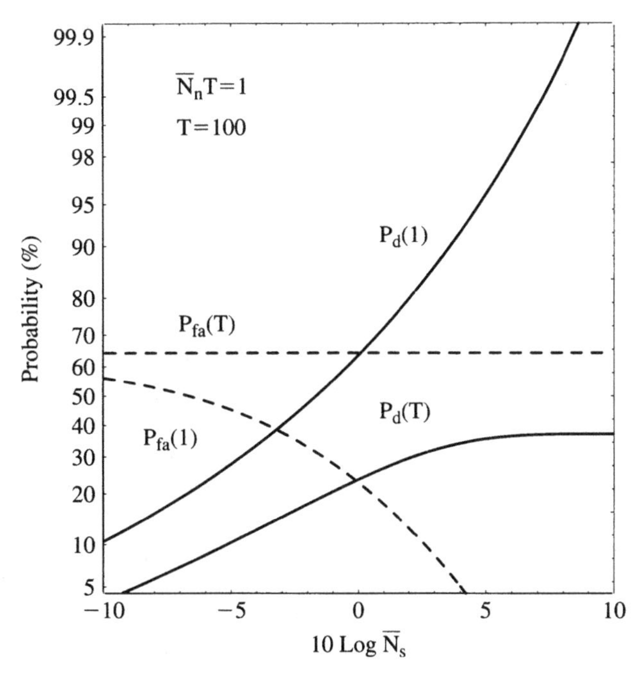

Figure 4-43. Single-pulse detection and false alarm probabilities versus signal photoelectron count $10\log\overline{N}_s$ for an APD detector operating in the Geiger mode against a specular (Poisson) target located in the front $P_x(1)$ and back $P_x(T)$ of a processing window containing $T = 100$ measurement intervals. The total background noise count is $\overline{N}_nT = 1$. (After R. Marino, et al., MIT Linc. Lab. Journ., Vol. 13, No. 2, 2002.)

Figure 4-43 shows the detection probabilities given by Eq. (4-220) and (4-221) and the false alarm probabilities given by Eqs. (4-222) and (4-223) for an assumed total noise photoelectron count within the processing window of $\overline{N}_nT =$ 1 with $T = 100$ intervals. It can be seen that when the signal resides in the first interval, improved detection and false alarm probabilities are obtained, especially at high signal photoelectron counts. This is a clear manifestation of the mutual exclusivity of the detection statistics whereby a high detection probability can be thought of as preventing or suppressing the occurrence of false alarms. A similar analysis can be performed in the case of rough targets.

4.9. DETECTION IN ATMOSPHERIC TURBULENCE

We saw in Chapter 3 that laser systems operating within the atmosphere may suffer from a variety of beam-perturbing effects caused by atmospheric turbulence. From a detection point of view, the most dominant effect is scintillation, that is, the random fluctuation in intensity of the received beam. The statistics governing these fluctuations in the weak turbulence limit have been shown to be log-normal in both amplitude and intensity. Here we wish to address the problem of target detection with a direct-detection receiver in the presence of atmospheric turbulence.

4.9.1. Poisson Signal in Turbulence

To simplify the problem, we focus on detection with a point receiver assuming a collimated beam that has propagated over an effective one-way path. By effective is meant that the path could be line of sight, as in the case of a free-space laser communications link, or folded as in the case of a bistatic laser radar operating against an infinite mirror target. These geometries are shown in Figure 4-44. In the latter geometry we assume no correlation of atmospheric inhomogeneities between the outgoing and return paths, an assumption that will be justified later in our comments about monostatic geometries.

Consider then a Poisson-distributed signal that has propagated through a turbulent atmosphere and is immersed in Poisson-distributed noise at the detector. The nonturbulent case was discussed in Section 4.5. Here it is again assumed that the Poisson background is fully incoherent, possibly being solar in origin, such that the atmosphere has negligible effect on its statistics. The discrete density function for the photoelectron count k may then be obtained by averaging the conditional Poisson statistics over all possible realizations of the atmosphere. We can therefore write, using Eq. (4-61),

$$q_{s+n}(k) = \int_{-\infty}^{\infty} q(k \mid \ell) p(\ell)\, d\ell \tag{4-225}$$

where ℓ is the log intensity defined by $I = \langle I \rangle e^{\ell}$ and $I = W/A_d \tau$ (cf. Section 3.4.2.4). From Eq. (4-62), we have

$$q(k \mid \ell) = \frac{\left(\overline{N}_s e^{\ell} + \overline{N}_n\right)^k e^{-(\overline{N}_s e^{\ell} + \overline{N}_n)}}{k!} \tag{4-226}$$

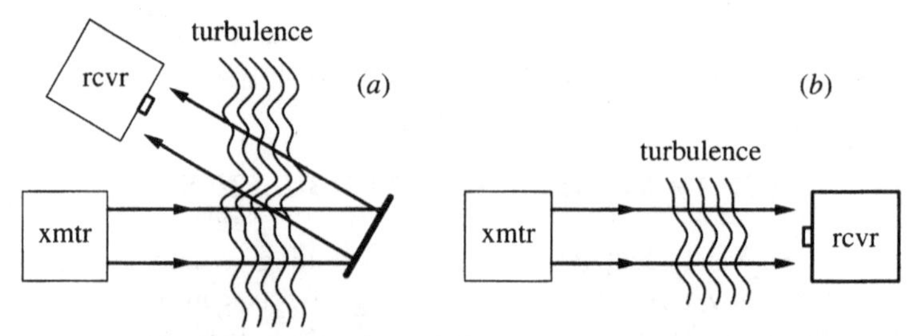

Figure 4-44. Equivalent propagation geometries for direct-detection systems operating in a turbulent atmosphere for (*a*) a bistatic geometry with an infinite mirror target and (*b*) a one-way communications link. In both cases a point receiver is assumed that is smaller than the correlation length ρ_0 of the atmosphere.

where $N_s \equiv \eta W/h\nu$ represents the mean number of signal photoelectrons received during the measurement interval τ. It is assumed that the count interval is short compared to the shortest time scales in the atmospheric fluctuations so that no averaging occurs.

The expression for $p(\ell)$ is simply the log-normal probability density function given by Eq. (3-224), that is,

$$p(\ell) = \frac{1}{\sqrt{2\pi\sigma_\ell^2}} e^{-(\ell+\sigma_\ell^2/2)^2/2\sigma_\ell^2} \tag{4-227}$$

where σ_ℓ^2 is the variance of log intensity.

Inserting Eqs. (4-226) and (4-227) into Eq. (4-225) yields the probability density function for a Poisson signal plus noise in the presence of turbulence. We have

$$q_{s+n}(k) = \frac{1}{k!\sqrt{2\pi\sigma_\ell^2}} \int_{-\infty}^{\infty} (\overline{N}_s e^\ell + \overline{N}_n)^k e^{-(\ell+\sigma_\ell^2/2)^2/2\sigma_\ell^2 - (\overline{N}_s e^\ell + \overline{N}_n)} d\ell \tag{4-228}$$

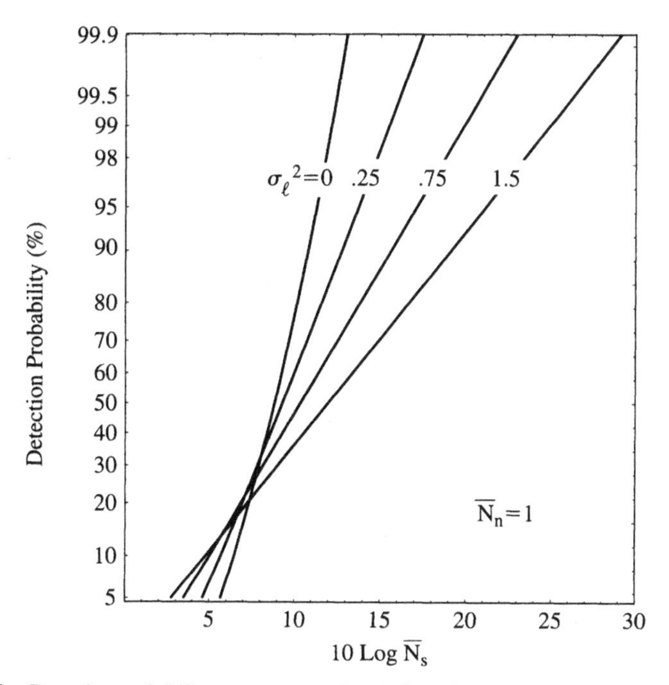

Figure 4-45. Detection probability versus mean signal photoelectron count for a Poisson signal in weak turbulence. The curves represent the performance achievable with the geometries shown in Figure 4-44 with the variance of log intensity σ_ℓ^2 as a parameter. The crossover of the curves at low detection probabilities is due to the finite probability of detecting a peak of the intensity fluctuations.

The false alarm probability is again given by Eq. (4-58), that is,

$$P_{\text{fa}} \geq 1 - \sum_{k=0}^{k_{\text{th}}-1} \frac{\left(\overline{N}_n\right)^k e^{-\overline{N}_n}}{k!} \qquad k_{\text{th}} = \text{integer} \tag{4-229}$$

and the detection probability by

$$P_d = 1 - \sum_{k=0}^{k_{\text{th}}-1} q_{s+n}(k) \tag{4-230}$$

Numerically integrating Eq. (4-228), Eq. (4-230) yields the point detection statistics shown in Figure 4-45. The reason for the crossover of the curves at low detection probabilities is similar to that observed in the case of area detection of rough targets (cf. Section 4.5). The reader may recall that in that case the effect was due to the finite probability that at *low* values of M, the system would detect a peak of the scattered spatial distribution. In the present case, the effect is due to the finite probability that at *high* values of σ_ℓ^2, the receiver may detect a peak of the atmospheric temporal fluctuations.

Comment 4-11. Some insight into the effects that scintillation can have on system performance may be obtained by considering the frequency of occurrence of missed detections. This is best accomplished by transforming Eq. (4-227) to intensity variables, which yields

$$p(I/\langle I\rangle) = \frac{1}{\sqrt{2\pi\sigma_\ell^2}} \frac{\langle I\rangle}{I} e^{-(\ln I/\langle I\rangle + \sigma_\ell^2/2)^2/2\sigma_\ell^2} \tag{4-231}$$

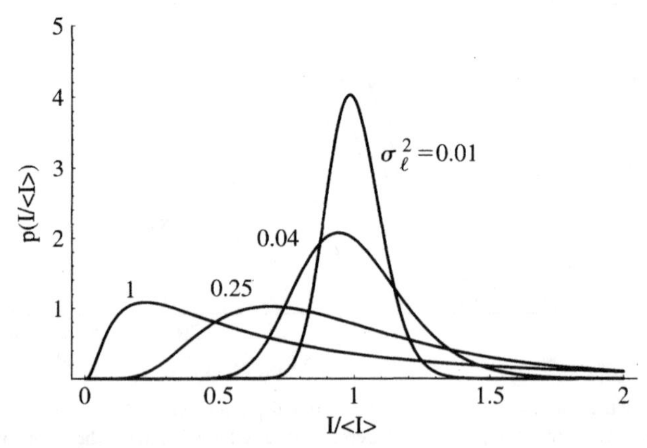

Figure 4-46. Log intensity probability density function for several values of the variance of log intensity.

Equation (4-231) is plotted in Figure 4-46 for several values of σ_ℓ^2. It can be seen that for large values of σ_ℓ^2, the most probable value of intensity (i.e., the mode) is less than the mean value. Temporally, this results in the signal being less than the mean more often than it is greater than the mean. This can be better understood using a semilog plot of the cumulative distribution function for the log intensity. Integrating Eq. (4-231), we have

$$F(I/\langle I\rangle) = \frac{1}{2}\left\{1 + \operatorname{erf}\left[\frac{\sigma_\ell^2/2 + \ln(I/\langle I\rangle)}{\sqrt{2\sigma_\ell^2}}\right]\right\} \tag{4-232}$$

Equation (4-232) is shown in Figure 4-47, where the abscissa represents the fraction of the time that I is less than the ordinate. Note that all curves are straight lines, which is characteristic of a log-normal distribution function. It can be seen that for $\sigma_\ell^2 = 1.5$, $I < \langle I\rangle$ about 75% of the time. Thus, for any nonzero value of σ_ℓ^2, the intensity at any point in the beam has an increased probability of being less than the mean intensity $\langle I\rangle$. This, of course, implies that deep fades may occur which can extend over many measurement intervals and during which the signal fails to cross threshold. On the other hand, since the turbulence-induced fluctuations are assumed to be a nondissipative process, Eq. (4-232) also implies that the occasional fluctuations above the mean must be sufficiently large as to maintain energy conservation. Such high peak excursions above the mean are clearly evident in the simulated log-intensity data of Figure 4-48.

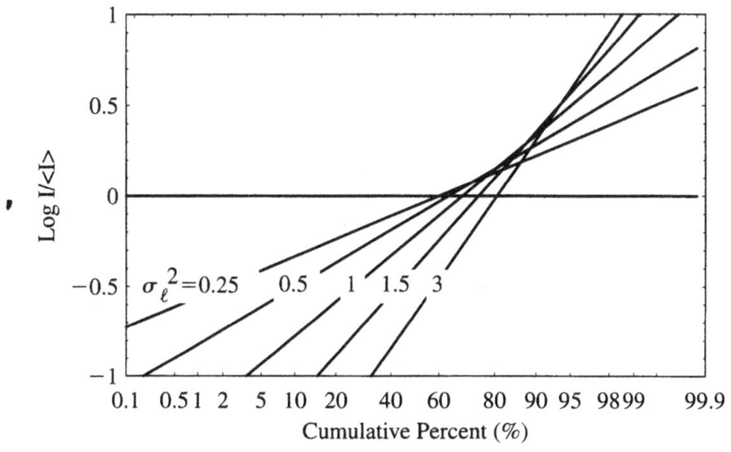

Figure 4-47. Cumulative distribution function for the log-intensity distribution showing the percentage of time in which the log intensity is less than the ordinate. Note that as σ_ℓ^2 increases, the percent of the time that I is less than $\langle I\rangle$ also increases. The fact that the curves are straight lines is characteristic of a log-normal distribution.

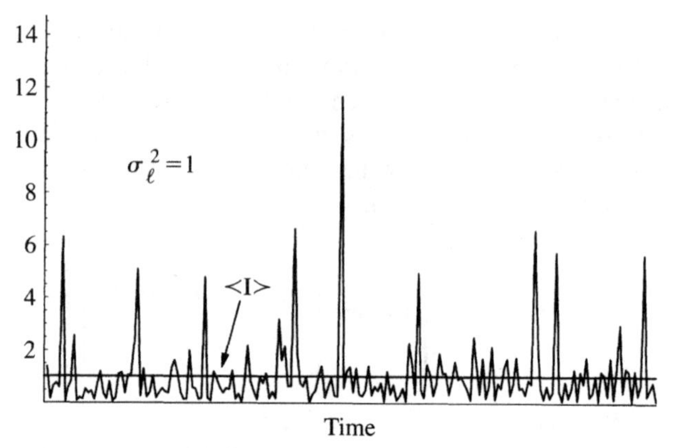

Figure 4-48. A 200-point sample of simulated log-intensity data. Note the occasional "high flyers" that occur due to the long tail of the distribution when σ_ℓ^2 is large.

■

Thus far we have considered the impact of atmospheric scintillation on point receivers (i.e., receivers that are small compared to the transverse correlation length ρ_0 of the turbulence-induced speckle). Consider now the statistics of large apertures, or equivalently, long ranges, either of which can result in the condition $D > \rho_0$, where D is the aperture diameter and ρ_0 is the transverse correlation length of the received field. In this case, the spatial averaging of uncorrelated regions of the signal beam results in a reduction of the variance of intensity in accordance with Eqs. (3-241) and (3-242):

$$\sigma_N^2 = \Theta(e^{\sigma_\ell^2} - 1) \equiv e^{\sigma_\Theta^2} - 1 \tag{4-233}$$

or

$$\sigma_\Theta^2 = \ln[\Theta(e^{\sigma_\ell^2} - 1) + 1] \tag{4-234}$$

where σ_N^2 is the normalized variance of intensity and Θ is the aperture-averaging factor described in Section 3.4.2.5 for plane and spherical waves. Notice that when $\Theta = 1$ there is no spatial averaging and $\sigma_\Theta^2 = \sigma_\ell^2$, and when $\Theta = 0$ there is complete spatial averaging and $\sigma_\Theta^2 = 0$. Hence, for any given σ_ℓ^2, σ_Θ^2 may be obtained using the appropriate Θ factor for the range, aperture diameter, and wavelength under consideration. Replacing σ_ℓ^2 with σ_Θ^2 in Eq. (4-228) yields, as a reasonable approximation in the weak turbulence limit, the discrete density function for a Poisson signal plus noise in the presence of spatially averaged atmospheric turbulence. The detection statistics follow immediately from Figure 4-45 simply by replacing σ_ℓ^2 with σ_Θ^2. Note that log-normal statistics have been assumed here, a consequence of the unusual behavior of the probability density function with regard to the central limit theorem discussed in Chapters 1 and 3. Indeed, it has been shown experimentally[39] that the distribution function remains log normal even for apertures that contain a large numbers of correlation cells.

To extend the bistatic results discussed thus far to the case of a monostatic geometry, account must be taken of the correlation of atmospheric inhomogeneities between the outgoing and return paths. This is usually accomplished by using the extended Huygens–Fresnel formula[40] and the reciprocity theorem of Helmholtz to calculate the mean intensity at the receiver after propagation over a two-way path to the target. The results may be expressed in terms of a normalized mean radial intensity defined as $N(r) = \langle I(r)\rangle/\langle I\rangle$[41], where r is the distance from the optic axis at the receiver and $\langle I\rangle$ is the mean intensity of the total field. It can be shown using the methods above that the normalized mean *on-axis* intensity $N(0) = \langle I(0)\rangle/\langle I\rangle$ can significantly depart from unity depending on target type (i.e., mirror, corner reflector, or diffuse reflector), target range R, and target diameter d_t. Here $I(0)$ is the intensity in the immediate vicinity of the optic axis. When $N(0) > 1$, the effect is referred to as the *backscatter amplification effect*, which at first glance appears to violate energy conservation. However, it must be realized that the effect holds only for strictly backward reflections that reside within a small cone given approximately by ρ_{tr}/R, where ρ_{tr} is a measure of the correlation distance between transmitted and received beams in the aperture plane. Reflected rays lying outside this cone can exhibit $N(0) < 1$ such that energy is conserved overall. Hence the effect averages to unity amplification for large receiver apertures, such as those employed in aperture-averaged direct-detection systems.

For the sake of brevity we will not solve the extended Huygen–Fresnel integrals here since they are quite complex mathematically and somewhat outside the intended scope of discussion, but will instead, use some of the limiting forms of the key equations presented in Reference 41 to demonstrate the application to detection theory. For a laser system with an aperture diameter D, the results are dependent on whether the incident wave at the target is a plane or spherical wave. This dependence may be expressed in terms of the aperture and target Fresnel numbers $\Omega = kD^2/4R$ and $\Omega_t = kd_t^2/4R$, respectively, where $k = 2\pi/\lambda$. For example, the incident wave may be considered a plane wave if $\Omega \gg 1$ (i.e., the target is in the near field of the aperture), or a spherical wave if $\Omega \ll 1$ (i.e., the target is in the far field of the aperture). Similar interpretations may be made for the aperture lying in the far field ($\Omega_t \ll 1$) or near field ($\Omega_t \gg 1$) of the target.

Consider the case of a flat mirror that has either a very small ($\Omega_t \ll 1$) or a very large ($\Omega_t \gg 1$) Fresnel number and lies in the far field of the illuminating beam ($\Omega \ll 1$). In this case, the mean on-axis intensity may be shown to be $N(0) = \exp(\sigma_\ell^2)$, where σ_ℓ^2 is the *spherical wave* variance of log intensity given by the second of Eqs. (3-223). Now it is easy to show using Eqs. (3-224) and (3-226) that if $N(0) = \exp(\sigma_\ell^2)$, then $I(0)/\langle I\rangle = \exp(2\ell)$. Thus in the limit of $D \ll \rho_0$, the detection statistics for a very small or a very large flat mirror target may be obtained in the same manner as has been outlined in this section but with Eq. (4-226) [and therefore Eq. (4-228)] conditioned on knowledge of $\exp(2\ell)$ rather than $\exp(\ell)$. Figure 4-49 shows the resulting detection statistics, where it can be seen that the performance is significantly degraded compared to the bistatic results. Note that if sufficient aperture averaging is employed (i.e., $D \gg \rho_0$), the on-axis amplification is averaged out and the bistatic results of Figure 4-45 apply.

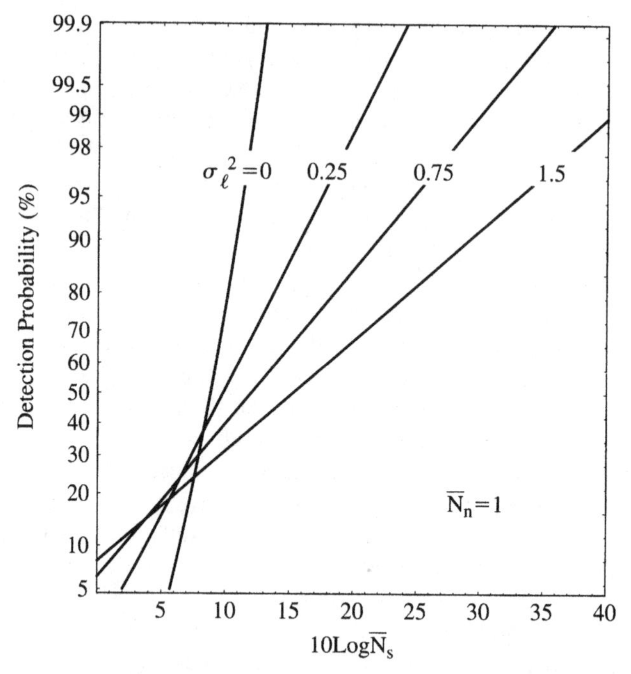

Figure 4-49. Detection probability versus mean signal photoelectron count for a Poisson signal in weak turbulence with a receiver that is small compared to ρ_0. The curves represent the performance of a monostatic laser radar geometry with an incident spherical wave at the target ($\Omega \ll 1$). The target is assumed to be a flat mirror that has either a very small ($\Omega_t \ll 1$) or a very large ($\Omega_t \gg 1$) Fresnel number. Note the scale change compared to Figure 4-45.

4.9.2. Bose–Einstein Signal in Turbulence

In the case of a diffuse target we begin once again with a bistatic geometry and a receiver aperture that is small compared to the transverse correlation length ρ_0 of the atmosphere. Let us also assume a condition of zero background noise and an aperture diameter that is also small compared to the correlation diameter of the target-induced speckle, as indicated in Figure 4-50. The appropriate density function representing the statistics of a single speckle of a negative binomial target has been shown to be the Bose-Einstein probability density function given by Eq. (4-85). In this case it must be conditioned on knowledge of the log-intensity random variable ℓ. We therefore write

$$q_s(k \mid \ell) = \frac{1}{1+\overline{N}_s e^{\ell}} \left(\frac{\overline{N}_s e^{\ell}}{1+\overline{N}_s e^{\ell}} \right)^k \tag{4-235}$$

Inserting Eqs. (4-235) and (4-227) into Eq. (4-225) results in the probability density function for a pure Bose–Einstein signal in a turbulent atmosphere,

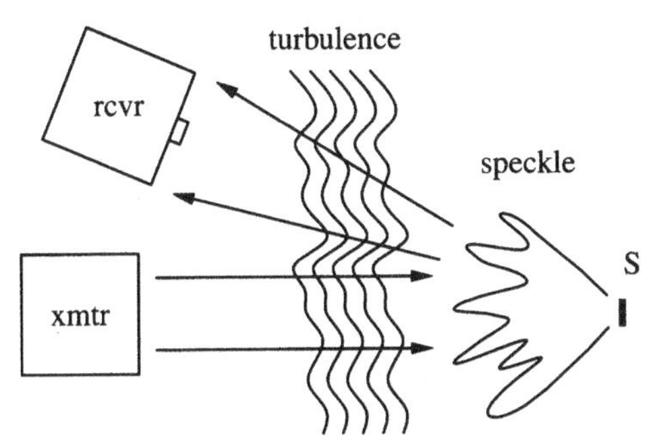

Figure 4-50. Bistatic geometry of a direct-detection laser system operating against a small Lambertian target S while in the presence of a turbulent atmosphere. The receiver aperture is assumed small compared to the correlation length ρ_0 of the atmosphere.

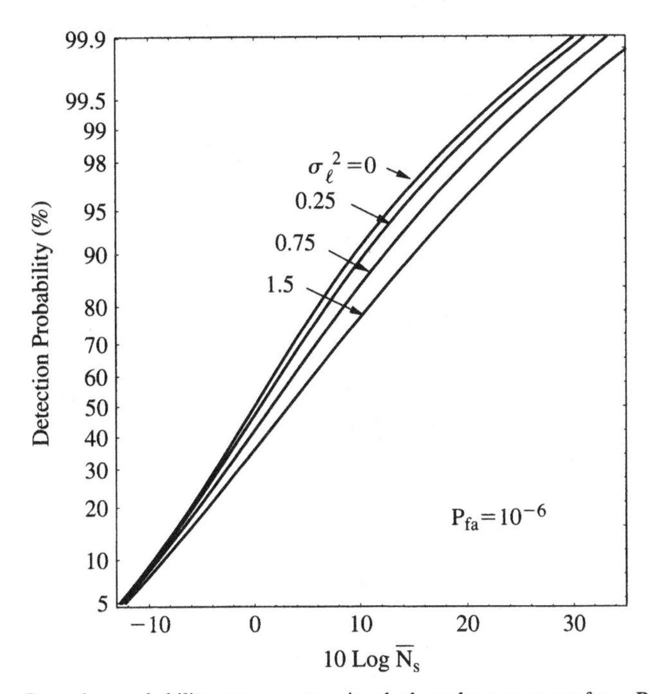

Figure 4-51. Detection probability versus mean signal photoelectron count for a Bose–Einstein signal in weak turbulence. The curves represent the performance achievable with the geometry shown in Figure 4-50 with the variance of log intensity σ_ℓ^2 as a parameter.

that is,

$$q_s(k) = \frac{1}{\sqrt{2\pi\sigma_\ell^2}} \int_{-\infty}^{\infty} \frac{1}{1+\overline{N}_s e^\ell} \left(\frac{\overline{N}_s e^\ell}{1+\overline{N}_s e^\ell} \right)^k e^{-(\ell+\sigma_\ell^2/2)^2/2\sigma_\ell^2} d\ell \qquad (4\text{-}236)$$

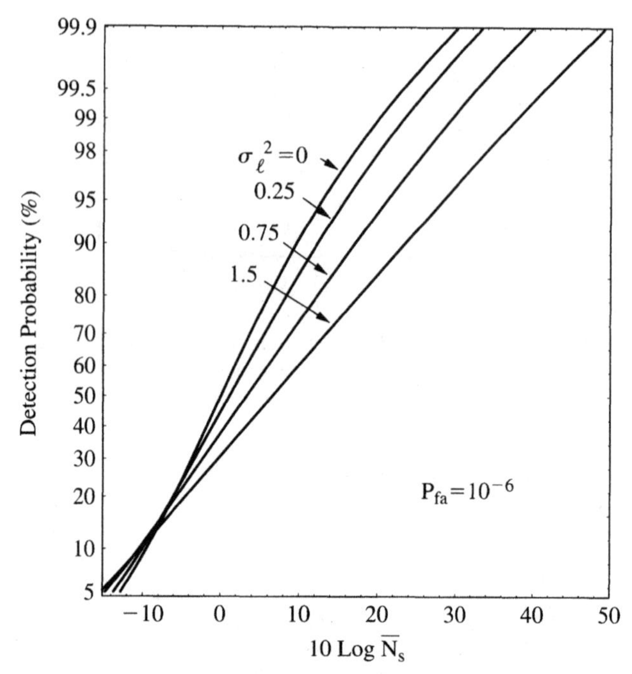

Figure 4-52. Detection probability versus mean signal photoelectron count for a Bose–Einstein signal in weak turbulence. The curves represent the performance achievable with a monostatic geometry and an incident spherical wave at the target ($\Omega \ll 1$). The target is assumed to be Lambertian with a Fresnel number $\Omega_t \ll 1$. Note the scale change compared to Figure 4-51.

Since there are no noise photoelectrons (i.e., $\overline{N}_n = 0$), a threshold photoelectron count of $N_t = 1$ is used. Summing over k as in Eq. (4-230) while numerically integrating Eq. (4-236) yields the detection probabilities shown in Figure 4-51. Note that when $\sigma_\ell^2 = 0$, the resulting curve corresponds to the $M = 1$ curve of Figure 4-14.

In the case of a diffuse target in a monostatic far-field geometry ($\Omega \ll 1$), it can be shown that the backscatter amplification effect is independent of the target size and is given once again by $N(0) = \exp(\sigma_\ell^2)$. However, to obtain the conditions for Bose–Einstein statistics, target dimensions that are small compared to the beam will be assumed. Thus the detection statistics follow in the same manner as discussed in the flat mirror case, that is, by conditioning $q_s(k \mid \ell)$ in Eq. (4-235) on knowledge of $\exp(2\ell)$ rather than $\exp(\ell)$. The resulting detection statistics are shown in Figure 4-52, where once again significant performance degradation is predicted compared to the bistatic case.

4.10. DETECTION IN ATMOSPHERIC CLUTTER

It is frequently the case that laser systems must contend with backscatter from particulates in the medium in which the transmitted beam propagates. That medium may be the atmosphere, the ocean, or some other medium that contains suspended particles. If the particles are on the order of a wavelength or larger in size, a certain fraction of the scattered radiation will be backscattered into the receiver. The effect is most pronounced in monostatic systems and is usually referred to as *optical clutter.* As we have seen, some systems are designed specifically to use *atmospheric backscatter* as the desired signal, as for example in DIAL and wind measurement systems. However, in most other systems it constitutes an undesirable form of clutter that can affect the detection statistics significantly. In Chapter 2 we saw how to account for backscatter in the signal-to-noise ratio. Here we consider it from the point of view of detection and false alarm probabilities.

In Sections 4.4 and 4.5 we developed expressions for the detection statistics associated with fully developed speckle in the presence of Poisson noise. This noise consisted of the random arrival times of the photoemmissive process as well as that due to background solar radiation or dark current. Both of these noise processes are incoherent and therefore follow Poisson statistics. However, in the case of optical clutter due to atmospheric backscatter, the radiation incident on the scattering medium is coherent and therefore produces a *clutter speckle* similar to that generated by a rough surface. This assumes, of course, that the coherence time τ_c of the transmitted waveform is longer than the measurement interval τ over which energy is collected (or, equivalently, that the coherence length $L_c = c\tau_c$ is longer than a range cell in the range processing window). If it is not, the energies from all coherent intervals within a range cell must be summed on an intensity basis.

For the moment, let us assume that the clutter is coherent during the measurement interval and that the signal plus background noise is described by the Kummer distribution. When clutter is present and the signal is absent, the clutter functions as the signal, so that the false alarm statistics are also described by the Kummer distribution. When written in terms of the Kummer distribution, the false alarm probability, may be written (cf. Section 4.5),

$$P_{\mathrm{fa}} \geq 1 - \sum_{k=0}^{k_{\mathrm{th}}-1} \left(\frac{M_C}{M_C + \overline{N}_C} \right)^{M_C} \left(\frac{\overline{N}_C}{M_C + \overline{N}_C} \right)^{k} \frac{e^{-\overline{N}_n}}{\Gamma(M_C)} \frac{\Gamma(k + M_C)}{\Gamma(k+1)} \, {}_1F_1(\alpha, \beta; z) \tag{4-237}$$

where $\overline{N}_C$ and $\overline{N}_n$ are the mean clutter and background photoelectron counts and M_C is the clutter speckle count. In general, Eq. (4-237) implicitly contains an inverse-square range dependence for $\overline{N}_C$ (cf. Section 2.9.4) as well as an inverse range dependence for the clutter speckle count M_C (cf. Section 3.3.3). As discussed in Section 2.9.4, this last effect means that aperture averaging of the clutter speckle remains relatively constant in the far field of the aperture, where the clutter signal is weakest, but improves in the near field, where the clutter signal is strongest.

If the clutter that resides in the range cell of the target is sufficiently low that it can be neglected, the detection probability is clearly given by Eq. (4-107), with

$q_{s+n}(k; \overline{N}_s, \overline{N}_n, M)$ given by Eq. (4-77). However, if the clutter in the range cell of the target cannot be neglected, Eq. (4-77) is still applicable, since, by assumption, both the target and the clutter are within the coherence length of the transmitted waveform. The resultant signal plus clutter speckle is therefore added on an amplitude basis so that the resultant density function must be modified via the replacement $\overline{N}_s \rightarrow \overline{N}_s + \overline{N}_C$.

The detection probability for a negative binomial signal immersed in Poisson noise, with the false alarm probability given by Eq. (4-237), is shown in Figure 4-53 for a relatively low value of clutter (i.e., $\overline{N}_C = 2$), a clutter speckle count of $M_C = 4$, and a background count of $\overline{N}_n = 1$. For simplicity, a hard target has been assumed that fills the beam completely such that the clutter contribution to the signal can be neglected in the expression for the detection probability. (Indeed, in this case there are no clutter contributions to range cells at or beyond the target.) The detection probability for the case of $M = \infty$ is given by Eq. (4-60) using Eq. (4-58). Compared with Figure 4-15, it can be seen that significantly larger signals are required in the presence of clutter to achieve the same detection probability, a consequence of the increased threshold level to

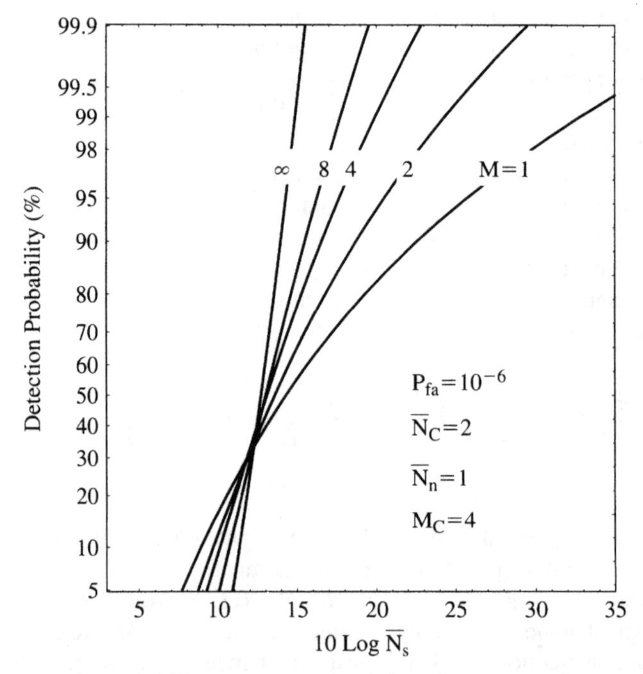

Figure 4-53. Detection probability versus mean signal photoelectron count for a negative binomial signal immersed in Poisson noise and atmospheric clutter (backscatter). The false alarm probability is $P_{fa} = 10^{-6}$, corresponding to a mean photoelectron count for the clutter of $\overline{N}_C = 2$ and an assumed clutter speckle count of $M_C = 4$. The mean photoelectron count for the background radiation is set at $\overline{N}_n = 1$ for both the signal and clutter while the clutter, count is set at $\overline{N}_C = 0$ in calculating the detection probability.

achieve the same false alarm probability. Similar procedures can be applied to other receiver classes, but these are left to the reader.

4.11. POLARIZATION DIVERSITY

Optical receivers that exhibit *polarization diversity* are capable of accepting partially polarized radiation of arbitrary orientation. Partial or fully depolarized signal returns can be generated to varying degrees, depending on the transmitting source characteristics, the degree of target roughness, the incident angle to the surface, and the reflectivity of the individual scatterers. With regard to the last effect, depolarization from multiple scattering events in either a flat surface geometry or a volumetric geometry is suppressed for low-reflectivity scatterers. (For example, a 10% reflectivity would result in only 1% depolarization for two events.) In this section we consider the impact of unpolarized optical signals on the detection statistics of polarization diverse optical receivers.

Calculation of the single-point probability density function for a speckle pattern consisting of partially polarized light is relatively straightforward.[42] With no loss of generality, we assume that the laser transmitter is linearly polarized in the y-direction. For clarity of definitions, we define the state $\langle I_x \rangle \neq \langle I_y \rangle$ as corresponding to a partially polarized field, $\langle I_x \rangle = \langle I_y \rangle = \langle I \rangle / 2$ as a fully depolarized field, and $\langle I_y \rangle = \langle I \rangle$, $\langle I_x \rangle = 0$ as a fully polarized field, where $\langle I \rangle$ is the mean intensity of the total scattered field. Since any partially polarized field can be resolved into uncorrelated orthogonal components through a coordinate rotation, selection of a coordinate system aligned with the illuminating beam polarization automatically uncouples the fields. Hence the depolarized component will appear in the plane orthogonal to the incident plane of polarization, in this case the x-direction, and will be uncorrelated with the scattered field in the original polarization direction. Since the underlying fields for each of the polarization components are uncorrelated circular complex Gaussian random variables, they are independent and the point density functions for the intensities are negative exponential distributions given by [cf. Eq. (4-18)]

$$p_x(I) = \frac{e^{-I/\langle I_x \rangle}}{\langle I_x \rangle} \quad \text{and} \quad p_y(I) = \frac{e^{-I/\langle I_y \rangle}}{\langle I_y \rangle} \tag{4-238}$$

Since I_x and I_y are statistically independent, the statistics of the sum of the two polarizations can be obtained from the characteristic function

$$G_p(iv) = \left[1 - iv\frac{\langle I \rangle}{2}(1 + P)\right]^{-1} \left[1 - iv\frac{\langle I \rangle}{2}(1 - P)\right]^{-1} \tag{4-239}$$

where P is the *degree of polarization* of the optical field, defined by

$$P = \frac{\langle I_x \rangle - \langle I_y \rangle}{\langle I_x \rangle + \langle I_y \rangle} \tag{4-240}$$

and $\langle I_x \rangle = \langle I \rangle(1 + P)/2$, $\langle I_y \rangle = \langle I \rangle(1 - P)/2$, and $\langle I_x \rangle + \langle I_y \rangle = \langle I \rangle$. The integrated intensity is obtained by Fourier-inverting Eq. (4-239). This leads to

$$p(I) = \frac{1}{\langle I \rangle P}\{e^{-[2/(1+P)](I/\langle I \rangle)} - e^{-[2/(1-P)](I/\langle I \rangle)}\} \tag{4-241}$$

If, on the other hand, we take the limit of Eq. (4-241) as $P \to 0$, we obtain

$$p(I) = \frac{4}{\langle I \rangle^2} I e^{-2I/\langle I \rangle} \tag{4-242}$$

Equation (4-242) corresponds to the χ^2 function with four degrees of freedom, $p(x) = (x/4)e^{-x/2}$, under the change of variables $x = 4I/\langle I \rangle$ and, as discussed in Chapter 5, is used in radar theory to represent the statistics of a one-dominant plus Rayleigh target model.

To proceed, we wish to find the probability density function for the integrated intensity corresponding to Eqs. (4-241) and (4-242). Following the procedures used in Section 4.3 for derivation of the gamma distribution, the characteristic function for the partially polarized case becomes

$$G_{px}^{M_x}(iv)G_{py}^{M_y}(iv) = \left(1 - iv\frac{\langle I_x \rangle}{M_x}\right)^{-M_x}\left(1 - iv\frac{\langle I_y \rangle}{M_y}\right)^{-M_y} \tag{4-243}$$

Here, for the sake of generality, the speckle counts in the two orthogonal polarizations are assumed to be different. This is generally the case for arbitrary angles of incidence and viewing directions relative to the planes of incidence and reflection. However, for those geometries in which the degree of polarization P of the scattered wave approaches zero, it is reasonable to expect that the two orthogonal speckle counts would become similar if not identical, especially for a homogeneous surface having uniform roughness in all directions. Assuming that this is the case, the characteristic function for the integrated intensity of an unpolarized field becomes

$$G_p^M(iv) = \left(1 - iv\frac{\langle I_s \rangle}{2M}\right)^{-2M} \tag{4-244}$$

where $M_x = M_y = M$ and we have let $\langle I_x \rangle = \langle I_y \rangle = \langle I_s \rangle/2$.

The inverse transformation of Eq. (4-244) offers a useful bound to the partially polarized problem. The Fourier inversion results in

$$p(I) = \begin{cases} \left(\dfrac{2M}{\langle I_s \rangle}\right)^{2M} \dfrac{I^{2M-1}e^{-2MI/\langle I_s \rangle}}{\Gamma(2M)} & I \geq 0 \\ 0 & I < 0 \end{cases} \tag{4-245}$$

which is again the *gamma distribution* but with M replaced by $2M$. The mean and variance are given by

$$\langle I \rangle = \langle I_s \rangle = \langle I_x \rangle + \langle I_y \rangle \tag{4-246}$$

and

$$\text{var}(I) = \frac{\langle I_s \rangle^2}{2M} \tag{4-247}$$

Equations (4-246) and (4-247) are consistent with our original assumption of equal means for the two polarizations. Notice that the variance is halved due to the additional diversity of the two polarizations. [cf. Eqs. (4-30) and (4-31)]. Thus we see that the integrated intensity of a completely unpolarized, fully developed speckle field can be obtained with the parameter replacement $M \to 2M$ in the fully polarized expression. It therefore follows that the detection statistics associated with fully developed unpolarized speckle can also be obtained via this replacement. For example, in the case of a pair of polarized and unpolarized fields of equal mean total intensity that produce the same signal-to-noise ratio in a polarization diverse direct detection receiver, the unpolarized field will result in better detection performance than will the polarized field. This follows from the fact that unpolarized radiation provides twice the speckle diversity of polarized radiation, thereby enhancing the detection statistics accordingly.

Given the similarity in the derivations of the detection probabilities for the various receiver classes considered thus far, it is safe to say that the above replacement rule can be applied in all cases as long as the fields are completely depolarized. A couple of examples should demonstrate this. In the first example, a partially developed speckle field will be assumed to be fully *depolarized.* In the second example, a partially developed speckle field will consist of a *fully depolarized* diffuse component and a *fully polarized* specular component.

In the first example, both the specular and diffuse components of the speckle field are depolarized, so that we can write for the characteristic function of the two polarizations oriented along the x and y directions, using Eq. (4-115) for each polarization,

$$\begin{aligned} G_p(iv) &= G_{px}(iv)G_{py}(iv) \\ &= \frac{e^{2iv\lambda_x/(1-2iv)}}{1-2iv}\frac{e^{2iv\lambda_y/(1-2iv)}}{1-2iv} \end{aligned} \tag{4-248}$$

where $\lambda_x = I_{cx}/\langle I_x \rangle$ and $\lambda_y = I_{cy}/\langle I_y \rangle$ are the x and y components of the normalized specular intensities, and $\langle I_x \rangle + \langle I_y \rangle = \langle I_s \rangle$ and $\langle I_{cx} \rangle + \langle I_{cy} \rangle = \langle I_c \rangle$. (Since the specular components are adding on an intensity basis, we must assume that they are uncorrelated, as is indeed the case for orthogonally polarized beams.)

In the most general case of a nonhomogeneous surface, the number of speckles M_x and M_y generated for the x and y polarizations will be different. Therefore,

$$G_p^M(iv) = G_{PX}^{M_X}(iv)G_{PY}^{M_Y}(iv)$$

$$= \left[\frac{e^{2iv\lambda_x/(1-2iv)}}{1-2iv} \right]^{M_X} \left[\frac{e^{2iv\lambda_y/(1-2iv)}}{1-2iv} \right]^{M_Y} \tag{4-249}$$

Assume for simplicity that $\langle I_x \rangle = \langle I_y \rangle = \langle I_s \rangle/2$ and $\langle I_{cx} \rangle = \langle I_{cy} \rangle = \langle I_c \rangle/2$. We can therefore let $\lambda_x = \lambda_y \equiv \lambda$, so that for a homogeneous surface, we have

$$G_p^M(iv) = \left[\frac{e^{2iv(\lambda_x+\lambda_y)/(1-2iv)}}{(1-2iv)^2} \right]^M$$
$$= \frac{e^{2iv\lambda(2M)/(1-2iv)}}{(1-2iv)^{2M}} \tag{4-250}$$

But this is just the characteristic function for fully polarized partially developed speckle, Eq. (4-117), with M replaced by $2M$. We can therefore immediately write the inverse transformation in the form of Eq. (4-121) with M replaced by $2M$, that is,

$$p(I) = \begin{cases} \dfrac{2M}{\langle I_s \rangle} \left(\dfrac{I}{I_c} \right)^{(1/2)(2M-1)} e^{-2M(I+I_c)/\langle I_s \rangle} I_{2M-1} \left(\dfrac{4M\sqrt{II_c}}{\langle I_s \rangle} \right) & I \geq 0 \\ 0 & I < 0 \end{cases} \tag{4-251}$$

Here we see that the speckle diversity has been increased by a factor of 2 in the same manner as for depolarized fully developed speckle. From the form of Eq. (4-251) we see that the detection probability may be obtained from the fully polarized expression given by Eq. (4-132) and Figures 4-22 and 4-23 with the replacement $M \to 2M$. In such a case, $\overline{N}_s$ and N_c correspond to the *total* diffuse and specular photoelectron counts, respectively.

Consider now the second case of polarization, which is preserved for the specular component and not for the diffuse component. Assuming complete depolarization of the diffuse components, the two orthogonal components can again be assumed to be statistically independent. Therefore, the characteristic function for the single-point probability density function of the total scattered field becomes the product of the characteristic functions for a fully polarized fully developed speckle field and an orthogonally polarized partially developed speckle field. Using the χ^2 variables of Section 4.6.1, we have

$$G_p(iv) = \frac{1}{1-2iv} \frac{e^{2iv\lambda/(1-2iv)}}{1-2iv} = \frac{e^{2iv\lambda/(1-2iv)}}{(1-2iv)^2} \tag{4-252}$$

where since there is no specular component in the x-direction, $\lambda_y = I_{cy}/\langle I_y \rangle \equiv 2I_c/\langle I_s \rangle \equiv \lambda$ and where once again $\langle I_x \rangle = \langle I_y \rangle = \langle I_s \rangle/2$ for the diffuse components. Assuming also the case of a homogeneous surface, we arrive at the characteristic function for M speckles,

$$G_p^M(iv) = \frac{e^{2iv\lambda M/(1-2iv)}}{(1-2iv)^{2M}} \tag{4-253}$$

We can put this in the same form as Eq. (4-248) by redefining the parameter $\lambda \equiv 2\lambda'$. Then

$$G_p^M(iv) = \frac{e^{2iv\lambda'(2M)/(1-2iv)}}{(1-2iv)^{2M}} \tag{4-254}$$

The inverse transformation follows immediately in the form of Eq. (4-251) but with $\lambda \equiv 2I_c/\langle I_s\rangle$ replaced by $\lambda' \equiv I_c/\langle I_s\rangle$, that is,

$$p(I) = \begin{cases} 2^{M+1/2}\dfrac{M}{\langle I_s\rangle}\left(\dfrac{I}{I_c}\right)^{(1/2)(2M-1)} & \\ \quad \times e^{-M(2I+I_c)/\langle I_s\rangle} I_{2M-1}\left(\dfrac{2\sqrt{2}M\sqrt{II_c}}{\langle I_s\rangle}\right) & I \geq 0 \\ 0 & I < 0 \end{cases} \tag{4-255}$$

Thus we see by comparing Eqs. (4-255) and (4-121) that our general replacement rule $M \to 2M$ does not hold in this case. This is because the specular component was not depolarized and therefore did not add to the diversity of the field.

4.12. MULTIPLE UNCORRELATED SIGNALS

So far we assumed coherent illumination of the target by the source. In that case, various scattered fields generated by different regions of the target are coherent with respect to each other and are therefore summed on an amplitude basis. The question arises as to the resultant detection statistics if the source has a finite coherence volume, that is, one in which the transverse coherence length L_c or coherence time τ_c is small compared to the target dimensions d or measurement interval τ, respectively. In each case, the coherently illuminated regions L_c and times τ_c generate their own speckle pattern, with a mean speckle size determined by the size of the coherently illuminated region on the target. When several of these coherent intervals are integrated, either spatially or temporally, the superposition of the fields at the receiver must be done on an intensity basis rather than an amplitude basis since the fields are uncorrelated.

It has been shown[43] that any arbitrary set of partially correlated intensities may be described at a point in space by the sum of the component intensities, that is,

$$I = \sum_{j=1}^{m} I_j \tag{4-256}$$

If the component intensities are correlated (i.e., $\langle I_j I_k \rangle \neq 0$) and we let the corresponding set of complex fields be given by $A_1, A_2, \ldots, A_m$, where $I_j = A_j^2$ represents a basis vector in amplitude space, a unitary transformation can always be found which transforms this vector to a new vector $A'_1, A'_2, \ldots, A'_m$, where $I'_j = A'^2_j$, such that all $\langle I'_j I'_k \rangle = 0$. This procedure is mathematically equivalent to the rotation of a coordinate system in order to decouple the orthogonal components of a partially polarized beam and ensures that the total intensity remains constant (i.e., $I = I'$). The component amplitudes A'_j are circular complex Gaussian random variables and, since they are uncorrelated, are described at any given point by negative exponential probability density functions. Thus the problem can always be reduced to one of finding the probability density function for the sum of the transformed intensities:

$$I = \sum_{j=1}^{m} I'_j \tag{4-257}$$

Once again we employ the method characteristic functions and write for the corresponding single-point probability density functions of the individual component intensities (while dropping the prime notation),

$$G_j(iv) = (1 - iv\langle I_j \rangle)^{-1} \tag{4-258}$$

At this point it must be decided whether the individual $\langle I_j \rangle$ are distinct or are the same. If they are different, it may be shown that

$$p(I) = \sum_{j=1}^{m} \frac{\langle I_j \rangle^{m-2} e^{-mI/\langle I_j \rangle}}{\prod_{p=1, p\neq j}^{m} (\langle I_j \rangle - \langle I_p \rangle)} \tag{4-259}$$

If they are the same, which is the immediate case of interest, we proceed as before (cf. Section 4.3) and note that for a single illuminating beam, conservation of energy requires that for m uncorrelated regions on the surface, $\langle I_j \rangle \to \langle I \rangle / m$ in the component characteristic functions, that is,

$$G_j(iv) = \left(1 - \frac{iv\langle I \rangle}{m}\right)^{-1} \tag{4-260}$$

Thus the characteristic function for the single-point intensity is

$$\begin{aligned} G_m(iv) &= \prod_{j=1}^{m} \left(1 - \frac{iv\langle I \rangle}{m}\right)^{-j} \\ &= \left(1 - \frac{iv\langle I \rangle}{m}\right)^{-m} \end{aligned} \tag{4-261}$$

From the form of Eq. (4-261), we know that the inverse transformation is given by the gamma distribution. Thus we can immediately write for the single-point probability density function of m uncorrelated speckle fields [cf. Eq. (4-32)],

$$p(W) = \begin{cases} \left(\dfrac{m}{\overline{W}}\right)^m \dfrac{(W)^{m-1} e^{-mW/\overline{W}}}{\Gamma(m)} & W \geq 0 \\ 0 & W < 0 \end{cases} \tag{4-262}$$

where $\overline{W}$ is the mean energy of the total speckle field. To obtain the integrated energy over a finite aperture, assuming M speckles generated at the receiver per uncorrelated region, we return to Eq. (4-261) and write

$$G_m^M(iv) = \left(1 - \frac{iv\langle I\rangle}{mM}\right)^{-mM} \tag{4-263}$$

The Fourier inversion of Eq. (4-263) is also a gamma distribution. Thus

$$p(W_0) = \begin{cases} \left(\dfrac{mM}{\overline{W}}\right)^{mM} \dfrac{(W_0)^{mM-1} e^{-mMW_0/\overline{W}}}{\Gamma(mM)} & W_0 \geq 0 \\ 0 & W_0 < 0 \end{cases} \tag{4-264}$$

where W_0 is the integrated energy and $\overline{W_0} \equiv \overline{W}$. Comparing Eq. (4-264) to Eq. (4-32), it can be seen that the specialized case of the sum of m uncorrelated speckle fields results in an identical expression for the integrated energy as that obtained for a single fully correlated speckle field under the parameter replacements $M \to mM$. As in the case of polarization diversity, which added two degrees of freedom to the speckle diversity, uncorrelated speckle fields are seen to add m degrees of freedom to the speckle diversity. As a consequence, the various detection statistics calculated for a single correlated speckle field are also applicable to multiple uncorrelated fields, given the parameter replacement $M \to mM$. Partially uncorrelated speckle fields may be solved in a similar manner but are considerable more complex and therefore are not discussed here.

REFERENCES

[1]J. W. Goodman, "Some effects of target-induced scintillation on optical radar performance," *Proc. IEEE*, Vol. 53, No. 11, Nov. 1963, p. 1688.

[2]P. Beckmann and A. Spizzichino, *The Scattering of Electromagnetic Waves from Rough Surfaces*, Artech House, Norwood, MA, 1987.

[3]L. Mandel, "Fluctuations of photon beams: the distribution of photoelectrons," *Proc. Phys. Soc. (London)*, Vol. 74, No. 475, 1959, pp. 233–243.

[4]J. W. Goodman, "Laser speckle and related phenomena," in *Topics in Applied Physics*, 2nd ed., Vol. 9, ed. J. C. Dainty, Springer-Verlag, New York, 1984, pp. 9–74.

[5]S. O. Rice, "Mathematical analysis of random noise," in *Selected Papers on Noise and Stochastic Processes*, ed. N. Wax, Dover, New York, 1954, p. 98.

[6]R. Barakat, "First-order probability densities of laser speckle patterns observed through finite-size scanning apertures," *Opt. Acta*, Vol. 20, No. 9, 1973, p. 729.

[7]J. C. Dainty, "Detection of images immersed in speckle noise," *Opt. Acta*, Vol. 18, No. 5, 1971, p. 327.

[8]A. A. Scribot, "First-order probability density functions of speckle measured with a finite apertures," *Opt. Commun.*, Vol. 11, No. 3, 1974, p. 238.

[9]Ibid., Ref. 4, p. 57.

[10]E. M. Purcell, "The question of the correlation between photons in coherent light rays," *Nature*, Vol. 178, Dec. 29, 1956, pp. 1449–1450.

[11]L. Mandel, "Fluctuations of photon beams and their correlations," *Proc. Phys. Soc. (London)*, Vol. 72, No. 1, Pt. 6, 1958, pp. 1037–1048.

[12]J. W. Goodman, *Statistical Optics*, John Wiley & Sons, New York, 1985, p. 95.

[13]Ibid., Ref. 1.

[14]M. Abramowitz and I. A. Stegun, Editors, *Handbook of Mathematical Functions*, Applied Mathematics Series 55, U.S. Department of Commerce, National Bureau of Standards, Washington, DC, June 1964, p. 504.

[15]G. R. Osche, K. N. Seeber, D. S. Young, and F. Lok, "Laser radar cross-section estimation from high resolution image data," *Appl. Opt.*, Vol. 31, No. 14, May 10, 1992, p. 2452.

[16]J. C. Dainty, "Coherent addition of a uniform beam to a speckle pattern," *J. Opt. Soc. Am.*, Vol. 62, Apr. 1972, p. 595.

[17]κ has been given various names by various authors. Abramowitz and Stegun (ref. 14) refer to it as a scattering ratio, whereas Goodman (ref. 4) refers to it as a beam ratio. However, coherence parameter, as suggested by Clifford and Hill (ref. 37, Chapter 3) seems best to convey the related physics and is adopted throughout this book.

[18]G. R. Osche, "Single and multiple-pulse detection statistics associated with partially developed speckle," *Appl. Opt.*, Vol. 39, No. 24, Aug. 20, 2000, p. 4255. (*Errata*: A $j!$ is missing in the denominator of the sum terms in Eqs. 29, 31, 43, and 44, due to a typographical error. Also, Eq. 30 should contain the term ${}_1F_1$, not ${}_2F_1$.)

[19]Ibid., ref. 14, p. 375, Eq. 9.6.10.

[20]T. Li and M. C. Teich, "Bit-error rate for a lightwave communication system incorporating an erbium-doped fibre amplifier," *Electron Letters*, Vol. 27, No. 7, 28 Mar 1991, p. 598.

[21]G. S. Mecherle, "Signal speckle effects on optical detection with additive Gaussian noise," *J. Opt. Soc. Am.*, Vol. 1, No. 1, Jan. 1984, p. 68.

[22]R. J. McIntyre, "The distribution of gains in uniformly multiplying avalanche photodiodes: theory," *IEEE Trans. Electron Devices*, Vol. ED-19, No. 6, June 1972, p. 703.

[23]J. Conradi, "The distribution of gains in uniformly multiplying avalanche photodiodes: experimental," *IEEE Trans. Electron Devices*, Vol. ED-19, No. 6, June 1972, p. 713.

[24]P. P. Webb, R. J. McIntyre, and J. Conradi, "Properties of avalanche photodiodes," *RCA Rev.*, Vol. 35, June 1974, p. 234.

[25]B. E. A Saleh, M. M. Hayat, and M. C. Teich, "Effect of dead space on the excess noise factor and time response of avalanche photodiodes," *IEEE Trans. Electron Devices*, Vol. 37, No. 9, Sept. 1990, p. 1976.

[26]S. A. Plimmer, J. P. R. David, D. S. Ong, and K. F. Li, "A simple model for avalanche multiplication including deadspace effects," *IEEE Trans. Electron Devices*, Vol. 46, No. 4, Apr. 1999, p. 769.

[27]Ibid., ref. 24.

[28]P. Balaban, "Statistical evaluation of the error rate of the fiberguide repeater using importance sampling," *Bell Syst. Tech. J.*, July–Aug. 1976, pp. 745–766.

[29]D. G. Youmans, "Avalanche photodiode detection statistics for direct detection laser radar," *SPIE*, Vol. 1633, Laser Radar VII, 1992.

[30]R. E. Burgess, "Homophase and heterophase fluctuations in semiconducting crystals," *Discuss Faraday Soc.*, Vol. 28, 1959, pp. 151–158.

[31]L. Mandel, "Image fluctuations in cascade intensifiers," *Br. J. Appl. Phys.*, Vol. 10, 1959, p. 233.

[32]M. C. Teich and B. E. Saleh, "Effects of random deletion and additive noise on bunched and antibunched photon-counting statistics," *Opt. Lett.*, Vol. 7, No. 8, Aug. 1982, p. 365.

[33]H. Dautet, P. Deschamps, B. Dion, A. D. MacGregor, D. MacSween, R. J. McIntyre, C. Trottier, and P. Webb, "Photon counting techniques with silicon avalanche photodiodes," *Appl. Opt.*, Vol. 32, No. 21, July 20, 1993, pp. 3894–3900.

[34]A. Lacaita, F. Zappa. S. Cova, and P. Lovati, "Single-photon detection beyond 1 μm: performance of commercially available InGaAs/InP detectors," *Appl. Opt.*, Vol. 35, No. 16, June 1, 1996, pp. 2986–2996.

[35]A. L. Lacaita, P. A. Francese, and S. D. Cova, "Single-photon optical-time-domain reflectometer at 1.3 μm with 5-cm resolution and high sensitivity," *Opt. Lett.*, Vol. 18, No. 13, July 1, 1993, pp. 1110–1112.

[36]S. Cova, M. Ghioni, A. Lacaita, C. Samori, and F. Zappa, "Avalanche photodiodes and quenching circuits for single-photon detection," *Appl. Opt.*, Vol. 35, No. 12, Apr. 20, 1996, pp. 1956–1976.

[37]EG&G data sheet on silicon avalanche photodiodes.

[38]R. M. Marino, M. A. Albota, B. F. Aull, G. G. Fouche, R. M. Heinrichs, D. G. Kocher, J. Mooney, M. O'Brien B. E. Player, B. C. Willard, and J. J. Zayhowski, "A three-dimensional imaging laser radar with Geiger-mode avalanche photodiode arrays," to be published in the MIT Lincoln Laboratory Journal, Vol. 13, No. 2, 2002.

[39]D. L. Fried, G. E. Mevers, and M. P. Keister, Jr., "Measurements of laser-beam scintillation in the atmosphere," *J. Opt. Soc. Am.*, Vol. 57, No. 6, June 1967, p. 787.

[40]J. H. Shapiro, B. A. Capron, and R. C. Harney, "Imaging and target detection with a heterodyne-reception optical radar," *Appl. Opt.*, Vol. 20, No. 19, Oct. 1, 1981, pp. 3292–3313.

[41]V. A. Banakh and V. L. Mironov, *Lidar in a Turbulent Atmosphere*, Artech House, Norwood, MA, 1987.

[42]Ibid., ref. 4.

[43]J. W. Goodman, "Probability density function of the sum of *N* partially correlated speckle patterns," *Opt. Commun.*, Vol. 13, No. 3, Mar. 1975, p. 244.

5

Single-Pulse Coherent Detection Statistics

5.1. INTRODUCTION

The coherent-detection receiver differs from the direct-detection receiver in several important respects. First, it cannot perform spatial averaging of random speckle fields as can the direct-detection receiver of Chapter 4. This is a consequence of the coherent mixing process on the detector, which is constrained by the mixing theorem for maximum heterodyne efficiency. Second, since the coherent receiver can measure frequency and phase as well as amplitude, the transmitted waveform need not be a simple pulse as in the direct-detection case, but can also be a CW-based waveform such as the frequency-modulated and pulse-compression waveforms. This ability to accommodate a multiplicity of waveforms greatly expands the versatility and usefulness of the coherently detected system. As discussed in Section 1.4, there are two fundamentally different configurations for mixing of the signal and local oscillator in a coherent detection system, depending on the degree of correlation established between the two beams. These are the *heterodyne* and *homodyne* configurations. Differences in the correlation properties between the signal and local oscillator for these two configurations lead to different constraints on the available waveforms and associated signal processing such that each configuration offers unique capabilities that are particularly suited for certain applications.

Figure 5-1 shows a generic receiver and signal processor architecture. If the signal frequency is unknown, spectrum analysis must be performed to locate the signal frequency in the processing band and to center that frequency in the narrow passband of the receiver. Despite the very high frequencies of optical radar systems ($\sim 10^{14}$ Hz), the IF frequencies can, in general, be confined to the RF spectrum similar to any coherent radar system. However, this is not always the case if target velocities are high and Doppler frequencies move into the microwave regime. In such cases, additional optical or microwave down-conversion techniques may be required. Ultimately, however, the received signal must pass through a narrowband filter at some IF that is compatible with the bandwidths of available signal processing technologies.

It is assumed below that the receiver functions as a matched filter for the waveform being processed. We know from linear filter theory that it is the bandwidth of the narrowest filter in the receiver chain that determines the noise bandwidth

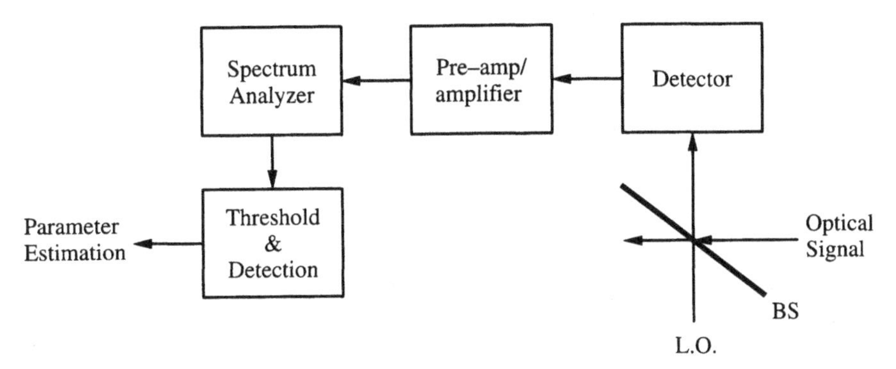

Figure 5-1. Top-level receiver and signal processor block diagram for coherent detection of optical radiation.

in the signal-to-noise ratio (SNR) expression of the receiver and that for shot-noise-limited performance, this is given by SNR $= \eta P_r / h\nu B$, where B is the IF bandwidth. This expression assumes that the local oscillator power on the detector can be made sufficiently high as to dominate all other competing noise sources, such as dark current and thermal noise (cf. Chapter 3). Thresholding is usually performed using an appropriate constant false alarm (CFAR) technique, especially in the presence of clutter. Finally, postdetection processing usually takes the form of parameter estimates in a data processor to extract range, Doppler, and signal amplitude. These fundamental data may then be tagged with image pixel locations in a scanning system in order to construct various types of imagery or detection locations in object space.

The chapter begins by considering the two fundamental target types already considered in the direct-detection theory of Chapter 4, namely, specular and rough targets. In radar terminology these are usually referred to as nonfluctuating and fluctuating targets, respectively. The *nonfluctuating target* is characterized by a constant-amplitude signal with an unknown phase that is uniformly distributed over the interval $0 \to 2\pi$. By *constant amplitude* is meant that the signal does not change when taken over an ensemble of such targets. A retroreflector or a specular surface in a nonturbulent atmosphere and having a nonvarying reflectivity would represent such a target. A raw laser beam directed at a receiver, such as in a free-space communications system, would also constitute a constant-amplitude signal, assuming once again no random perturbations due to the intervening atmosphere.

The *fluctuating target* corresponds to a signal having a random amplitude and phase when taken over an ensemble of targets. The amplitude of the signal is assumed to obey Rayleigh statistics, whereas the phase is assumed to be uniformly distributed over the interval $0 \to 2\pi$. We saw in Chapter 4 that a Rayleigh-distributed signal amplitude (or, equivalently, an exponentially distributed signal intensity) corresponds to the fully developed speckle statistics of a rough target. Independent speckle realizations may be obtained either by rotating the target through angles greater than $\sim\lambda/d$, where d is the spot size on the target, or by translating the beam spot beyond the correlation length of the surface. Rayleigh fluctuation statistics also arise in the case of soft targets such as the atmosphere, either as a result of scanning or temporal decorrelation of the atmosphere. Later in the chapter, these fundamental models are extended to partially developed speckle fields and propagation in turbulent atmospheres.

Finally, it should be pointed out that in the limit of a strong local oscillator signal, the noise properties of the optical heterodyne receiver assume the characteristics of white circular Gaussian statistics that are essentially identical to those of conventional radar receivers. As a consequence, the detection statistics are also identical, so that there is a close correspondence between coherent radar and optical theories.

5.2. CONSTANT-AMPLITUDE SIGNAL IN GAUSSIAN NOISE

The simplest case of coherent detection is that of a single pulse of constant but unknown amplitude and unknown phase. This is a common occurrence for optical systems where target variations on the order of an optical wavelength can randomize the signal phase over the full interval $(0, 2\pi)$. We shall begin by assuming a *narrowband receiver* with an output noise voltage waveform given by

$$f(t) = r(t)\cos[\omega_c t + \theta(t)] \tag{5-1}$$

where $r(t)$ is the time-dependent *envelope*, $\omega_c = 2\pi f_c$ is the center frequency of the narrowband filter, and $\theta(t)$ is the time-dependent phase. To obtain the statistics for the signal envelope, the signal must first be split into in-phase (I) and quadrature (Q) components. The optimum signal processing architecture to accomplish this is shown in Figure 5-2. Equation (5-1) therefore becomes

$$f(t) = x(t)\cos\omega_c t - y(t)\sin\omega_c t \tag{5-2}$$

where $x(t) = r(t)\cos\theta(t)$ and $y(t) = r(t)\sin\theta(t)$ represent zero-mean white Gaussian noise variables that are described by the density functions $p_n(x)$

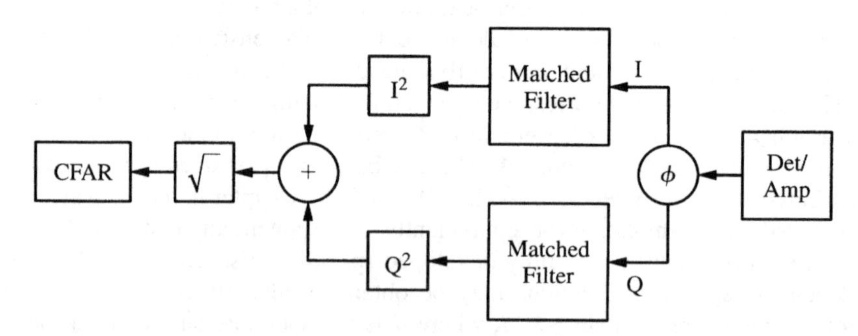

Figure 5-2. Optimal receiver architecture for coherent detection of optical signals.

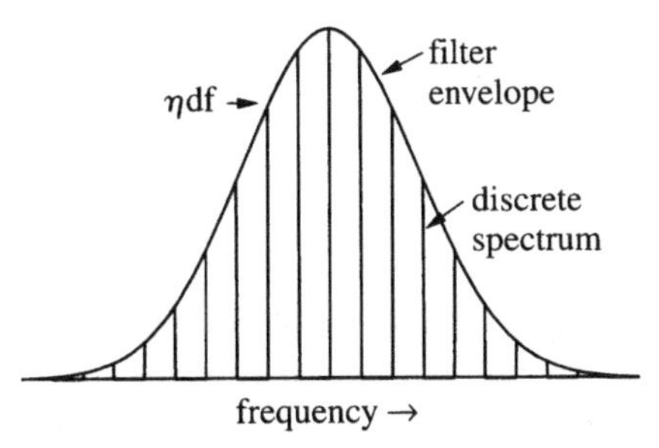

Figure 5-3. Discrete randomly phased sine-wave approximation to the continuous noise spectrum of a narrowband receiver. The amplitude of each frequency component is given by the product of the power spectral density η and the spectral width of each component *df*.

and $p_n(y)$. They are related to the envelope and phase through the relations $r^2(t) = x^2(t) + y^2(t)$ and $\tan\theta(t) = y(t)/x(t)$, respectively.

The statistics of $x(t)$ and $y(t)$ follow from the fact that the continuous noise spectrum of a narrowband receiver may be modeled by a discrete spectrum of n randomly phased sinusoids as shown in Figure 5-3, each associated with a power spectral density η and a bandwidth df. In the limit as $n \to \infty$ and $df \to 0$, the fluctuation statistics of the discrete spectrum approaches that of a zero-mean white Gaussian noise process in accordance with the central limit theorem (cf. Section 1.2.6). Since the center frequency of a narrowband receiver is much greater than the bandwidth (i.e., $\omega_c \gg B$, where $B = 1/2t_p$ is the filter bandwidth matched to the pulse width t_p), the quadrature components may also be considered statistically independent. The joint statistics of the I and Q components of the noise can therefore be written as

$$p_n(x, y) = p_n(x)p_n(y) = \frac{e^{-(x^2+y^2)/2\sigma^2}}{2\pi\sigma^2} \tag{5-3}$$

Here the second moment of $p_n(x)$ or $p_n(y)$ corresponds to the total noise power at the input to the detector and is given by $\langle x^2\rangle = \langle y^2\rangle = \sigma^2$. We wish to find the corresponding statistics of the envelope and phase of the noise as represented by Eq. (5-3). To do this, the random variables x and y need to be transformed to variables that correspond to the envelope and phase of the waveform, namely, r and θ, where $x = r\cos\theta$ and $y = r\sin\theta$. Performing the usual transformation results in

$$p_n(r, \theta) = \frac{re^{-r^2/2\sigma^2}}{2\pi\sigma^2} \tag{5-4}$$

Lacking a precise definition of the distribution function for the phase, we can invoke the minimax criterion outlined in Section 1.3.4 and assume the least favorable distribution, which is a uniform distribution. Thus the marginal probability density function for the envelope alone is obtained by integrating Eq. (5-3) from $0 \to 2\pi$. Hence

$$p_n(r) = \int_0^{2\pi} \frac{re^{-r^2/2\sigma^2}}{2\pi\sigma^2} d\theta$$
$$= \frac{re^{-r^2/2\sigma^2}}{\sigma^2} \quad (5\text{-}5)$$

Equation (5-5) is, once again, the Rayleigh distribution. Its derivation may be recognized as mathematically identical to that which was used in developing the statistics of the amplitude of a fully developed speckle field in Chapter 4. From Comment 1-3, the first two moments and the variance are given by $\langle r \rangle = \sigma\sqrt{\pi/2}$, $\langle r^2 \rangle = 2\sigma^2$, and $\sigma_r^2 = (2 - \pi/2)\sigma^2$, respectively. From a receiver point of view, the mean is to be interpreted as the dc noise voltage, the second moment as the total (dc plus ac) noise power, and the variance as the fluctuation power about the mean. Note also that the *mean noise power* at the output of the envelope detector in the absence of a signal is $\langle r^2/2 \rangle = \sigma^2$.

When a signal is inserted into the receiver, it is added to the noise amplitude as shown in Figure 5-4 and is thereby corrupted by that noise. By corrupted is meant that both the shape and peak amplitude are modified by the additive noise. Therefore, Eqs. (5-1) and (5-2) must be modified to include these effects. This is most conveniently accomplished by adding a constant but unknown signal voltage a_s to the random variable x in Eq. (5-2) such that

$$f(t) = x'(t)\cos\omega_c t - y(t)\sin\omega_c t \quad (5\text{-}6)$$

where $x'(t) = x(t) + a_s$. In terms of x', Eq. (5-3) becomes

$$p_{s+n}(x', y) = \frac{e^{-[(x'-a_s)^2+y^2]/2\sigma^2}}{2\pi\sigma^2} \quad (5\text{-}7)$$

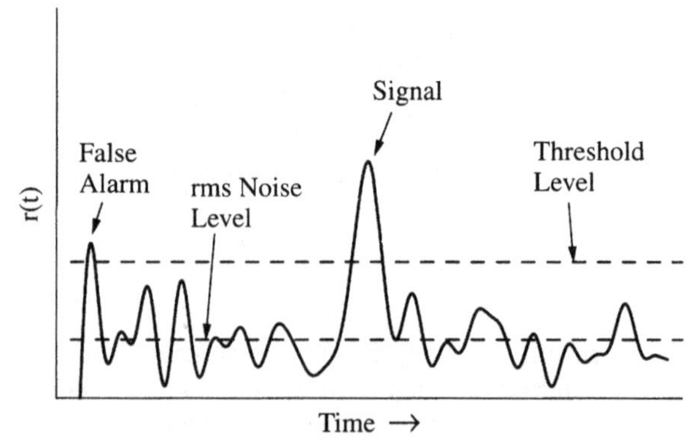

Figure 5-4. Relationship of the rms noise and threshold levels in a sample waveform of a narrow-band receiver.

where the voltage envelope is now given by

$$r(t) = \sqrt{[x(t) + a_s]^2 + y^2(t)} \tag{5-8}$$

and the phase by

$$\theta(t) = \tan^{-1} \frac{y(t)}{x(t) + a_s} \tag{5-9}$$

Changing the variables in Eq. (5-7) to r and θ yields

$$p_{s+n}(r, \theta) = \frac{re^{-[(r^2+a_s^2)-2ra_s \cos\theta]/2\sigma^2}}{2\pi\sigma^2} \tag{5-10}$$

where $x' = r\cos\theta$, $y = r\sin\theta$, and the subscript $s + n$ indicates that the random variable represents the signal plus noise.

Since the phase is still unknown, the probability density function for the voltage envelope is once again obtained by integrating over all θ, resulting in

$$p_{s+n}(r) = \frac{re^{-(r^2+a_s^2)/2\sigma^2}}{\sigma^2} I_0\left(\frac{ra_s}{\sigma^2}\right) \tag{5-11}$$

where

$$I_0\left(\frac{ra_s}{\sigma^2}\right) = \frac{1}{2\pi}\int_0^{2\pi} e^{ra_s \cos\theta/\sigma^2}\, d\theta \tag{5-12}$$

is the modified Bessel function of the first kind, zeroth order. Equation (5-11) is known as the *Rician*[1] *probability density function* and is shown in Figure 5-5 for small, intermediate, and large signals. The Rician distribution represents the *marginal* probability density function, hence the voltage envelope, of a constant

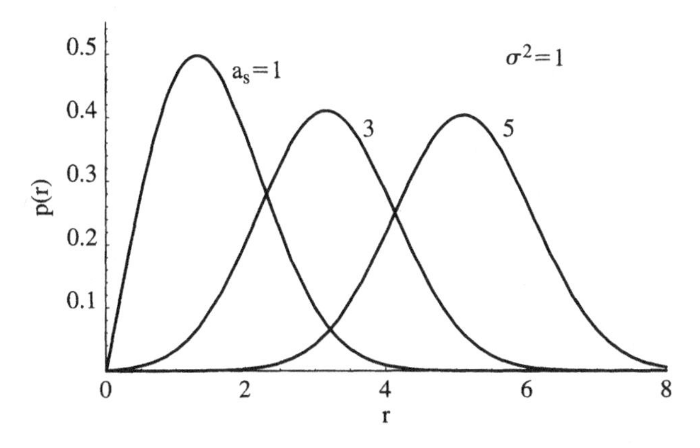

Figure 5-5. Rician probability density function.

signal of unknown phase embedded in the Gaussian noise of a narrowband receiver.

Two limiting cases of Eq. (5-11) are worth mentioning. The first is for small signals (i.e., $a_s \to 0$), in which case $I_0(x) \to 1$ and we obtain the Rayleigh distribution, as expected. The second is for large signals for which we make use of the approximation

$$I_0(z) \approx \frac{e^z}{\sqrt{2\pi z}} \tag{5-13}$$

and write $p_{s+n}(r)$ as

$$p_{s+n}(r) \approx \frac{r}{\sqrt{2\pi r a_s \sigma^2}} e^{-(r-a_s)^2/2\sigma^2} \tag{5-14}$$

Since most of the contribution to $p_{s+n}(r)$ is in the vicinity of $r = a_s$, a consequence of the exponential, we can let $r = a_s$ and obtain

$$p_{s+n}(r) \approx \frac{1}{\sqrt{2\pi \sigma^2}} e^{-(r-a_s)^2/2\sigma^2} \tag{5-15}$$

which is a Gaussian distribution. These limiting distributions are clearly evident in Figure 5-5. From Eq. (5-15) it is easy to see that the mean of the Rician distribution approaches a_s when $a_s^2/2\sigma^2 \gg 1$, while the variance approaches σ^2.

The first moment of the Rician distribution is relatively complex and therefore not very useful. On the other hand, the second moment is relatively simple and is given by $\langle r^2 \rangle = 2\sigma^2 + a_s^2$. This may be rearranged to yield the mean power out of the envelope detector given by $\langle r^2/2 \rangle = \sigma^2 + a_s^2/2$. It is easy to show that the term $a_s^2/2$ corresponds to the *mean signal power* of a constant-amplitude harmonic signal of the form $s = a_s \cos(\omega t + \phi)$ by taking the time average of s^2.

The probability of detecting a target in the presence of receiver noise can be calculated using the Neyman–Pearson criterion of Chapter 1. In this case the a priori probability density function associated with the hypothesis H_0 that noise alone is present is given by Eq. (5-5), while the hypothesis H_1 that signal plus noise is present is given by Eq. (5-11). Thus the hypothesis H_1 is chosen if $r \geq t$ and H_0 otherwise. Therefore,

$$P_d = \int_t^\infty p_{s+n}(r; a_s)\, dr \tag{5-16}$$

and

$$P_{\text{fa}} = \int_t^\infty p_n(r)\, dr \tag{5-17}$$

Performing the integration in the latter results in

$$P_{\text{fa}} = e^{-t^2/2\sigma^2} \tag{5-18}$$

from which the threshold level t may be obtained, given a desired false alarm probability P_{fa}. Inserting Eq. (5-11) into Eq. (5-16) and integrating yields the detection probability,

$$P_d = \int_t^\infty \frac{re^{-(r^2+a_s^2)/2\sigma^2}}{\sigma^2} I_0\left(\frac{ra_s}{\sigma^2}\right) dr$$

$$= Q\left(\frac{a_s}{\sigma}, \frac{t}{\sigma}\right) \tag{5-19}$$

where

$$Q(\alpha, \beta) = \int_\beta^\infty ze^{-(z^2+\alpha^2)/2} I_0(\alpha z)\, dz \tag{5-20}$$

is the well-known *Marcum Q-function.*[2] Equation (5-19) can only be solved numerically or by using Marcum's table of Q-functions.[3] In the former case it is convenient to put Eq. (5-19) in the form

$$P_d = 1 - \int_0^t \frac{re^{-(r^2+a_s^2)/2\sigma^2}}{\sigma^2} I_0\left(\frac{ra_s}{\sigma^2}\right) dr \tag{5-21}$$

which restricts the integration limits to finite values.

Advantage can be taken of the Gaussian approximation at high signal levels to obtain an approximate closed-form expression having a more intuitive form. Substituting Eq. (5-15) into Eq. (5-16) yields

$$P_d \approx \frac{1}{\sqrt{2\pi\sigma^2}} \int_t^\infty e^{-(r-a_s)^2/2\sigma^2} dr \tag{5-22}$$

Letting $y = (r - a_s)/\sqrt{2\sigma^2}$, Eq. (5-22) becomes

$$P_d \approx \int_{t'}^\infty e^{-y^2} dy \tag{5-23}$$

where $t' = (t - a_s)/\sqrt{2}\,\sigma$. Thus

$$P_d \approx \frac{1}{2}\left[1 + \text{erf}\left(\frac{a_s}{\sqrt{2}\,\sigma} - \frac{t}{\sqrt{2}\,\sigma}\right)\right] \tag{5-24}$$

It should be kept in mind that Eq. (5-24) is valid only for $t/\sigma \gg 1$ and $a_s/\sigma \gg 1$.

The reader may have noted that the previous derivations could just as well have been done in terms of power variables. In that case we treat a_s as a constant and let $X = r^2$, $S = a_s^2$, and $N = \sigma^2$. Thus Eq. (5-11) becomes

$$p_{s+n}(X) = \begin{cases} \dfrac{1}{2N} e^{-(X+S)/2N} I_0\left(\dfrac{\sqrt{XS}}{N}\right) & X \geq 0 \\ 0 & X < 0 \end{cases} \tag{5-25}$$

where S and N are the *peak signal* and the *rms noise* powers at the input to the detector, respectively. In coherent radar terminology, Eq. (5-25) is referred to as the *power form* of the Rician distribution. In optical speckle theories, it is referred to as the *modified Rician distribution*, which was introduced in Eq. (4-110) for a partially developed speckle field.

To be consistent with the convention introduced in Chapter 4, detection probabilities will be expressed in terms of mean signal-to-noise power ratios. Thus, since the mean signal power of a constant-amplitude harmonic signal at the input to an envelope detector is given by $\overline{S} = a_s^2/2$ and the mean noise power by $N = \sigma^2$, the mean threshold-to-noise and signal-to-noise power ratios at the input to the detector become $\text{TNR} = \overline{T}/N = t^2/2\sigma^2$ and $\text{SNR} = \overline{S}/N = a_s^2/2\sigma^2$, respectively. The detection statistics corresponding to Eqs. (5-18) and (5-19) are therefore

$$P_{\text{fa}} = e^{-\text{TNR}} \tag{5-26}$$

$$P_d = 1 - \frac{1}{2N}\int_0^T e^{-(X/2N)-\text{SNR}} I_0\left(\sqrt{\frac{2X\ \text{SNR}}{N}}\right) dX$$

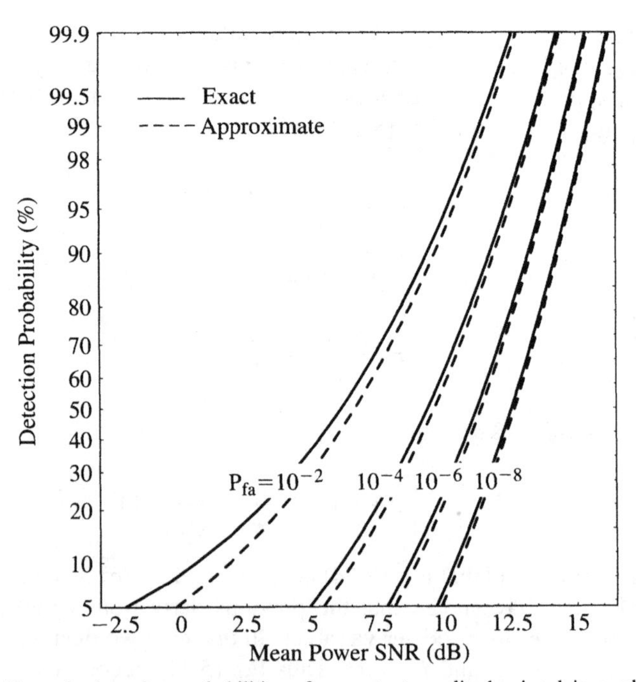

Figure 5-6. Example detection probabilities of a constant-amplitude signal in a white Gaussian noise environment plotted as a function of the mean signal-to-noise power ratio, SNR(dB), and several values of false alarm probability P_{fa}. Dashed curves represent the large-signal Gaussian approximation to the Rician distribution.

$$= 1 - \int_0^{T/2N} e^{-X'-\mathrm{SNR}} I_0(2\sqrt{X'\,\mathrm{SNR}})\,dX' \tag{5-27}$$

where $X' = X/2N$.

In the high signal limit, the power form of Eq. (5-24) assumes the particularly simple form

$$P_d \approx \tfrac{1}{2}[1 + \mathrm{erf}(\sqrt{\mathrm{SNR}} - \sqrt{\mathrm{TNR}})] \tag{5-28}$$

It should be noted that the error function is an odd function, so that the order of the terms in its argument is important.

Figure 5-6 shows an example of the numerically integrated detection statistics generated by Eq. (5-27) plotted as a function of the mean input power SNR for several values of false alarm probability, P_{fa}. Superimposed in the figure are plots of Eq. (5-28) (dashed curves) shown for comparison where Eq. (5-18) in the form $\mathrm{TNR} = -\ln P_{\mathrm{fa}}$ is used to calculate the threshold-to-noise ratio. Notice that excellent agreement is obtained for high signals and high detection probabilities.

Comment 5-1. It is frequently the case that the output signal and noise properties of a narrowband receiver are measured in terms of voltage rather than power, such as with a linear spectrum analyzer. In such cases, care must be exercised in calculating the power signal-to-noise ratio from measured voltage signal-to-noise ratio data since their relationship is not always obvious. Let us denote the *mean power signal-to-noise ratio* out of an envelope detector as SNR_{po} and the *mean voltage signal-to-noise ratio* as SNR_{vo}. The former can be defined as the mean signal-plus-noise power divided by the noise variance in the absence of a signal [i.e., $\mathrm{SNR}_{po} = \langle r^2/2\rangle/\mathrm{var}(r)$]. On the other hand, the latter can be defined as the mean signal-plus-noise voltage out of the detector divided by the square root of the variance, or standard deviation, in the absence of a signal [i.e., $\mathrm{SNR}_{vo} = \langle r\rangle/\sqrt{\mathrm{var}(r)}$]. Thus, using the Gaussian high signal limit for the Rician probability density function as a simple example [i.e., Eq. (5-15)], the power signal-to-noise ratio out of the detector becomes $\mathrm{SNR}_{po} = 1 + a_s^2/\sigma^2 \approx a_s^2/\sigma^2$ and the voltage signal-to-noise ratio becomes $\mathrm{SNR}_{vo} = (a_s/\sigma)/(2 - \pi/2)^{1/2}$. Hence $\mathrm{SNR}_{po} = (2 - \pi/2)\mathrm{SNR}_{vo}^2$ and $\mathrm{SNR}_{po} \neq \mathrm{SNR}_{vo}^2$, as is frequently assumed. This inequality is due primarily to the fact that receiver noise has been added to the signal at the output. At the input to the envelope detector, the mean power signal-to-noise ratio is given by $\mathrm{SNR}_{pi} = a_s^2/2\sigma^2$, while the mean voltage signal-to-noise ratio is $\mathrm{SNR}_{vi} = a_s/\sqrt{2}\,\sigma$. Thus, at the input, $\mathrm{SNR}_{pi} = \mathrm{SNR}_{vi}^2$ since noise has yet to be added to the signal. In the following sections, the subscripts o and i are dropped, unless specifically required for clarity, and the terms SNR and SNR_v will be used to denote the mean power and voltage signal-to-noise ratios at the *input* to the receiver, respectively. ■

5.3. RAYLEIGH FLUCTUATING SIGNAL IN GAUSSIAN NOISE

The previous case of a constant-amplitude signal will now be generalized to the case of a fluctuating signal. In this case the amplitude of the signal is also a random variable and is described by Rayleigh statistics. In Section 4.2 it was shown that the Rayleigh distribution describes the single-point statistics of a fully developed speckle field, which in the intensity domain becomes negative exponential. It can also be shown that the negative exponential distribution corresponds to a χ^2-distribution with two degrees of freedom. It might be recalled that the Rayleigh fluctuating target is one in which there are many small randomly phased scatterers of similar cross section contributing to the total radiation field. Optically, this corresponds to any surface that has an rms height roughness which is on the order of a wavelength or larger. Hence a flat paint is considered rough in the visible portion of the spectrum, but a glossy paint is not.

To find the probability density function for the signal plus noise of a Rayleigh fluctuating signal in Gaussian noise, Eq. (5-11) must be integrated over all possible realizations of the target signal:

$$p_{s+n}(r) = \int_0^\infty p(r \mid a_s)p(a_s)\,da_s \tag{5-29}$$

where

$$p(r \mid a_s) = \frac{re^{-(r^2+a_s^2)/2\sigma^2}}{\sigma^2} I_0\left(\frac{ra_s}{\sigma^2}\right) \tag{5-30}$$

and

$$p(a_s) = \frac{a_s}{a_o^2} e^{-a_s^2/2a_o^2} \tag{5-31}$$

Here we have indicated that a_s is a random variable by letting $p(r; a_s) \to p(r \mid a_s)$ in Eq. (5-30). Since a_s is a random variable, we define the mean signal power as $a_o^2 = \langle a_s^2/2 \rangle$, where the brackets indicate averaging over all possible realizations of the target signal and the factor of 2 indicates a harmonic signal. Setting $dp(a_s)\,da_s = 0$ and solving for a_s yields a_o for the most probable value of the signal.

Comment 5-2. As indicated in Chapter 4, the various target models follow from the gamma distribution, which also corresponds to the χ^2-distribution with $2n$ degrees of freedom.[4] For example, starting with the gamma distribution, $a^n x^{n-1} \exp(-ax)/\Gamma(n)$, and letting $n = 1$, we have $a\exp(-ax)$, which is the negative exponential distribution. Letting $n = 2$ results in $a^2 x \exp(-ax)$, which is the one-dominant-plus-Rayleigh distribution introduced by Swerling (cf. Section 5.4). Higher-order target models follow accordingly. ■

Inserting Eqs. (5-30) and (5-31) into Eq. (5-29) and rearranging yields

$$p_{s+n}(r) = \frac{re^{-r^2/2\sigma^2}}{\sigma^2 a_o^2} \int_0^\infty a_s e^{-(a_s^2/2)(1/\sigma^2+1/a_o^2)} I_0\left(\frac{ra_s}{\sigma^2}\right) da_s \tag{5-32}$$

Changing the variable of integration to $y = a_s/\sigma$ gives

$$p_{s+n}(r) = \frac{re^{-r^2/2\sigma^2}}{a_o^2} \int_0^\infty ye^{-y^2/2a_o^2(\sigma^2+a_o^2)} I_0\left(\frac{ry}{\sigma}\right) dy \tag{5-33}$$

The integral in Eq. (5-33) is a tabulated integral[5] given by

$$\int_0^\infty e^{-pt} I_0(2\sqrt{\alpha t})\, dt = p^{-1} e^{\alpha/p} \qquad \text{Re } p > 0 \tag{5-34}$$

Letting $t = y^2$, $p = (\sigma^2 + a_o^2)/2a_o^2$, and $\alpha = r^2/4\sigma^2$, the integral in Eq. (5-33) evaluates to

$$\int_0^\infty f(y)\, dy = \frac{a_o^2}{\sigma^2 + a_o^2} e^{r^2 a_o^2/2\sigma^2(\sigma^2+a_o^2)} \tag{5-35}$$

We therefore obtain after some minor algebra

$$\begin{aligned} p_{s+n}(r) &= \frac{re^{-r^2/2\sigma^2}}{a_o^2} \frac{a_o^2}{\sigma^2 + a_o^2} e^{r^2 a_o^2/2\sigma^2(\sigma^2+a_o^2)} \\ &= \frac{r}{\sigma^2 + a_o^2} e^{-r^2/2(\sigma^2+a_o^2)} \end{aligned} \tag{5-36}$$

This is an interesting result since it demonstrates the additive nature of Gaussian processes. It is an expression of the fact that a receiver with a Rayleigh-distributed noise amplitude, when convolved with a target signal that is also Rayleigh distributed, results in a Rayleigh distribution for the signal plus noise. We see this again in the case of the Rician target model.

The detection probability follows in the usual manner,

$$\begin{aligned} P_d &= \int_t^\infty \frac{r}{\sigma^2 + a_o^2} e^{-r^2/2(\sigma^2+a_o^2)} dr \\ &= e^{-t^2/2(\sigma^2+a_o^2)} \end{aligned} \tag{5-37}$$

However, we know from Eq. (5-18) that $t = \sigma\sqrt{2 \ln P_{\text{fa}}^{-1}}$, which leads to the useful result that the detection probability can simply be expressed in terms of the false alarm probability and the mean input power signal-to-noise ratio, that is

$$\begin{aligned} P_d &= e^{\ln P_{\text{fa}}/(1+a_o^2/\sigma^2)} \\ &= P_{\text{fa}}^{1/(1+\text{SNR})} \end{aligned} \tag{5-38}$$

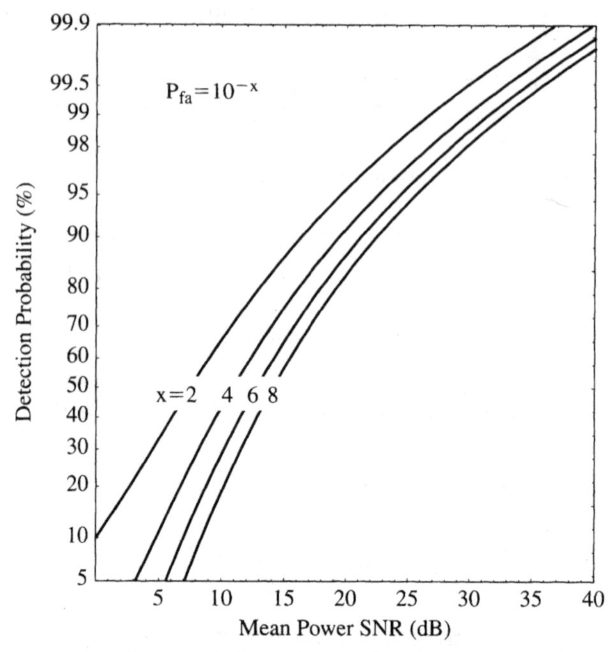

Figure 5-7. Example detection probabilities of a Rayleigh fluctuating signal detected in a white Gaussian noise environment plotted as a function of the mean signal-to-noise power ratio, SNR(dB), and for several values of false alarm probability P_{fa}.

Here we have used the fact that the mean signal power at the input to the receiver is, from Eq. (5-31), $\overline{S} = \langle a_s^2/2 \rangle = a_o^2$ and the total (ac + dc) mean noise power is $N = \langle r^2/2 \rangle = \sigma^2$ such that the mean signal-to-noise power ratio is $\text{SNR} = \overline{S}/N = a_o^2/\sigma^2$.

Equation (5-38) is a very simple and useful expression that yields exact results for single-pulse detection probabilities, given a desired false alarm probability and a mean SNR at the input to the receiver. An example of the detection statistics associated with Eq. (5-38) is shown in Figure 5-7 for several values of P_{fa}.

Comment 5-3. Recalling the discussion in Comment 5-1, it is straightforward to show that $\text{SNR}_{po} \neq \text{SNR}_{vo}^2$ is also true for the case of a *Rayleigh fluctuating signal* in a narrowband receiver. With $\langle r \rangle$ and $\langle r^2 \rangle$ representing the first two moments of Eq. (5-36) and var(r) representing the noise variance in the absence of a signal calculated from Eq. (5-5), we have

$$\text{SNR}_{po} = \frac{\langle r^2/2 \rangle}{\text{var}(r)} = \frac{1 + a_o^2/\sigma^2}{2 - \pi/2} \tag{5-39}$$

and

$$\text{SNR}_{vo} = \frac{\langle r \rangle}{\sqrt{\text{var}(r)}} = \frac{\sqrt{1 + a_o^2/\sigma^2}}{4/\pi - 1} \tag{5-40}$$

which results in

$$\mathrm{SNR}_{po} = \frac{2}{\pi}\mathrm{SNR}_{vo}^2 \tag{5-41}$$

It is important to note that such comparisons are very dependent on the definitions of the noise term in the denominator of SNR_{po}. For example, one might choose to let the noise power be given by $\langle r^2/2\rangle = \sigma^2$ rather than the variance, which would lead to

$$\mathrm{SNR}_{po} = \left(\frac{4}{\pi} - 1\right)\mathrm{SNR}_{vo}^2 \tag{5-42}$$

The difference between Eqs. (5-41) and (5-42) depends on whether only the ac or fluctuating noise power is included in the definition of SNR_{po}, as in the former case, or whether the total ac plus dc noise power is included, as in the latter case [cf. discussion following Eq. (5-5)]. Thus, failing to consider either Eq. (5-41) or Eq. (5-42) in calculations of the power signal-to-noise ratio from measured voltage data using a linear spectrum analyzer[6] can lead to overestimates of 2 dB or 5.63 dB, respectively, depending on the noise definition employed. In either case, we find that $\mathrm{SNR}_{po} < \mathrm{SNR}_{vo}^2$. ■

5.4. ONE-DOMINANT-PLUS-RAYLEIGH SIGNAL IN GAUSSIAN NOISE

The one-dominant-plus-Rayleigh probability density function was proposed by Swerling to account for the fluctuating radar signals observed with aircraft targets. Measurements showed that aircraft radar cross sections tended to consist of one or more large-amplitude components together with many low-level Rayleigh fluctuating components. The proposed signature model was that of a χ^2-distribution with four degrees of freedom. In the power domain, the probability density function for the signal is written

$$p(S) = \frac{4S}{\langle S\rangle^2} e^{-2S/\langle S\rangle} \tag{5-43}$$

and in the amplitude domain

$$p(a_s) = \frac{9a_s^3}{2a_o^4} e^{-3a_s^2/2a_o^2} \tag{5-44}$$

where $S = a_s^2$ and $\langle S\rangle = \langle a_s^2\rangle = 4a_o^2/3$. The probability density functions given by Eqs. (5-43) and (5-44) have been generally accepted in the radar community

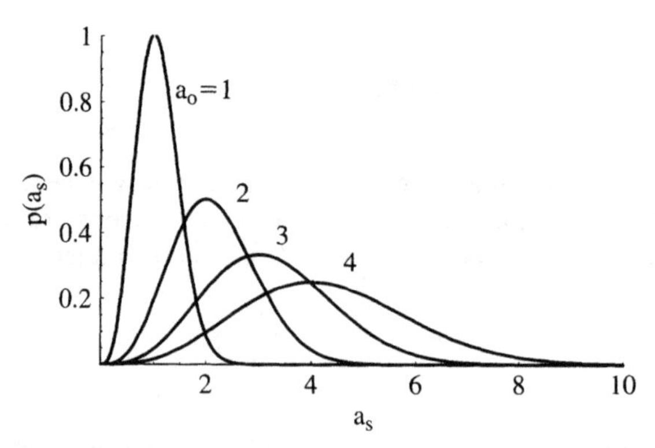

Figure 5-8. Probability density function for the signal of a one-dominant-plus-Rayleigh target model.

despite the fact that they do not represent mathematically the intended model (cf. Section 5.5 for a discussion of the latter point).

Equation (5-44) represents a one-dominant-plus-Rayleigh signal voltage with a most probable value of a_o, obtained by solving for a_s in $dp(a_s)/da_s = 0$, and a mean signal power $\langle a_s^2/2\rangle = 2a_o^2/3$. Figure 5-8 shows a plot of Eq. (5-44) for several values of the parameter a_o. The probability density function can be obtained once again by inserting Eqs. (5-44) and (5-30) into Eq. (5-29) and integrating. Thus

$$p(r) = \frac{9re^{-r^2/2\sigma^2}}{2\sigma^2 a_o^4} \int_0^\infty a_s^3 e^{-a_s^2(1/\sigma^2+3/a_o^2)/2} I_0\left(\frac{ra_s}{\sigma^2}\right) da_s \tag{5-45}$$

The integral can be evaluated using the identity[7]

$$\int_0^\infty t^3 e^{-\alpha t^2} I_0(\beta t)\, dt = \frac{1}{2\alpha^2}\left(1+\frac{\beta^2}{4\alpha}\right) e^{\beta^2/4\alpha} \tag{5-46}$$

Letting $p = [(a_o^2+3\sigma^2)/(2\sigma^2 a_o^2)]^{1/2}$ and $\alpha = r/\sigma^2$, we obtain

$$p(r) = \frac{9r}{(a_o^2+3\sigma^2)^2}\left[\sigma^2 + \frac{r^2a_o^2}{2(a_o^2+3\sigma^2)}\right] e^{-3r^2/2(a_o^2+3\sigma^2)} \tag{5-47}$$

Equation (5-47) is shown graphically in Figure 5-9 for $\sigma = 1$ and several values of a_o.

The detection probability for the one-dominant-plus-Rayleigh signal becomes

$$P_d = \int_t^\infty p(r)\, dr$$

$$= \frac{9}{(a_o^2+3\sigma^2)^2}\int_t^\infty r\left[\sigma^2 + \frac{r^2a_o^2}{2(a_o^2+3\sigma^2)}\right] e^{-3r^2/2(a_o^2+3\sigma^2)} dr \tag{5-48}$$

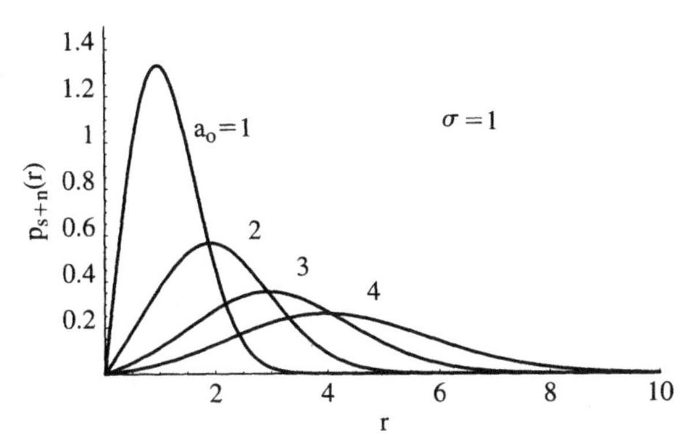

Figure 5-9. Probability density function for the signal plus noise of a one-dominant-plus-Rayleigh target model.

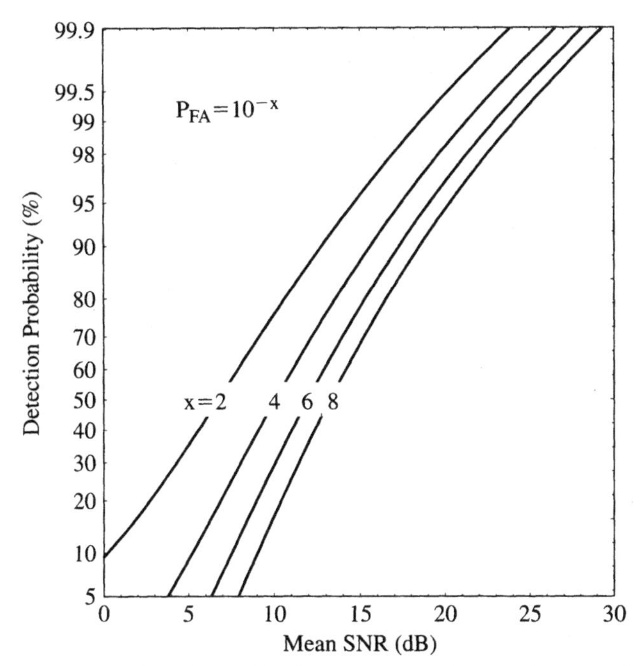

Figure 5-10. Single-pulse detection probability versus mean power signal-to-noise ratio for a one-dominant-plus-Rayleigh target model.

Changing variables to $\rho = r\sqrt{3/b}$, where $b = (a_o^2 + 3\sigma^2)$, the integral reduces to

$$P_d = \frac{3}{b}\int_{t\sqrt{3/b}}^{\infty} \rho\left(\sigma^2 + \frac{a_o^2\rho^2}{6}\right) e^{-\rho^2/2} d\rho \tag{5-49}$$

This can be solved directly, yielding

$$P_d = \frac{3}{a_o^2 + 3\sigma^2}\left[\sigma^2 + \frac{a_o^2}{3} + \frac{t^2}{2(1 + 3\sigma^2/a_o^2)}\right]e^{-3t^2/2(a_o^2+3\sigma^2)} \tag{5-50}$$

The false alarm probability is again given by Eq. (5-18). As in the Rayleigh signal model, we can substitute for the threshold $t = \sigma\sqrt{2\ln P_{\text{fa}}^{-1}}$ in Eq. (5-50) to obtain the detection probability as a function of the false alarm probability. Thus

$$P_d = \frac{1}{1 + \text{SNR}/2}\left(1 + \frac{\text{SNR}}{2} - \frac{\ln P_{\text{fa}}}{1 + 2/\text{SNR}}\right)e^{\ln P_{\text{fa}}/(1+\text{SNR}/2)} \tag{5-51}$$

where $\text{SNR} = \langle a_s^2\rangle/2\sigma^2 = 2a_o^2/3\sigma^2$ is the mean power signal-to-noise ratio at the input to the envelope detector. Equation (5-50) is plotted in Figure 5-10 for the case $\sigma = 1$ and several values of false alarm probability, P_{fa}.

5.5. RICIAN SIGNAL IN GAUSSIAN NOISE

Historically, the one-dominant-plus-Rayleigh distribution was accepted, in part, because of its ability to yield tractable expressions for the various models, especially for multiple-pulse integration. However, it is easy to show that Eq. (5-43) represents the probability density function for the incoherent summation of two equal-intensity fully developed speckle patterns [cf. Section 4.11, specifically Eq. (4-242)], which suggests that the name *one dominant plus Rayleigh* is somewhat of a misnomer. Indeed, since the one-dominant-plus-Rayleigh distribution is an attempt to model a target having one large specular component plus many low-level Rayleigh or diffuse components, it would seem to be more appropriate to use the partially developed speckle model discussed in Section 4.5. This idea was originally suggested by Scholefield[8] in 1967 from a radar standpoint and will be given some renewed interest in this section.

In Section 4.5 the complex surface was considered in the context of a direct-detection receiver in which aperture averaging can occur. Here the situation must be limited statistically to a single speckle and a single polarization state at the aperture, a consequence of the mixing theorem for heterodyne detection (cf. Section 2.6). It might be remembered that a complex surface is defined as one that exhibits both specular and diffuse components in the scattered radiation, the resultant speckle pattern being referred to as *partially developed*. It might also be recalled that the associated intensity distribution is one for which the point density function differs from that of a negative exponential distribution. Once again, the assumption is made that the coherence length of the transmitted waveform is large compared to the surface depth and that the illuminating beam is coherent in a plane normal to the direction of propagation.

The probability density function $p(r)$ may be calculated using either the voltage or power representation. We pursue the voltage representation by starting, as in Section 5.4, with Eq. (5-29) and with the conditional density function $p(r \mid a_s)$ as given by Eq. (5-30). However, the density function for the signal is now given by the Rician distribution given by Eq. (5-11), with a_s representing the random variable and a_c representing the coherent component of the signal amplitude. Thus

$$p(a_s) = \frac{a_s}{a_o^2} e^{-(a_s^2+a_c^2)/2a_o^2} I_0\left(\frac{a_s a_c}{a_o^2}\right) \tag{5-52}$$

Inserting Eqs. (5-52) and (5-30) into Eq. (5-29), we obtain after some rearrangement

$$\begin{aligned} p_{s+n}(r) &= \int_0^\infty p(r \mid a_s) p(a_s)\, da_s \\ &= \frac{r}{\sigma^2 a_o^2} e^{-(r^2/2\sigma^2 + a_c^2/2a_o^2)} \\ &\quad \times \int_0^\infty a_s e^{-(a_s^2/2)[(\sigma^2+a_o^2)/\sigma^2 a_o^2]} I_0\left(\frac{r a_s}{\sigma^2}\right) I_0\left(\frac{a_c a_s}{a_o^2}\right) da_s \end{aligned} \tag{5-53}$$

Letting $p = \frac{1}{2}[(\sigma^2 + a_o^2)/\sigma^2 a_o^2]$, $\alpha = r/\sigma^2$ and $\beta = a_c/a_o^2$ in the integral, Eq. (5-53) can be written as

$$p_{s+n}(r) = K \int_0^\infty a_s e^{-p a_s^2} I_0(\alpha a_s) I_0(\beta a_s)\, da_s \tag{5-54}$$

where

$$K = \frac{r}{\sigma^2 a_o^2} e^{-(r^2/2\sigma^2 + a_c^2/2a_o^2)} \tag{5-55}$$

The integral in Eq. (5-54) can be found in various tables of integrals[9] and is given by

$$\begin{aligned} &\int_0^\infty e^{-pt} I_\nu(\sqrt{2at}) I_\nu(\sqrt{2bt})\, dt \\ &\quad = p^{-1} e^{(a+b)/2p} I_\nu\left(\frac{\sqrt{ab}}{p}\right) \qquad \operatorname{Re} \nu > -1, \quad \operatorname{Re} p > 0 \end{aligned} \tag{5-56}$$

Letting $t = a_s^2$, $a = \alpha^2/2$, and $b = \beta^2/2$, Eq. (5-54) becomes for the case of $\nu = 0$,

$$p_{s+n}(r) = K \frac{\sigma^2 a_o^2}{\sigma^2 + a_o^2} e^{(r^2 a_o^2/2\sigma^2 + a_c^2 \sigma^2/2a_o^2)/(\sigma^2+a_o^2)} I_0\left(\frac{r a_c}{\sigma^2 + a_o^2}\right) \tag{5-57}$$

Substituting Eq. (5-55) for K finally yields

$$p_{s+n}(r) = \frac{r}{\sigma^2 + a_o^2} e^{-(r^2+a_c^2)/2(\sigma^2+a_o^2)} I_0\left(\frac{r a_c}{\sigma^2 + a_o^2}\right) \tag{5-58}$$

Thus we find that a Rician signal convolved with Gaussian noise results in a Rician density function for the signal plus noise with a variance equal to the sum of the variances of the Gaussian processes. This is similar to the Rayleigh signal in Gaussian noise discussed earlier where the variances also added. Note also that when $a_c \to 0$, the Rayleigh distribution of Eq. (5-36) is obtained corresponding to a Rayleigh fluctuating signal in Gaussian noise. On the other hand, when $a_o \to 0$, the density function becomes that of a Rician distribution given by Eq. (5-11), corresponding to a constant-amplitude signal in Gaussian noise.

We can write Eq. (5-58) in terms of a coherence parameter κ, as was done for the direct-detection case of partially developed speckle (cf. Section 4.6.2), and the input power signal-to-noise ratio (SNR). Expressed in terms of amplitude variables, we have

$$\kappa = \frac{a_c^2}{2a_o^2} \tag{5-59}$$

and

$$\text{SNR} = \frac{a_o^2 + a_c^2/2}{\sigma^2} = \frac{a_o^2}{\sigma^2}(1+\kappa) \tag{5-60}$$

which results in

$$p_{s+n}(r) = \frac{r(1+\kappa)}{\sigma^2(1+\kappa+\text{SNR})} e^{-[r^2(1+\kappa)/2\sigma^2+\kappa\text{SNR}]/(1+\kappa+\text{SNR})} \times I_0\left[\frac{r\sqrt{2\text{SNR}\kappa(1+\kappa)}}{\sigma(1+\kappa+\text{SNR})}\right] \tag{5-61}$$

The Rician distribution was shown in Section 5.2 to approach a Gaussian distribution at high signal-to-noise ratios and a Rayleigh distribution at low signal-to-noise ratios. The limiting cases above are clearly evident in Figures 5-11 and 5-12, where Eq. (5-61) is plotted for low and high values of κ, respectively.

The detection statistics follow in the usual manner. The false alarm probability is given by Eq. (5-18),

$$P_{\text{fa}} = e^{-t^2/2\sigma^2} \tag{5-62}$$

and the detection probability by

$$P_d = 1 - \int_0^t p_{s+n}(r)\,dr \tag{5-63}$$

Equation (5-63), with $p_{s+n}(r)$ given by Eq. (5-61), is numerically integrated and plotted in Figure 5-13 for the case of $\kappa = 1$ and several values of false alarm probability.

Figure 5-14 shows the effect of varying the coherence parameter for a fixed false alarm probability. It can be seen that the statistics approach the fully

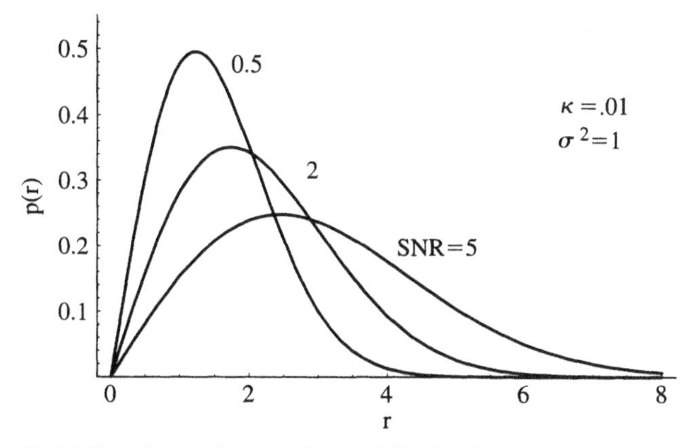

Figure 5-11. Probability density function for partially developed speckle in a white Gaussian noise environment of variance $\sigma^2 = 1$ for the case of $\kappa = 0.01$ and several values of the power signal-to-noise ratio. Here the coherence parameter κ is defined as the ratio of specular to diffuse optical power in the signal received. Note that the distribution corresponds to a Rayleigh distribution in this limit.

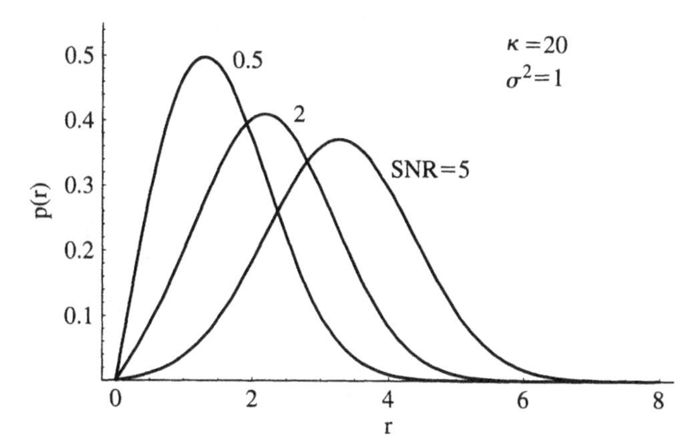

Figure 5-12. Probability density function for partially developed speckle in a white Gaussian noise environment of variance $\sigma^2 = 1$ for the case of $\kappa = 20$ and several values of power signal-to-noise ratio. Here the coherence parameter κ is defined as the ratio of specular to diffuse optical power in the signal received. Note that the distribution corresponds to a Rician distribution in this limit.

developed speckle case of a pure Rayleigh target in the limit of $\kappa \to 0$ and that of a specular target when $\kappa \to \infty$. It can also be seen that the detection probabilities reverse at low values of detection probability. This is a well-known manifestation of the finite probability of receiving a peak of a Rayleigh fluctuating signal at low signal-to-noise ratio together with the finite probability of receiving a fade at high signal-to-noise ratios. It also occurs in direct-detection systems when the statistics transition from Rayleigh fluctuating to specular. In

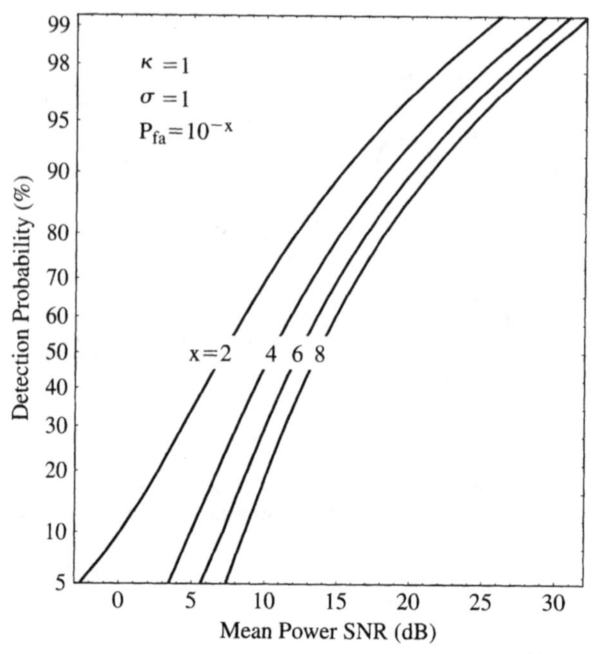

Figure 5-13. Single-pulse detection probabilities for partially developed speckle coherently detected in a white Gaussian noise environment versus the mean power signal-to-noise ratio, SNR(dB), for several values of the false alarm probability P_{fa}. The coherence parameter $\kappa = 1$ and the mean noise power variance $\sigma^2 = 1$.

that case it has been shown (cf. Section 4.6.2) that the crossover may occur not only due to an increase in the coherence parameter κ of the target but also due to an increase in the speckle count M intercepted by the aperture.

5.6. DETECTION IN ATMOSPHERIC TURBULENCE

Previously considered effects of atmospheric turbulence on direct-detection optical systems are now reconsidered for the case of coherent detection. Since the coherent system responds to the received field rather than intensity, log-amplitude statistics are used to develop the relevant detection statistics. It was shown in Section 3.4.2.6 in discussions about the limitations of coherent detection systems operating in turbulent atmospheres that the effective receiver aperture is determined by the coherence length of the turbulent medium. In addition, the log-normal signal fluctuations can produce deep fades in the received signal, which translate into reduced detection probabilities for a given false alarm rate. It is the latter effect that is explored next.

As discussed in Chapter 4, there are a variety of conditions and parameters that can affect detection performance when operating in a turbulent atmosphere. This is especially the case for monostatic geometries where the irradiance fluctuations

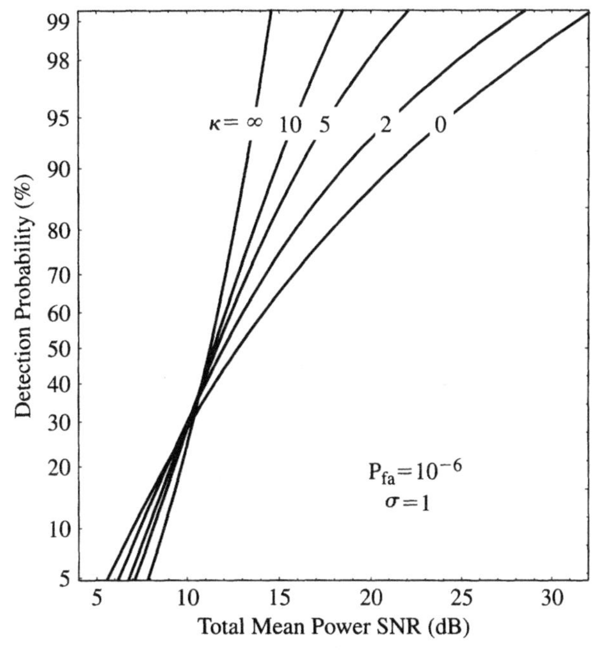

Figure 5-14. Single-pulse detection probabilities for partially developed speckle coherently detected in a white Gaussian noise environment versus mean power signal-to-noise ratio, SNR(dB), for several values of the coherence parameter κ. The false alarm probability $P_{fa} = 10^{-6}$ and the mean noise power variance $\sigma^2 = 1$. A Rician density function was used for the $\kappa = \infty$ case.

at the receiver depend on target and aperture size, target type and range, and whether the incident wave is plane or spherical. Each set of specifications has its own solution, so that for the sake of simplicity only a few representative examples are discussed to demonstrate the methodology.

5.6.1. Constant-Amplitude Signal in Weak Turbulence

The simplest arrangement that one might consider to demonstrate the effects of atmospheric turbulence on the detection statistics of a coherent laser system is that of a one-way path or, equivalently, a folded path using a uniformly reflecting mirror of infinite dimensions that is oriented at nonnormal incidence. With this idealization, which neglects any correlations in the refractive index between the outgoing and return paths, the system concept is limited to that of a bistatic radar geometry or a one-way communications link, as shown in Figure 5-15.

It will also be assumed that the receiver aperture is small compared to the saturation length of the beam. The saturation length was defined in Section 3.4.2.6 as $r_0 \approx 2.1\rho_0$, where ρ_0 is the transverse correlation length of the beam in the weak turbulence limit. As may be recalled, ρ_0 is defined for both plane waves

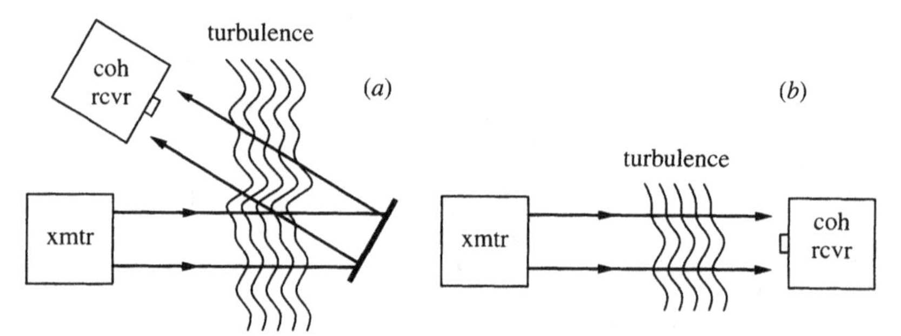

Figure 5-15. Equivalent propagation geometries for coherent detection systems operating in a weak turbulent atmosphere for (*a*) a bistatic laser radar system with an infinite mirror target, and (*b*) a one-way communications link. In both cases a point receiver is assumed that is much smaller than the correlation length ρ_0 of the atmosphere.

and spherical waves and for the near- and far-field cases,

$$
\begin{aligned}
&\rho_0 \approx \sqrt{\lambda R} && \text{near field} \\
&\left.\begin{aligned}
\rho_0^{(p)} &\approx (1.5k^2 C_n^2 R)^{-3/5} && \text{plane wave} \\
\rho_0^{(s)} &\approx (0.5k^2 C_n^2 R)^{-3/5} && \text{spherical wave}
\end{aligned}\right\} && \text{far field}
\end{aligned}
\tag{5-64}
$$

Under this assumption, only one correlation cell will, on the average, intercept the aperture at any instant of time such that no aperture averaging occurs. Hence a point density function is an appropriate representation for the statistics.

Calculation of the detection statistics associated with a specular target in a turbulent atmosphere requires that we once again solve an integral expression similar to Eq. (5-29). As noted earlier, there are two equivalent formulations that one may pursue in developing the probability density function given by Eq. (5-29), one in terms of power and the other in terms of amplitude or voltage. To be consistent with previous sections, we pursue the latter using the methodology employed by Shapiro et al.[10] but applied to the infinite mirror or one-way-path geometries discussed above. In all cases, effects due to beam expansion and beam bending are assumed negligible.

In a coherent detection system the voltage output from the detector is directly proportional to the signal field strength of the incident wave according to the relationship $a_s = \Re E_S E_{\text{LO}} \cos[(\omega_1 - \omega_2)t + \phi]$, where E_s is the signal field, E_{LO} the local oscillator field, and $\Re$ the detector responsivity. Since by definition the log-amplitude random variable is normalized to the mean value [i.e., $\chi = \ln(E_s/E_o) = \ln(ka_s/k\langle a_s^2\rangle) = \ln(a_s/\langle a_s^2\rangle)$, where k is a constant of proportionality that includes $\Re$], we can change the conditional variable in Eq. (5-29) to χ, obtaining

$$p_{s+n}(r) = \int_{-\infty}^{\infty} p(r \mid \chi) p(\chi)\, d\chi \tag{5-65}$$

As was pointed out in Chapter 3, the log-amplitude χ is related to the log intensity ℓ by

$$\frac{I}{\langle I \rangle} = \frac{a_s^2}{\langle a_s^2 \rangle} = e^{\ell} \equiv e^{2\chi} \tag{5-66}$$

Dividing the numerator and denominator of the second term in Eq. (5-66) by 2 to ensure consistency with the notation of Section 5.3, we obtain

$$a_s = \sqrt{2}\, a_o e^{\chi} \tag{5-67}$$

where $a_o^2 = \langle a_s^2/2 \rangle$ is again the mean signal power in the absence of turbulence. The probability density function for the log amplitude χ was discussed in Chapter 3 and is given by

$$p(\chi) = \frac{1}{\sqrt{2\pi\sigma_\chi^2}} e^{-(\chi+\sigma_\chi^2)^2/2\sigma_\chi^2} \tag{5-68}$$

The conditional Rician density function, Eq. (5-30), can be written in terms of χ by using Eq. (5-67) in the definition of the mean amplitude. Thus

$$p(r \mid \chi) = \frac{r}{\sigma^2} e^{-(r^2+2a_o^2 e^{2\chi})/2\sigma^2} I_0 \left(\frac{\sqrt{2}\, r a_o e^{\chi}}{\sigma^2} \right) \tag{5-69}$$

Inserting Eqs. (5-68) and (5-69) into Eq. (5-65) yields

$$p_{s+n}(r) = \frac{r e^{-r^2/2\sigma^2}}{\sigma^2 \sqrt{2\pi\sigma_\chi^2}} \int_{-\infty}^{\infty} e^{-(\chi+\sigma_\chi^2)^2/2\sigma_\chi^2 - a_o^2 e^{2\chi}/\sigma^2} I_0 \left(\frac{\sqrt{2}\, r a_o e^{\chi}}{\sigma^2} \right) d\chi \tag{5-70}$$

The detection probability therefore becomes

$$\begin{aligned} P_d &= \int_t^{\infty} p_{s+n}(r)\, dr \\ &= \frac{1}{\sigma^2 \sqrt{2\pi\sigma_\chi^2}} \int_t^{\infty} r e^{-r^2/2\sigma^2} \int_{-\infty}^{\infty} e^{-(\chi+\sigma_\chi^2)^2/2\sigma_\chi^2 - a_o^2 e^{2\chi}/\sigma^2} I_0 \left(\frac{\sqrt{2}\, r a_o e^{\chi}}{\sigma^2} \right) d\chi\, dr \end{aligned} \tag{5-71}$$

Interchanging the order of integration results in

$$P_d = \frac{1}{\sigma^2 \sqrt{2\pi\sigma_\chi^2}} \int_{-\infty}^{\infty} e^{-(\chi+\sigma_\chi^2)^2/2\sigma_\chi^2} Q \left(\frac{\sqrt{2}\, a_o e^{\chi}}{\sigma}, \frac{t}{\sigma} \right) d\chi \tag{5-72}$$

where the inner integral has been recognized as Malcom's Q-function. Now, from Section 5.3, we know that in the absence of turbulence $a_o^2/\sigma^2 \equiv \text{SNR}$ and $t^2/2\sigma^2 \equiv \text{TNR}$ so that Eq. (5-72) can be written

$$P_d = \frac{1}{\sigma^2\sqrt{2\pi\sigma_\chi^2}} \int_{-\infty}^{\infty} e^{-(\chi+\sigma_\chi^2)^2/2\sigma_\chi^2} Q(\sqrt{2\text{SNR}}e^\chi, \sqrt{2\text{TNR}})\, d\chi \tag{5-73}$$

where SNR and TNR are the mean power signal-to-noise and threshold-to-noise ratios, respectively.

Equation (5-73) is solved numerically and plotted in Figure 5-16 for a probability of false alarm of $P_{\text{fa}} = 10^{-7}$ and several values of σ_χ^2. It can be seen that performance is degraded as σ_χ^2 increases while the curves converge to the case of a constant-amplitude signal when $\sigma_\chi^2 \to 0$. The degradation of performance is, of course, a direct manifestation of the log-amplitude signal statistics that, as we have shown above, favors signal levels that are predominantly less than the mean. The reason for the crossover of the curves at low detection probabilities is similar to that discussed for direct detection of rough targets in Chapter 4. In that case, the spatial distribution of speckle resulted in the enhancement of

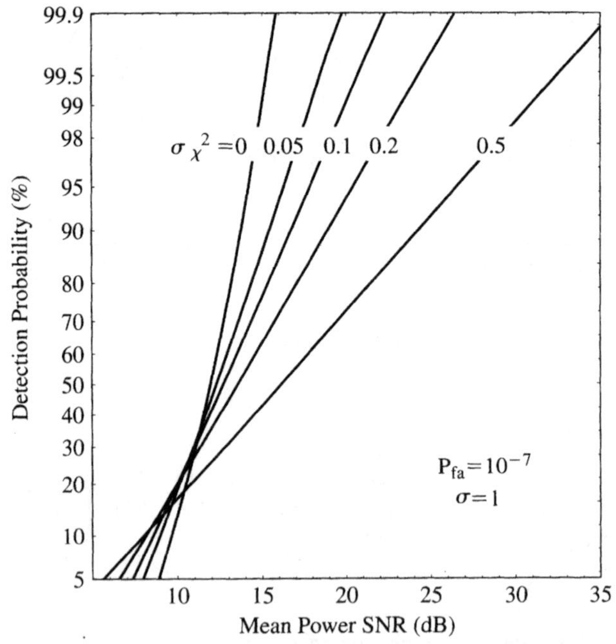

Figure 5-16. Detection probability versus mean signal photoelectron count for a constant-amplitude signal in weak turbulence. The curves represent the performance achievable with the geometries shown in Figure 5-15 with the variance of log amplitude σ_χ^2 as a parameter. The crossover of the curves at low detection probabilities is due to the finite probability of detecting a peak of the intensity fluctuations.

the lower detection probabilities at low values of speckle number M due to the finite probability that the receiver may sit on a speckle peak. In the present case, it is the temporal distribution of the log-amplitude fluctuations which has resulted in the enhancement of the lower detection probabilities at high scintillation levels, a consequence of the finite probability that a high-amplitude pulse may be received.

Shapiro, Capron, and Harney have considered the case of a *monostatic* geometry with aperture and target dimensions that are smaller than the transverse correlation length ρ_0 of the propagating field. In that case, it is easy to show that inclusion of the backscatter amplification effect discussed in Section 4.9.1 leads to the same density function as Eq. (5-72), but conditioned on $\exp(2\chi)$ rather than $\exp(\chi)$. Figure 5-17 shows the corresponding detection statistics for this case, where it can be seen that performance is again significantly degraded compared to the bistatic case (cf. Section 4.9.1).

Under the assumption that the statistics remain log normal at high turbulence levels, the effects of saturation of scintillation may be taken into account in Eq. (5-73) using the empirical formula given by Eq. (3-295), which is repeated

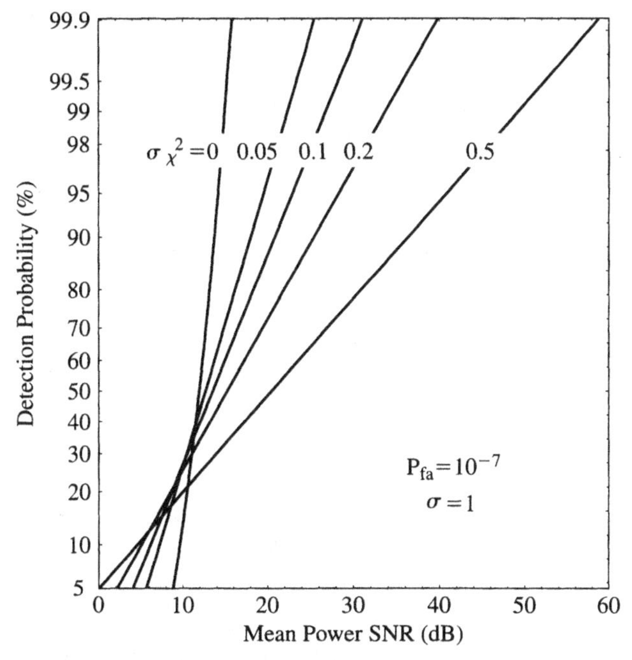

Figure 5-17. Single-pulse detection probabilities for a monostatic laser radar versus mean power signal-to-noise ratio for the case of a constant-amplitude signal in weak turbulence. An incident spherical wave is assumed ($\Omega \ll 1$) with a flat mirror target that has either a very small ($\Omega_t \ll 1$) or a very large ($\Omega_t \gg 1$) Fresnel number. Note the scale change compared to Figure 5-16. (After J. H. Shapiro, B. A. Capron, and R. C. Harney, *Appl. Opt.*, Vol. 20, No. 19, Oct. 1, 1981.)

below for reference,

$$\sigma_{\chi s}^2 = \frac{\sigma_\chi^2}{1 + A\sigma_\chi^{2B}} \tag{5-74}$$

Here A and B are parameters determined by a fit to measured data.

5.6.2. Rayleigh Fluctuating Signal in Weak Turbulence

Detection of rough targets in a turbulent atmosphere requires that both the Rayleigh fluctuations of the target and the log-normal fluctuations of the atmosphere be taken into account.[11] The simplest geometry that one might consider is that of a bistatic system operating against a rough target that has a diameter d_t which is small compared to both the far-field beam diameter ω_t. This implies an incident spherical wave at the target with a Fresnel number $\Omega \ll 1$ and a corresponding target Fresnel number $\Omega_t < 1$. It is also assumed that the receiver aperture diameter D is small compared to the transverse correlation length ρ_0 of the received beam. The geometry is shown in Figure 5-18. The assumption of $\Omega_t < 1$ necessarily implies that the surface-induced speckle size is large compared to the receiver aperture diameter, which, together with the condition $D < \rho_0$, ensures that the heterodyne phase-matching conditions required by the antenna theorem are satisfied. Assuming a homogeneous and isotropic medium with no correlation of inhomogeneities between the outgoing and return paths, the signal fluctuations may be viewed as resulting from a log-normal modulation of the irradiance of the Rayleigh distributed target (or equivalently, an effective log-normal modulation of the target reflectivity).

Thus, we again begin with Eq. (5-65),

$$p_{s+n}(r) = \int_{-\infty}^{\infty} p(r \mid \chi) p(\chi)\, d\chi \tag{5-75}$$

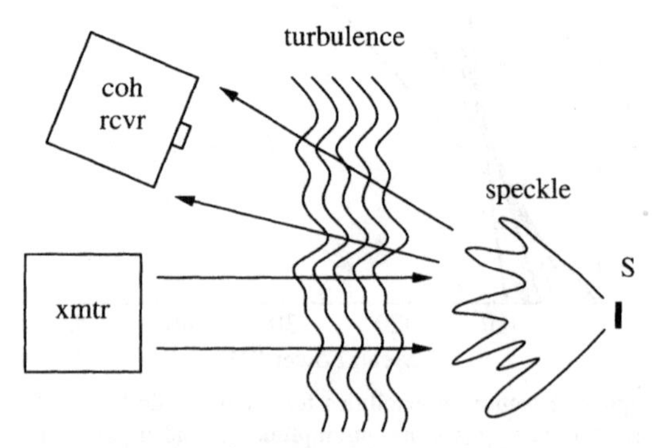

Figure 5-18. Bistatic geometry of a coherent detection laser radar operating against a Lambertian target S of arbitrary dimensions while in the presence of a weak turbulent atmosphere. The receiver aperture is assumed small compared to the correlation length ρ_0 of the atmosphere.

where $p(\chi)$ is given Eq. (5-68) and $p(r \mid \chi)$ is the conditional probability density function representing the envelope of the signal. The latter is obtained by first applying a log-normal modulation to the mean signal power in Eq. (5-31) (i.e., we let $a_o^2 \rightarrow a_o^2 e^{2\chi}$), obtaining

$$p(a_s) = \frac{a_s}{a_o^2 e^{2\chi}} e^{-a_s^2/2a_o^2 e^{2\chi}} \tag{5-76}$$

Using Eq. (5-76) instead of Eq. (5-31) in a derivation identical to that leading up to Eq. (5-36) (cf. Section 5.3) yields

$$p(r \mid \chi) = \frac{r}{\sigma^2 + a_o^2 e^{2\chi}} e^{-r^2/2(\sigma^2 + a_o^2 e^{2\chi})} \tag{5-77}$$

where χ is the log-normal random variable. Inserting Eqs. (5-77) and (5-68) into Eq. (5-75) results in a probability density function that represents a small Rayleigh distributed target in a turbulent atmosphere, that is

$$p_{s+n}(r) = \frac{r}{\sqrt{2\pi\sigma_\chi^2}} \int_{-\infty}^{\infty} \frac{e^{-(\chi+\sigma_\chi^2)^2/2\sigma_\chi^2 - r^2/2(\sigma^2 + a_o^2 e^{2\chi})}}{\sigma^2 + a_o^2 e^{2\chi}} d\chi \tag{5-78}$$

The detection probability follows immediately as

$$P_d = \frac{1}{\sqrt{2\pi\sigma_\chi^2}} \int_t^{\infty} \int_{-\infty}^{\infty} \frac{r e^{-(\chi+\sigma_\chi^2)^2/2\sigma_\chi^2 - r^2/2(\sigma^2 + a_o^2 e^{2\chi})}}{\sigma^2 + a_o^2 e^{2\chi}} d\chi \, dr \tag{5-79}$$

Interchanging the order of integration yields

$$P_d = \frac{1}{\sqrt{2\pi\sigma_\chi^2}} \int_{-\infty}^{\infty} e^{-(\chi+\sigma_\chi^2)^2/2\sigma_\chi^2} \int_t^{\infty} \frac{r e^{-r^2/2(\sigma^2 + a_o^2 e^{2\chi})}}{\sigma^2 + a_o^2 e^{2\chi}} dr \, d\chi \tag{5-80}$$

As in Section 5.3, the integral over r can be solved exactly, resulting in

$$\int_t^{\infty} f(r) \, dr = e^{-t^2/2(\sigma^2 + a_o^2 e^{2\chi})} = P_{\text{fa}}^{1/(1+\text{SNR} e^{2\chi})} \tag{5-81}$$

where $t^2/2\sigma^2 = -\ln P_{\text{fa}}$ and $\text{SNR} = a_o^2/\sigma^2$ is the input power signal-to-noise ratio in the absence of turbulence. The detection probability therefore becomes

$$P_d = \frac{1}{\sqrt{2\pi\sigma_\chi^2}} \int_{-\infty}^{\infty} e^{-(\chi+\sigma_\chi^2)^2/2\sigma_\chi^2} P_{\text{fa}}^{1/(1+\text{SNR} e^{2\chi})} d\chi \tag{5-82}$$

Equation (5-82) is plotted in Figure 5-19 as a function of SNR and a false alarm probability of $P_{\text{fa}} = 10^{-7}$, where it can be seen that the curves approach the pure

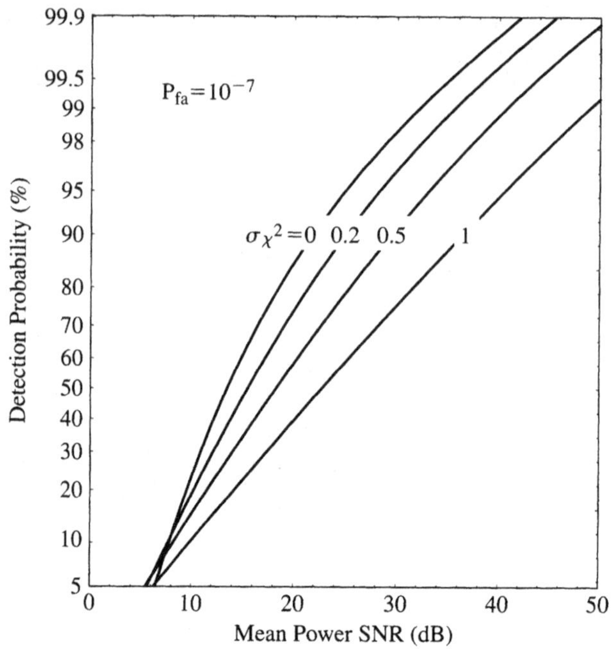

Figure 5-19. Single-pulse detection probabilities for a bistatic laser radar versus mean power signal-to-noise ratio for the case of a Rayleigh-distributed signal in a weak turbulent atmosphere. The curves represent the performance achievable with the geometry shown in Figure 5-18 with the variance of log amplitude σ_χ^2 as a parameter.

Rayleigh target case for $\chi = \sigma_\chi = 0$ (i.e., a nonturbulent atmosphere). It can also be seen that the curves begin to cross over at very low values of detection probability where the probability of detecting a peak of the intensity fluctuations dominates the statistics.

A monostatic geometry must take into account the correlation between outgoing and return paths, despite the randomizing effects of the rough surface.[12] As discussed in Section 4.9.2, the backscatter amplification effect for the diffuse target case is independent of target size, so that in terms of the variance of log amplitude, the normalized intensity becomes $\langle I/\langle I\rangle\rangle = \exp(4\sigma_\chi^2) > 1$, where $I/\langle I\rangle = e^{4\chi}$ represents the irradiance fluctuations at the receiver. Thus, letting the conditional variable in Eq. (5-77) be given by $e^{4\chi}$ rather than $e^{2\chi}$, we obtain the detection performance shown in Figure 5-20. Once again it can be seen that performance is severely degraded relative to the bistatic case.

5.7. COHERENT VERSUS NONCOHERENT PERFORMANCE

In Section 2.10 we saw that coherent detection offers a sensitivity advantage over direct detection, even when the latter is operated in the shot noise limit. It was

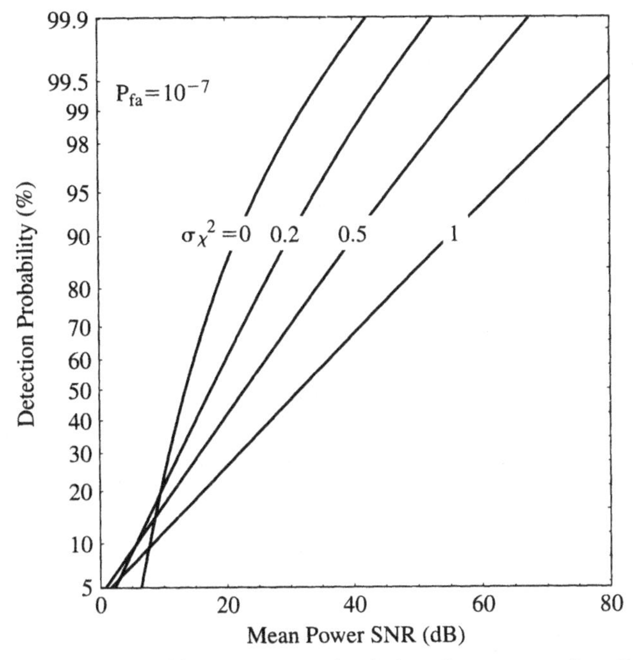

Figure 5-20. Detection probability versus mean signal photoelectron count for a Rayleigh-distributed signal in a weak turbulent atmosphere. The curves represent the case of a monostatic geometry with an incident spherical wave ($\Omega \ll 1$) and a Lambertian target that has a Fresnel number in the range $0 < \Omega_t < \infty$. Note the scale change compared to Figure 5-19.

also pointed out that this performance advantage for coherent detection becomes even greater when receiver noise begins to dominate. However, from a detection point of view, this conclusion does not necessarily hold, due primarily to the different detection statistics associated with the two processes but also because aperture averaging can be performed with direct detection but not with coherent detection. The latter is, of course, a consequence of the limitations imposed by the mixing theorem (cf. Chapter 2).

A direct comparison of coherent and direct-detection performance is complicated by the fact that the noise in a direct-detection receiver depends, in part, on the signal [cf. Eq. (2-276)]. Indeed, a rigorous analysis should compare detection probabilities as a function of the power signal-to-noise ratio for the two receiver types, using common aperture sizes and system losses. However, for the present discussion a more general formulation that emphasizes the key features of each receiver type that ultimately determine the relative performance advantages should be sufficient. To accomplish this, we compare the detection statistics of the two receiver types as a function of a common independent variable that is related to the signal-to-noise ratio,[13] namely, $10 \log \overline{N}_s$, where $\overline{N}_s$ is the mean signal photoelectron count during the measurement interval. For a coherent receiver matched to the received waveform, it is easy to see that $\overline{N}_s$ corresponds to the power signal-to-noise ratio through the identity $\overline{N}_s = \eta W/h\nu$, where $W = P/B$ is the energy collected during the measurement interval. For a direct-detection receiver, it simply corresponds to the log of the photoelectron count as has been used throughout Chapter 4. (It could also be identified with the shot-noise limited expression obtained at high signal levels [cf. Eq. (2-284)], although this identification would be somewhat restrictive for the present analysis.)

As an example of such a comparison, consider the detection statistics given by Figures 5-6 and 5-7 and 4-14 and 4-15. To make a proper comparison, a fixed false alarm probability of $P_{\rm fa} = 10^{-6}$ is chosen. Thus the $P_{\rm fa} = 10^{-6}$ nonfluctuating (NF) and fluctuating (F) curves of Figures 5-6 and 5-7, respectively, are superimposed on the data from Figure 4-14 and shown in Figure 5-21. Note that the $M = 1$ and ∞ curves represent Rayleigh fluctuating (rough) and nonfluctuating (specular) targets, respectively, for the direct-detection receiver and are therefore direct analogs of the F and NF curves for the coherent case (where $M \equiv 1$). It can be seen that for $\overline{N}_n = 0$ and fluctuating targets, the direct-detection receiver appears to outperform the corresponding coherent detection receiver even for $M = 1$. This performance advantage can be mitigated somewhat by increasing the false alarm probability of the coherent receiver while noting that $\overline{N}_n = 0$

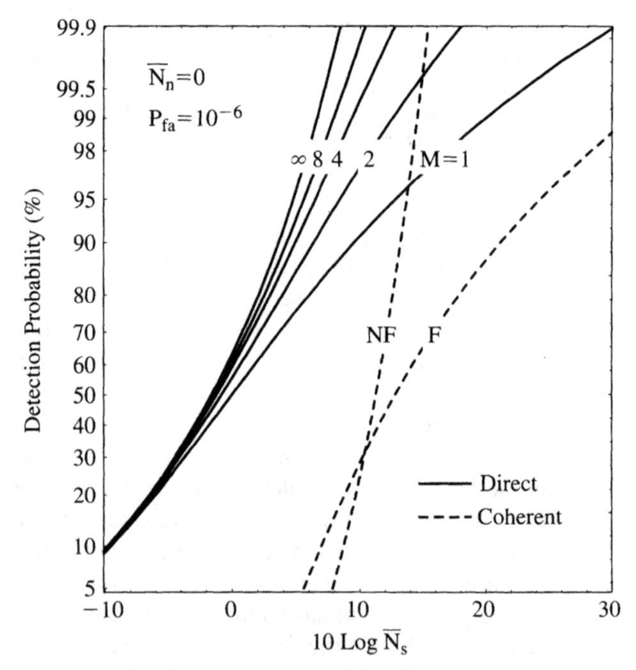

Figure 5-21. Comparison of coherent and direct detection probability versus the log of the mean signal photoelectron count $\overline{N}_s$ with a mean background photoelectron count $\overline{N}_n = 0$ and a false alarm probability $P_{\rm fa} = 10^{-6}$. The fluctuating (F) and nonfluctuating (NF) curves for the direct-detection receiver correspond to the $M = 1$ and $M = \infty$ cases, respectively.

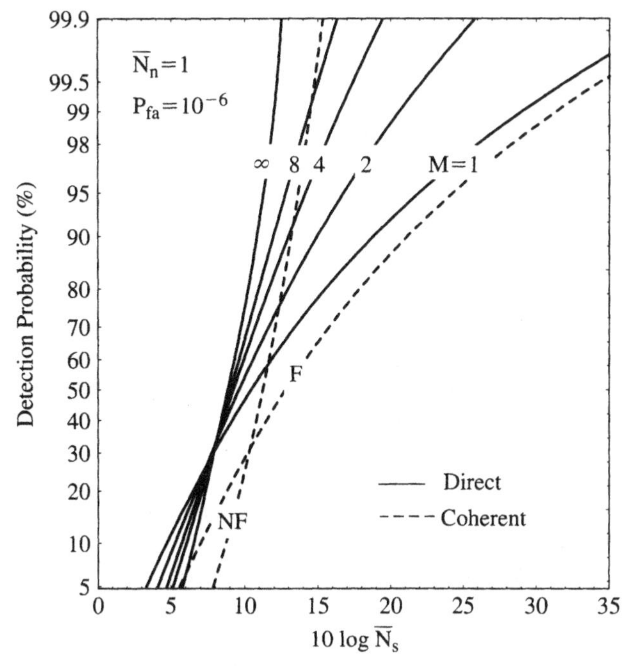

Figure 5-22. Comparison of coherent and direct-detection probability versus the log of the mean signal photoelectron count $\overline{N}_s$ with a mean background photoelectron count $\overline{N}_n = 1$ and a false alarm probability $P_{fa} = 10^{-6}$. The fluctuating (F) and nonfluctuating (NF) curves for the direct-detection receiver correspond to the $M = 1$ and $M = \infty$ cases, respectively.

implies a *quantum limit* for the direct-detection receiver that is independent of the false alarm probability (cf. Chapter 4).

The results shown in Figure 5-21 are very sensitive to the background noise $\overline{N}_n$ in the direct-detection receiver. This is shown in Figure 5-22 for the case of $\overline{N}_n = 1$ photoelectron counts. Note that the relative performance advantage of the direct-detection receiver has been reduced considerably. Further increases in $\overline{N}_n$ would clearly lead to a performance advantage for the coherent receiver. However, it is interesting that even when $\overline{N}_n$ is of such a magnitude that the coherent receiver outperforms the direct-detection receiver with $M = 1$, aperture averaging may still provide an advantage for the latter.

As indicated earlier, the performance comparisons above cannot be taken too seriously since the results do not include all of the noise and loss mechanisms experienced by real systems. For example, coherent receivers require a single polarization for optimum performance so that depolarization of the signal by the target may introduce a loss mechanism not experienced by unpolarized direct-detection receivers. Other system variables can also be considered which may further complicate the trade-off process, such as high receiver noise and multiple-pulse integration.

REFERENCES

[1]S. O. Rice, "Mathematical analysis of random noise," *Bell Syst. Tech. J.*, Vol. 23, July 1944 and Vol. 24, Jan. 1945.

[2]J. Marcum, "Studies of target detection by pulsed radar," *IEEE Trans. Inform. Theory*, Vol. IT-6, No. 2, April 1960.

[3]J. Marcum, *Table of Q-Functions*, Memo RM-339, RAND Corporation, Santa Monica, CA, Jan. 1, 1950.

[4]R. S. Berkowitz, Editor, *Modern Radar: Analysis, Evaluation, and System Design*, John Wiley & Sons, New York, 1965, p. 172.

[5]A. Erdelyi, Editor, *Tables of Integral Transforms*, Vol. I, McGraw-Hill Book Company, New York, 1954, p. 197, Eq. 14.

[6]J. H. Shapiro, "Precise comparison of experimental and theoretical SNRs in CO_2 laser heterodyne systems: comments," *Appl. Opt.*, Vol. 24, May 1, 1985, p. 1245.

[7]I. S. Gradshteyn and I. M. Ryzhik, *Tables of Integrals, Series, and Products*, Academic Press, San Diego, CA, 1965, p. 716, Eq. 6.631. This reference expresses the integral in terms of the Bessel functions $J_\nu(\beta x)$, but it may also be written in terms of the modified Bessel functions by using the relationship $I_\nu(\beta x) = i^{-\nu} J(-i\beta x)$.

[8]P. H. R. Scholefield, "Statistical aspects of ideal radar targets," *Proc. IEEE*, Apr. 1967, p. 587.

[9]Ibid., ref. 5, Eq. 22.

[10]J. H. Shapiro, B. A. Capron, and R. C. Harney, "Imaging and target detection with a heterodyne-reception optical radar," *Appl. Opt.*, Vol. 20, No. 19, Oct. 1, 1981, pp. 3292–3313.

[11]Ibid., Ref. 10.

[12]V. A. Banakh and V. L. Mironov, *Lidar in a Turbulent Atmosphere*, Artech House, Norwood, MA, 1987.

[13]J. W. Goodman, "Comparative performance of optical-radar detection techniques," *IEEE Trans. Aerosp. Electron. Syst.*, Vol. AES-2, No. 5, Sept. 1966, pp. 526–535.

6

Multiple-Pulse Detection

6.1. INTRODUCTION

Systems that lack sufficient signal-to-noise ratio on a single-pulse basis can benefit significantly using multiple-pulse integration. Unlike wide-beam search radar systems that have the luxury of long dwell times to perform integration, narrow-beam laser radar systems are frequently driven to detection schemes that rely on single-pulse detection to achieve the requisite scan coverage. However, there are many laser applications that require multiple-pulse integration to achieve the necessary sensitivity to perform the intended task. These include, among others, systems that operate against soft targets, such as Doppler wind-sensing systems, or search and track systems that employ low-power, high-PRF laser transmitters. Improved signal-to-noise ratio helps not only detection probabilities but also range and Doppler measurement accuracies and is therefore a desirable feature to implement if the cost is not too great.

System efficiency can also be affected by the integration process, either through the fact that the laser transmitter efficiency depends upon the PRF or through the losses or gains inherent in the detection statistics. The latter effect is characterized by a loss factor known as *integration loss*, which will be described for each target/receiver class presented. Below, a few representative single-pulse detection cases discussed in Chapters 4 and 5 are generalized to the case of integrated pulse trains to demonstrate the general methodology involved as well as to point out some fundamental differences between fluctuating and nonfluctuating target detection statistics.

Historically, radar detection theory has considered two types of integrated pulse trains, those in which there is statistical independence from scan to scan, each scan consisting of a group of correlated pulses, and those in which there is complete statistical independence from pulse to pulse throughout the entire pulse train. These models were first introduced by Swerling and are discussed in Section 6.3 for the coherent detection receiver. In Section 6.2, in which we address the direct-detection receiver, we consider only the latter model, that is, uncorrelated pulse trains. Also, it should be noted that, consistent with the notation of previous chapters, lowercase subscripts on the detection and false alarm probabilities represent single-pulse probabilities, while uppercase subscripts represent integrated or multiple-pulse probabilities.

6.2. DIRECT-DETECTION SYSTEMS

Direct-detection optical systems are almost exclusively pulsed systems. This follows from the two facts that the direct-detection signal-to-noise ratio is peak power dependent and that phase is not available to process more complex waveforms, such as frequency modulation or pulse compression. The basic receiver and signal processing architecture for a multiple-pulse direct-detection system is shown in Figure 6-1. Here it is assumed that the modern digital signal processor sums the A/D samples from a train of n pulses and subsequently applies some type of (optimal or nonoptimal) constant false alarm rate (CFAR) thresholding or filtering to the summed data to detect the received pulse. Since the signal improves linearly with the number of pulses n and the noise increases as $\sqrt{n}$, an overall improvement in signal-to-noise ratio of approximately $\sqrt{n}$ is generally realizable. Thresholding can also be performed prior to the summing process, however. In this case, each pulse is detected independently and a binary detection algorithm applied to the n pulses. In the following paragraphs, examples of both types of integration will be given, including a performance comparison between the two approaches. We will find that they differ only slightly in performance and that the selection of either one depends more on implementation complexity and cost than on performance. It is assumed throughout the chapter that the integrated pulses are independent since any correlation from pulse to pulse will cause a departure from the idealized $\sqrt{n}$. This has been demonstrated,[1] for example, in direct-detection differential absorption lidar (DIAL) systems that operate in statistically nonstationary atmospheres.

6.2.1. Poisson Signal in Poisson Noise

Consider a train of constant-amplitude pulses received from a specular target. Each pulse corresponds to the case of a Poisson signal immersed in a Poisson background as given by Eq. (4-58):

$$q_{s+n}(k; \overline{N}_{s+n}) = \frac{(\overline{N}_{s+n})^k e^{-\overline{N}_{s+n}}}{k!} \tag{6-1}$$

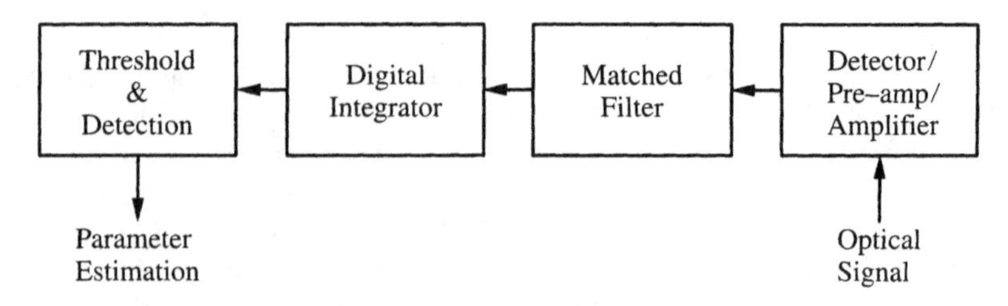

Figure 6-1. Block diagram of a coherent detection system with noncoherent integration.

where $\overline{N}_{s+n} = \overline{N}_s + \overline{N}_n$. The characteristic function for the Poisson distribution was shown to be

$$G(i\upsilon) = e^{\overline{N}_{s+n}(e^{i\upsilon}-1)} \tag{6-2}$$

If we now assume n independent, constant-amplitude pulses, we can write for the integrated pulse train

$$\begin{aligned} G_k(i\upsilon) &= G_{k1}(i\upsilon)Gk_2(i\upsilon)\cdots G_{kn}(i\upsilon) \\ &= \prod_{j=1}^{n} G_{ki}(i\upsilon) \end{aligned} \tag{6-3}$$

where j stands for the jth pulse in the train. If we assume that all $\overline{N}_{s+n}$ are equal (i.e., the mean signal and the mean noise counts are constant during the n-pulse train), Eq. (6-3) reduces to

$$G_k(i\upsilon) = e^{n\overline{N}_{s+n}(e^{i\upsilon}-1)} \tag{6-4}$$

By inspection, the discrete density function for the integrated pulse train can be written as

$$q_{s+n}^{(n)}(k; \overline{N}_s, \overline{N}_n) = \frac{n^k(\overline{N}_s + \overline{N}_n)^k e^{-n(\overline{N}_s+\overline{N}_n)}}{k!} \tag{6-5}$$

Using the methods of Comment 4-2, it is straightforward to show that the mean and variance of Eq. (6-5) are both equal to $n(\overline{N}_s + \overline{N}_n)$.

The detection statistics can be calculated in the same manner as the single-pulse case. We have

$$P_{\text{FA}}^{(n)} \geq 1 - \sum_{k=0}^{k_{\text{th}}-1} \frac{(n\overline{N}_n)^k e^{-n\overline{N}_n}}{k!} \tag{6-6}$$

and

$$P_D^{(n)} = 1 - \sum_{k=0}^{k_{\text{th}}-1} p_{s+n}^{(n)}(k; \overline{N}_s, \overline{N}_n) \tag{6-7}$$

The detection probability $P_D^{(n)}$ is shown in Figure 6-2 for several values of n with $\overline{N}_n = 1$ and $P_{\text{FA}}^{(n)} \cong 10^{-6}$. For reference, the $n = 1$ curve is identical with the $\overline{N}_n = 1$ curve shown in Figure 4-11. The advantage in detection probability with increasing n is clearly evident. As mentioned earlier, a $\sqrt{n}$ improvement in SNR is also realized. This is most easily seen from the definition of the voltage SNR given by

$$\text{SNR}_V = \frac{\langle k \rangle}{\sqrt{\text{var}(k)}} = \frac{n\overline{N}_s}{\sqrt{n(\overline{N}_s + \overline{N}_n)}} = \frac{\sqrt{n}\overline{N}_s}{\sqrt{\overline{N}_s + \overline{N}_n}} \tag{6-8}$$

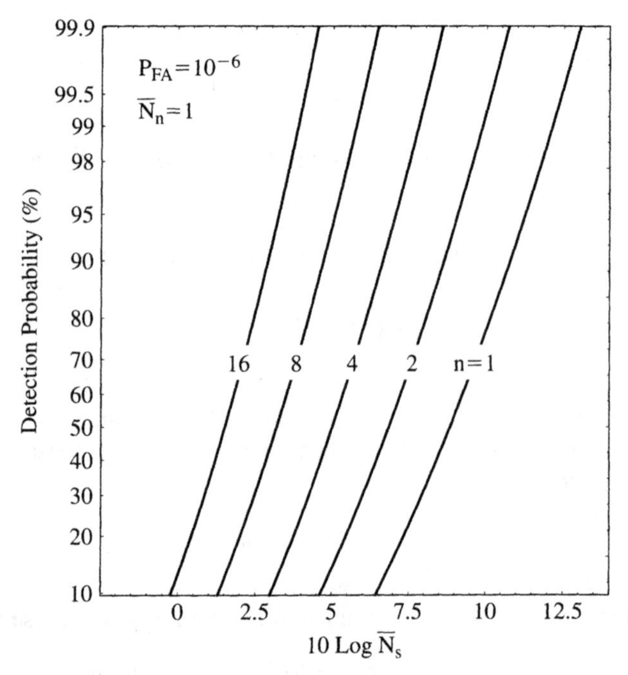

Figure 6-2. Integrated detection probability for a Poisson signal in Poisson noise versus mean single-pulse photoelectron count $10 \log \overline{N}_s$ with a mean noise count of $\overline{N}_n = 1$ and a false alarm probability $P_{FA} = 10^{-6}$. The case of $n = 1$ is the same as in Figure 4-11.

Comment 6-1. It should be noted that the noise variance increases as $\sqrt{n}$ with integration. This seems counterintuitive since one would expect that the noise fluctuations would be reduced with integration. However, one must realize that this reduction is only *relative* to the signal, which has increased in amplitude by n. Thus, for a constant-amplitude signal and a limited dynamic range signal processor with variable gain, the noise floor appears to compress as n increases. ■

The advantage in SNR noted above comes, however, at a cost in efficiency. The efficiency loss is directly attributable to the statistics of the noncoherent integration process and in radar terminology is referred to as *integration loss*. It is usually defined for an n-pulse integration process as

$$L = 10 \log \left(\frac{n\text{SNR}_n}{\text{SNR}_c} \right) \text{dB} \tag{6-9}$$

where SNR_n is the single-pulse signal-to-noise ratio required to achieve a given detection and false alarm probability when integrating n pulses and SNR_c is the signal-to-noise ratio required to achieve the same detection parameters with coherent integration. Thus, if coherent integration is performed, $n\text{SNR}_n = \text{SNR}_c$ and $L = 0$ dB. However, for noncoherent integration, one suspects that $L > 0$, since it requires, in general, more signal energy to achieve a given detection probability with direct detection than with coherent detection. Also, it seems reasonable to expect that L cannot be less than zero since coherent integration is generally assumed to be the more efficient detection process. We will find that this is not always the case.

The conclusions above can be in error when specific target fluctuations are considered. Indeed, it will be shown in the following sections that either a direct or coherent detection system operating against a specular target results in $L > 0$. However, that same system operating against a rough target can result in $L < 0$ over a finite range of the integration parameter n. These results have important implications for system design. This follows from the fact that for a given detection and false alarm probability, L less than zero implies that less total energy is required per dwell (integration period) using an n-pulse waveform than that required by a single pulse of the same total energy. Thus, advantages in system power and weight can be achieved through a judicious choice of integration parameters, such as PRF and dwell time, if the target statistics support an $L < 0$ regime.

For equal noise contributions to SNR_n and SNR_c of Eq. (6-9), the expression for L can be rewritten in terms of energy per dwell:

$$L = 10\log\left(\frac{n\overline{E}_{np}}{\overline{E}_{sp}}\right) \tag{6-10}$$

where $\overline{E}_{np}$ is the individual mean pulse energy in the integrated n-pulse train and $\overline{E}_{sp}$ is the energy of a single pulse ($n = 1$) required to achieve the same detection statistics. Note that a single pulse of energy $\overline{E}_{sp}$ may always be considered as a coherent sum of n pulses of energy $\overline{E}_{sp}/n$. In terms of the photoelectron count, Eq. (6-10) becomes

$$L = 10\log\left(\frac{n\overline{N}_{np}}{\overline{N}_{sp}}\right) \tag{6-11}$$

where the relationship $N = \eta E/h\nu$ was used. Note once again that if $\overline{N}_{sp} = n\overline{N}_{np}$, $L = 0$ dB.

To calculate L for a specific detection statistic, such as the Poisson distribution of Eq. (6-5), one must select a P_D and a P_{FA} using Eqs. (6-6) and (6-7) and evaluate L for each value of n. The result of such a process is illustrated in Figure 6-3, where it can be seen that L is positive for all $n > 0$. As stated earlier, this condition leads to a more energy-efficient system for single-pulse detection than for n-pulse detection. A $10\log\sqrt{n}$ straight line representing the performance achievable with a video integrator (e.g., a visual observer on an integrating display) is also shown and corresponds to the slope of the curves at

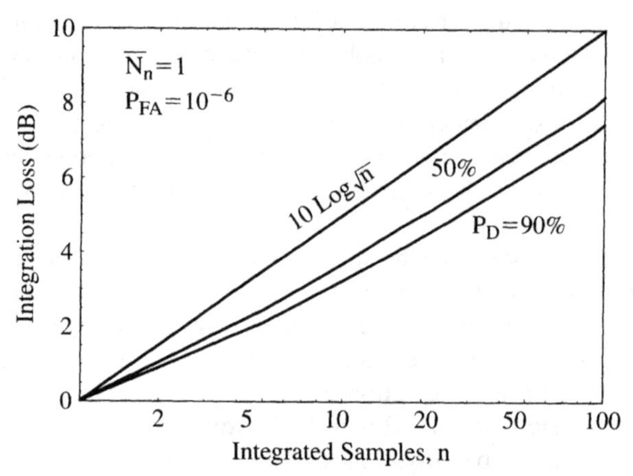

Figure 6-3. Integration loss for an integrated Poisson signal in Poisson noise for a noise count of $\overline{N}_n = 1$, a false alarm probability $P_{\text{FA}} = 10^{-6}$, and detection probabilities $P_D = 0.9$ and 0.5.

large n. The x-axis corresponds to the integration loss of a coherent integrator which, by definition, is $L = 0$ dB. We conclude from the figure that for the receiver class under consideration, noncoherent integration results in less integration loss than that for a video integrator but somewhat greater loss than a coherent integrator or a single pulse of the same total energy.

Comment 6-2. Better insight into the bounding lines $L = 10\log\sqrt{n}$ and $L = 0$ in Figure 6-3 can be obtained as follows. If we represent the mean single-pulse photoelectron count $\overline{N}_{sp}$ as the coherent sum of n pulses having counts $\overline{N}_{sp}/n$, and assume a $\sqrt{n}$ performance for the noncoherent integrator, the total count of the integrator is $\sqrt{n}(\overline{N}_{sp}/n)$ and, from the definition of integration loss,

$$L = 10\log\left(\frac{n\sqrt{n}(\overline{N}_{sp}/n)}{n\overline{N}_{sp}/n}\right) = 10\log\sqrt{n} \tag{6-12}$$

A similar thought process for coherent integration results in $L = 0$ dB for all n. ■

6.2.2. Negative Binomial Signal in Poisson Noise

The previous case of a constant-intensity signal immersed in Poisson noise will now be generalized to a fluctuating signal immersed in Poisson noise. In this case, each pulse in the integrated pulse train experiences a new and independent realization of the target-induced speckle. We saw in Section 4.5 that the single-pulse case resulted in a discrete density function consisting of the product of a negative binomial distribution and a confluent hypergeometric distribution.

Indeed, we also found that the corresponding characteristic function consisted of the product of the characteristic functions of negative binomial and Poisson distributions [i.e., Eq. (4-104)], which is repeated here for convenience:

$$G(i\upsilon) = p^r(1 - qe^{i\upsilon})^{-r}e^{-qz(1-e^{i\upsilon})} \tag{6-13}$$

We can build on this result[2] to obtain the characteristic function for an n-pulse negative binomial signal immersed in Poisson noise. Hence, for n independent, constant-amplitude pulses, we write

$$\begin{aligned} G_k(i\upsilon) &= G_{k1}(i\upsilon)G_{k2}(i\upsilon)\cdots G_{kn}(i\upsilon) \\ &= p^{\mathcal{R}}(1 - qe^{i\upsilon})^{-\mathcal{R}}e^{-\overline{\mathcal{N}}_n(1-e^{i\upsilon})} \end{aligned} \tag{6-14}$$

where $\mathcal{R} = nr$ and $\overline{\mathcal{N}}_n = nqz$. Since Eq. (6-14) is identical in form to the single-pulse characteristic function under the parameter replacements $r \to \mathcal{R}$ and $qz \to \overline{\mathcal{N}}_n$, the inverse transformation of Eq. (6-14) should yield an identical discrete density function but with the foregoing parameter replacements. Thus, with $r = M$ and $z = \overline{N}_n/q$, we let $\mathcal{M} = nM$ and $\overline{\mathcal{N}}_n = n\overline{N}_n$ obtaining

$$q_{s+n}^{(n)}(k; \overline{N}_s, \overline{\mathcal{N}}_n, \mathcal{M}) = p^{\mathcal{M}}q^k \frac{e^{-\overline{\mathcal{N}}_n}}{\Gamma(\mathcal{M})} \frac{\Gamma(k + \mathcal{M})}{\Gamma(k+1)} {}_1F_1\left(\alpha, \beta'; \frac{\overline{\mathcal{N}}_n}{q}\right) \tag{6-15}$$

where $\alpha = -k$ and $\beta' = -k - \mathcal{M} + 1$. Notice that neither p nor q is modified in this procedure. The detection statistics follow immediately by summing Eq. (6-15) over the appropriate limits.

Using Eq. (6-14) to calculate the moments, the mean and variance become the expected generalizations of Eqs. (4-105) and (4-106), that is,

$$\langle k \rangle = \frac{nrq}{p} + n\overline{N}_n = n(\overline{N}_s + \overline{N}_n) \tag{6-16}$$

and

$$\text{var}(k) = \frac{nrq}{p^2} + n\overline{N}_n = n\overline{N}_s\left(1 + \frac{\overline{N}_s}{M}\right) + n\overline{N}_n \tag{6-17}$$

The detection statistics may be calculated in the usual manner using Eqs. (6-6) for the false alarm probability and Eq. (6-15) in Eq. (6-7) for the detection probability. Figures 6-4 and 6-5 show the resulting detection statistics for the cases of $M = 1$ and 2, respectively, with $\overline{N}_n = 1$ and $P_{\text{FA}} = 10^{-6}$.

Integration loss for the Kummer distribution is shown in Figure 6-6. We find that the large exponentially distributed intensity fluctuations associated with a single speckle realization at the aperture ($M = 1$) result in $L < 0$ over a wide range of the integration parameter n when P_D is large. As noted above, this

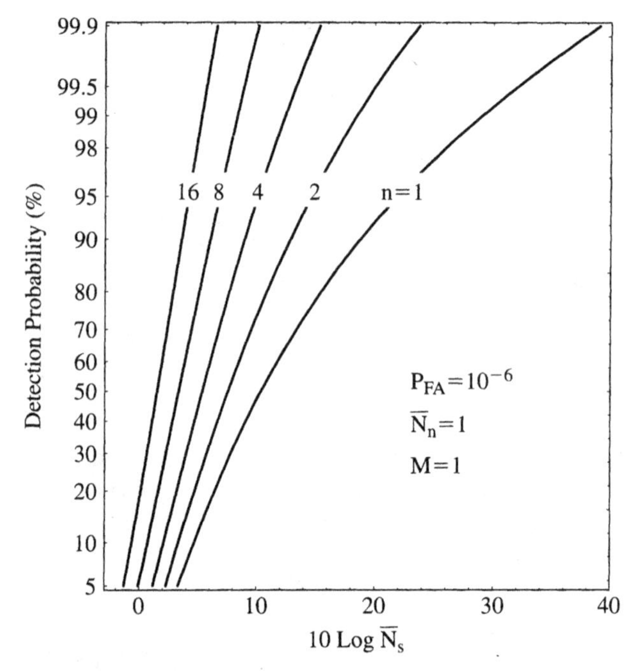

Figure 6-4. Integrated detection probability for a negative binomial signal in Poisson noise versus mean single-pulse photoelectron count $10 \log \overline{N}_s$ with a noise count of $\overline{N}_n = 1$, a false alarm probability $P_{FA} = 10^{-6}$, and a speckle count of $M = 1$. The case of $n = 1$ is the same as the $M = 1$ curve in Figure 4-15.

implies that for the receiver class being considered, the process of noncoherent integration prior to thresholding and detection allows for a more energy-efficient system than a coherently integrated pulse train (or a single pulse) of the same total energy. This advantage is quickly diminished, however, as M increases or as the detection probability is decreased.

6.2.3. Noncentral Negative Binomial Signal in Poisson Noise

Using procedures similar to that used in Section 6.2.2, the characteristic function for the discrete density function of a single pulse in partially developed speckle given by Eq. (4-130) becomes[3]

$$G(iv) = p^M (1 - qe^{iv})^{-M} e^{-\overline{N}_n(1-e^{iv})} e^{-(pN_c/q)[1-p(1-qe^{iv})^{-1}]} \tag{6-18}$$

Once again we assume n statistically independent, constant-amplitude pulses and obtain

$$G(iv) = p^{\mathcal{M}} (1 - qe^{iv})^{-\mathcal{M}} e^{-\overline{\mathcal{N}}_n} (1 - e^{iv}) e^{-(p\mathcal{N}_c/q)\{1-p[1-q\exp(iv)]^{-1}\}} \tag{6-19}$$

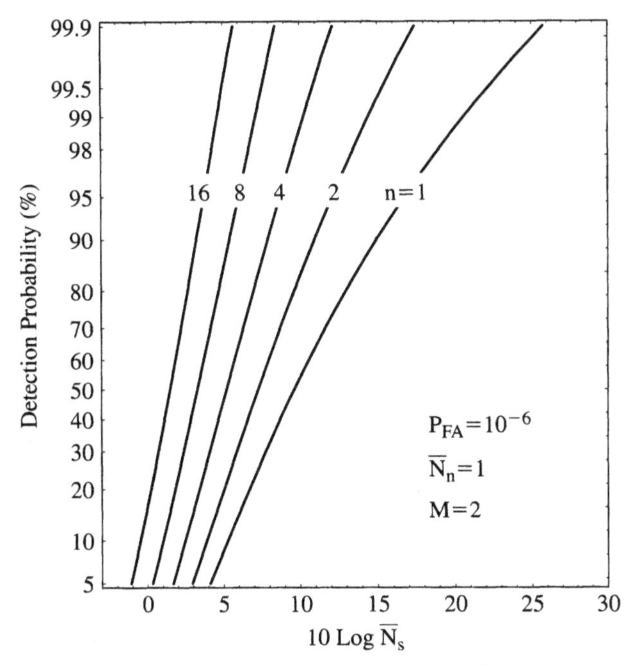

Figure 6-5. Integrated detection probability for a negative binomial signal in Poisson noise versus mean single-pulse photoelectron count $10 \log \overline{N}_s$ with a noise count of $\overline{N}_n = 1$, a false alarm probability $P_{\text{FA}} = 10^{-6}$ and a speckle count of $M = 2$. The case of $n = 1$ is the same as the $M = 2$ curve in Figure 4-15. (From G. R. Osche, *Appl. Opt.*, Vol. 39, No. 24, Aug. 20, 2000.)

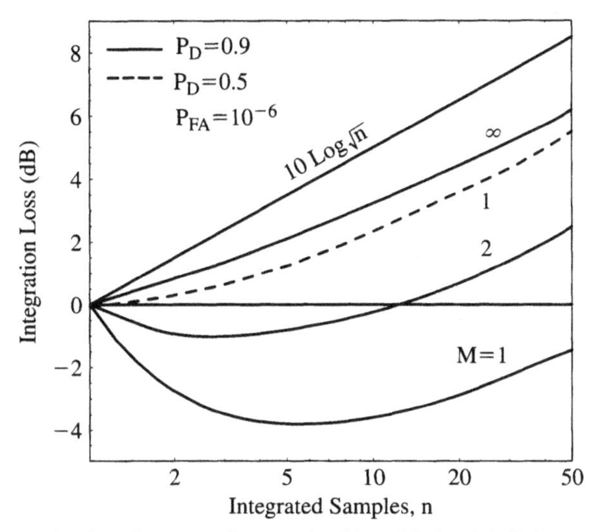

Figure 6-6. Integration loss for an n-pulse negative binomial signal in Poisson noise for a noise count of $\overline{N}_n = 1$, a false alarm probability $P_{\text{FA}} = 10^{-4}$, and detection probabilities $P_D = 0.9$ and 0.5. M represents the speckle count for the signal received. (From G. R. Osche, *Appl. Opt.*, Vol. 39, No. 24, Aug. 20, 2000.)

where $\mathcal{N}_c = nN_c$. The inverse transformation assumes the form of Eq. (4-130) so that

$$q_{s+n}^{(n)}(k) = p^{\mathcal{M}} q^k e^{-(\overline{\mathcal{N}}_n + p\mathcal{N}_c/q)} \sum_{j=0}^{\infty} \left(\frac{p^2 \mathcal{N}_c}{q}\right)^j \frac{\Gamma(\mathcal{M}+k+j)}{j!\,\Gamma(\mathcal{M}+j)\Gamma(k+1)}\, {}_1 \\ \times F_1\left(\alpha, \beta_j'; \frac{\overline{\mathcal{N}}_n}{q}\right) \tag{6-20}$$

where $\beta_j' = -k - \mathcal{M} - j + 1$. Expressing Eq. (6-20) in terms of the coherence parameter $\kappa = N_c/\overline{N}_s$ yields

$$q_{s+n}^{(n)}(k) = p^{\mathcal{M}} q^k e^{-(\overline{\mathcal{N}}_n + \kappa\mathcal{M})} \sum_{j=0}^{\infty} (\kappa p \mathcal{M})^j \frac{\Gamma(\mathcal{M}+k+j)}{j!\,\Gamma(\mathcal{M}+j)\Gamma(k+1)}\, {}_1 \\ \times F_1\left(\alpha, \beta_j'; \frac{\overline{\mathcal{N}}_n}{q}\right) \tag{6-21}$$

Figure 6-7 shows the effect of changing κ on the multiple-pulse detection statistics, where detection and false alarm probabilities again follow from Eqs. (6-6) and (6-7). A single value of $n = 4$ has been chosen, but the trend applies to any n. It can be seen from the figure that for a given M and $\overline{N}_n$, the detection curves converge to the corresponding Poisson limit as $\kappa \to \infty$ and to the fully developed speckle results as $\kappa \to 0$. These limits are consistent with the $n = 1$ results of Figures 4-22 and 4-23. In the present case, the n-pulse Poisson limit is given by

$$q_{s+n}^{(n)}(k) = \frac{(\overline{\mathcal{N}}_n + \overline{\mathcal{N}}_s + \mathcal{N}_c)^k}{k!} e^{-(\overline{\mathcal{N}}_n + \overline{\mathcal{N}}_s + \mathcal{N}_c)} \tag{6-22}$$

where $\overline{\mathcal{N}}_s = n\overline{N}_s$.

Integration loss for the case of partially developed speckle can be understood by referring to Figure 6-6 for the case of $\kappa = 0$. Clearly, the $M = 1, \kappa = 0$ curve represents a lower bound for the set of $\kappa > 0$ curves, while we have already shown that in the limit of $\kappa \to \infty$, the performance of the Poisson distribution given by Eq. (6-22) is obtained. Hence all κ curves are contained within these bounds, and we see that if κ is large enough, even the $M = 1$ curve can have $L > 0$ for all n.

6.2.4. Parabolic Cylinder Signal in Gaussian Noise

Consider now the Gaussian receiver operating against a rough target as discussed in Section 4.7. There it was shown that the probability density function was

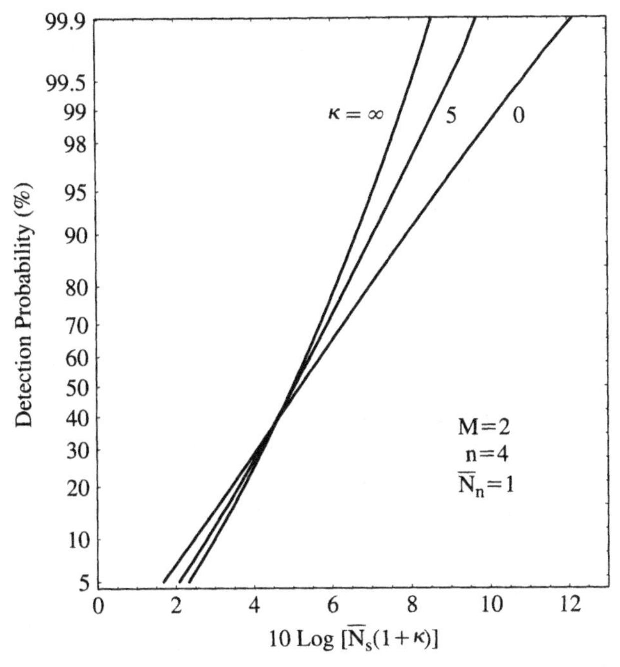

Figure 6-7. Integrated detection probability of a noncentral negative binomial signal in Poisson noise for the case of several values of the coherence parameter κ and $n = 4$ pulses. $\kappa = 0$ corresponds to the fully developed speckle case while $\kappa = \infty$ corresponds to a specular surface. (From G. R. Osche, *Appl. Opt.*, Vol. 39, No. 24, Aug. 20, 2000.)

described by a parabolic cylinder function having a characteristic function given by Eq. (4-156), which we repeat here for convenience:

$$G_k(i\upsilon) = \left(\frac{a}{a - i\upsilon}\right)^M e^{-(1/2)\sigma^2\upsilon^2} \tag{6-23}$$

Once again, $a = M/\overline{N}_s$. Assuming n independent, constant-amplitude pulses, the n-pulse characteristic function becomes

$$\begin{aligned} G_k(i\upsilon) &= G_{k1}(i\upsilon)G_{k2}(i\upsilon)\cdots G_{kn}(i\upsilon) \\ &= \left[\frac{\prod_{j=1}^n a_j}{\prod_{j=1}^n (a_j - i\upsilon)}\right]^M e^{-(1/2)\sum_{j=1}^n \sigma_j^2\upsilon^2} \end{aligned} \tag{6-24}$$

which, under the assumption of equal mean energies per pulse, becomes

$$G_k(i\upsilon) = \left(\frac{a}{a - i\upsilon}\right)^{\mathcal{M}} e^{-(1/2)\sigma_n^2\upsilon^2} \tag{6-25}$$

where $\mathcal{M} = nM$ and $\sigma_n^2 = \sum_{i=1}^{n} \sigma_i^2$. If the pulses are not identically distributed, the simplification of Eq. (6-25) could not be assumed and more complex procedures would have to be invoked.

The probability density function representing n integrated pulses having identical parabolic density distributions with equal mean values for the statistical variables can then be obtained from Eq. (6-25) by comparing with the original single-pulse probability density function given by Eq. (4-144), yielding

$$p_{s+n}^{(n)}(k; a, \mathcal{M}, \sigma_n) = \frac{(a\sigma_n)^{\mathcal{M}}}{\sqrt{2\pi\sigma_n^2}} e^{-(k^2/4\sigma_n^2 + ak/2 - a^2\sigma_n^2/4)} D_{-\mathcal{M}}\left(a\sigma_n - \frac{k}{\sigma_n}\right) \tag{6-26}$$

where, again, $a = M/\overline{N}_s$.

Equation (6-26) is identical to the single-pulse density function, Eq. (4-144), under the replacements $\sigma \to \sigma_n$ and $M \to \mathcal{M}$. Once again, since σ and M are parameters, it is reasonable to assume that the single-pulse detection probability, Eq. (4-180), can be generalized with similar replacements. Thus

$$\begin{aligned} P_D^{(n)} &= \frac{1}{2}\operatorname{erfc}\left(\frac{k_{\text{th}}}{\sqrt{2}\sigma_n}\right) + \frac{(a\sigma_n)^{\mathcal{M}-1}}{\sqrt{2\pi}} e^{a^2\sigma_n^2/4 - ak_{\text{th}}/2 - k_{\text{th}}^2/4\sigma_n^2} \\ &\times \left[D_{-\mathcal{M}}\left(a\sigma_n - \frac{k_{\text{th}}}{\sigma_n}\right) + \sum_{j=1}^{\mathcal{M}-1} (a\sigma_n)^{-j} D_{-(\mathcal{M}-j)}\left(a\sigma_n - \frac{k_{\text{th}}}{\sigma_n}\right)\right] \end{aligned} \tag{6-27}$$

where $\sigma_n = \sqrt{n}\,\sigma$ and $\mathcal{M} = nM$.

The false alarm probability is obtained from an n-pulse generalization of Eq. (4-161) for the zero-mean Gaussian noise probability density function. This follows in a straightforward manner, that is

$$G_n^{(n)}(iv) = e^{-(1/2)\sigma_n^2 v^2} \tag{6-28}$$

where, again, $\sigma_n^2 = \sum_{i=1}^{n} \sigma_i^2 = n\sigma^2$. The inverse transformation leads to the well-known zero-mean Gaussian noise probability density function:

$$p_n^{(n)}(k; 0) = \frac{e^{-k^2/2\sigma_n^2}}{\sqrt{2\pi\sigma_n^2}} \tag{6-29}$$

The false alarm probability is then

$$\begin{aligned} P_{\text{FA}}^{(n)} &= \frac{1}{\sqrt{2\pi\sigma_n^2}} \int_{k_{\text{th}}}^{\infty} e^{-k^2/2\sigma_n^2}\, dk \\ &= \frac{1}{2}\left[1 - \operatorname{erf}\left(\frac{k_{\text{th}}}{\sqrt{2n}\,\sigma}\right)\right] \end{aligned} \tag{6-30}$$

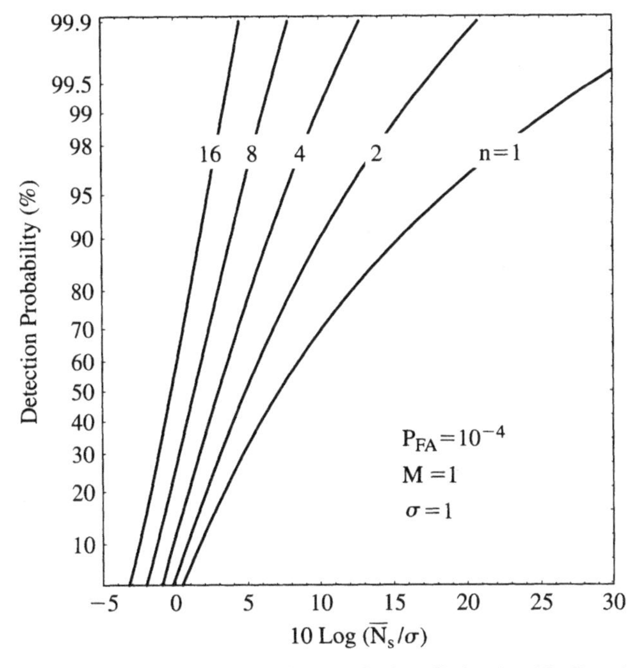

Figure 6-8. Integrated detection probability for a parabolic cylinder signal in Gaussian noise versus the single-pulse signal-to-noise ratio expressed in decibels. The integrated false alarm probability $P_{FA} = 10^{-4}$, a noise count $\sigma = 1$, and a speckle count of $M = 1$. The case of $n = 1$ is the same as the $M = 1$ curve in Figure 4-27.

Probability plots of Eq. (6-27) are shown in Figures 6-8 and 6-9 for several values of n with $M = 1$ and $M = 2$, $\sigma = 1$ and $P_{FA} = 10^{-4}$. It can be seen that considerable improvement in detection probability is realized with integration, especially at high values of P_D.

Integration loss for the Gaussian receiver operating against a fluctuating target is calculated in the same manner as for the previous cases. Using Eqs. (6-7), (6-8), and (6-30), L is shown as a function of n for $\sigma = 1$ in Figure 6-10. As in the Kummer distribution, there exists a regime of $L < 0$ characteristic of negative exponential intensity fluctuations.

6.3. COHERENT DETECTION SYSTEMS

The Swerling models are a set of assumptions about the rate of target fluctuations relative to the PRF of a coherent radar system that employs noncoherent integration. The models were introduced by Marcum[4] and by Swerling[5] to aid the system designer in selecting the optimum, or nearly optimum, detection strategy for coherent radar based on prior knowledge of the target statistics. Application of these models to coherent laser systems is a common design practice within the

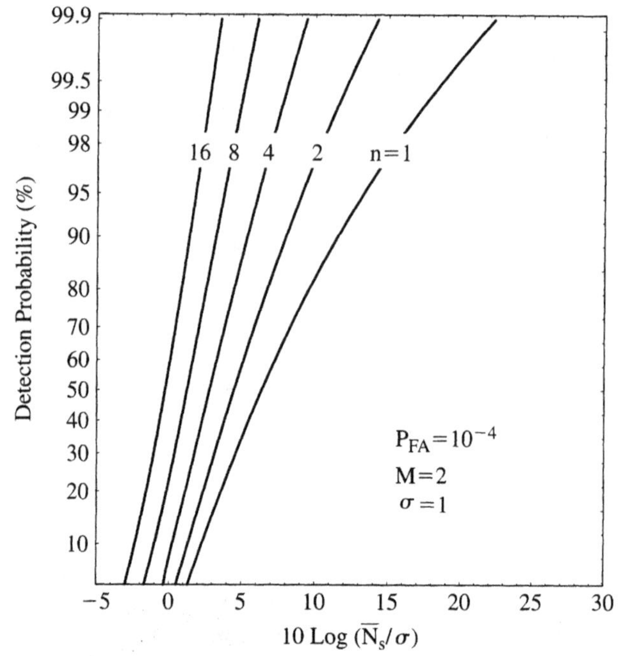

Figure 6-9. Integrated detection probability for a parabolic cylinder signal in Gaussian noise versus single-pulse signal-to-noise ratio expressed in decibels. The false alarm probability $P_{FA} = 10^{-4}$, a noise count $\sigma = 1$, and a speckle count of $M = 2$. The case of $n = 1$ is the same as the $M = 2$ curve in Figure 4-27.

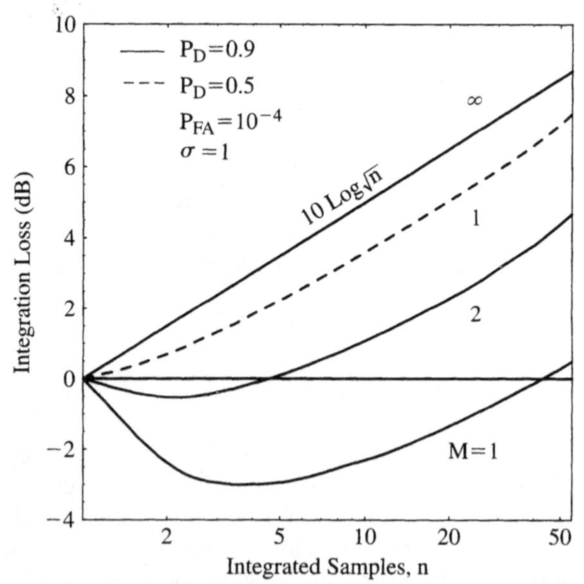

Figure 6-10. Integration loss for an n-pulse parabolic cylinder signal in Gaussian noise for a noise count of $\sigma = 1$, a false alarm probability $P_{FA} = 10^{-4}$, and detection probabilities $P_D = 0.9$ and 0.5. M is the speckle count at the receiver.

electro-optic community based on well-established measurements[6,7] that confirm the Rayleigh nature of amplitude fluctuations from either rough hard (surface) targets or soft (volume) targets. Swerling introduced four models for fluctuating targets, to which was added a fifth representing the nonfluctuating case.

Swerling Case 0. This model assumes a constant, unknown amplitude with random phase from pulse to pulse throughout the integration period.

Swerling Case I. This model assumes a constant, unknown amplitude with random phase from pulse to pulse within a scan but random fluctuations in amplitude from scan to scan for a Rayleigh fluctuating target. Restated, each scan consists of a constant-amplitude noncoherent pulse train similar to Swerling case 0.

Swerling Case II. This model assumes random fluctuations in amplitude and phase from pulse to pulse (and hence scan to scan) for a Rayleigh fluctuating target.

Swerling Case III. This model is the same as case I except that the target is a one-dominant-plus-Rayleigh fluctuating target.

Swerling Case IV. This model is the same as case II except that the target is a one-dominant-plus-Rayleigh fluctuating target.

Since the Swerling III and IV models are based on the one-dominant plus Rayleigh signal model and we have shown in Section 5.5 that this model does not properly represent a specular target immersed in a diffuse background, they will not be discussed here. Instead, multiple-pulse statistics will be developed in Section 6.3.4 for the Rician signal model corresponding to partially developed speckle.

Some definitions are warranted here. The word *scan* is used in the signal processing context to mean one integration period of, say, n pulses and is not related to angle scan. Hence the assumption in Swerling's case I is that the target decorrelation time is slow compared to the interpulse period but fast compared to the scan or integration time. This clean division of regimes is seldom true, of course, but it provides a reasonable and intuitive model from which one might estimate the best detection strategy.

The general assumption in the Swerling models is that the pulses are incoherent relative to each other. If this is not the case, the pulse train may be considered coherent with a fixed phase relationship from pulse to pulse, albeit with an unknown initial phase. But such a coherent pulse train can be optimally processed with a filter bandwidth matched to the full pulse train, not to an individual pulse. Thus, when applying the model to laser systems generating coherent pulse trains, such as those employing mode- or phase-locked transmitters, each *pulse train* may be considered a Swerling pulse.

One should also keep in mind that the Swerling models and their associated detection statistics (P_D, P_{FA}, etc) assume an idealized stationary white Gaussian

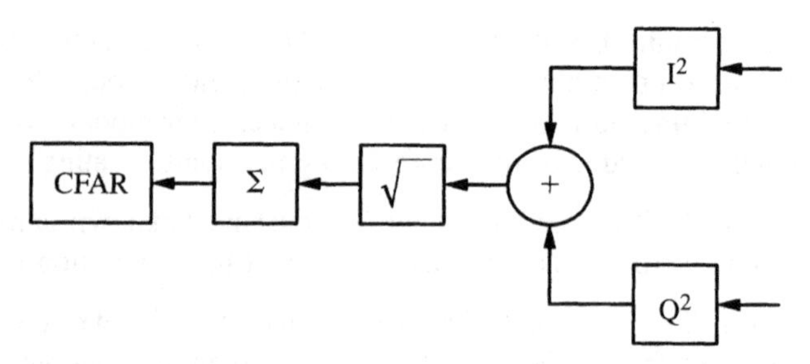

Figure 6-11. Receiver architecture for a multiple-pulse coherent detection receiver.

noise environment, which is generally realizable only in a clutter-free environment. Coherent laser systems operating in the shot-noise-limited regime via a strong local oscillator usually satisfy these requirements. However, the presence of rain, atmospheric backscatter, or other volume clutter effects can significantly alter the noise characteristics of the receiver, so that the Swerling models must always be considered limited in their range of applicability.

The optimum architecture for multiple-pulse integration in a coherent detection system is identical to that shown in Figure 5-2 with the addition of a summing device after the envelope processing. This is shown in Figure 6-11.

6.3.1. Swerling Case 0 Model

Consider a train of constant, but unknown, amplitude pulses each with random carrier phase as shown in Figure 6-12. The problem at hand is to develop detection statistics for the noncoherent integration of such a pulse train. Clearly, we must begin with the probability density function for a single pulse within that train and then proceed to develop the corresponding density function for n pulses. We do this by using the method of characteristic functions.

In Section 5.2 we developed the probability density function for a single pulse obtained from a matched filter, Eq. (5-25), which was rewritten in terms of the signal plus noise power at the output as

$$p_{s+n}(X_j;\, S) = \frac{1}{N} e^{-(X_j+S)/N} I_0 \left(\frac{2\sqrt{X_j S}}{N} \right) \tag{6-31}$$

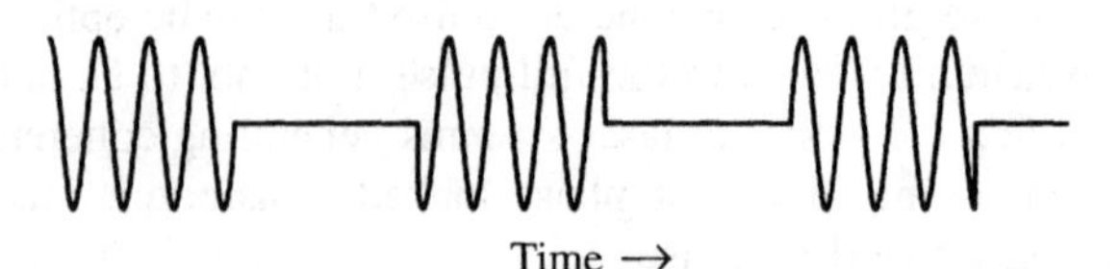

Figure 6-12. Constant amplitude signals with random initial phase assumed in Swerling case 0 model.

where X_j is the power variable of the jth pulse. Since Eq. (6-31) is identical in form to Eq. (4-110), its characteristic function should also be identical in form. Hence, identifying Eq. (6-31) as a noncentral χ^2 distribution with two degrees of freedom, we can write

$$p(y_j;\lambda) = \tfrac{1}{2}\, e^{-(y_j/2)-\lambda} I_0\left(\sqrt{2y_j\lambda}\right) \tag{6-32}$$

where $\lambda = S/N$ and $y_j = 2x_j/N$ for the jth pulse. The characteristic function is then given by

$$G(iv) = \frac{e^{-\lambda[1-(1-2iv)^{-1}]}}{1-2iv} \tag{6-33}$$

Now we wish to find the probability density function for the random variable Y representing n integrated pulses, where $Y = \sum_{j=1}^{n} Y_j$. Assuming random phases and therefore statistical independence from pulse to pulse, we have

$$\begin{aligned} G^{(n)}(iv) &= \prod_{i=1}^{n} \frac{e^{-\lambda[1-(1-2iv)^{-1}]}}{1-2iv} \\ &= \frac{e^{-n\lambda[1-(1-2iv)^{-1}]}}{(1-2iv)^n} \end{aligned} \tag{6-34}$$

Using the series representation of Eq. (6-34) given by Eq. (4-116) together with the procedures outlined in Section 4.6.1, the inverse transformation becomes

$$\begin{aligned} p_{s+n}^{(n)}(Y;S) &= \frac{1}{2\pi}\int_{-\infty}^{\infty} G^{(n)}(iv)e^{-ivY}\,dv \\ &= 2^{-(1/2)(n+1)}\left(\frac{Y}{n\lambda}\right)^{(1/2)(n-1)} e^{-(Y/2)-n\lambda} I_{n-1}(\sqrt{2n\lambda Y}) \end{aligned} \tag{6-35}$$

where $I_{n-1}(z)$ is the modified Bessel function of the first kind, order $n-1$. In terms of the signal and noise power at the output of the matched filter, Eq. (6-35) becomes, after transforming variables,

$$p_{s+n}^{(n)}(X;S) = \frac{1}{N}\left(\frac{X}{nS}\right)^{(1/2)(n-1)} e^{-(X+nS)/N} I_{n-1}\left(\frac{2\sqrt{nSX}}{N}\right) \tag{6-36}$$

where $X = NY/2$ and $\lambda = S/N$. Equation (6-36) represents the probability density function for an integrated train of n constant-amplitude, randomly phased pulses (or samples, as the case may be).

The mean and variance of $p_{s+n}^{(n)}(X;S)$ are readily shown to be

$$\langle X\rangle = n(S+N) \tag{6-37}$$

and

$$\operatorname{var}(X) = nN(2S+N) \tag{6-38}$$

Note that the integration process has increased both the mean and variance of the distribution by n.

We can calculate the false alarm probability by letting $S \to 0$ in $p_{s+n}^{(n)}(X; S)$ and integrating from the threshold level T to infinity. With the help of Eq. (4-125),

$$\lim_{z \to 0} \{I_{M-1}(z)/z^{M-1}\} = [2^{M-1}(M-1)!]^{-1} \tag{6-39}$$

we obtain for the noise density function

$$\begin{aligned} p_n^{(n)}(X; 0) &= \frac{1}{N}\left(\frac{X}{nS}\right)^{(1/2)(n-1)} \exp -\frac{X}{N} \frac{(2\sqrt{nXS}/N)^{n-1}}{2^{n-1}(n-1)!} \\ &= \frac{\alpha^n X^{n-1} e^{-\alpha X}}{\Gamma(n)} \end{aligned} \tag{6-40}$$

where $\alpha = 1/N$. But this is simply the well-known gamma distribution. The false alarm probability is therefore

$$\begin{aligned} P_{\text{FA}}^{(n)} &= \int_T^\infty p_n^{(n)}(X; 0)\, dX \\ &= \frac{1}{N} \int_T^\infty \frac{(X/N)^{n-1} e^{-X/N}}{\Gamma(n)} dX \\ &= \frac{1}{\Gamma(n)} \int_{T/N}^\infty Z^{n-1} e^{-Z} dZ \end{aligned} \tag{6-41}$$

where $Z = X/N$ and $T =$ threshold power level. The integral in Eq. (6-41) is identifiable as the incomplete gamma function $\Gamma(n, Z)$, so that

$$P_{\text{FA}} = \frac{\Gamma(n, T/N)}{\Gamma(n)} \tag{6-42}$$

From the definition of the gamma function and the incomplete gamma function, it is not too difficult to show that for $n = 1$, P_{FA} reduces to the single-pulse false alarm probability $P_{\text{fa}} = \exp(-T/N)$ derived in Section 5.2.

The detection probability for Swerling case 0 is given by the integral

$$\begin{aligned} P_D^{(n)} &= 1 - \int_0^T p_n^{(n)}(X, S)\, dX \\ &= 1 - \frac{1}{N} \int_0^T \left(\frac{X}{nS}\right)^{(1/2)(n-1)} e^{-(X+nS)/(N)} I_{n-1}\left(\frac{2\sqrt{nXS}}{N}\right) dX \end{aligned} \tag{6-43}$$

Equation (6-43) can be shown to be related to the Toronto function that has been tabulated by Heatley[8] (cf. Appendix A). However, with the advent of modern analysis tools, it is more convenient to use a numerical integration routine. The

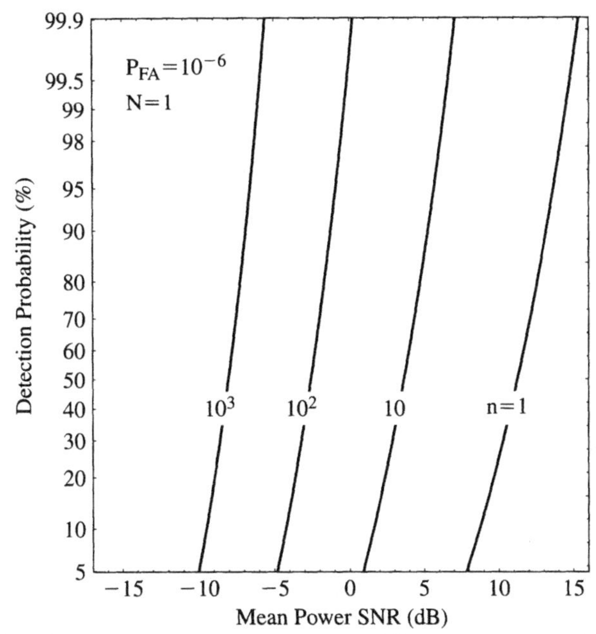

Figure 6-13. Swerling case 0 integrated detection probabilities versus mean single-pulse power signal-to-noise ratio, SNR(dB), for the case of $P_{FA} = 10^{-6}$ and $N = 1$. Plots are shown for several representative values of the number of pulses integrated, n.

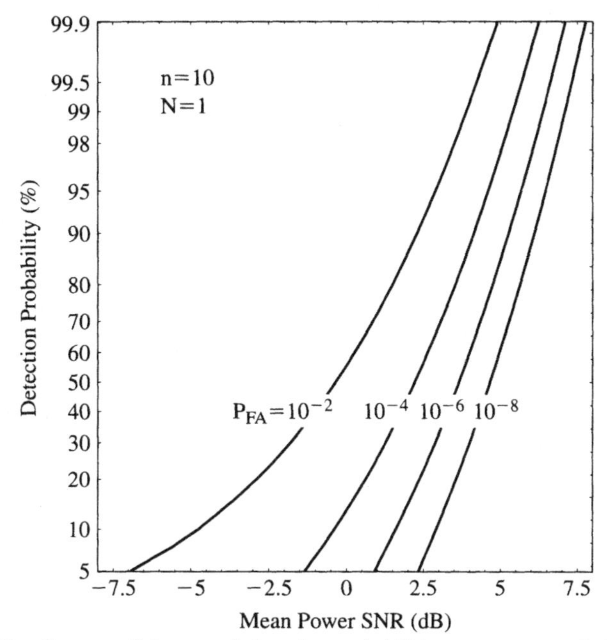

Figure 6-14. Swerling case 0 integrated detection probabilities versus mean single-pulse power signal-to-noise ratio, SNR(dB), for the case of $n = 10$ pulses and $N = 1$. Plots are shown for several representative values of the false alarm probability P_{FA}.

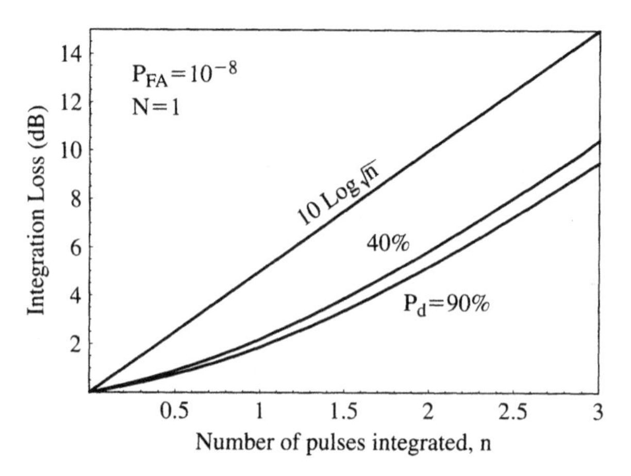

Figure 6-15. Integration loss for the Swerling Case 0 model.

result is shown in Figure 6-13 where detection probability is plotted as a function of mean power SNR for a false alarm probability $P_{FA} = 10^{-6}$ and several values of n. Figure 6-14 shows the family of false alarm curves associated with the $n = 10$ case. Integration loss for the Swerling Case 0 is shown in Figure 6-15.

Comment 6-3. The Swerling statistics frequently exhibit gamma-related functions. Historically, radar detection statistics have been expressed in terms of Pearson's incomplete gamma function, defined as

$$I(u, p) = \frac{1}{\Gamma(p+1)} \int_0^{u\sqrt{p+1}} t^p e^{-t}\, dt \tag{6-44}$$

A simple change in variables shows that $I(u, p)$ is related to the more standard incomplete gamma function $\Gamma(a, z)$ through the relationship

$$I(u, p) = \frac{\gamma(p+1, u\sqrt{p+1})}{\Gamma(p+1)} = 1 - \frac{\Gamma(p+1, u\sqrt{p+1})}{\Gamma(p+1)} \tag{6-45}$$

Thus for example, Eq. (6-42) for the false alarm probability may be written in the equivalent forms

$$\begin{aligned} P_{FA} &= \frac{\Gamma(n, T/N)}{\Gamma(n)} = 1 - \frac{\gamma(n, T/N)}{\Gamma(n)} \\ &= 1 - I\left(n-1, \frac{T}{\sqrt{n}N}\right) \end{aligned} \tag{6-46}$$

■

6.3.2. Swerling Case I Model

The Swerling case I model assumes a constant-amplitude, noncoherent (i.e., randomly phased) pulse train during each scan period with a Rayleigh distribution in amplitude from scan to scan. This is shown graphically in Figure 6-16. But we have already developed the probability density function for a train of constant-amplitude, randomly phased pulses, namely Eq. (6-36). Hence the Swerling case I problem becomes one of generalizing Eq. (6-36) to the case of the signal also being a random variable and convolving the resultant density function with the density function for that signal. To do this, it is convenient to transform back to the original voltage representation by letting $r = \sqrt{2X}$, $\sigma = \sqrt{N}$, and $a_s = \sqrt{2S}$, and write

$$p_{s+n}(r) = \int_0^\infty p(r \mid a_s) p(a_s)\, da_s \tag{6-47}$$

where

$$p(r \mid a_s) = \frac{r}{\sigma^2} \left(\frac{r^2}{na_s^2} \right)^{(1/2)(n-1)} e^{-(r^2+na_s^2)/2\sigma^2} I_{n-1} \left(\frac{ra_s\sqrt{n}}{\sigma^2} \right) \tag{6-48}$$

and

$$p(a_s) = \frac{a_s}{a_o^2} e^{-a_s^2/2a_o^2} \tag{6-49}$$

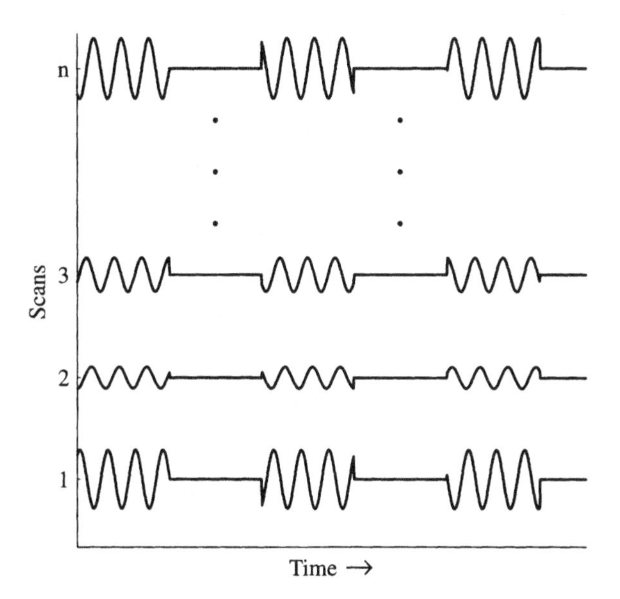

Figure 6-16. Sequence of processing scans for the Swerling case I model.

Inserting Eqs. (6-48) and (6-49) into Eq. (6-47) and rearranging yields

$$p_{s+n}(r) = K \int_0^\infty a_s^{2-n} e^{-ba_s^2/2} I_{n-1}\left(\frac{ra_s\sqrt{n}}{\sigma^2}\right) da_s \tag{6-50}$$

where

$$K = \frac{n^{(1/2)(1-n)} r^n e^{-r^2/2\sigma^2}}{\sigma^2 a_o^2} \quad \text{and} \quad b = \frac{\sigma^2 + na_o^2}{\sigma^2 a_o^2} \tag{6-51}$$

The integral in Eq. (6-50) may be found in various tables of integrals[9] and evaluates to

$$\int_0^\infty f(a_s)\, da_s = b^{n-2}\left(\frac{nr^2}{\sigma^4}\right)^{(1-n)/2} \times \exp\left[\frac{nr^2}{2b\sigma^4} \frac{\gamma(n-1, nr^2/2b\sigma^4)}{\Gamma(n-1)}\right] \tag{6-52}$$

where $\gamma(n-1, z)$ is the *alternate* incomplete gamma function related to the gamma and incomplete gamma function through the relation $\gamma(n-1, z) = \Gamma(n-1) - \Gamma(n-1, z)$. Substituting Eqs. (6-51) and (6-52) into Eq. (6-50) yields, after some algebra,

$$p_{s+n}^{(n)}(r) = \frac{(\sigma^2 + na_o^2)^{n-2}}{(na_o^2)^{n-1}} r \times \exp\left[-\frac{r^2}{2(\sigma^2 + na_o^2)} \frac{\gamma(n-1, (na_o^2/2\sigma^2)[r^2/(\sigma^2 + na_o^2)])}{\Gamma(n-1)}\right] \tag{6-53}$$

where again the superscript indicates an n-pulse integration. In terms of the power variable $X = r^2/2$, Eq. (6-53) can be transformed to read

$$p_{s+n}^{(n)}(X) = \frac{d^{n-2}}{na_o^2} e^{-X/na_o^2 d} \frac{\gamma(n-1, X/\sigma^2 d)}{\Gamma(n-1)} \tag{6-54}$$

where $d = 1 + \sigma^2/na_o^2$.

The false alarm probability is the same as that for the Swerling case 0 model [i.e., Eq. (6-42)], while the detection probability becomes

$$P_D = \int_T^\infty p_{s+n}^{(n)}(X)\, dX = \frac{d^{n-2}}{na_o^2 \Gamma(n-1)} \int_T^\infty e^{-X/na_o^2 d} \gamma\left(n-1, \frac{X}{\sigma^2 d}\right) dX \tag{6-55}$$

Equation (6-55) is relatively straightforward to solve.[10] Using the integral definition of the alternative incomplete gamma function given by

$$\gamma(a, z) = \int_0^z t^{a-1} e^{-t}\, dt \tag{6-56}$$

in Eq. (6-55), we have

$$P_D = \frac{d^{n-2}}{na_o^2\Gamma(n-1)} \int_T^\infty dX e^{-X/na_o^2 d} \int_0^{X/\sigma^2 d} dt\, t^{n-2} e^{-t} \tag{6-57}$$

The two-dimensional integration given by Eq. (6-57) represents the area under the curve $t = X/\sigma^2 d$ and the x-axis between $x = T$ and infinity. However, it can also be represented as the sum of the areas shown in Figure 6-17 as regions 1 and 2. Thus

$$P_D = \frac{d^{n-2}}{na_o^2\Gamma(n-1)} \left[\int_0^{T/\sigma^2 d} dt\, t^{n-2} e^{-t} \int_t^\infty dX e^{-X/na_o^2 d} + \int_{T/\sigma^2 d}^{0} dt\, t^{n-2} e^{-t} \int_{t\sigma^2 d}^\infty dX\, e^{-X/na_o^2 d} \right] \tag{6-58}$$

Evaluating the integrals over X results in

$$P_D = \frac{d^{n-2}(na_o^2 d)}{na_o^2\Gamma(n-1)} \left(e^{-T/na_o^2 d} \int_0^{T/\sigma^2 d} dt\, t^{n-2} e^{-t} + \int_{T/\sigma^2 d}^{0} dt\, t^{n-2} e^{-dt} \right) \tag{6-59}$$

This can be rewritten as

$$P_D = \frac{d^{n-1}}{\Gamma(n-1)} \left[e^{-T/na_o^2 d} \gamma(n-1, T/\sigma^2 d) + \frac{1}{d^{n-1}} \Gamma(n-1, T/\sigma^2 d) \right] \tag{6-60}$$

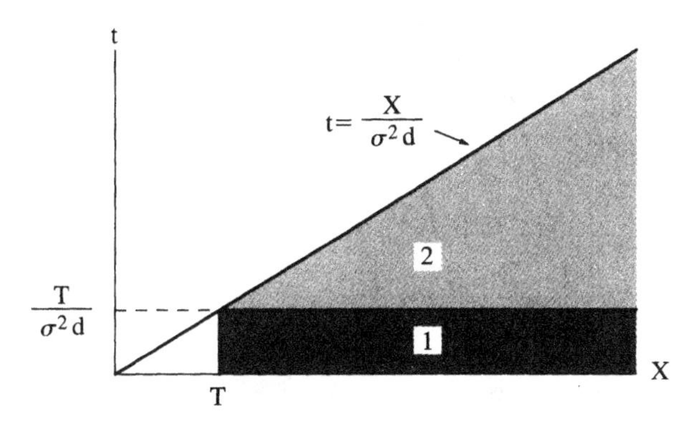

Figure 6-17. Regions of integration for evaluating the integral in Eq. (6-55).

or

$$P_D = \frac{\Gamma(n-1, T/\sigma^2)}{\Gamma(n-1)} + d^{n-1} e^{-T/na_o^2 d} \frac{\gamma(n-1, T/\sigma^2 d)}{\Gamma(n-1)} \tag{6-61}$$

where $d = 1 + \sigma^2/na_o^2$.

Figure 6-18 shows Eq. (6-61) plotted as a function of the mean single-pulse power signal-to-noise ratio given by SNR $= a_o^2/\sigma^2$ with $\sigma = 1$, expressed in decibels, for several values of n and a false alarm probability $P_{FA} = 10^{-6}$. Figure 6-19 shows one set of false alarm curves for the case of $n = 10$.

Integration loss for Swerling case I is shown in Figure 6-20. Here we have defined the integration loss L as the ratio $n\text{SNR}_n/\text{SNR}_1$, where SNR_n is the power signal-to-noise ratio required for n pulses with the prescribed detection probability and false alarm probability, and SNR_1 is the single-pulse signal-to-noise ratio required for the same set of detection statistics. It can be seen that L is positive for all n, implying that single-pulse detection is always more efficient than multiple-pulse detection when Swerling case I targets are being detected. The $10 \log \sqrt{n}$ curve corresponds to a video integrator and the x-axis a coherent integrator.

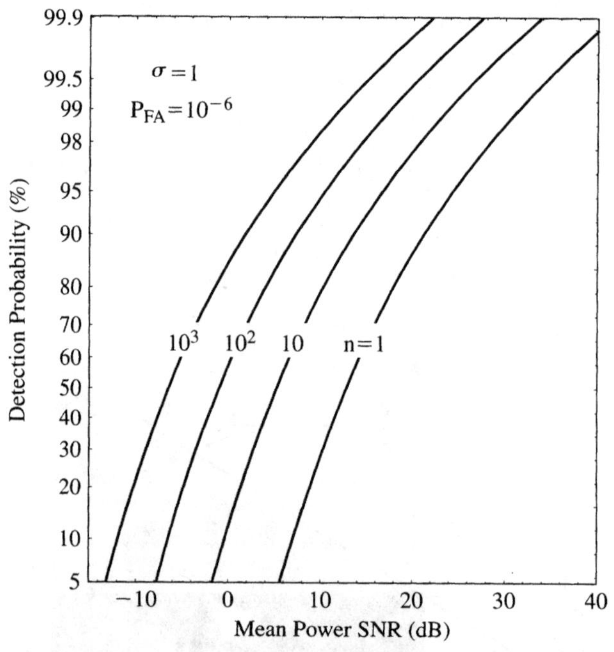

Figure 6-18. Swerling case I integrated detection probabilities versus mean single-pulse power signal-to-noise ratio, SNR(dB), for the case of $P_{FA} = 10^{-6}$ and $\sigma = 1$. Plots are shown for several representative values of the number of pulses integrated, n.

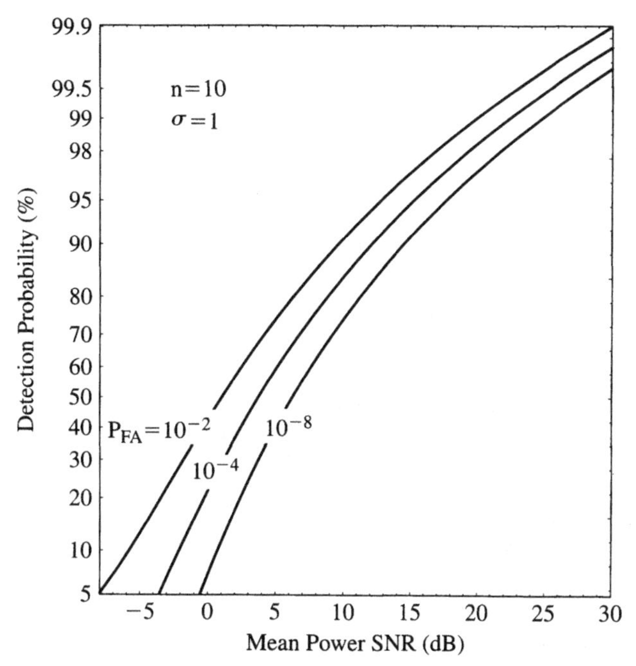

Figure 6-19. Swerling case I integrated detection probabilities versus mean single-pulse power signal-to-noise ratio, SNR(dB), for the case of $n = 10$ pulses and $\sigma = 1$. Plots are shown for several representative values of the false alarm probability P_{FA}.

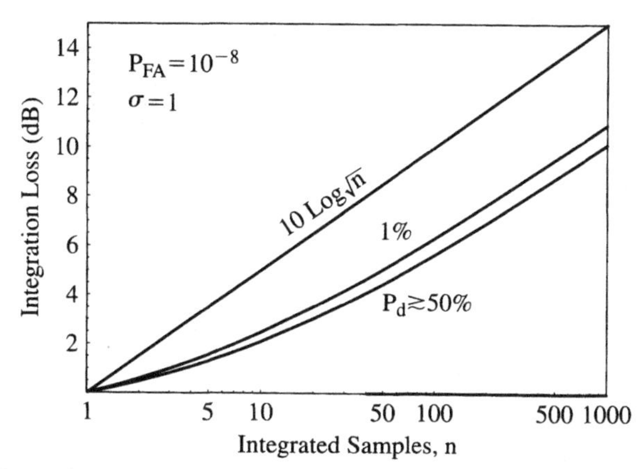

Figure 6-20. Integration loss for a Swerling case I target. The probability of false alarm is $P_{FA} = 10^{-8}$ and $\sigma = 1$.

6.3.3. Swerling Case II Model

The Swerling case II target model assumes a signal that is Rayleigh distributed in amplitude and uniformly distributed in phase from pulse to pulse as shown in

Figure 6-21. In this case, we start with Eq. (5-36) for the single-pulse probability density function of a Rayleigh fluctuating signal obtained from a matched filter, that is,

$$p(r) = \beta r e^{-\beta r^2/2} \tag{6-62}$$

where $\beta = (\sigma^2 + a_o^2)^{-1}$. Expressing Eq. (6-62) in terms of the power variable $X = r^2/2$, we obtain

$$p(X) = \beta e^{-\beta X} \tag{6-63}$$

As expected, this is the negative exponential distribution with parameter β^{-1}. The characteristic function was developed in Eq. (4-22), where it was shown to be given by

$$G(iv) = \frac{1}{1 - iv/\beta} \tag{6-64}$$

For n independent pulses,

$$G^{(n)}(iv) = \frac{1}{(1 - iv/\beta)^n} \tag{6-65}$$

from which the inverse transformation yields the gamma distribution (cf. Appendix B), namely

$$p(X) = \frac{\beta^n X^{n-1} e^{-\beta X}}{\Gamma(n)} \tag{6-66}$$

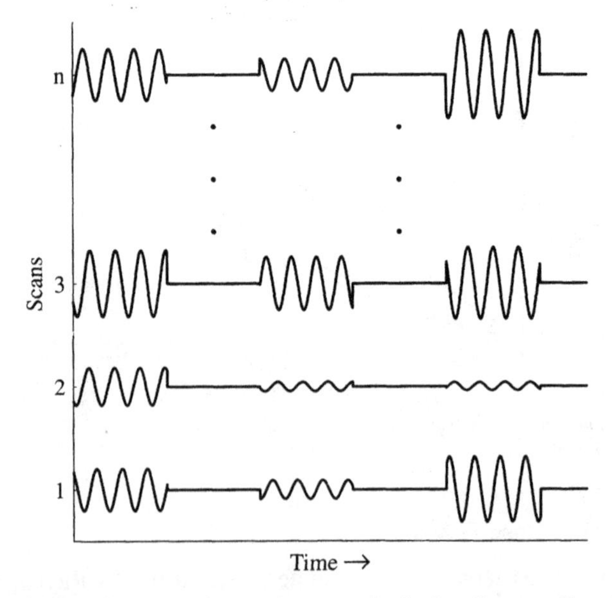

Figure 6-21. Sequence of processing scans for the Swerling case II model.

The detection probability for a Swerling case II target is therefore

$$\begin{aligned} P_D &= \int_T^\infty \frac{\beta^n X^{n-1} e^{-\beta X}}{\Gamma(n)} \, dX \\ &= \int_{\beta T}^\infty \frac{Y^{n-1} e^{-Y}}{\Gamma(n)} \, dY \\ &= \frac{\Gamma(n, \beta T)}{\Gamma(n)} \end{aligned} \tag{6-67}$$

where the variable of integration $Y = \beta X$ was used. Note that for a single pulse, $n = 1$, and

$$P_d = \int_{\beta T}^\infty e^{-y} \, dy = e^{-T/\sigma^2 + a_o^2} = e^{-t^2/2(\sigma^2 + a_o^2)} \tag{6-68}$$

which is identical to Eq. (5-37).

Once again, the false alarm probability is the same as those for Swerling cases 0 and I [i.e., Eq. (6-42)]. The reason for this is that all the receivers have the same noise properties, only the signal characteristics being different from model to model: thus for completeness,

$$P_{\mathrm{FA}} = \frac{\Gamma(n, T/\sigma^2)}{\Gamma(n)} \tag{6-69}$$

Equations (6-67) and (6-69) were used to generate the detection probabilities shown in Figures 6-22 and 6-23 as a function of the mean single-pulse power signal-to-noise ratio defined as SNR $= a_o^2/\sigma^2$ expressed in decibels and with $\sigma = 1$.

Integration loss for Swerling case II is shown in Figure 6-24. Once again integration loss L is defined as the ratio $n\mathrm{SNR}_n/\mathrm{SNR}_1$, where SNR_n is the power signal-to-noise ratio required for n pulses with the prescribed detection probability and false alarm probability, and SNR_1 is the single-pulse signal-to-noise ratio required for the same set of detection statistics. In this case $L < 0$ for large detection probabilities over a broad range of integrated pulses n. In this region of the curve we can conclude that it is always more efficient to detect using multiple pulses than a single pulse, a consequence of the fluctuating signal of the Swerling case II target. The $10 \log \sqrt{n}$ curve again corresponds to the case of a video integrator and the x-axis a coherent integrator.

6.3.4. Rician Signal Model

Section 5.5 addressed the Rician signal model in Gaussian noise for the case of a single observation. There it was found that an analytic solution for the signal plus noise resulted in the form of a Rician probability density function. To obtain the

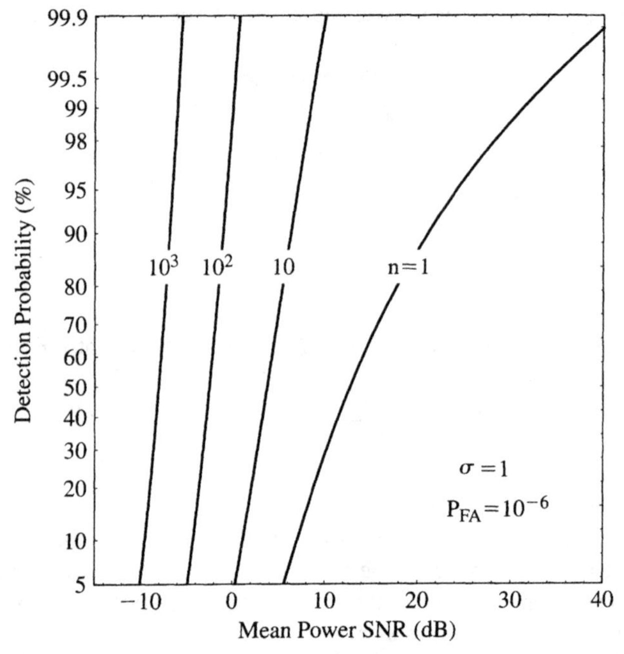

Figure 6-22. Swerling case II integrated detection probabilities versus mean single-pulse power signal-to-noise ratio, SNR(dB), for the case of $P_{FA} = 10^{-6}$ and $\sigma = 1$. Plots are shown for several representative values of the number of pulses integrated, n.

multiple-pulse statistics for this model, it is convenient to transform Eq. (5-52) to the power domain so that advantage can be taken of results obtained in Chapter 4. Thus, expressing Eq. (5-25) in terms of *mean* power variables rather than *peak* power variables, we can immediately write

$$p_{s+n}(X) = \begin{cases} \dfrac{1}{N_n} e^{-(X+X_c)/N_n} I_0\left(\dfrac{2\sqrt{XX_c}}{N_n}\right) & X \geq 0 \\ 0 & X < 0 \end{cases} \tag{6-70}$$

where $X = r^2/2$, $X_c = a_c^2/2$, and $N_n = \sigma^2 + a_o^2$. But Eq. (6-70) is just the modified Rician distribution given by Eq. (4-110), which also corresponds to a noncentral χ^2-distribution with two degrees of freedom. Following the derivation in Eqs. (4-115) through (4-121), the characteristic function for n pulses, when expressed in terms of χ^2 variables, is found to be

$$G^{(n)}(i\upsilon) = \frac{e^{2i\upsilon n\lambda_n/(1-2i\upsilon)}}{(1-2i\upsilon)^n} \tag{6-71}$$

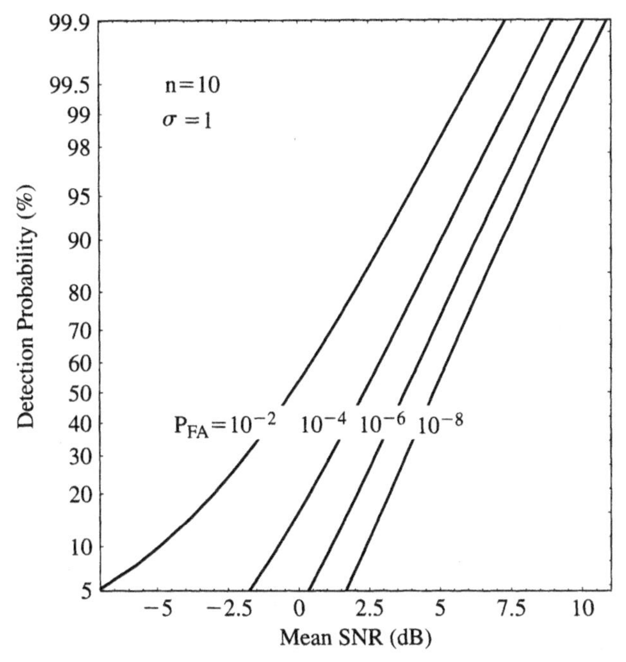

Figure 6-23. Swerling case II integrated detection probabilities versus mean single-pulse power signal-to-noise ratio, SNR(dB), for the case of $n = 10$ pulses and $\sigma = 1$. Plots are shown for several representative values of the false alarm probability P_{FA}.

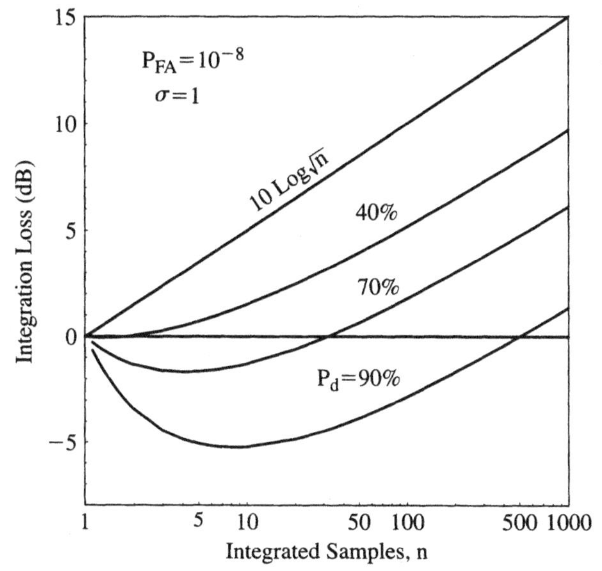

Figure 6-24. Integration loss for a Swerling case II target. The probability of false alarm is $P_{FA} = 10^{-8}$ and $\sigma = 1$.

where $\lambda_n = X_c/N_n$. The inverse transformation on Eq. (6-71) then becomes[11]

$$p_{s+n}^{(n)}(X) = \begin{cases} \dfrac{n}{N_n}\left(\dfrac{X}{X_c}\right)^{(1/2)(n-1)} e^{-n(X+X_c)/N_n} I_{n-1}\left(\dfrac{2n\sqrt{XX_c}}{N_n}\right) & X \geq 0 \\ 0 & X < 0 \end{cases} \tag{6-72}$$

which is identical in form to Eq. (4-121).

To maintain the connection with the single-pulse case of Chapter 5, Eq. (6-72) will be transformed back to the voltage representation, yielding

$$p_{s+n}^{(n)}(r) = \begin{cases} \dfrac{nr}{\sigma^2 + a_o^2}\left(\dfrac{r^2}{a_c^2}\right)^{(1/2)(n-1)} e^{-n(r^2+a_c^2)/(\sigma^2+a_o^2)} I_{n-1}\left(\dfrac{nra_c}{\sigma^2 + a_o^2}\right) & r \geq 0 \\ 0 & r < 0 \end{cases} \tag{6-73}$$

This can be written in terms of the coherence parameter κ and the input power signal-to-noise ratio (SNR) in the same manner as in Section 5.5 for the single-pulse case. Thus we let

$$\kappa = \frac{a_c^2}{2a_o^2} \tag{6-74}$$

and

$$\mathrm{SNR} = \frac{a_o^2 + a_c^2/2}{\sigma^2} = \frac{a_o^2}{\sigma^2}(1+\kappa) \tag{6-75}$$

obtaining

$$p_{s+n}^{(n)}(r) = \frac{nr(1+\kappa)}{\sigma^2(1+\kappa+\mathrm{SNR})}\left[\frac{r^2(1+\kappa)}{2\kappa\sigma^2\mathrm{SNR}}\right]^{(1/2)(n-1)} \times e^{-[nr^2(1+\kappa)/2\sigma^2 + n\kappa\mathrm{SNR}]/(1+\kappa+\mathrm{SNR})} I_0\left[\frac{nr\sqrt{2\mathrm{SNR}\kappa(1+\kappa)}}{\sigma(1+\kappa+\mathrm{SNR})}\right] \tag{6-76}$$

The detection statistics are given by

$$P_{\mathrm{FA}} = \frac{\Gamma(n, t^2/2\sigma^2)}{\Gamma(n)} \tag{6-77}$$

and

$$P_D = 1 - \int_0^t p_{s+n}^{(n)}(r;\kappa)\,dr \tag{6-78}$$

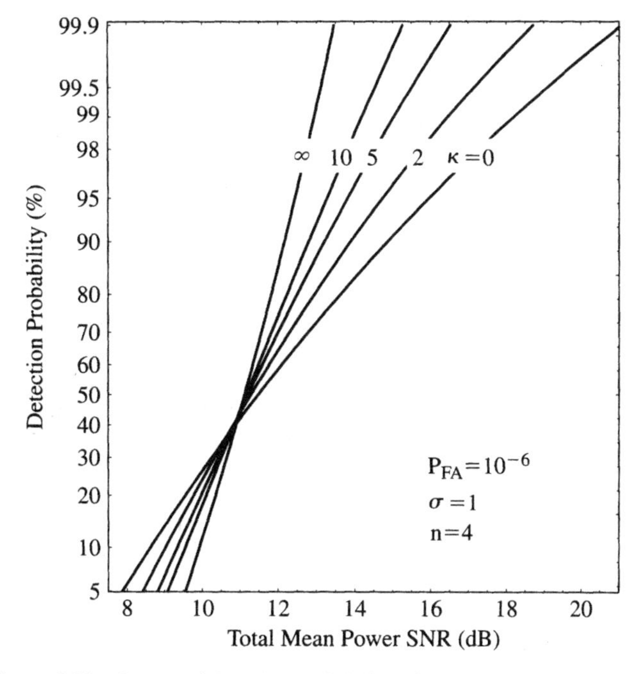

Figure 6-25. Integrated detection probabilities for the Rician signal model.

Numerical integration of Eq. (6-78) is plotted in Figure 6-25 for several values of κ with $\sigma = 1$ and $n = 4$. As in the case of Figure 5-14, the curves are seen to crossover at low values of detection probability for any given n when plotted with κ as a parameter and converge to the Swerling case II curves for $\kappa = 0$ and to the constant-amplitude case for $\kappa = \infty$. Given these limiting cases, integration loss for the Rician signal model will lie between the bounds given by Figures 6-15 and 6-24, depending on the value of κ.

6.4. BINARY INTEGRATION

The integration processes discussed thus far represent summing procedures that are performed prior to thresholding and detection. Other schemes based on algorithmic techniques are also possible. One such scheme is the binary integration algorithm, also referred to as the *coincidence detection* or *double-thresholding algorithm*. With this approach, thresholding and detection are done on a pulse-to-pulse basis followed by a detection criteria based on the number of pulses m detected out of a total number n transmitted. Algorithmic approaches such as this offer the potential advantages of reduced signal processing hardware and better matches to target characteristics, although signal processing speeds must match the full PRF of the transmitter.

The probability P that *exactly* j out of n pulses will be detected when a signal is present is given by the binomial distribution,

$$P_d(j:n) = \binom{n}{j} P_d^j (1 - P_d)^{n-j} \tag{6-79}$$

where

$$\binom{n}{j} = \frac{n!}{j!(n-j)!} \tag{6-80}$$

are the binomial coefficients that represent the number of ways in which j successes out of n trials can occur, P_d represents the single-pulse probability of detection, and $(1 - P_d)$ is the single-pulse probability of a miss.

A detection is declared when the number of single-pulse detections, j, satisfies the condition $m \leq j \leq n$; hence the overall detection probability P_D is given by the cumulative binomial distribution

$$\begin{aligned} P_D(m:n) &= \sum_{j=m}^{n} \binom{n}{j} P_d^j (1 - P_d)^{n-j} \\ &= 1 - \sum_{j=0}^{m-1} \binom{n}{j} P_d^j (1 - P_d)^{n-j} \end{aligned} \tag{6-81}$$

Similarly, the probability that exactly j false alarms will occur out of n trials when no signal is present is given by

$$P_{\text{fa}}(j:n) = \binom{n}{j} P_{\text{fa}}^j (1 - P_{\text{fa}})^{n-j} \tag{6-82}$$

The total false alarm probability is then the sum over all possible false events, that is,

$$\begin{aligned} P_{\text{FA}}(m:n) &= \sum_{j=m}^{n} \binom{n}{j} P_{\text{fa}}^j (1 - P_{\text{fa}})^{n-j} \\ &= 1 - \sum_{j=0}^{m-1} \binom{n}{j} P_{\text{fa}}^j (1 - P_{\text{fa}})^{n-j} \end{aligned} \tag{6-83}$$

where P_{fa} is the single-pulse false alarm probability and $(1 - P_{\text{fa}})$ is the single-pulse probability of no false alarms.

Equation (6-83) can be greatly simplified when it is realized that $P_{\text{fa}} \ll 1$ (typically, 10^{-3} or less); hence

$$P_{\text{FA}}(m:n) \approx \binom{n}{m} P_{\text{fa}}^m \tag{6-84}$$

Equations (6-80) and (6-82) constitute the basic algorithms for the binary integration procedure, given P_d and P_{fa} as obtained from the probability density functions for the signal plus noise and the noise, respectively. The algorithm is frequently referred to as the *m-out-of-n algorithm* because of its relationship to binomial statistics.

Schwartz[12] has shown that an optimum m exists, which will be denoted as m_{opt}, that minimizes the required single-pulse signal-to-noise ratio for any given n. He showed that for a constant-amplitude radar signal, m_{opt} follows closely a relationship given by $b\sqrt{n}$, where b is a constant approximately equal to 1.5. However, in general, the constant b depends on the noise characteristics of the target/receiver system such that each case must be evaluated independently.

As an example of the application of binary integration to the direct-detection optical receiver, consider the case of the Poisson signal and noise statistics associated with specular targets. To determine the optimum m, we first choose $n = 8$ and 16 as two representative examples, and let the mean noise photoelectron count $\overline{N}_n = 1$, and the false alarm probability $P_{FA} = 10^{-6}$ for comparison with Figures 4-11 and 6-2. Since $P_{fa} \ll 1$, we use Eq. (6-84) to calculate the single-pulse false alarm probability for each value of m, that is,

$$P_{fa} = \left[\frac{m!(n-m)!}{n!} P_{FA}\right]^{1/m} \tag{6-85}$$

The corresponding threshold values k_{th} may be obtained from Eq. (4-61), that is,

$$P_{fa} \geq 1 - \sum_{k=0}^{k_{th}-1} \frac{(\overline{N}_n)^k e^{-\overline{N}_n}}{k!} \tag{6-86}$$

Having obtained k_{th}, the corresponding single-pulse detection probability given by Eq. (4-62),

$$P_d = 1 - \sum_{k=0}^{k_{th}-1} \frac{(\overline{N}_s + \overline{N}_n)^k e^{-(\overline{N}_s + \overline{N}_n)}}{k!} \tag{6-87}$$

can then be used in Eq. (6-81) to find the mean signal photoelectron count $\overline{N}_s$ that produces the desired integrated detection probability P_D for each value of m.

Figures 6-26 and 6-27 show plots of the required single-pulse photoelectron count $\overline{N}_s$, expressed in decibels, versus m for the cases of $n = 8$ and $n = 16$. Minima are seen to occur in the figures at $m_{opt} = 4$ and 8, respectively, which are exactly $n/2$. Notice that the data points are relatively flat in the vacinity of m_{opt} when n is large so that the choice of m_{opt} is not critical in such cases. The fact that an optimum m exists follows from the fact that large signals are required to overcome the larger thresholds (smaller P_{fa} values) that occur for small m, while large signals are again required to achieve a high percentage of detections as m approaches n.

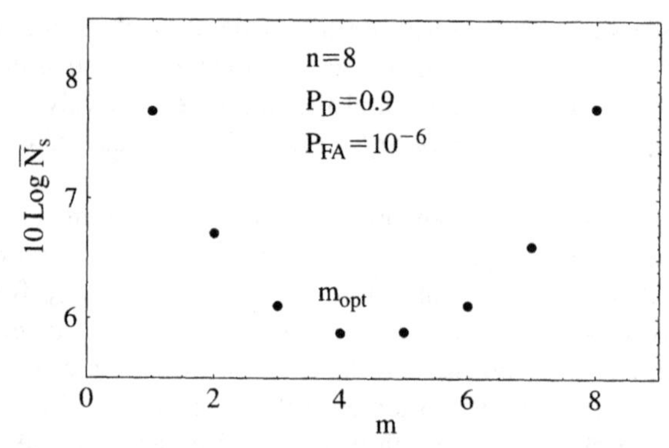

Figure 6-26. Optimum m for a binary integrated Poisson signal from a specular target. $M_{\text{opt}} = 4$ for $n = 8$ samples.

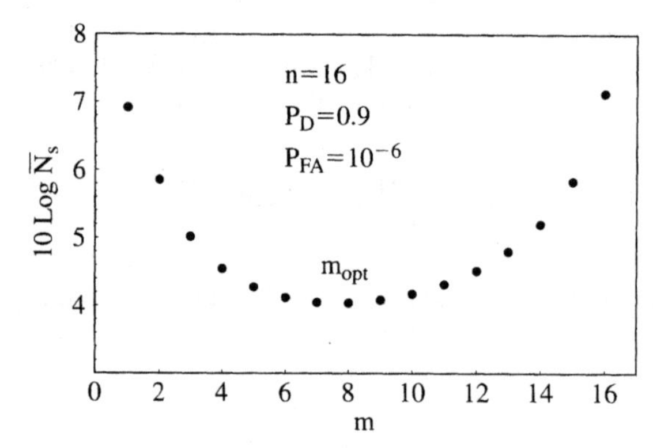

Figure 6-27. Optimum m for a binary integrated Poisson signal in Poisson noise. $M_{\text{opt}} = 8$ for $n = 16$ samples.

The question now arises as to the signal-to-noise ratio performance of binary detection compared to conventional integration procedures. Figure 6-28 shows this comparison wherein the $n = 16$ curve of Figure 6-2 is compared to the corresponding 8-out-of-16 curve for the binary detection case. It can be seen that a little over 1 dB of additional signal is required for the binary detection process. This is consistent with calculations of other target and receiver configurations, including coherent detection receivers, where degradations of 1 to 2 dB are typical.

6.4.1. Application to Geiger-Mode APD Detectors

As was mentioned in Section 4.8.3, background radiation and dark noise are sources of false alarms for Geiger-mode APD detectors. The achievable dark-count rates of one per second in silicon devices generally do not present a problem for laser systems having processing windows on the order of microseconds or less.

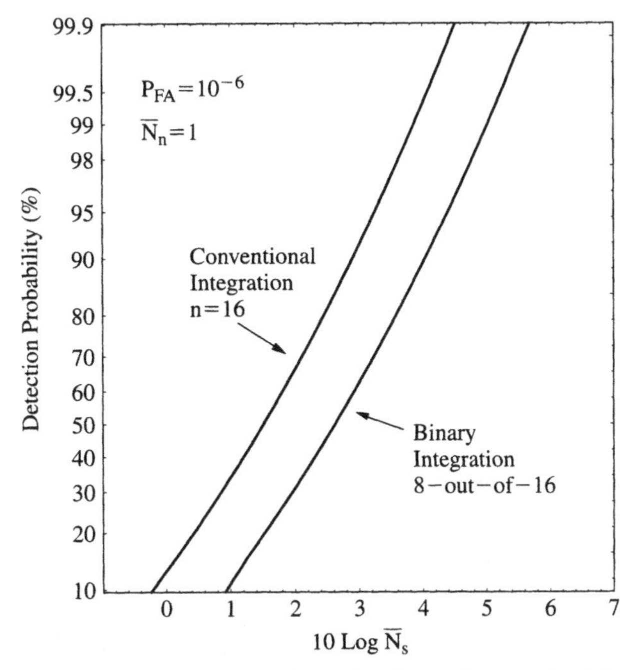

Figure 6-28. Performance comparison of binary detection and conventional integration for a Poisson signal in Poisson noise for the case of $n = 16$ samples.

However, background radiation can present a problem for daytime use despite the availability of ultra-narrowband (<1 nm) optical filters. Further reduction in the false alarm probability can be achieved by using various multiple-pulse integration techniques. In this section we show that coincidence detection offers a convenient method for achieving acceptable target-to-background discrimination for a variety of applications.[13]

Consider a set of n processing windows each with T measurement intervals as shown in Figure 6-29. If we transmit n pulses and set a coincidence threshold for detecting at least m out of n pulses that occur *in the same measurement interval*, we can use Eqs. (6-81) and (6-83) to predict the overall detection and false alarm probabilities. Restricting the analysis once again to the case of the signal residing in either the first or last measurement intervals of the processing window, Eq. (6-81) can be written as

$$P_D(m:n) = 1 - \sum_{j=0}^{m-1} \binom{n}{j} P_d^j(1,T)[1 - P_d(1,T)]^{n-j} \tag{6-88}$$

where $P_d(1, T)$ represents the single-pulse detection probabilities for the first and last measurement intervals, respectively, as given by Eqs. (4-220) and (4-221). On the other hand, there are $t = T - 1$ potential measurement intervals that can

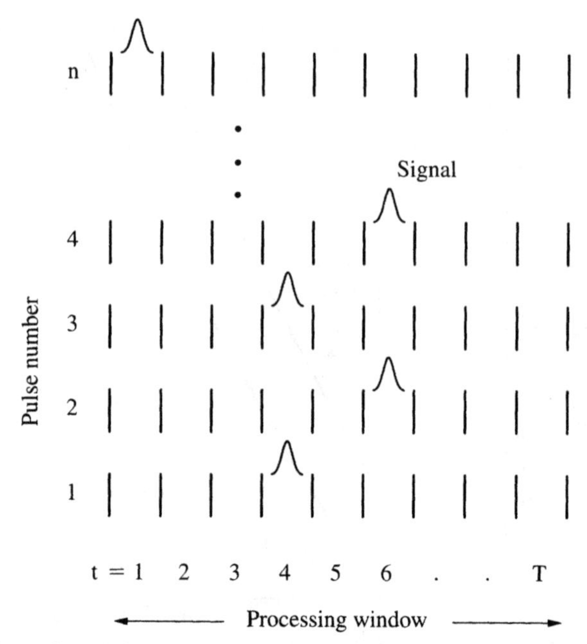

Figure 6-29. Processing windows and measurement intervals for a Geiger-mode APD receiver and processor using n-pulse coincidence detection.

generate a false alarm, so that with the notation of Section 4.8.3 we must write

$$P_{\mathrm{FA}}(m:n) = \sum_{t=1}^{T-1} \left\{ 1 - \sum_{j=0}^{m-1} \binom{n}{j} P_{\mathrm{fa}}^{j}(t;1,T)[1 - P_{\mathrm{fa}}(t;1,T)]^{n-j} \right\} \tag{6-89}$$

where

$$\begin{aligned} P_{\mathrm{fa}}(t;1) &= P_m \times (P_{0\mathrm{fa}})^{t-1} \times P_{\mathrm{fa}} \\ &= e^{-\overline{N}_{s+n}} e^{-\overline{N}_n(t-1)}(1 - e^{-\overline{N}_n}) \\ &= e^{-\overline{N}_s} e^{-\overline{N}_n t}(1 - e^{-\overline{N}_n}) \end{aligned} \tag{6-90}$$

and

$$\begin{aligned} P_{\mathrm{fa}}(t;T) &= P_{\mathrm{fa}} \times (P_{0\mathrm{fa}})^{t-1} \\ &= (1 - e^{-\overline{N}_n})(e^{-\overline{N}_n})^{t-1} \\ &= (1 - e^{-\overline{N}_n})e^{-\overline{N}_n(t-1)} \end{aligned} \tag{6-91}$$

It is easy to show that summing Eqs. (6-90) and (6-91) over all measurement intervals other than the signal interval yields the single-pulse false alarm probabilities given by Eqs. (4-222) and (4-223); that is,

$$P_{\text{fa}}(1, T) = \sum_{t=1}^{T-1} p_{\text{fa}}(t; 1, T) \tag{6-92}$$

Now since it is also possible for noise detections to satisfy the threshold requirement of m-out-of-n detections in a common time interval, these additional possibilities must be taken into account in calculating the overall detection statistics. A little thought shows that if $m/n > 1/2$, n even, there is only one possible way for an m-out-of-n detection to occur in a single time interval, while if $m/n = 1/2$, n even, there are two possible ways, and so on. For example, in the case of $m : n = 2 : 3$, pulses 1 and 2, 2 and 3, or 1 and 3 may occur in the same time interval. In the case of $m : n = 2 : 4$, the signal may reside in 1 and 3 and noise in 2 and 4, each in a separate interval, and so on. Similar analysis can be done for n odd.

In the case of $m/n > 1/2$, n even, Eqs. (6-88) and (6-89) may be used directly since there are no other possibilities. However, in the case of $m/n = 1/2$, n even, corrections need to be subtracted from Eq. (6-88) and (6-89) to account for the additional possibilities that $m : n$ noise detections may also satisfy the threshold requirements and result in a *no-detection*, which we denote as P_N. The fact that multiple detections must be classified as no-detections follows from the ambiguity in identifying which detection constitutes the signal. Since all possibilities have been accounted for, we can write $P_D + P_{\text{FA}} + P_N = 1$.

The correction factors may be obtained using multinomial statistics, a generalization of binomial statistics that is described in Comment 6-4 below. Thus the probability that both the signal interval and a noise interval will satisfy the threshold requirements is given by

$$C_D(m, m : n) = \frac{n!}{m!\,m!} \sum_{t=1}^{T-1} P_d^m(t; 1, T) P_{\text{fa}}^m(t; 1, T) \tag{6-93}$$

where the factorial fraction is the multinomial coefficient. Similarly, for the false alarm probability,

$$C_{\text{FA}}(m, m : n) = C_D(m, m : n) + \frac{n!}{m!\,m!} \sum_{t=1}^{T-1} \left[P_{\text{fa}}^m(t; 1, T) \sum_{t=1}^{T-1} P_{\text{fa}}^m(t; 1, T) \right] \tag{6-94}$$

where $C_D(m, m : n)$ is given by Eq. (6-93).

Equations (6-88) through (6-94) are used to plot the integrated detection and false alarm ($\times 10$) probabilities in Figure 6-30 for the case of $m : n = 2 : 4$, $\overline{N}_n T = 1$, and $T = 100$. It can be seen that substantial improvement has been obtained for both statistics compared to the single-pulse probabilities shown in Figure 4-43.

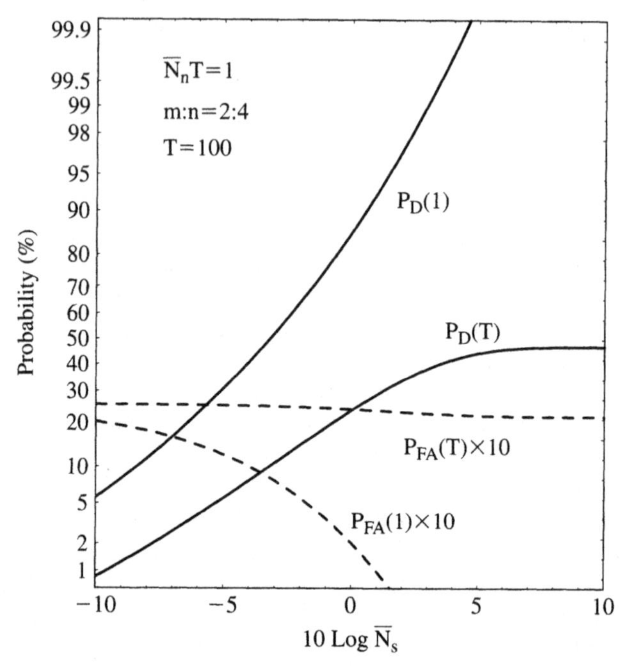

Figure 6-30. Detection and false alarm probabilities versus the mean signal photoelectron count $10 \log \overline{N}_s$ for a Geiger mode detector using an $m : n = 2 : 4$ coincidence detection algorithm with $T = 100$ measurement intervals per processing window. Note that the false alarm curves have been multiplied by a factor of 10. (After R. Marino, et al., MIT Linc. Lab. Journ., Vol. 13, No. 2, 2002.)

Comment 6-4. The binomial distribution describes the probability that x Bernoulli events will occur given a probability P for a single event. There are situations in which more than two mutually exclusive events can occur, as, for example, the faces on a die for which there are six possible (in this case, equally probable) outcomes for each trial. Thus if there are k distinct outcomes with associated probabilities $P_1, P_2, \ldots, P_k$, where $\sum_{j=1}^{k} P_j = 1$, for n independent trials and x_1 occurrences of outcome 1, x_2 occurrences of outcome $2, \ldots, x_k$ occurrences of outcome k, the probability of a specific sequence of outcomes is

$$P_1^{x_1} P_2^{x_2} \cdots P_k^{x_k} \tag{6-95}$$

where $\sum_{j=1}^{k} x_j = n$. Now there are

$$\frac{n!}{x_1! x_2! \cdots x_k!} \tag{6-96}$$

equally likely ways in which such a sequence can occur. Thus the probability of *exactly* x_1 occurrences of outcome 1, x_2 occurrences of outcome $2, \ldots, x_k$

occurrences of outcome k in n trials is

$$C(x_1, x_2, \ldots, x_k : n) = \frac{n!}{x_1!x_2!\cdots x_k!} P_1^{x_1} P_2^{x_2} \cdots P_k^{x_k} \tag{6-97}$$

where $x_1, x_2, \ldots, x_j = 0, 1, 2, \ldots, n$ and $0 \le P_1, P_2, \ldots, P_j \le 1$.

For example, in the case of the single die, the probability of rolling a 6 six times out of six tries is

$$C(0, 0, 0, 0, 0, 6 : 6) = \frac{6!}{(0!)^5 6!} \left(\frac{1}{6}\right)^{0\times 5} \left(\frac{1}{6}\right)^6 = 0.0000214 \tag{6-98}$$

whereas the probability of each face appearing once (in any order) in the six tries is

$$C(1, 1, 1, 1, 1, 1 : 6) = \frac{6!}{(1!)^6} \left(\frac{1}{6}\right)^6 = 0.015 \tag{6-99}$$

Clearly, the uniform distribution yields the higher probability, as one would expect. ■

REFERENCES

[1]N. Menyuk, D. K. Killinger, and C. R. Menyuk, "Limitations of signal-averaging due to temporal correlation in laser remote-sensing measurements," *Appl. Opt.*, Vol. 21, No. 18, Sept. 15, 1982, p. 3377.

[2]G. R. Osche, "Single- and multiple-pulse noncoherent detection statistics for partially developed speckle," *Appl. Opt.*, Vol. 39, No. 24, Aug. 20, 2000, pp. 4255–4262.

[3]Ibid., Ref. 2.

[4]J. I. Marcum, "*A statistical theory of target detection by pulsed radar*," Memo RM-754, RAND Corporation, Santa Monica, CA, Dec. 1947. Also, *IRE Trans. Inf. Theory*, Vol. IT-6, No. 2, Apr. 1960, pp. 59–144.

[5]P. Swerling, "*Probability of detection for fluctuating targets*," Memo RM-1217, RAND Corporation, Santa Monica, CA, Mar. 1954. Also, *IRE Trans. Inf. Theory*, Vol. IT-6, No. 2 Apr. 1960, pp. 269–308.

[6]R. M. Hardesty, R. J. Keeler, M. J. Post, and R. A. Ricter, "Characteristics of coherent lidar returns from calibration targets and aerosols," *Appl. Opt.*, Vol. 20, No. 21, Nov. 1, 1981, p. 3763.

[7]J. Y. Wang and P. A. Pruitt, "Laboratory target reflectance measurements for coherent laser radar applications," *Appl. Opt.*, Vol. 23, No. 15, Aug. 1, 1984, p. 2559.

[8]A. H. Heatley, "A short table of the Toronto functions," *Trans. R. Soc. Canada*, Vol. 37 (Sec. III), 1943, pp. 13–29.

[9]M. Abramowitz and I. A. Stegun, Editors, *Handbook of Mathematical Functions*, Applied Mathematics Series 55, U.S. Department of Commerce, National Bureau of standards,Washington, DC, June 1964, p. 486, Eq. 11.4.2.8.

[10]The author is indebted to F. Horrigan for developing this simple integral solution while proofreading the manuscript.

[11]P. H. R. Scholefield, "Statistical aspects of ideal radar targets," *Proc. IEEE*, Apr. 1967, p. 587.

[12]M. Schwartz, "A coincidence procedure for signal detection," *IEEE Trans. Inf. Theory*, Dec. 1956, p. 135.

[13]R. M. Marino, M. A. Albota, B. F. Aull, G. G. Fouche, R. M. Heinrichs, D. G. Kocher, J. Mooney, M. O'Brien, B. E. Player, B. C. Willard, and J. J. Zayhowski, "A three-dimensional imaging laser radar with Geiger-mode avalanche photodiode arrays," to be published in the *MIT Lincoln Laboratory Journal*, Vol. 13, No. 2, 2002.

Appendix A

Advanced Mathematical Functions

In this appendix we present a brief overview of some of the key features of several of the less commonly known mathematical functions referred to in the text. Sufficient detail is presented to allow the reader some degree of versatility in working the mathematics, but for more comprehensive treatments, the reader is referred to the excellent handbooks by Abramowitz and Stegun,[1] Gradshteyn and Ryzhik,[2] or Erdelyi.[3] The Toronto function can be found in the work of Heatley.[4]

A.1. DIRAC DELTA AND UNIT STEP FUNCTIONS

The Dirac delta function is defined by the following properties

$$\delta(x) = 0 \qquad x \neq 0 \tag{A-1}$$

$$\int_{-\infty}^{\infty} \delta(x)\, dx = 1 \tag{A-2}$$

and

$$\int_{-\infty}^{\infty} f(x)\delta(x)\, dx = f(0) \tag{A-3}$$

where $f(x)$ is assumed to be continuous at $x = 0$. It should be noted that the integral expression, Eq. (A-3), is the defining equation for the delta function, not the function itself. In point of fact, the delta function itself does not exist in the usual sense of a function, as evidenced by the fact that it is indeterminate (i.e., infinite) at $x = 0$. It is more properly defined as the limit of a sequence of integrals given by

$$\lim_{n\to\infty} \int_{-\infty}^{\infty} \delta_n(x) f(x)\, dx = f(0) \tag{A-4}$$

where $\delta_n(x)$ is any well-behaved continuous function in the interval $(-\infty, \infty)$. An example of such a function is the Gaussian function shown in Figure A-1 and given by

$$\delta_n(x) = \frac{n}{\sqrt{\pi}} e^{-n^2x^2} \tag{A-5}$$

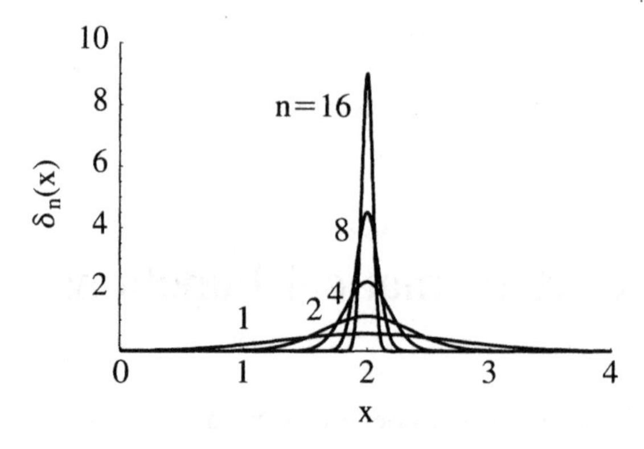

Figure A-1. Gaussian approximation to the Dirac delta function as $n \to \infty$.

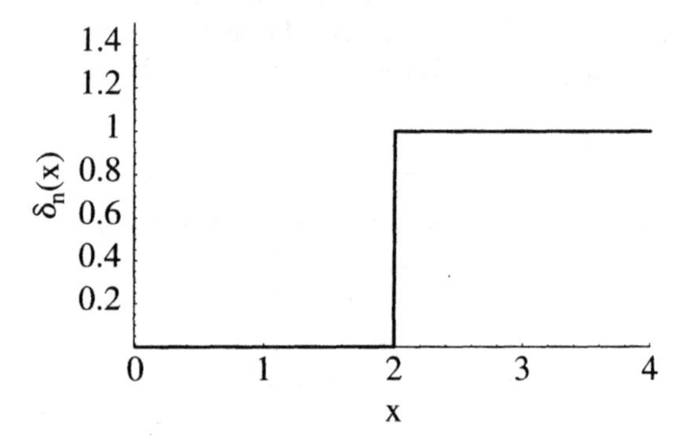

Figure A-2. Unit step function.

Equation (A-3) may also be defined for $x \neq 0$ as

$$\int_{-\infty}^{\infty} f(x)\delta(x - x')\,dx = f(x') \tag{A-6}$$

The unit step function $U(x - x')$, shown in Figure A-2, is defined as

$$U(x - x') = \begin{cases} 0 & x < x' \\ 1 & x \geq x' \end{cases} \tag{A-7}$$

It is related to the delta function through the relationships

$$\frac{dU(x - x')}{dx} = \delta(x - x')$$

$$U(x - x') = \int_{-\infty}^{\infty} \delta(x - x')\,dx \tag{A-8}$$

A.2. GAMMA FUNCTION

The gamma function is defined through integral and infinite product expressions. The two most useful for this work are the integral expression of Euler,

$$\Gamma(z) = \int_0^\infty e^{-z} t^{z-1} dt \qquad \text{Re } z > 0 \tag{A-9}$$

and the product expression by Gauss,

$$\Gamma(z) = z^{-1} \prod_{n=1}^{\infty} \left[\left(1 + \frac{1}{n}\right)^z \left(1 + \frac{z}{n}\right)^{-1} \right] \tag{A-10}$$

Several useful functional relationships are satisfied by the gamma function. For example, if we integrate Eq. (A-9) by parts, with $U = e^{-t}$ and $dv = t^{s-1}$, we obtain

$$\Gamma(z+1) = z\Gamma(z) \tag{A-11}$$

If we let $z = 1$, Eq. (A-9) becomes

$$\Gamma(1) = \int_0^\infty e^{-t} dt = 1 \tag{A-12}$$

so that with $n =$ positive integer in Eq. (A-11),

$$\begin{aligned} \Gamma(n+1) &= n\Gamma(n) = n(n-1)\Gamma(n-1) \\ &= n(n-1)(n-2)\cdots\Gamma(1) \\ &= n! \end{aligned} \tag{A-13}$$

More generally, we can write

$$\begin{aligned} \Gamma(z+n) &= (z+n-1)\Gamma(z+n-1) \\ &= (z+n-1)(z+n-2)\cdots z\Gamma(z) \end{aligned} \tag{A-14}$$

from which it follows that

$$\begin{aligned} \frac{\Gamma(z)}{\Gamma(z-n)} &= (z-1)(z-2)\cdots(z-n) \\ &= (-1)^n \frac{\Gamma(-z+n+1)}{\Gamma(-z+1)} \end{aligned} \tag{A-15}$$

and

$$\frac{\Gamma(-z+n)}{\Gamma(-z)} = (-1)^n z(z-1)\cdots(z-n+1)$$
$$= (-1)^n \frac{\Gamma(z+1)}{\Gamma(z-n+1)} \qquad \text{(A-16)}$$

Other useful expressions follow from Eq. (A-10):

$$\Gamma(z)\Gamma(-z) = -z^{-2}\prod_{n=1}^{\infty}\left(1-\frac{z^2}{n^2}\right)^{-1} \qquad \text{(A-17)}$$

But since

$$\sin(\pi z) = \pi z\prod_{n=1}^{\infty}\left(1-\frac{z^2}{n^2}\right) \qquad \text{(A-18)}$$

we can write

$$\Gamma(z)\Gamma(-z) = -\frac{\pi}{z}\csc(\pi z) \qquad \text{(A-19)}$$

or

$$\Gamma(\tfrac{1}{2}+z)\Gamma(\tfrac{1}{2}-z) = \pi\sec(\pi z) \qquad \text{(A-20)}$$

With $z = 0$ in Eq. (A-20) and $z = \frac{1}{2}$ in Eq. (A-19), we obtain the well-known identities

$$\Gamma(\tfrac{1}{2}) = \sqrt{\pi} \quad \text{and} \quad \Gamma(-\tfrac{1}{2}) = -2\sqrt{\pi} \qquad \text{(A-21)}$$

There are many functions related to the gamma function. We list only a few. The incomplete gamma function and its complement are defined by

$$\Gamma(a,z) = \int_a^{\infty} e^{-t}t^{z-1}dt \qquad \text{Re } a > 0$$
$$\gamma(0,a) = \int_0^{a} e^{-t}t^{z-1}dt \qquad \text{Re } a > 0 \qquad \text{(A-22)}$$

Clearly,

$$\Gamma(z) = \Gamma(a,z) + \gamma(0,a) \qquad \text{(A-23)}$$

The regularized incomplete gamma function is defined as

$$Q(a,z) = \frac{\Gamma(a,z)}{\Gamma(a)} = \frac{\Gamma(z)-\gamma(0,a)}{\Gamma(a)} \qquad \text{(A-24)}$$

The asymptotic form for the gamma function is

$$\Gamma(az+b) \sim \sqrt{2\pi}e^{-az}(az)^{az+b-1/2} \qquad |\arg z| < \pi, \quad a > 0 \qquad \text{(A-25)}$$

A.3. CONFLUENT HYPERGEOMETRIC FUNCTION

There are four functions: the confluent hypergeometric functions, Kummer's function, ${}_1F_1(a, b; z)$, its associated solution $U(a, b; z)$, and two Whittaker functions, $M_{k,n}(x)$ and $W_{k,n}(x)$. Many of the well-known functions of mathematical physics can be expressed in terms of these functions, indeed, may be considered as subsets of these functions. The word *confluent* is used to indicate a confluence or merging of two regular singularities at b and ∞ in the more general differential equation of Gauss. The word *hypergeometric* arises from the fact that any solution consisting of an ascending power series of the type

$$1 + x + x^2 + \cdots + x^n$$

is referred to as hypergeometric when coefficients are inserted. We are concerned here with Kummer's function as it relates to the detection statistics in Chapter 4.

Kummer's differential equation is

$$x\frac{d^2y}{dx^2} + (b - x)\frac{dy}{dx} - ay = 0 \tag{A-26}$$

Its solutions are confluent hypergeometric functions, written as

$$\begin{aligned} {}_1F_1(a, b; x) &= 1 + \frac{a}{b}x + \frac{a(a+1)}{b(b+1)}\frac{x^2}{2!} + \frac{a(a+1)(a+2)}{b(b+1)(b+2)}\frac{x^3}{3!} + \cdots \\ &\equiv \sum_{n=0}^{\infty} \frac{(a)_n}{(b)_n}\frac{x^n}{n!} \end{aligned} \tag{A-27}$$

where the $(a)_n$ and $(b)_n$ are Pochhammer symbols, defined by

$$\begin{aligned} (z)_0 &= 1 \\ (z)_n &= z(z+1)(z+2)\cdots(z+n-1) = \frac{\Gamma(z+n)}{\Gamma(z)} \end{aligned} \tag{A-28}$$

The series functions ${}_1F_1$ converge absolutely for all values of a, b, and x, real or complex, excluding $b = 0, -1, -2, \ldots, -n$, where n is an integer. Other notations seen in the literature for ${}_1F_1$ are $M(a, b; x)$ and $\Phi(a, b; x)$.

The integral representation is given by

$${}_1F_1(a, b; x) = \frac{\Gamma(b)}{\Gamma(a)\Gamma(b-a)} \int_0^1 e^{xu} u^{a-1}(1-u)^{b-a-1} du \qquad \text{Re } b > \text{Re } a > 0 \tag{A-29}$$

This expression can be verified by expanding e^{xu} in powers of x.

Several useful functional relationships exist for ${}_1F_1$. Kummer's transformation, also known as Kummer's first theorem, is given by

$${}_1F_1(a, b; x) = e^x {}_1F_1(b - a, b; -x) \tag{A-30}$$

Since the series given by Eq. (A-27) converges absolutely, it can be differentiated term by term, yielding

$$\frac{d}{dx}{}_1F_1(a,b;x) = \frac{a}{b}\sum_{n=1}^{\infty}\frac{(a+1)_{n-1}x^{n-1}}{(b+1)_{n-1}(n-1)!}$$
$$= \frac{a}{b}{}_1F_1(a+1,b+1;x) \qquad \text{(A-31)}$$

and in general,

$$\frac{d^n}{dx^n}{}_1F_1(a,b;x) = \frac{(a)_n}{(b)_n}{}_1F_1(a+n,b+n;x) \qquad \text{(A-32)}$$

Relationships can also be derived for contiguous functions [i.e., functions that differ from ${}_1F_1(a,b;x)$ by an integer in a or b]. Thus

$$(b-a){}_1F_1(a-1,b;x) + (2a-b+x){}_1F_1(a,b;x) - a{}_1F_1(a+1,b;x) = 0 \qquad \text{(A-33)}$$

and

$$(a-b+1){}_1F_1(a,b;x) - a{}_1F_1(a+1,b;x) - (b-1){}_1F_1(a,b-1;x) = 0 \qquad \text{(A-34)}$$

Several other such relationships can be derived from these two.

Relationships to some other well-known functions of interest here are shown in Table A-1.

Table A-1. Connections of ${}_1F_1(a,b;x)$ to other functions

a	b	x	Relation	Function
$\frac{1}{2}$	$\frac{3}{2}$	$-x^2$	$\frac{\sqrt{\pi}}{2x}\,\text{erf}\,(x)$	Error integral
$-n$	$\frac{1}{2}$	$\frac{x^2}{2}$	$\frac{n!}{(2n)!}\left(\frac{1}{2}\right)^{-n} H_{2n}(x)$	Hermite polynomials
$\pm a$	$\pm a$	x	e^x	Exponential
$-n$	$\alpha+1$	x	$\frac{n!}{(\alpha+1)_n}L_n^{(\alpha)}(x)$	Laguerre polynomials
$n+\frac{1}{2}$	$2n+1$	$2ix$	$\Gamma(n+1)e^{ix}\left(\frac{z}{2}\right)^{-n} J_n(x)$	Bessel functions

A.4. PARABOLIC CYLINDER FUNCTIONS

The parabolic cylinder functions, also called the *Weber–Hermite functions*, are solutions to the differential equation

$$\frac{d^2y}{dx^2} + \left(\nu + \frac{1}{2} - \frac{1}{4}z^2\right) y = 0 \tag{A-35}$$

They can be defined in terms of the confluent hypergeometric functions through the expression

$$\begin{aligned} D_\nu(z) = 2^{\nu/2} e^{-z^2/2} \Bigg[& \frac{\Gamma(1/2)}{\Gamma((1-\nu)/2)} {}_1F_1\left(-\frac{\nu}{2}, \frac{1}{2}; \frac{z^2}{2}\right) \\ & + \frac{z}{\sqrt{2}} \frac{\Gamma(-1/2)}{\Gamma(-\nu/2)} {}_1F_1\left(\frac{1-\nu}{2}, \frac{3}{2}; \frac{z^2}{2}\right)\Bigg] \end{aligned} \tag{A-36}$$

For ν an integer, $D_\nu(z), D_\nu(-z), D_{-\nu-1}(iz),$ and $D_{-\nu-1}(-iz)$ all satisfy Eq. (A-35). A commonly used integral representation is

$$D_\nu(z) = \frac{e^{-z^2/4}}{\Gamma(-\nu)} \int_0^\infty x^{-(\nu+1)} e^{-(x^2/2)-zx}\, dx \qquad \text{Re } \nu < 0 \tag{A-37}$$

Recursion formulas include

$$\begin{aligned} D_{\nu+1}(z) - zD_\nu(z) + \nu D_{\nu-1}(z) &= 0 \\ \frac{d}{dz} D_\nu(z) + \tfrac{1}{2} z D_\nu(z) - \nu D_{\nu-1}(z) &= 0 \\ \frac{d}{dz} D_\nu(z) - \tfrac{1}{2} z D_\nu(z) + D_{\nu+1}(z) &= 0 \end{aligned} \tag{A-38}$$

The parabolic cylinder functions are connected with the Hermite polynomials through the expression

$$D_\nu(z) = 2^{-\nu/2} e^{-z^2/4} H_\nu\left(\frac{z}{\sqrt{2}}\right) \tag{A-39}$$

Equations (A-39) and (A-36) can be used to generate the first two functions having zero and negative parameters. Thus

$$\begin{aligned} D_0(z) &= e^{-z^2/4} \\ D_{-1}(z) &= \sqrt{\frac{\pi}{2}} e^{z^2/4} \left[1 - \text{erf}\left(\frac{z}{\sqrt{2}}\right)\right] \\ &= \sqrt{\frac{\pi}{2}} e^{z^2/4} \,\text{erfc}\left(\frac{z}{\sqrt{2}}\right) \end{aligned} \tag{A-40}$$

where

$$\text{erf}\,(z) = \frac{2}{\sqrt{\pi}} \int_0^z e^{-t^2} dt \tag{A-41}$$

and

$$\text{erfc}\,(z) = \frac{2}{\sqrt{\pi}} \int_z^\infty e^{-t^2} dt \tag{A-42}$$

are the error function and the error function complement, respectively. The first of Eqs. (A-38) together with Eqs. (A-40) can be used to generate parabolic cylinder functions recursively to any order. The first five functions are listed below. Figure A-3 shows these five functions plotted as a function of z.

$$
\begin{aligned}
D_0(z) &= e^{-z^2/4} \\
D_{-1}(z) &= \sqrt{\frac{\pi}{2}} e^{z^2/4} \text{ erfc}\left(\frac{z}{\sqrt{2}}\right) \\
D_{-2}(z) &= \sqrt{\frac{\pi}{2}} e^{z^2/4}\left[\sqrt{\frac{2}{\pi}} e^{-z^2/2} - z \text{ erfc}\left(\frac{z}{\sqrt{2}}\right)\right] \\
D_{-3}(z) &= \frac{1}{2}\sqrt{\frac{\pi}{2}} z e^{-z^2/4}\left[\frac{1+z^2}{z} e^{z^2/2} z \text{ erfc}\left(\frac{z}{\sqrt{2}}\right) - 1\right] \\
D_{-4}(z) &= \frac{1}{3}\left(1 - \frac{1}{2}\sqrt{\frac{\pi}{2}} z^2\right) e^{-z^2/4} - \frac{z}{3}\sqrt{\frac{\pi}{2}} e^{z^2/4} \text{ erfc}\left(\frac{z}{\sqrt{2}}\right)\left[1 + \frac{1}{2}(1+z^2)\right]
\end{aligned}
\tag{A-43}
$$

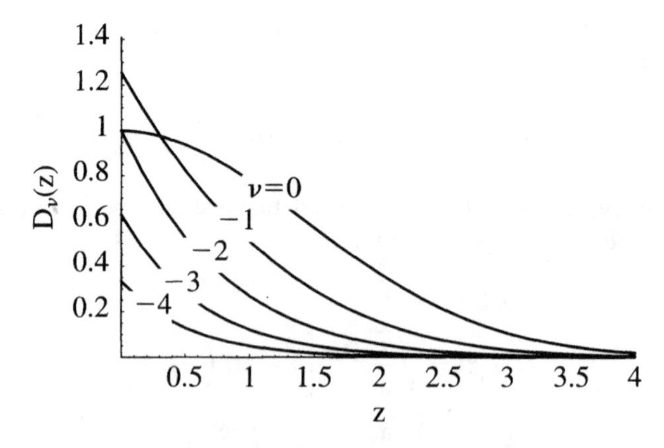

Figure A-3. Parabolic cylinder functions of order $\nu = 0, -1, -2, -3,$ and -4.

A.5. TORONTO FUNCTION

The Toronto functions are defined by the expression

$$T_B(m, n, r) = 2r^{n-m+1}e^{-r^2}\int_0^B t^{m-n}e^{-t^2}I_n(2rt)\,dt \tag{A-44}$$

REFERENCES

[1]M. Abramowitz and I. A. Stegun, Editors, *Handbook of Mathematical Functions, Applied Mathematics Series 55*, U.S. Department of Commerce, National Bureau of Standards, Washington, DC, June 1964.

[2]I. S. Gradshteyn and I. M. Ryzhik, *Tables of Integrals, Series and Products*, 4th ed., Academic Press, San Diego, CA, 1965.

[3]H. Bateman manuscript, ed. A. Erdelyi, *Higher Transcendental Functions*, Vols. 1–3, McGraw-Hill Book Company, New York, 1953.

[4]A. H. Heatley, "A short table of the Toronto functions," *Trans. R. Soc. Canada*, Vol. 37 (Sec. III), 1943, p. 13–29.

Appendix B

Additional Derivations

B.1. GAMMA DISTRIBUTION

Consider a characteristic function of the form

$$G^{(n)}(iv) = \frac{1}{(1 - iv/a)^n} \tag{B-1}$$

where a is a parameter. The corresponding probability density function is given by

$$\begin{aligned} p(x) &= \frac{1}{2\pi} \int_{-\infty}^{\infty} G_x^{(n)}(iv) e^{-ivx} dv \\ &= \frac{1}{2\pi} \int_{-\infty}^{\infty} \frac{e^{-ivx}}{(1 - iv/a)^n} dv \end{aligned} \tag{B-2}$$

A pole is seen to exist at $v = -ia$ so that a contour integral over the lower half of the complex plane yields a nonzero solution, as shown in Figure B-1. Thus, in terms of the complex variable z, the contour integral may be written

$$\frac{1}{2\pi} \oint f(z)\, dz = -i \text{ Re } s(z_0) \tag{B-3}$$

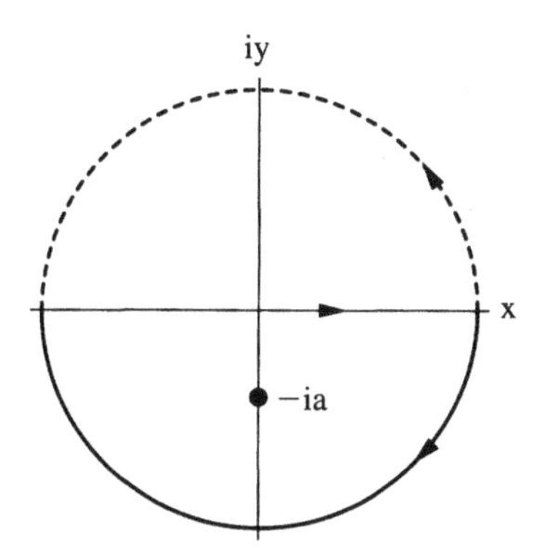

Figure B-1. Contour diagram used in derivation of the gamma distribution.

where

$$f(z) = \frac{e^{-ixz}}{(1 - iz/a)^n} \tag{B-4}$$

and Re $s(z_0)$ is the residue of $f(z)$ with value z_0. The residue may be calculated from

$$\text{Re } s(z_0) = \frac{1}{(m-1)!} \left\{ \frac{d^{m-1}}{dz^{m-1}} [(z - z_0)^m f(z)] \right\}_{z=z_0} \tag{B-5}$$

Thus

$$\begin{aligned}
\text{Re } s(-ia) &= \frac{1}{(n-1)!} \left\{ \frac{d^{n-1}}{dz^{n-1}} [(z + ia)^n f(z)] \right\}_{z=-ia} \\
&= \frac{1}{(n-1)!} \left\{ \frac{d^{n-1}}{dz^{n-1}} \left[(z + ia)^n \frac{e^{-ixz}}{(1 - iz/a)^n} \right] \right\}_{z=-ia} \\
&= i \frac{a^n x^{n-1} e^{-ax}}{\Gamma(n)}
\end{aligned} \tag{B-6}$$

Note that we have generalized n to noninteger values by letting $(n-1)! \to \Gamma(n)$. Inserting Eq. (B-6) into Eq. (B-3) results in

$$p(x) = \begin{cases} \dfrac{a^n x^{n-1} e^{-ax}}{\Gamma(n)} & x \geq 0 \\ 0 & x < 0 \end{cases} \tag{B-7}$$

which is the gamma distribution.

B.2. BURGESS VARIANCE THEOREM

The Burgess variance theorem[1,2] is applicable to any particle interaction in which a single primary gives rise to g secondary or multiplied particles. Let $q(k)$ be the discrete probability density function of k primary particles, and let $P(g)$ be the probability of obtaining g secondaries from any one primary. If $q(m \mid k)$ is the conditional probability that k primaries produce m secondaries, assuming statistical independence and stationarity for the process,

$$\begin{aligned}
q(m \mid k) = &\sum_{g_1=0}^{\infty} \sum_{g_2=0}^{\infty} \cdots \sum_{g_{n-1}=0}^{\infty} P(g_1) P(g_2) \cdots P(g_{n-1}) \\
&\times P(m - g_1 - g_2 - \cdots - g_{n-1})
\end{aligned} \tag{B-8}$$

where $P(g) = 0$ for $g < 0$.

The conditional mean is therefore

$$\langle m_k \rangle = \sum_{m=0}^{\infty} mq(m \mid k)$$
$$= \sum_{g'=0}^{\infty} \sum_{g_1=0}^{\infty} \cdots \sum_{g_{n-1}=0}^{\infty} (g' + g_1 + \cdots + g_n) P(g') P(g_1) \cdots P(g_{n-1}) \quad \text{(B-9)}$$

where $g' = m - g_1 - g_2 - \cdots - g_{n-1}$. Thus

$$\langle m \rangle = \sum_{k=0}^{\infty} \langle m_k \rangle q(k)$$
$$= \langle k \rangle \langle g \rangle \quad \text{(B-10)}$$

The conditional mean square follows similarly:

$$\langle m_k^2 \rangle = \sum_{m=0}^{\infty} m^2 q(m \mid k)$$
$$= \sum_{g'=0}^{\infty} \sum_{g_1=0}^{\infty} \cdots \sum_{g_{n-1}=0}^{\infty} (g' + g_1 + \cdots + g_n)^2 P(g') P(g_1) \cdots P(g_{n-1})$$
$$= k \langle g^2 \rangle + k(k-1) \langle g \rangle^2 \quad \text{(B-11)}$$

The second moment is therefore

$$\langle m^2 \rangle = \sum_{k=0}^{\infty} m_k^2 q(k)$$
$$= \langle k \rangle \langle g^2 \rangle + (\langle k^2 \rangle - \langle k \rangle) \langle g \rangle^2 \quad \text{(B-12)}$$

Hence

$$\text{var}(m) = \langle m^2 \rangle - \langle m \rangle^2$$
$$= \langle g \rangle^2 \ \text{var}(k) + \langle k \rangle \ \text{var}(g) \quad \text{(B-13)}$$

REFERENCES

[1]R. E. Burgess, "Homophase and heterophase fluctuations in semiconducting crystals," *Discuss. Faraday Soc.*, Vol. 28, 1959, pp. 151–158.

[2]L. Mandel, "Image fluctuations in cascade intensifiers," *Br. J. Appl. Phys.*, Vol. 10, 1959, pp. 233–234.

Index

ABCD ray transfer matrix, 79
Additive nature of Gaussian processes, 329
Airy pattern, 73, 98, 116
Alternate incomplete gamma function, 373
Angular uncertainty, beam spread, 95
Aperture averaging
  and integrated intensity, 174
  factor, 204
  in direct detection, 150, 302
  incoherent detection, 150, 302
Aperture transfer function, 174, 233
Area receiver, 234
Area statistics of speckle field, *see* Integrated statistics
Array theorem, 104
Atmospheric clutter, 306
Atmospheric turbulence
  and beam wander, 212
  and conservative passive additive, 185
  and one-way communication links, 298, 339
  as a homogeneous medium, 184
  as an inhomogeneous medium, 185
  beam-wave theories of, 184
  bistatic geometry in, 298, 339
  coherent detection in, 338
  constant amplitude signal in, 339
  correlation function of, 196
  direct detection in, 297
  frozen-in model, 183
  inertial sub-range, 185
  inner and outer scales of turbulence, definition of, 185
  isotropy, 185
  Kolmogorov theory of turbulence, 185–186
  monostatic geometry in, 302, 306, 343
  normalized correlation function for, 196
  plane wave theory of, 190
  power spectral density of the refractive index fluctuations, 191
    Kolmogorov spectrum, 193, 200
    modified von Karman spectrum, 194, 200
    Tatarski spectrum, 193
  Rayleigh fluctuating signal in, 344
  refraction index structure function, 186
  refractive index structure constant, 186
  Rytov method, 188
  Rytov parameter, 196
  saturation of scintillation, 217
  saturation length, 211
  spherical wave theory of, 190
  strong turbulence theory, 217
  structure constant, definition of, 185
  supersaturated regime, 217
  Tatarski theory of, 187
  transverse correlation length of, 201, 339
  turbulence-induced speckle, 182
  turbulent eddies, 183, 221
  two-dimensional spectral density, 3–67
  two-thirds law, 185
  universal theory of turbulence, 218
  variance
    of amplitude and intensity, 201–202
    of log amplitude and intensity, 201–202
  wave structure function, 199
  weak turbulence theory, 187
Autocorrelation function, 42
  of the intensity, 175, 177
Avalanche photodiode (APD), 61, 140, 276
  Mcintyre distribution, probability density function for, 278
  WMC distribution
    for specular targets, 279
    for diffuse targets, 288
Avalanche processes, 140–141
Average value, *see* Statistical parameters

Background radiation, 143
Backpropagated local oscillator, 115, 118
Backscatter amplification effect, 303, 306, 343
Beam wander, *see* Atmospheric turbulence
Beam modes or wavefunctions, 75
Bernoulli trials, 7, 296, 388
Bessel functions
  integral representation of, 70
  of the first kind, zeroth order, 33
Binary detection, 47
Binary integration algorithm, 381
  and Poisson signal in Poisson noise, 383
Binary optical communications system, 52
Binomial distribution, 7, 249, 382

Binomial expansion, 246, 264
Bistatic reflectance, 168
Bit error rate (BER), 53
Bivariate or second order statistics, 16
Born approximation, 188
Bose–Einstein distribution
 as limit of negative binomial distribution, 250
 definition of, 250
 for signal in turbulence, 304
Bulk dark current, 141
Bunching and anti-bunching statistics, 64, 290
Burgess variance theorem, 289, 402

Campbell's theorem, 139
Capture efficiency, 120
Center of mass, statistical analog of, 10
Central-limit theorem, 30
Centrally obscured aperture, 73
Characteristic function
 definition of, 27
 for chi-squared distribution, 35, 260
 for Gaussian distribution, 29, 267
 for Kummer's distribution, 254
 for integrated intensity, 235
 for negative binomial distribution, 252
 for negative exponential distribution, 235
 for noncentral chi-squared distribution, 35, 260
 for parabolic-cylinder distribution, 271
 for Poisson distribution, 242
 for Poisson signal in Poisson noise, 243
Chi-squared distribution, probability density function for, 34
Clutter sources, 142
Clutter speckle, 307
Coherence
 area, 154
 complex coherence factor, 153, 180
 complex degree of, 153, 157
 complex degree of spatial coherence, 153
 complex degree of temporal coherence, 153
 degree of coherence, 153–154, 157, 177
 time, 154
Coherence parameter, 220, 265, 336
Coherent detection statistics
 for constant amplitude signal in Gaussian noise, 320
 for constant amplitude signal, Gaussian approximation, 324–325
 for constant amplitude signal in turbulence, 339
 for one-dominant plus Rayleigh signal, 331
 for Rayleigh fluctuating signal in Gaussian noise, 328
 for Rayleigh fluctuating signal in turbulence, 344
 for Rician signal in Rician noise, 334
 in terms of power variables, 325
Coherent or heterodyne detection, 2, 62–66, 318–319
Coherent state, 64
Coherent-detection signal-to-noise ratio
 with truncated beams, 125
 with untruncated beams, 117
Coincidence detection
 for APD Geiger-mode detectors, 384
  correction factors for, 387
  measurement intervals for, 386
Coincidence detection algorithm, *see* Binary integration algorithm
Collimated beams, 71, 103, 123, 128
Comparison of coherent and direct detection statistics, 346
Complementary error function, 269
Complex beam parameter, definition of, 76
Complex number, definition of, 27
Complex scalar field amplitude, 3–4
Complex surface, 4–39, 5–22
Complex wave amplitude, 69
Computer generated images of
 Airy pattern, 73
 fully developed speckle, 228
 Gaussian beam speckle, 109
  aperture distribution for, 108
 partially developed speckle, 258
 two-beam interference pattern, 106
Conditional probability, definition of, 17–18
Constant amplitude signal in Gaussian noise, 320
Contour integration, 31, 236, 259, 401
Contrast of a speckle pattern, 259
Convex sum, definition of, 36
Convolution integral, 27
Convolution theorem, 27
Correlation area, 178
Correlation coefficient, *see* Statistical parameters
Cumulative binomial distribution, 382
Cumulative distribution function, definition of, 9

Dark current, 140
Dark noise, 267
Dead time, in Geiger mode detectors, 292
Decision theory
 a posteriori probabilities, 50
 a priori probabilities, 50
 alternative hypothesis, 48

average cost, 77
Bayes decision rule, 54
Bayes risk, 60
Bayes rule, 18
binary or signal detection, 47
composite hypotheses, 48
detection probability, definition of, 49
error costs, 54
error of the first kind, 49
as type I error, 54
error of the second kind, 49
as type II error, 54
false alarm probability, definition of, 49
false alarm, definition of, 49
hypothesis testing, 48
ideal observer test, 52, 79
information parameters, 47
least favorable statistics, 60
likelihood functions, 51
likelihood ratio, 51
minimax criterion, 60
miss probability, definition of, 49
missed detection, 54
Neyman–Pearson decision rule, 58
null hypothesis, 48
parameter estimation, 47
simple hypotheses, 48
test statistic, 50
threshold level, 51–52
Degree of polarization, 309
Density function, definition of, 6–7
Detector optimization, 134–136
Detector responsivity, 65, 146
Determinant, 24
Differential absorption lidar (DIAL), 2
Diffraction, far-field distribution of a circular aperture, 73
Diffraction
definition of, 68
extended Huygens–Fresnel formula, 303
Fraunhofer formula, 71
Fresnel–Kirchhoff formula, 69
Huygens–Fresnel integral formula, 70, 98
Huygens–Fresnel principle, example of, 74
inclination factor, 69
limit, 75, 93
near and far fields, 72
secondary Huygens' wavelets, 69
theory of, 68
Dirac delta function, 7
Direct detection, 2, 64, 227
receiver, 227–228
signal processor, 227–228
Direct-detection statistics
for a Geiger-mode APD detector
and rough targets, 294
and specular targets, 293
and multiple time-bins, 295
for a noncentral negative binomial signal in Poisson noise, 265–266
for a negative binomial signal in Poisson noise, 254–256
for a parabolic-cylinder signal in Gaussian noise, 275–276
for a Poisson signal in Poisson noise, 244
for an APD detector and a rough target, 288–292
for an APD detector and a specular target, 278–288
in the high signal limit, 281
with the WMC approximation, 279
in atmospheric clutter, 306
in atmospheric turbulence, 297
for depolarized fields, 309–313
for multiple uncorrelated signals, 313–315
Direct-detection signal-to-noise ratio
with truncated beams, 131
with untruncated beams, 129
Discrete Fourier coefficients, 40
Discrete Fourier transform, 285
Distribution function, definition of, 9
Doppler frequency, 63
Double-thresholding algorithm, *see* Binary integration algorithm
Doubly-stochastic Poisson process, 245

Effective aperture, 74
Effective ionization rate ratios, 142
Energy spectral density, 40
Ensemble autocorrelation function, 42
Erbium doped fiber amplifier (EDFA), 266
Ergodic hypothesis, 38, 184, 234–235
Excess noise, 140, 277
Exponential fading, 232
Extinction coefficient, 148
Extremely strong turbulence, 219

Fano factor, 290
First order statistics, *see* Univariate statistics
Fluctuating and nonfluctuating targets, 319
Focused CW coherent system, 124
Fourier optics, 104
Fourier integral transforms, definition of, 28
Fourier–Stieltjes transform, 44–46, 191
Free-space optical communications links
coherent detection in, 64
coherent detection in turbulence, 339–344
direct detection in turbulence, 298–300

Frequencies of occurrence, 6
Frequency bins, 6
Fringe visibility, 154

Gamma distribution
  and partially polarized light, 310–311
  and uncorrelated speckle fields, 315
  as a chi-squared distribution, 236
  as limit of noncentral chi-squared distribution, 261
  cumulative distribution function of, 238–239
  for the integrated intensity, 236
  in generalized notation, 236
  probability density function for, 31, 236, 268
Gamma function, definition of, 393
Gaussian beam
  focal range of, 82
  focal volume of, 83
  radius of curvature of, 76–77
  radius of, 76–77
  range resolution of, 83
  Rayleigh range of, 77
  theory of, 75–78
  waist, 76
Gaussian distribution
  and the central limit theorem, 30–33
  and the parabolic-cylinder distribution, 267
  as limiting form of the Poisson distribution, 243
  as limiting form of the Rician distribution, 324
  as limiting form of the WMC distribution, 281–282
  definition of, 7
  two-dimensional form, 24, 230, 321
Geiger-mode APD detector, 62, 291
  dead time of, 292
  measurement intervals or time bins, 294
  multiple-pulse detection statistics for, 385–388
  passive and active quenching circuits for, 292
  photoelectron detection probability, 291
  photon detection efficiency of, 291
  probability of avalanche production in, 294
    mutual exclusivity of, 294
    statistical independence of, 294
  processing windows for, 294, 386
  single-pulse detection statistics for, 296–297
Noncentral or generalized negative-binomial distribution, 264
Golf example, 6, 11, 20
Green's function
  in surface scattering theory, 162
  in turbulence theory, 189–191

Hanbury Brown and Twiss experiment, 158
Hard targets, 3
Heisenberg uncertainty principle, 64, 95
Helmholtz scalar wave equation, 69, 75, 160, 187
Heterodyne detection, 62, 318
  applications of, 2–3
Heuristic models of strong turbulence, 218
Histogram, 6
Hole and electron ionization rates, 141
Homodyne detection
  definition of, 63–64
  in quantum optics, 64
Huygens' wavelets, 69

I–K distribution, 218, 263
  moments of, 222
Impulse function, 7
Inclination factor, 69. *See also* Diffraction
Incomplete Gamma function, 239
In-phase (I) and quadrature (Q) components, 320
Integrated energy and intensity, 233
Integrated statistics, *see* Summed statistics
Integration loss
  definition of, 354–355
  for negative binomial signal in Poisson noise, 359
  for noncentral negative binomial signal in Poisson noise, 360
  for parabolic-cylinder signal in Gaussian noise, 364
  for Poisson signal in Poisson noise, 356
  for Rician signal model, 381
  for Swerling Case 0, 370
  for Swerling Case I, 375
  for Swerling Case II, 379
Interference, 152–154
Intermediate frequency, 66
Inverse steradian power reflectivity, 119, 144, 168

Jacobian, 24
Johnson or thermal noise, 136–138
Joint cumulative distribution function, 20

K-distribution, 219–220
Karhunan–Loeve expansion, 238
Kirchhoff or physical optics approximation, 160
Kummer's confluent hypergeometric function, 247
Kummer's distribution. *See also* Negative binomial signal in Poisson noise
  characteristic function for, 254

moments of, 254
probability density function for, 247–248
Kummer's transformation, 267, 395

L'Hopital's rule, 274
Lambert's cosine law, 169, 171–173
Lambertian reflector, 164
Lambertian surface, 170–171
Laser cavities or resonators, 75
Laser radar applications, 1–3
Limb darkening, 158
Log-amplitude correlation function, 196
Log-amplitude random variable, 202
probability density function for, 202, 341, 345
Log-intensity random variable, 202
probability density function for, 202, 299
Log-normally modulated negative exponential distribution, 221
Log-normal Rician distribution, 221, 263
moments of, 222
Log-normal statistics
amplitude distribution for, 202
cumulative distribution for the intensity, 301
intensity distribution for, 202, 221, 300

Marcum's Q-function, 325
Matched filter theory, 45
Maxwell's equations, 187
Mean noise power, 322
Mean power signal-to-noise ratio, 327
Mean signal power, 324
Mean-square noise current, 138
Mean voltage signal-to-noise ratio, 327
Measurement area for integrated speckle, 178
Minimum beam spread, 95
Mixing or heterodyne efficiency, 112
Mixing theorem, 112
Modified Bessel function of the first kind, 221, 261
Modified Rician distribution
moments of, 257–258
probability density function for, 257, 263, 325–326
Moment generating function, 28
Moment theorem, 176
M-out-of-n algorithm, *see* Binary integration algorithm
Multimode beams, 84
Multinomial statistics, 387
Multiple pulse detection theory
coherent detection
for Swerling Case 0, 366
for Swerling Case I, 371
for Swerling Case II, 375
for Rician signal model, 377
direct detection
for a negative binomial signal in Poisson noise, 356
for a noncentral negative binomial signal in Poisson noise, 358
for a parabolic-cylinder signal in Gaussian noise, 360
for a Poisson signal in Poisson noise, 352
Mutual coherence function, 151, 198, 210
Mutual intensity, 152
Mutually exclusive random variables
definition of, 4–5
in APD Geiger mode statistics, 295

Narrowband receiver, 320–321
Natural light, 151
Negative binomial distribution
and Bose–Einstein distribution, 250
characteristic function for, 252
definition of, 248–250
moments of, 252–253
probability density function for, 248
Negative exponential distribution
definition of, 232
moments of, 231
and the intensity of a speckle field, 232
and extremely strong turbulence, 219
Neyman–Pearson criterion, 58
Noise equivalent power, 120, 146
Noise spectral densities, 146
Noncentral chi-squred distribution
characteristic function for, 35
noncentrality factor for, 36, 260
probability density function for, 36, 261
series form of, 36, 260
Noncentral or generalized negative binomial distribution, 264–265
Noncentral negative binomial signal in Poisson noise, 264,
probability density function for, 264–265
Noncoherent illumination, 227
Nondiffraction-limited performance, 75
Nyquist derivation of thermal noise, 136

Observation, statistical, 4
Offset-homodyne, 63
One-dominant plus Rayleigh distribution
as an incoherent sum of two speckle fields, 310, 334
probability density function for, 332
Optical modulation techniques, 64
Optical preamplifiers, 266–267

Outcomes of a random event, 4
Overlap integral for circular apertures, 179

Parabolic-cylinder distribution
  characteristic function for, 271
  moments of, 271
  probability density function for, 268
Parabolic-cylinder function, 268, 397
Parabolic-cylinder signal in Gaussian noise, 267
  detection probability for, 275
Parabolic-cylinder signal, 268
Parameter estimation, 47, 227
Paraxial approximation, 70
Parseval's theorem, 40
Partially coherent light, 154
Partially developed speckle, 256
Pascal distribution, and the negative binomial distribution, 250
Peak signal power, 325
Pearson's incomplete Gamma function, 370
Phase structure function, 199
Photodetector, 61, 239
Photoelectric effect, 61, 239
Photoelectron detection probability, 291
Photoemissive processes, 233, 239
Photomultiplier tube, 140
Photon bunching and anti-bunching, 64, 290
Planck's radiation law, 137
Poisson distribution
  characteristic function of, 242
  limiting forms of, 243
  moments of, 242
  probability density function of, 241
Poisson probabilities, 36
Poisson signal in atmospheric turbulence, 298
Poisson signal in Poisson noise
  characteristic function of, 243
  probability density function for, 244, 266
Polarization diversity, 227, 309
Potential temperature, 186
Power signal-to-noise ratio, 145
  compared to voltage signal-to-noise ratio, 327, 330
Power spectral density
  definitions, 38–42
  of a turbulent field, 193
Primary carriers, 141, 277
Probability density function
  definition of, 6
  conditional, 17–18
  continuous and discrete, 7
  joint, 17–18
  marginal, 19
  two-dimensional, 17
Probability, 6
Processing scan, 365
Pulse oximeter, 3
Pupil function, 174, 233

Quantum limit
  for shot noise limited receivers, 147
  for the direct detection receivers, 349
Quantum well APD, 140

Radiance, 172
Random variable, 4
Randomly phased sinusoids, 32, 321
Range or time bins, 294, 386
Rayleigh distribution
  definition of, 13
  in thresholding example, 53
  representing receiver noise, 322
  representing speckle amplitude, 231
Rayleigh and exponential signal fading, 228, 232
Rayleigh fluctuating signal, 328
Rayleigh–Rice approximation, 160
Real-beam, definition of, 96
Receiver class, 227
Reciprocity theorem of Helmholtz, 303
Reflectivity per steradian, 172
Replacement rule for depolarized fields, 311
Resolved and unresolved targets, 68
Riccati equation, 188
Rice–Nakagami distribution, 221, 263
Rician distribution
  for constant-amplitude signal in Gaussian noise, 323
  limiting cases of, 324
  mean noise power of, 322
  mean signal power of, 324
  moments of, 324
  power form of, 325–326
Rician signal in Gaussian noise, probability density function for, 335
Rician signal model
  relationship to one-dominant plus Rayleigh signal, 334
  relationship to partially developed speckle, 336
Rms noise power, 326
Rotating rough surface, 234–235
Rough surface, definition of, 160

Second order or bivariate statistics, 16
  in coherence theory, 151
Semiclassical radiation theory, 240

Shot noise, 138
  limited performance, 147
Signal envelope, 320
Signal-to-noise ratio
  of truncated coherent detection systems, 125–128
  of truncated direct-detection systems, 131–134
    and detector optimization, 134–136
  of untruncated coherent detection systems, 120, 122
    far-field limiting cases, 121
    soft or volume targets, 123–124
  of untruncated direct-detection systems, 131
    far-field limiting cases, 131
Simulated far-field pattern, 71
Soft or volume target, 123–125
Solar or background radiation, 143
Spatial and temporal coherence, 153–154
Spatial autocorrelation function, 43
Spatial Fourier transform
  and Fraunhofer formula, 72
  in Fourier optics, 104
Spatial homogeneity, 184
Spatial power spectral density, 43
Speckle
  averaging, 229, 255, 265, 276
  diversity, 227, 313, 315
  pattern, 109, 228, 258
Spectral filter, 143
Spectral irradiance, 143
Spectral radiance, 143
Specular reflection, 159
Squeezed states, 64
Stationary processes
  in increments, 37
  in random events, 36–37
  in the ergodic sense, 38
  in the strict sense, 37
  in the wide-sense, 37
  in turbulent media, 38, 184
Statistical decision theory, *see* Decision theory
Statistical dependence and independence, 4
Statistical event, 4
Statistical parameters
  correlation coefficient, 21, 175
  correlation diameter, 174
  correlation radius, 180
  covariance, 21
  cross-correlation function, 151
  degrees of freedom, 34
  examples of, 13–14
  expected value, 10–11
  joint events, 4
  joint probability, 17
  kurtosis or fourth central moment, 13
  mean, 10
  median, 12
  mode or most probable value, 12
  moments, 10
  normalized covariance, 21
  second moment, 11
  skewness or third central moment, 12
  standard deviation, 12
  variance or second central moment, 12, 21
Statistical realization of a random event, 38
Statistical sample
  sample function, 38
  sample space, 4
  sample waveform, 38
Statistical stationarity, *see* Stationary processes
Statistical trial, 4
Stellar interferometry, 158
Summed statistics
  of a Gaussian signal in Gaussian noise, 267
    characteristic function for, 271
    probability density function for, 268
  of fully developed speckle, 234
    characteristic function for, 235
    probability density function for, 236
  of partially developed speckle, 259,
    characteristic function for, 260
    probability density function for, 261
Surface dark current, 141
Surface height correlation distance, 166
Surface scatter, 159
Swerling models, 365

Target and turbulence induced speckle, 150
Temporal coherence
  factor, 153
  function, 153
Time autocorrelation function, 42
Top-hat aperture distribution, 88
Toronto function, 368, 399
Transformation of random variables
  one-dimensional, 22
  multi-valued functions, 23
  two-dimensional, 24
Transverse coherence length, 200
Transverse electromagnetic modes (TEM), 86
Turbulent atmosphere, *see* Atmospheric turbulence
Two-parameter atmospheric models, 222–224

Uncorrelated speckle cells, 260
Uncorrelated speckle fields, 313
Univariate statistics, 4

Van Cittert–Zernike theorem, 155–159, 177
Variance, *see* Statistical parameters
Variance, as a measure of beam width, 93
Variance of turbulence parameters, *see* Atmospheric turbulence
Virtual local oscillator, 118
Volume or soft target, 123–125

Wave structure function, 199
Weighted average, as a statistical moment, 10
White light, *see* Natural light
White noise, 138
Wiener–Khintchine theorem, 42, 45, 154
WMC distribution for APD detectors, 279

Young's two-beam experiment, 152